Canadian
Environmental History

Canadian Environmental History

Essential Readings

Edited by David Freeland Duke

Canadian Scholars' Press Inc.
Toronto

Canadian Environmental History: Essential Readings
Edited by David Freeland Duke

First published in 2006 by
Canadian Scholars' Press Inc.
180 Bloor Street West, Suite 801
Toronto, Ontario
M5S 2V6

www.cspi.org

Canadian Scholars' Press gratefully acknowledges financial support for our publishing activities from the Government of Canada through the Book Publishing Industry Development Program (BPIDP).

Library and Archives Canada Cataloguing in Publication

    Canadian environmental history : essential readings / edited by David Freeland Duke. Includes bibliographical references.

ISBN 1-55130-310-8
    1. Human ecology--Canada--History--Textbooks. 2. Environmental sciences--Textbooks. I. Duke, David Freeland, 1967-

HC120.E5C3536 2006   333.720971   C2006-904361-2

Cover design by Aldo Fierro
Cover photo: "B.C. Rain-Forest." Reprinted by permission of the National Archives of Canada / PA-40959.
Page design and layout by Brad Horning

06 07 08 09 10   5 4 3 2 1

Printed and bound in Canada by Marquis Book Printing Inc.

This book is printed on 100% post-consumer recycled paper.

Canadä

Table of Contents

PART III
BIOLOGY AND IMPERIALISM IN
NORTH AMERICAN ENVIRONMENTAL HISTORY

PART IV
PRE-INDUSTRIAL RESOURCES
AND THE CHANGING CULTURE OF NATURE

PART V
INDUSTRIALIZATION

PART VI
SUSTAINABILITY AND CONSERVATION

Preface

It is unusual to acknowledge another book, a potential competitor, in one's preface. However, I am deeply indebted to Chad and Pam Gaffield's now out-of-print classic, *Consuming Canada: Readings in Environmental History*. At the time of its publication, more than a decade ago, the editors of *Consuming Canada* wrote that historians in Canada had only recently begun to undertake serious research in the environmental history of their country. The perception was that Canadian historiography was lagging behind that of the United States, that its focus on the histories of the two founding cultures, or on the history of Canada's regions, was masking broader themes that played very significant roles in the formation of Canadian history. Today we may claim with justification that Chad and Pam Gaffield's argument that the Canadian historical community ignored environmental history can be laid to rest—and that it is in no small measure as a result of the publication of *Consuming Canada* that we may do so.

There is no doubt that Canadian historiography is now marked by the distinctive presence of environmental history. The flagship journal of the field, *Environmental History*, is in 2006 publishing a special edition devoted entirely to Canadian environmental history, and the journal *Urban History Review* recently published a similar edition (34, no. 1 [Fall 2005]) dealing with the theme of urban environmental history in Canada. The environmental history community is blossoming in this country, as demonstrated by the creation of the Network in Canadian History and Environment (NiCHE) in the autumn of 2004. Conference panels and themes are increasingly devoted to aspects of Canadian environmental history, and the interactions among scholars working in the field are growing rapidly.

It is still difficult to teach Canadian environmental history, however, or more difficult than it should be. Although resources, both primary and secondary, are growing in number all the time, the latter tend to be quite specialized in nature and are often difficult to employ in the classroom. The Gaffields' *Consuming Canada* has long been the standard reader, especially in junior classes, but it is now out of print. Furthermore, much work of tremendous importance has been undertaken and published in the decade since *Consuming Canada* appeared, and it is occasionally difficult to integrate this material into undergraduate courses. I have heard it suggested that Canadian environmental history doesn't yet belong in university or high school curricula because it remains underdeveloped and characterized by a lack of interpretive secondary sources. I hope that this reader can serve to undermine that suggestion. It is my intention that it will be an introduction for students to this remarkable, energetic, and rapidly growing field, and that it can act as an easily accessible collection of scholarship that demonstrates both the breadth and sophistication of environmental history in Canada today. I have tried to select sources that reflect the field's breadth and its recent growth—more than half of the items in this text are less than a decade old. There is a freshness, energy, and quality apparent in this scholarship that builds upon rather than replaces the work of earlier trailblazers. This synergy between older and new scholarship bodes well for the future of Canadian environmental history.

In this volume's initial form I had intended, somewhat grandly, to create a collection that tied together an earlier focus of Canadian historiography with the challenge offered by environmental history. I wanted to develop a collection that focused on the environmental histories of Canada, with *regions as the binding theme*. However, I was dissuaded from the regional approach by four excellent reviews—academics teaching, writing, and researching in the area. The anonymous authors of those reviews each offered a remarkable series of themes that should be incorporated into the collection and did so on the basis of clear and sophisticated arguments. I realized then, on the ground

so to speak, what I had long known intellectually—that environmental history really *is* a "big tent" that shelters a very broad group of scholars and perspectives indeed.

In revamping this volume, as a result of the reviewers' input, I turned away from the regional approach and chose instead to develop a collection that is based on a concise, *chronological* framework. This framework, I believe, has created a text that can serve as a structure upon which teachers and students can investigate the directions of Canadian environmental history and which provides a gateway for the investigation of specialized topics. In addition, I have tried to provide useful further readings on those topics, together with Web sites that can provide access to primary sources and multimedia presentations on the subjects under investigation.

I am deeply grateful to the reviewers for their perspectives. The collection is much stronger than it would have been otherwise.

I am especially indebted to my editor at Canadian Scholars' Press, Megan Mueller, whose energy and reserves of patience concerning this project have been seemingly inexhaustible. My ability to keep deadlines is not well evolved, but Megan pushed gently but firmly to keep me on track—approximately. Her genuine enthusiasm for the idea of environmental history in Canada and its importance for students is infectious and deeply perceptive.

I also owe a significant debt of gratitude to my students in my Canadian environmental history seminar at Acadia University, the group that is test-driving these readings. Their suggestions and responses have pushed me to think about and present the articles here in new ways that I otherwise would have overlooked. As everyone who teaches the discipline knows, students in environmental history courses are, quite simply, the best there is: They tend to self-select for the courses (how many undergraduate programs or degrees require environmental history as a core component?) and they bring an enthusiasm to class that most teachers would long to have.

Last, but not least, I want to thank my wife, Laura, who suffers the unusual hours and working weekends, and especially the last-minute rushes, with patience and understanding. I rely on her more than she knows, but less than she would like. It's for that reason that this book is dedicated to her; in a real sense she has contributed a great deal to its completion.

A NOTE FROM THE PUBLISHER

Thank you for selecting *Canadian Environmental History: Essential Readings*, edited by David Freeland Duke. The editor and publisher have devoted considerable time and careful development (including meticulous peer reviews) to this book. We appreciate your recognition of this effort and accomplishment.

TEACHING FEATURES

This volume distinguishes itself on the market in many ways. One key feature is the book's well-written and comprehensive part openers, which help to make the readings all the more accessible to undergraduate students. The part openers add cohesion to the section and to the whole book. The themes of the book are very clearly presented in these section openers.

The general editor, David Freeland Duke, has also greatly enhanced the book by adding pedagogy to close and complete each section. As noted by the editor in the Preface, each part ends with critical thinking questions, detailed annotated further readings, and annotated relevant Web sites.

Introduction to Environmental History

What is environmental history? This is a question that has often been directed at practitioners in the field ever since it developed as a discrete area of historical inquiry. Historians who separate themselves from the mainstream of the discipline, as environmental historians did in the 1970s, often do so either because they find mainstream inquiries too pedestrian or too restrictive, or because they are part of a generational shift that allows junior historians to approach historical problems with a fresh perspective not easily encompassed by older methodologies or outlooks. This phenomenon is by no means restricted to history—it has happened, and continues to happen, in all intellectual activities from the arts to the natural sciences. Obviously the nature of knowledge is not static, and as it changes, so too do the forms of inquiry that allow investigators to uncover and understand new knowledge, categorize it, and disseminate it to the larger community. Furthermore, new areas of scholarly investigation can appear and grow in response to larger social phenomena—witness the dramatic growth of women's and gender history as part of the rise of the feminist movement after the Second World War (or, for that matter, the tremendous technological advances of the war itself—in aviation, computing, radar, medicine, chemistry, and, of course nuclear technology—all consequences of the exigencies of survival in the face of a serious threat). Environmental history as a discipline is no different. Although the roots of environmental consciousness are both deep and old (reverence for nature is probably an impulse as ancient as humanity itself), the appearance of a group of professional scholars who identified themselves as belonging to a school of thought called "environmental history" dates only from the 1970s. It is, therefore, a very new field indeed.

The rise of this school came about as a consequence of a variety of factors. One was the spread of popular environmental consciousness on a broad scale in the 1960s. Partly a response to the publication of Rachel Carson's seminal *Silent Spring* in 1962, and partly a result of the generational tensions created as the baby boom generation came of age, in the 1960s there was a growing uneasiness that the industrial activities that characterized modern society—and which provided that society's material wealth—were producing significant negative environmental consequences. There were fears concerning nuclear fallout from atomic weapons tests (and growing concerns about nuclear power), evidence of increasingly obvious damage produced by acid rain, worries about penetration of the natural environment by artificial compounds and their effects there, well-publicized environmental crises such as the ignition of the Cuyahoga River and the "death" of Lake Erie, and more.[1] The environmental activism spurred on by these events affected the new generation of historians who increasingly came to focus on environmental questions in their analyses.

The rise of ecology at roughly the same time was another contributing factor in the emergence of environmental history. Ecologists turned away from limited investigations of components of the natural world and instead increasingly broadened their focus to encompass the webs that bound entire communities of species together into ecosystems. It was only a small intellectual step for ecologists to begin to include the activities of humans as another species as factors in shaping the web of life. As long as scientific investigations remained focused on just the natural world, history could play little or no part. But when ecology wove the analysis of human activities into a larger examination of the physical environment, historians—those engaged in the study of human activities in the recoverable past—could become involved.

Lastly, there were investigative themes that emerged from within history itself. These propelled the forms of analysis that came to characterize environmental history. Chief among these was the rise of the *Annales* school, whose founders, Lucien Febvre and Marc Bloch, were each acutely conscious of the importance of environmental factors on the development and evolution of human societies. As Donald Worster points out in the first of the articles in this section, environmental historians owe a great debt to the *Annalistes* for their development of the concept of *la longue durée* (the long duration) in history. The long duration requires us to reorient our perspective, to move it away from the short-term event as the engine of historical change, and to find it instead in the slow rhythms of humanity's relations with nature.

If the *Annales* school encouraged historians to adjust their temporal lens, then another historical school called upon them to expand their geographical focus. World history, increasingly popular after the Second World War, found new perspectives not in the histories of nation-states or peoples, but in studying the relationships between them. Emulating the epistemology of the ecologists, perhaps unconsciously, world historians argued that a thorough understanding of historical phenomena could be gained only by studying the ties that bound geographical areas together. Thus for world historians, the North Atlantic Triangle, for example, was a crucial object of study for investigation of the civilizations of West Africa, the colonization of the Americas, or the industrialization of Europe, and any attempt to understand any one of these geographical areas or historical processes without reference to the ties that bound them together was necessarily incomplete.

All of these processes—a growing environmental consciousness, the new epistemology of ecology, the environmental and long-duration emphasis of the *Annales* school, and the geographic breadth and systemic breadth of world history—coalesced as the intellectual foundation of environmental history. The four historians presented in this section were junior scholars in the 1970s and are now among the leaders in the field today. Each in his or her own way, and repeatedly throughout their careers, have had to answer the question posed at the beginning of this introduction: What is environmental history? Their answers are found in these readings and are daunting in their breadth.

One thing is common to each of the authors' contributions, and that is their discussion of the nature of power in the human–environment relationship. For Carolyn Merchant, human power is often exercised along gendered lines. For Douglas Weiner it is often ideological; for Donald Worster it is economic; and for William Cronon it is political. This raises the question: To what extent should we emphasize power as a central axis of analysis in environmental history? Are there any pitfalls inherent in this approach?

Another point worth considering is raised by Douglas Weiner toward the end of his death-defying attempt to define environmental history, when he notes that environmental historians have become "socially 'credentialed' interpreters of the world." To a certain extent this is what Cronon suggested environmental historians should in fact aspire to in his piece, written more than a decade earlier. But if it is true, as Weiner says, that society at large increasingly looks to environmental history and its practitioners for directions and prescriptions concerning current environmental politics and practices, then what burdens consequently accumulate on the shoulders of environmental historians? How should we deal with the social responsibilities—of activism, perhaps even of partisanship—that come with social credentials?

NOTE

1. Ralph Lutts, "Chemical Fallout: Rachel Carson's *Silent Spring*, Radioactive Fallout and the Environmental Movement," *Environmental Review* 9 (Fall 1985): 210–225.

Chapter One
Doing Environmental History

Donald Worster

In the old days, the discipline of history had an altogether easier task. Everyone knew that the only important subject was politics and the only important terrain was the nation-state. One was supposed to investigate the connivings of presidents and prime ministers, the passing of laws, the struggles between courts and legislatures, and the negotiations of diplomats. That old, self-assured history was actually not so old after all—a mere century or two at most. It emerged with the power and influence of the nation-state, reaching a peak of acceptance in the nineteenth and early twentieth centuries. Often its practitioners were men of intensely nationalistic feelings, who were patriotically moved to trace the rise of their individual countries, the formation of political leadership in them, and their rivalries with other states for wealth and power. They knew what mattered, or thought they did.

But some time back that history as "past politics" began to lose ground, as the world evolved toward a more global point of view and, some would say, toward a more democratic one. Historians lost some of their confidence that the past had been so thoroughly controlled or summed up by a few great men acting in positions of national power. Scholars began uncovering long submerged layers, the lives and thoughts of ordinary people, and tried to reconceive history "from the bottom up." Down, down we must go, they maintained, down to the hidden layers of class, gender, race, and caste. There we will find what truly has shaped the surface layers of politics. Now enter still another group of reformers, the environmental historians, who insist that we have got to go still deeper yet, down to the Earth itself as an agent and presence in history. Here we will discover even more fundamental forces at work over time. And to appreciate those forces we must now and then get out of parliamentary chambers, out of birthing rooms and factories, get out of doors altogether, and ramble into fields, woods, and the open air. It is time we bought a good set of walking shoes, and we cannot avoid getting some mud on them.

So far this extending of the scope of history to include a deeper and broader range of subjects has not challenged the primacy of the nation-state as the proper territory of the historian. Social, economic, and cultural history are all still commonly pursued within national boundaries. Thus, to an extent that is quite extraordinary among the disciplines of learning, history (at least for the modern period) has tended to remain the insular study of the United States, Brazil, France, and the rest. Such a way of organizing the past has the undeniable virtue of preserving some semblance of order in the face of a threatening chaos—some way of synthesizing all the layers and forces. But at the same time it may set up obstacles to new inquiries that do not neatly fit within national borders, environmental history among them. Many of the issues in this new field defy a narrow nationality: the wanderings of Tuareg nomads in the African Sahel, for instance, or the pursuit of the great whales through all the world's oceans. Other environmental themes, to be sure, have developed strictly within the framework of single-nation politics, [...] but not all have done so, and in the history that will be written tomorrow, fewer and fewer will be.

Environmental history is, in sum, part of a revisionist effort to make the discipline far more inclusive in its narratives than it has traditionally been. Above all, it rejects the conventional assumption that human experience has been exempt from natural constraints, that people are a separate and "supernatural" species, that the ecological consequences of their past deeds can be ignored. The old history could hardly deny that we have been living for a long while on this planet, but it assumed by its general disregard of that fact that we have not been and are not truly part of the planet. Environmental historians, on the other hand, realize that we can no longer afford to be so naive.

The idea of environmental history first appeared in the 1970s, as conferences on the global predicament were taking place and popular environmentalist movements were gathering momentum in several countries. It was launched, in other words, in a time of worldwide cultural reassessment and reform. History was hardly alone in being touched by that rising mood of public concern; scholarship in law, philosophy, economics, sociology, and other areas was similarly responsive. Long after popular interest in environmental issues crested and ebbed, as the issues themselves came to appear more and more complicated, without easy resolution, the scholarly interest continued to expand and take on greater and greater sophistication. Environmental history was, therefore, born out of a moral purpose, with strong political commitments behind it, but also became, as it matured, a scholarly enterprise that had neither any simple, nor any single, moral or political agenda to promote. Its principal goal became one of deepening our understanding of how humans have been affected by their natural environment through time and, conversely, how they have affected that environment and with what results.

One of the liveliest centres of the new history has been the United States, a fact that undoubtedly stems from the strength of American leadership in environmental matters. The earliest attempt to define the field was Roderick Nash's essay, "The State of Environmental History." Nash recommended looking at our entire surroundings as a kind of historical document on which Americans had been writing about themselves and their ideals. More recently, a comprehensive effort by Richard White to trace the development of the field credits the pioneering work of Nash and that of the conservation historian Samuel Hays, but also suggests that there were anticipations before them in the frontier and western school of American historiography (among such land-minded figures as Frederick Jackson Turner, Walter Prescott Webb, and James Malin). Those older roots became increasingly recalled as the field moved beyond Hays's politics of conservation and Nash's intellectual history to focus on changes in the environment itself and consider, once more, the environment's role in the making of American society.

Another centre of innovation has been France, particularly the historians associated with the journal *Annales*, who have been drawing attention to the environment for several decades now. That journal was founded in 1929 by two professors at the University of Strasbourg, Marc Bloch and Lucien Febvre. Both of them were interested in the environmental basis of society, Bloch through his studies of French peasant life and Febvre as a social geographer. The latter's protegé, Fernand Braudel, would also make the environment a prominent part of his historical studies, notably in his great work on the Mediterranean. For Braudel, the environment was the shape of the land—mountains, plains, seas—as an almost timeless element shaping human life over the long duration (*la longue dureé*). There was, he insisted, more to history than the succession of events in individual lives; on the grandest scale, there was history seen from the vantage of nature, a history "in which all change is slow, a history of constant repetition, ever-recurring cycles."

Like the frontier historians in the United States, the *Annalistes* in France found their environmental interests reanimated by the popular movements of the sixties and early seventies. In 1974, the journal devoted a special issue to "Histoire et Environnement." In a short preface Emmanuel Le Roy Ladurie, himself one of the leading lights in the field, gave this description (my translation) of the field's program:

Environmental history unites the oldest themes with the newest in contemporary historiography: the evolution of epidemics and climate, those two factors being integral parts of the human ecosystem; the series of natural calamities aggravated by a lack of foresight, or even by an absurd "willingness" on the part of the simpletons of colonization; the destruction of Nature, caused by soaring population and/or by the predators of industrial overconsumption; nuisances of urban and manufacturing origin, which lead to air or water pollution; human congestion or noise levels in urban areas, in a period of galloping urbanization.

Denying that this new history was merely a passing fashion, Le Roy Ladurie insisted that the inquiry had in truth been going on for a long time as part of a movement toward "histoire écologique."

Much of the material for environmental history has indeed been around for generations, if not for centuries, and is only being reorganized in the light of recent experience. It includes data on tides and winds, on ocean currents, on the position of continents in relation to each other, on the geological and hydrological forces creating our land and water base. It includes the history of climate and weather, as these have made for good or bad harvests, sent prices up or down, ended or promoted epidemics, led to population increase or decline. All these have been powerful influences over the course of history, and continue to be so, as when massive earthquakes destroy cities or starvation follows in the wake of drought or rivers determine the flow of settlement.

The fact that such influences continue in the late twentieth century is evidence of how far we are yet from controlling the environment to our complete satisfaction. In a somewhat different category are those living resources of the Earth, which the ecologist George Woodwell calls the most important of all: the plants and animals (and one might add the soil as a collective organism) that, in Woodwell's phrase, "maintain the biosphere as a habitat suitable for life." These resources have been far more susceptible to human manipulation than the abiotic ones, and at no point more so than today. But pathogens are also a part of that living realm, and they continue, despite the effectiveness of medicine, to be a decisive agency in our fate.

Put in the vernacular then, environmental history is about the role and place of nature in human life. By common understanding we mean by "nature" the non-human world, the world we have not in any primary sense created. The "social environment," the scene of humans interacting only with each other in the absence of nature, is therefore excluded. Likewise is the built or artifactual environment, the cluster of things that people have made and which can be so pervasive as to constitute a kind of "second nature" around them. That latter exclusion may seem especially arbitrary, and to an extent it is. Increasingly, as human will makes its imprint on the forest, on gene pools, on the polar ice cap, it may seem that there is no practical difference between "nature" and "artifact." The distinction, nonetheless, is worth keeping, for it reminds us that there are different forces at work in the world and not all of them emanate from humans; some remain spontaneous and self-generating. The built environment is wholly expressive of culture; its study is already well advanced in the history of architecture, technology, and the city. But with such phenomena as the forest and the water cycle, we encounter autonomous energies that do not derive from us. Those forces impinge on human life, stimulating some reaction, some defence, some ambition. Thus, when we step beyond the self-reflecting world of humankind to encounter the non-human sphere, environmental history finds its main theme of study.

There are three levels on which the new history proceeds, three clusters of issues it addresses, though not necessarily all in the same project, three sets of questions it seeks to answer, each drawing on a range of outside disciplines and employing special methods of analysis. The first deals with understanding nature itself, as organized and functioning in past times; we include both organic and inorganic aspects of nature, and not least the human organism as it has been a link in nature's food chains, now functioning as womb, now belly, now eater, now eaten, now a host for micro-organisms, now a kind of parasite. The second level in this history brings in the socio-economic realm as it interacts with the environment. Here we are concerned with tools and work, with the social relations that grow out of that work, with the various modes people have devised of producing goods from natural resources. A community organized to catch fish at sea may have very different institutions, gender roles, or seasonal rhythms than one raising sheep in high mountain pastures. Power to make decisions, environmental or other, is seldom distributed through a society with perfect equality, so locating the configurations of power is part of this level of analysis. Then, forming a third level for the historian is that more intangible and uniquely human type of encounter—the purely mental or intellectual, in which perceptions, ethics, laws, myths, and other structures of meaning become part of an individual's or group's dialogue with nature. People are constantly engaged in constructing maps of the world around them, in defining what a resource is, in determining which sorts of behaviour may be environmentally degrading and ought to be prohibited, and generally in choosing the ends of their lives. Though for the purposes of clarification, we may try to distinguish between these three levels of environmental study, in fact they constitute a single dynamic inquiry in which nature, social and economic organization, thought and desire are treated as one whole. And this whole changes as nature changes, as people change, forming a dialectic that runs through all of the past down to the present.

This in general is the program of the new environmental history. It brings together a wide array of subjects, familiar and unfamiliar, rather than setting up some new, esoteric specialty. From that synthesis, we hope, new questions and answers will come.

NATURAL ENVIRONMENTS OF THE PAST

The environmental historian must learn to speak some new languages as well as ask some new questions. Undoubtedly, the most outlandish language that must be learned is the natural scientist's. So full of numbers, laws, terms, and experiments, it is as foreign to the historian as Chinese was to Marco Polo. Yet, with even a smattering of vocabulary, what treasures are here to be understood and taken back home! Concepts from geology, pushing our notions of history back into the Pleistocene, the Silurian, the Precambrian. Graphs from climatology, on which temperatures

and precipitation oscillate up and down through the centuries, with no regard for the security of kings or empires. The chemistry of the soil with its cycles of carbon and nitrogen, its pH balances wavering with the presences of salts and acids, setting the terms of agriculture. Any one of these might add a powerful tool to the study of the rise of civilizations. Together, the natural sciences are indispensable aids for the environmental historian, who must begin by reconstructing past landscapes, learning what they were and how they functioned before human societies entered and rearranged them.

But above all it is ecology, which examines the interactions among organisms and between them and their physical environments, that offers the environmental historian the greatest help. This is so in part because, ever since Charles Darwin, ecology has been concerned with past as well as present interactions; it has been integral to the study of evolution. Equally significant, ecology is at heart concerned with the origins, dispersal, and organization of all plant life. Plants form by far the major portion of the Earth's biomass. All through history people have depended critically on them for food, medicine, building materials, hunting habitat, and a buffer against the rest of nature. Far more often than not, plants have been humans' allies in the struggle to survive and thrive. Therefore, where people and vegetation come together, more issues in environmental history cluster than anywhere else. Take away plant ecology and environmental history loses its foundation, its coherence, its first step.

So impressed are they with this fact that some scholars speak of doing, not environmental, but "ecological history" or "historical ecology." They mean to insist on a tighter alliance with the science. Some years back the scientist and conservationist Aldo Leopold projected such an alliance when he spoke of "an ecological interpretation of history." His own illustration of how that might work had to do with the competition among Native Indians, French and English traders, and American settlers for the land of Kentucky, pivotal in the westward movement. The canebrakes growing along Kentucky bottomlands were a formidable barrier to any agricultural settlement, but as luck would have it for the Americans, when the cane was burned and grazed and chopped out, bluegrass sprouted in its place. And bluegrass was all that any farmer, looking for a homestead and a pasture for his livestock, could want. American farmers entered Kentucky by the thousands, and the struggle was soon over. "What if," Leopold wondered, "the plant succession inherent in this dark and bloody ground had, under the impact of these forces, given us some worthless sedge, shrub, or weed?" Would Kentucky have become American property as and when it did?

Shortly after Leopold called for that merging of history and ecology, the Kansas historian James Malin brought out a series of essays leading to what he termed "an ecological reexamination of the history of the United States." He specially had in mind examining his native grasslands and the problem in adaptation they had set for Americans, as they had for the Indians before them. From the late nineteenth century on, White settlers, coming out of a more humid, wooded country, had tried to create a stable agriculture on the dry, treeless plains, but with only mixed results. Malin was impressed that they had succeeded in turning the land into prosperous wheat farms, but not before they had had to unlearn many of their old

agricultural techniques. Dissatisfied with traditional history, which did not give such matters any prominence, Malin found himself reading ecologists to find the right questions to ask. He read them with a certain freedom, as a source of inspiration rather than a set of rigid models. "The ecological point of view," he believed, "is valuable to the study of history; not under any illusion that history may thus be converted into a science, but merely as a way of looking at the subject matter and processes of history."

Those were alliances sought some 30 or 40 years back. Since then, as ecology has developed into a more rigorously mathematical science, with more elaborate models of natural processes, neither Malin's nor Leopold's casual sort of alliance has seemed adequate. Environmental historians have had to learn to read at a more advanced level, though they are still faced with Malin's problem of deciding just how scientific their history needs to be and which ideas in science can or ought to be adopted.

Today's ecology offers a number of angles for understanding organisms in their environment, and they all have their limits as well as uses in history. One might, for example, examine the single organism and its response to external conditions; in other words, study adaptation in individual physiological terms. Or one might track the fluctuations in size of some plant or animal population in an area, its rates of reproduction, its evolutionary success or failure, its economic ramifications. Although both sorts of inquiry may have considerable practical significance for human society, there is a third strategy that holds the most promise for historians needing to understand humans and nature in the composite.

When organisms of many species come together, they form communities, usually highly diverse in makeup, or as they are more commonly called now, ecosystems. An ecosystem is the largest generalization made in the science, encompassing both the organic and inorganic elements of nature bound together in a single place, all in active, reciprocating relationship. Some ecosystems are fairly small and readily demarcated, like a pond in New England, while others are sprawling and ill-defined, as large as the Amazonian rain forest or the Serengeti plain or even the whole Earth. All are commonly described, in language derived heavily from physical mechanics and cybernetics, as self-equilibrating, like a machine that runs on and on automatically, checking itself when it gets too hot, speeding up when it slows and begins to sputter. Outside disturbances may affect that equilibrium, throwing the machine temporarily off its regular rhythm, but always (or almost always) it returns to some steady state condition. The numbers of species constituting an ecosystem fluctuate around some determinable point; the flow of energy through the machine stays constant. The ecologist is interested in how such systems go on functioning in the midst of continual perturbations, and how and why they break down.

But right there occurs a difficult issue on which the science of ecology has reached no clear consensus. How stable are those natural systems and how susceptible to upset? Is it accurate to describe them as balanced and stable until humans arrive? And if so, then at what point does a change in their equilibrium become excessive, damaging or destroying them? Damage to the individual organism is easy enough to define: it is an impairment of health or, ultimately, it is death. Likewise, damage to a population is not very hard to determine, simply, when its numbers decline. But damage to whole ecosystems

is a more controversial matter. No one would dispute that the death of all its trees, birds, and insects would mean the death of a rain-forest ecosystem, or that the draining of a pond would spell the end of that system. But most changes are less catastrophic, and the degree of damage has no easy method of measurement.

The difficulty of determining ecosystem damage applies to changes worked by people as well as non-human forces. A South American tribe, for instance, may clear a small patch in the forest with their machetes, raise a few crops, and then let the field revert to forest. Such so-called swidden, or slash-and-burn, farming has usually been regarded as harmless to the whole ecosystem; eventually, its natural equilibrium is restored. But at some point, as this farming intensifies, the capacity of the forest to regenerate itself must be permanently impaired and the ecosystem damaged. What is that point? Ecologists are not sure and cannot give precise answers. For that reason the ecological historian more often than not ends up talking about people inducing "change" in the environment—"change" being a neutral and indisputable term—rather than doing "damage," a far more problematical concept.

Until recently the ruling authority in ecosystem science has been Eugene Odum, through the various editions of his popular textbook, *Fundamentals of Ecology*. Odum is a system man nonpareil, one who sees the entire realm of nature as hierarchically organized into systems and subsystems, all made up of parts that function harmoniously and homeostatically, the rhythm of each system rather resembling the eighteenth-century's watchlike nature that never missed a tick. That earlier version was supposed to reveal the contriving hand of its divine maker; Odum's, in contrast, is the spontaneous work of nature. But increasingly, ecologists are retreating from his picture of order. Led by paleoecologists, especially paleobotanists, who collect core samples from peat bogs and, through pollen analysis, try to reconstruct ancient environments, they are finding Odum's blueprint a bit static. Looking backward in time to the Ice Age and before, they are discovering plenty of disorder and upheaval in nature. Abstracted from time, the critics say, ecosystems may have a reassuring look of permanence; but out there in the real, the historical, world, they are more perturbed than imperturbable, more changing than not.

This scientific difference of opinion is partly over evidence, partly over perspective, like disputing whether a glass is half empty or half full. Stand back far enough, stand off in outer space as the British scientist James Lovelock has tried imaginatively to do, and the Earth still looks like a remarkably stable place, with organisms maintaining conditions highly suitable for life for over a billion years: all the gases in the atmosphere properly adjusted, fresh water and rich soil preserved in abundance, though evolution rages on and on, ice sheets come and go, and continents go drifting off in all directions. That may be how things look to the cosmic eyeball. Seen up close, however, the organic world may have a very different aspect. Stand on any given acre in North America and contemplate its past 1,000 years or so, even a single decade, and the conclusion ecologists are coming to these days is change, change, change.

There is a further unresolved problem in translating ecology into history. Few scientists have perceived people or human societies as being integral parts of their ecosystems. They leave them out as distractions, imponderables. But

people are what the historian mainly studies; consequently, his or her job is to join together what scientists have put asunder.

Human beings participate in ecosystems either as biological organisms akin to other organisms or as culture bearers, though the distinction between the two roles is seldom clear-cut. Suffice it here to say that, as organisms, people have never been able to live in splendid, invulnerable isolation. They breed, of course, like other species, and their offspring must survive or perish by the quality of food, air, and water and by the number of micro-organisms that are constantly invading their bodies. In these ways and more, humans have inextricably been part of the Earth's ecological order. Therefore, any reconstruction of past environments must include not only forests and deserts, boas and rattlesnakes, but also the human animal and its success or failure in reproducing itself.

HUMAN MODES OF PRODUCTION

Nothing distinguishes people from other creatures more sharply than the fact that it is people who create culture. Precisely what culture really is, however, is anybody's guess. There are literally scores of definitions. For preliminary purposes it can be said that the definitions tend to divide between those including both mental and material activities and those emphasizing mental activities exclusively, and that these distinctions between the mental and material correspond to the second and third levels of analysis in our environmental history. In this section we are concerned with the material culture of a society, its implications for social organization, and its interplay with the natural environment.

In any particular place nature offers the humans dwelling there a flexible but limited set of possibilities for getting a living. The Eskimos of the northern polar regions, to take an extreme case of limits, cannot expect to become farmers. Instead, they have ingeniously derived a sustenance, not by marshalling seed, plows, and draft animals of other, warmer latitudes, but through hunting. Their food choices have focused on stalking caribou over the tundra and pursuing bowhead whales among floating cakes of ice, on gathering blueberries in season and gaffing fish. Narrow though those possibilities are, they are the gift of technology as much as nature. Technology is the application of skills and knowledge to exploiting the environment. Among the Eskimos technology has traditionally amounted to fish hooks, harpoons, sled runners, and the like. Though constrained by nature, that technology has nonetheless opened up for them a nutritional field otherwise out of reach, as when a sealskin boat allowed them to venture farther out to sea in pursuit of prey. Today's Eskimos, invaded as they are by the instruments of more materially advanced cultures, have still more choices laid before them; they can, if they desire, import a supply of wheat and oranges by cargo plane from California. And they can forget how their old choices were made, surrender their uniqueness, their independence of spirit, their intimacy with the icy world. Much of environmental history involves examining just such changes, voluntary or imposed, in subsistence modes and their ramifications for people and the Earth.

As historians address these elemental issues of tools and sustenance, they soon become aware that there have been other disciplines at work here too, and for a long time. Among them is the discipline of anthropologists, and environmental historians have been reading their work with great interest. They have begun to search for clues from anthropologists to critical pieces of the ecological puzzle: What is the best way to understand the relation of human material cultures to nature? Is technology to be viewed as an integral part of the natural world, akin to the fur coat of the polar bear, the sharp teeth of the tiger, the fleet agility of the gazelle, all adaptive mechanisms functioning within ecosystems? Or should cultures be viewed as setting people apart from and outside of nature? Everything in the ecosystem, we are told by natural scientists, has a role and therefore an influence on the workings of the whole; conversely, everything is shaped by its presence in the ecosystem. Are cultures and the societies that create them also to be seen in that double position, both acting on and being acted on? Or are they better described as forming their own kind of "cultural systems" that mesh with ecosystems only in rare, isolated cases? Or, to make the puzzle more complicated still, do humans create with their technology a series of new, artificial ecosystems—a rice paddy in Indonesia or a carefully managed German forest—that require constant human supervision? There is, of course, no single or consistent set of answers to be given to such questions; but anthropologists, who are among the most wide-ranging and theory-conscious observers of human behaviour, can offer some provocative insights.

Anthropological thinking on such questions goes back well into the nineteenth century, but it has been particularly the last three or four decades that have seen the emergence of an ecological school (one with no settled curriculum, bearing such contending labels as cultural ecology, human ecology, ecological anthropology, and cultural materialism). The best guide to this literature is probably John Bennett's *The Ecological Transition*, though there are other useful surveys by Emilio Moran, Roy Ellen, Robert Netting, and others. Bennett defines the ecology school as the study of "how and why humans use Nature, how they incorporate Nature into Society, and what they do to themselves, Nature, and Society in the process." Some of these anthropologists have maintained that culture is an entirely autonomous and superorganic phenomenon, emerging apart from nature and understandable only in its own terms—or at least, as Bennett himself would have it, modern culture is trying to become so. Others, in contrast, have argued that all culture is, to some important degree, expressive of nature and ought not be rigidly set off in its own, self-contained sphere. Both positions are illuminating to the environmental historian, though for the historical era that is the main focus of this book, Bennett's is surely the more plausible one.

No one did more to found the ecological study of culture than Julian Steward, who published in 1955 his influential work, *Theory of Culture Change*, from which comes the idea of "cultural ecology." Steward began by examining the relationship between a people's system of economic production and their physical environment. He asked what resources they chose to exploit and what technology they devised for that work. This set of subsistence activities he called the "cultural core." Then he asked how such a system affected the behaviour of people toward one another, that is, how they organized themselves to produce their living. Social relations in turn shaped other aspects of culture. Some of the most interesting case studies for him were the great irrigation empires of the ancient world, in which large-scale control of water in arid environments led again and again to parallels in socio-political organization. Such regularities, he hoped, would suggest a general law of human evolution: not the old Victorian scheme that had all cultures moving along a single, fixed line of progress from hunting and gathering to industrial civilization, but rather one that explained the multilinear evolution of cultures, now diverging, now converging, now colliding with one another, with no end point in sight.

Steward's leadership in the new ecological approach inspired, directly or indirectly, a younger generation of field researchers who fanned out to all parts of the globe. John Bennett went to the Canadian prairies, Harold Conklin to the Philippines, Richard Lee to the !Kung Bushmen of Africa, Marshall Sahlins to Polynesia, Robert Netting to Nigeria to observe the hillside farmers there, Betty Meggers was off to the Amazon basin, Clifford Geertz to Indonesia, and there were still others. But above all, it has been Marvin Harris who has taken Steward's ideas and transformed them into a comprehensive and, some would complain, a highly reductive theory of the relationship between nature and culture. Like Steward, he has identified the "techno-environment" (i.e., the application of technology to environment) as providing the core of any culture, the main influence over how a people live with one another and think about the world. He has been even more rigidly deterministic than Steward was about that core. He has also been more interested in its dynamics. The techno-environmental system is not at all stable, he insists, certainly not forever. There is always the tendency to intensify production. It may come from population increase, climate change, or competition between states. Whatever the cause, the effect is always the same: depletion of the environment, declining efficiency, worsening living standards, pressures to move on—or if there is no new place to go, then pressure to find new tools, techniques, and resources locally, creating thereby another techno-environment. In other words, the degradation of the environment can be tragic, unhappy, or if people rise successfully to the challenge, it can mean the triumphant birth of a new culture. Harris calls this theory "cultural materialism." Clearly, it draws not only on Steward but on recent energy shortages, the present decline of a techno-environment based on fossil fuels, and the revival of Malthusian anxieties about world resource scarcity, though Harris would argue that a time of scarcity can also be a time of opportunity and revolution.

Marvin Harris has explicitly compared his theory of cultural materialism to that of Karl Marx, who gave the world "dialectical materialism," a view of history impelled forever forward by the struggle of one economic class to dominate another. The contrast between the two theories is emphatic: One sees change coming from the struggle of whole societies to exploit nature, with diminishing returns; the other points to internal conflicts within societies as the prime historical agency, with nature serving as a passive background. Perhaps, however, the distance between the two men is not hopelessly unbridgeable. One might put a little more Marxism into Harris by arguing that, among the factors leading to depletion and ecological disequilibrium, is competition between classes as well as states. Capitalists devise a social and technological order that makes them rich and elevates them to power. They set up factories for mass production. They drive the Earth to the point of breakdown with their technology, their management of

the labouring class, and their appetites. Subsistence gets redefined as endless want, endless consumption, endless competing for status. The system eventually self-destructs, and a new one takes its place. Similarly, we might improve Marxism by adding Harris's ecological factors to help explain the rise of classes and class conflict. Neither theory, taken alone, adequately accounts for the past. Together, they might work more effectively, each supplying the other's shortcomings. In so far as the course of history has been shaped by material forces, and hardly anyone would deny that they have indeed been important, we will undoubtedly need something like that merger of the two theories.

The modes of production are an endless parade of strategies, as complex in their taxonomies as the myriad species of insects thriving in the canopy of a rain forest or the brightly coloured fish in a coral reef. In broad terms, we may speak of such modes as hunting and gathering, agriculture, and modern industrial capitalism. But that is only the bare outline of any full taxonomy. We must also include, as modes, submodes, or variations on them, the history of cowboys herding cattle across a Montana grassland, of dark-skinned fishermen casting their nets on the Malabar coast, of Laplanders trailing after their reindeer, of Tokyo factory workers buying bags of rice and seaweed in a supermarket. In all these instances and more, the environmental historian wants to know what role nature had in shaping the productive methods and, conversely, what impact those methods had on nature.

This is the age-old dialogue between ecology and economy. Though deriving from the same etymological roots, the two words have come to denote two separate spheres, and for good reason: Not all economic modes are ecologically sustainable. Some last for centuries, even millennia, while others appear only briefly and then fade away, failures in adaptation. And ultimately, over the long stretch of time, no modes have ever been perfectly adapted to their environment, or there would be little history.

PERCEPTION, IDEOLOGY, AND VALUE

Humans are animals with ideas as well as tools, and one of the largest, most consequential of those ideas bears the name "nature." More accurately, "nature" is not one idea but many ideas, meanings, thoughts, feelings, all piled on top of one another, often in the most unsystematic fashion. Every individual and every culture has created such agglomerations. We may think we know what we are saying when we use the word, but frequently we mean several things at once and listeners may have to work at getting our meaning. We may suppose too that nature refers to something radically separate from ourselves, that it is "out there" someplace, sitting solidly, concretely, unambiguously. In a sense, that is so. Nature is an order and a process that we did not create, and in our absence it will continue to exist; only the most strident solipsist would argue to the contrary. All the same, nature is a creation of our minds too, and no matter how hard we may try to see what it is objectively, in and by and for itself, we are to a considerable extent trapped in the prison of our own consciousness and web of meanings.

Environmental historians have done some of their best work on this level of cultural analysis, studying the perceptions and values people have held about the non-human world. They have, that is, put people thinking about nature under

scrutiny. So impressed have they been by the enduring, pervasive power of ideas that sometimes they have blamed present environmental abuse on attitudes that go far back into the recesses of time: as far back as the book of Genesis and the ancient Hebraic ethos of asserting dominion over the Earth; or the Greco-Roman determination to master the environment through reason; or the still more archaic drive among patriarchal males to lord it over nature (the "feminine" principle) as well as women. The actual effects of such ideas, in the past or in the present, are extremely difficult to trace empirically, but that has not deterred scholars from making some very large claims here. Nor should it altogether. Perhaps we have too wildly exaggerated a notion of our mental prowess and its impact on the rest of nature. Perhaps we spend too much time talking about our ideas, neglecting to examine our behaviour. But however overblown some of these claims may be, it is certainly true that our ideas have been interesting to contemplate, and nothing among them has been more interesting than our reflections on other animals, plants, soils, and the entire biosphere that gave birth to us. So, for good reason, environmental history must include in its program the study of aspects of esthetics and ethics, myth and folklore, literature and landscape gardening, science and religion—must go wherever the human mind has grappled with the meaning of nature.

For the historian, the main object must be to discover how a whole culture, rather than exceptional individuals in it, perceived and valued nature. Even the most materially primitive society may have had quite sophisticated, complex views. Complexity, of course, may come from unresolved ambiguities and contradictions as well as from profundity. People in industrial countries especially seem to abound in these contradictions: They may chew up the land wholesale and at a frightful speed through real estate development, mining, and deforestation, but then turn around and pass laws to protect a handful of fish swimming in a desert spring. Some of this is simply confusion; some of it may be quite reasonable. Given the protean qualities of nature, the fact that the environment presents real dangers as well as benefits to people, this contradictoriness is inescapable. It has everywhere been true of the human reaction. Yet not a few scholars have fallen into the trap of speaking of "the Buddhist view of nature" or "the Christian view" or "the American Indian view," as though people in those cultures were all simple-minded, uncomplicated, unanimous, and totally lacking in ambivalence. Every culture, we should assume, has within it a range of perceptions and values, and no culture has ever really wanted to live in total harmony with its surroundings.

But ideas should not be left floating in some empyrean realm, free from the dust and sweat of the material world. They should be studied in their relations with those modes of subsistence discussed in the preceding section. Without reducing all thought and value to some material base, as though the human imagination was a mere rationalization of the belly's needs, the historian must understand that mental culture does not spring up all on its own. One way to put this relationship is to say that ideas are socially constructed and, therefore, reflect the organization of those societies, their techno-environments and hierarchies of power. Ideas differ from person to person within societies according to gender, class, race, and region. Men and women, set apart almost

everywhere into more or less distinctive spheres, have arrived at different ways of regarding nature, sometimes radically so. So too have slaves and their masters, factory owners and workers, agrarian and industrial peoples. They may live together or in close proximity, but still see and value the natural world differently. The historian must be alert to these differences and resist easy generalizations about the "mind" of a people or of an age.

Sometimes it is maintained that modern science has enabled us to rise above these material conditions to achieve for the first time in history an impersonal, transcultural, unbiased understanding of how nature works. The scientific method of collecting and verifying facts is supposed to deliver truth pure and impartial. Such confidence is naive. Few scholars writing the history of science today would accept it uncritically. Science, they would caution, has never been free of its material circumstances. Though it may indeed be a superior way of arriving at the truth, certainly superior in its capacity to deliver power over nature, it has nonetheless been shaped by the techno-environment and social relations of its time. According to historian Thomas Kuhn, science is not simply the accumulating of facts but involves fitting those facts into some kind of "paradigm," or model of how nature works. Old paradigms lose their appeal, and new ones rise to take their place. Although Kuhn does not himself derive those paradigm shifts from material conditions, other historians have insisted that there is a connection. Scientists, they say, do not work in complete isolation from their societies but reflect, in their models of nature, their societies, their modes of production, their human relations, their culture's needs and values. Precisely because of this fact, as well as the fact that modern science has had a critical impact on the natural world, the history of science has a part in the new environmental history.

Finally, the historian must confront the formidable challenge of examining ideas as ecological agents. We return to the matter of choices that people make in specific environments. What logic, what passion, what unconscious longings, what empirical understanding goes into those choices? And how are choices expressed in rituals, techniques, and legislation? Sometimes choices are made in the halls of national governments. Sometimes they are made in that mysterious realm of the Zeitgeist that sweeps across whole eras and continents. But some are also made, even in this day of powerful centralized institutions, by scattered households and farmsteads, by lumberjacks and fishing crews. We have not studied often or well enough the implementation of ideas in those microcosms.

Once again, it is anthropologists who have a lot to offer the historian seeking insight and method. One of the most intriguing pieces of fieldwork that comes from them bears directly on this question of ideas at work in the small setting. It comes out of a mountain valley in New Guinea, where the Tsembaga people subsist on taro, yams, and pigs. Published by Roy Rappaport under the title *Pigs for the Ancestors*, it exemplifies brilliantly how one might conceive of humans and their mental cultures functioning within a single ecosystem.

The Tsembaga appear in Rappaport's study as a population engaged in material relations with other components of their environment. Unlike their plant and animal congeners, however, they create symbols, values, purposes, and meanings, above all, religious meanings, out of the world around them. And that culture performs, though at points obscurely

and indirectly, an important function: It encourages the Tsembaga to restrain their use of the land and avoid its degradation. For long periods of time, up to 20 years, these people busy themselves raising pigs, which they accumulate as payment to their ancestral spirits for help in battles with their neighbouring enemies. Then at last, when they feel they have enough pigs to satisfy the spirits, a ritualistic slaughter ensues. Hundreds of the animals die and are consumed on behalf of the ancestors. Now, the debt paid, the Tsembaga are ready to go back to war, confident that they will have divine power on their side again. So their lives go round, year after year, decade after decade, in a ritualistic cycle of pig-raising, pig-slaughtering, dancing, feasting, and warring. The local explanation for this cycle is wholly religious, but the outside observer sees something else going on: an elaborate ecological mechanism at work, keeping the number of pigs under control and the people living in equilibrium with their surroundings.

In this forested valley Rappaport has found an example, assuming the validity of the study, of how a culture can take shape through addressing the problems of living within a peculiar ecosystem. The harmony between the two realms of nature and culture seems in this case to be nearly perfect. But the historian wants to know whether human populations are always as successfully adaptive as the Tsembaga. Moreover, are the people that the historian is most likely to study—people organized in advanced, complex societies, relating to nature through modern rituals, religions, and other structures of meaning and value—quite so successful? Rappaport ventures to suggest that the "ecological wisdom" embodied unconsciously in the New Guinea ritual cycle is by no means common. It is most likely to be found where the household is the primary unit of production, where people produce for immediate use rather than for sale and profit, and where "signs of environmental degradation are likely to be apparent quickly to those who can do something about them." Modern industrial societies, on the other hand, he finds culturally maladaptive. In them an economic and technological rationality has replaced the Tsembaga's ecological rationality. Rappaport's case is therefore of limited application elsewhere. Nor does it explain why a change in rationality has occurred, why cultures have drifted away from ecosystem harmony, why modern religion fails to restrain our environmental impact. Generally, anthropology bows out as those issues arise, retiring to its remote green valleys and leaving the historian to face the grinding, shrieking disharmonies of modernity alone.

As it tries to redefine the search into the human past, environmental history has, as indicated above, been drawing on a number of other disciplines, ranging from the natural sciences to anthropology to theology. It has resisted any attempt to put strict disciplinary fences around its work, which would force it to devise all its own methods of analysis, or to require all these overlapping disciplines to stay within their own discrete spheres. Each may have its tradition, to be sure, its unique way of approaching questions. But if this is an age of global interdependence, it is surely also the moment for some cross-disciplinary co-operation. Scholars need it, environmental history needs it, and so does the Earth.

One discipline not so far explicitly discussed is geography. Environmental historians have leaned on many geographers for insight, on those such as Michael Williams and Donald Meinig among presently active scholars, and from the recent past, names like Carl Sauer,

H.C. Darby, and Lucien Febvre. Over the last century scholars from the two disciplines have crossed into one another's territory often and found that they share much in temperament. Geographers, like historians, have tended to be more descriptive than analytical. Taking place rather than time as their focus, they have mapped the distribution of things, just as historians have narrated the sequence of events. Geographers have liked a good landscape just as historians have liked a good story. Both have shown a love of the particular and a resistance to easy generalizing — a quality that may be their common virtue and strength. But they also bear a neighbourly resemblance in their weaknesses, above all in their recurring tendency to lose sight of the elemental human–nature connection: historians when they have measured time only by elections and dynasties, geographers when they have tried to reduce the Earth and its complexities to the abstract idea of "space." Nature, the land, climate, ecosystems, these are the entities that have relevance. When and where geographers have talked about such forces, they have offered much in the way of information to the new history. More, it has pre-eminently been geographers who have helped us all see that our situation is no longer one of being shaped by environment; rather, it is increasingly we who are doing the shaping, and often disastrously so. Now the common responsibility of both disciplines is to discover why modern people have been so determined to escape the restraints of nature and what the ecological effects of that desire have been.

Put so comprehensively, with so many lines of investigation possible, it may seem that environmental history has no coherence, that it includes virtually all that has been and is to be. It may appear so wide, so complex, so demanding as to be impossible to pursue except in the most restricted of places and times: say, on a small, scarcely populated island well isolated from the rest of the world and then only for a period of six weeks. Historians of every sort will recognize that feeling of being engulfed by one's subject. No matter how inclusive or specialized one's perspective, the past seems these days like a vast buzzing confusion of voices, forces, events, structures, and relationships defying any coherent understanding. The French speak bravely of doing "total history." History is everything, they say, and everything has a history. True and noble that realization may be, but it does not give much ease of mind. Even delimiting some part of the totality as "environment" may seem to leave us with the still unmanageable burden of trying to write the history of "almost everything." Unfortunately, there is no feasible alternative open to us any longer. We did not make nature or the past; otherwise, we might have made them simpler. Now we are challenged to make some sense of them—and in this case, to make sense of their working intricately together.

Chapter Two
The Uses of Environmental History

William Cronon

When I first started teaching a lecture course on American environmental history at Yale over half a decade ago, I came to the end of the semester feeling that despite all the rough spots and gaps, it had gone as well as I could have expected. My ordinary practice on such occasions is to distribute teaching evaluations during the penultimate week of classes so I can read students' comments and report back to them on what they collectively see as the strengths and weaknesses of the course. When I did this for the new environmental history class, I was taken aback to discover that despite my students' enthusiasm for the course, the vast majority seemed profoundly depressed by what they had learned in it. I was unprepared for this reaction. What my students had apparently concluded from their encounter with my subject was that the American environment had gone from good to bad in an unrelentingly depressing story that left little or no hope for the future. Because my own feelings about the matter were not nearly so bleak, I had not intended to lead students to this dreary conclusion, and the more I thought about it, the more it seemed to me that I

had no right to end the course on such a note. Whether or not my students' sense of despair was justified, I did not think it was a particularly useful emotion, either personally or politically. To conclude that the environmental past teaches the hopelessness of the environmental future struck me as a profoundly disempowering lesson—albeit a potentially self-fulfilling one—and I felt that my responsibility both as a teacher and as someone who cares about the future must be to resist such a conclusion.

I therefore wrote a final lecture that ended the class on a deliberately upbeat note with a very personal set of reflections about lessons I had extracted from my study of environmental history—the morals I drew from its stories—and the reasons why I continue to remain hopeful despite all the apparent reasons for feeling otherwise. Leaving aside my own worries about the appropriateness of temporarily turning my lectern into the secular equivalent of a pulpit, I'm persuaded that it was the right thing to do, for my students seemed genuinely grateful for this unusual bout of sermonizing on my part. I still end my environmental history course with a similar lecture. And yet I also think there's something odd about an academic subject that seems to require such an antidote against despair. Certainly I've never felt the need for a comparable closing lecture in my classes on the history of the American West, where I suspect that a residue of frontier optimism and high spiritedness somehow combine with moral outrage and regional pride to produce more ambiguous lessons. Because I've also encountered this sense of despair not just among students but among readers as well, I think it's worth asking why environmental history seems regularly to provoke such a response. A more general way of framing the question is to ask how our study of the environmental past affects our sense of the environmental present and future. Perhaps the simplest way to put this is just to ask: What are the uses of environmental history?[1]

Do practitioners of environmental history have special reason to worry about their field's usefulness? Yes. Like the several other "new" histories born or re-energized in the wake of the 1960s—women's history, African-American history, Chicano history, gay and lesbian history, and the new social history generally—environmental history has always had an undeniable relation to the political movement that helped spawn it. The majority (but not quite all) of those who become environmental historians tend also to regard themselves as environmentalists. And so it is no accident that many of the most important works in the field approach their subjects with explicitly present-day concerns. Any number of environmental histories have clearly been framed to make contemporary political interventions. Roderick Nash's *Wilderness and the American Mind* has played a significant role in helping frame debates about wilderness protection in the three decades since its publication.[2]

Samuel Hays's *Conservation and the Gospel of Efficiency* and *Beauty, Health, and Permanence*, though less obviously partisan in their politics than Nash's book, speak just as powerfully to major trends in conservation and environmental politics in the twentieth century.[3] Among the most consistently interventionist of environmental historians has been Donald Worster, whose unflinching moral vision has never failed to produce works of history that are also passionately committed to change. *Nature's Economy* critiqued the twentieth-century evolution of ecological science by seeking to rehabilitate an older natural-history tradition that had fallen into disrepute with many modern ecologists, while *Dust Bowl* and *Rivers of Empire* located the origins of environmental degradation in capitalist world views and modes of production that are as alive in the present as they have been in the past.[4] Carolyn Merchant joined Worster in bringing an environmentalist perspective to the history of science, but combined it with a more feminist approach to argue in *The Death of Nature* that Western science has harmed nature and women in parallel ways; her *Radical Ecology*, though less historical, is still more activist in its efforts to intervene in contemporary political struggles.[5] Even scholars whose work has been less explicitly political have consciously sought to make it relevant to contemporary environmental concerns. Joel Tarr's many studies of pollution and waste streams have always aimed to address the concerns of contemporary policy makers, while Stephen Pyne's epic histories of fire have consistently tried to persuade present-day resource managers of the complexity of their task.[6] Pyne has even gone so far as to author a textbook on fire management practices.[7] And so on and on. The list of such interventions is long, and applies in varying degrees to the majority of historians who work in this field. So I think we can take it as a given that many if not most environmental historians aspire to contribute to contemporary environmental politics: they want their histories to be useful not just in helping us understand the past, but in helping us change the future.

THE PROBLEM OF AUDIENCE

How successful have we been at this? Or to put it a little less comfortably, just how useful have our contributions been so far? (For now, I leave aside the even more uncomfortable question of how useful we are being when so many who study our work apparently find in it a counsel of despair.) One way to start answering these questions is to think about the different audiences our work has been intended to address. Questions about whether environmental history is useful can only be answered—explicitly or implicitly— relative to the people or things we seek to reach and help. *Useful to whom?* Whom do we see as our chief audiences, and how do *they* define usefulness? These are among the most basic questions any writer or teacher can ask. Each of our different audiences in some sense represents a different occasion for usefulness, with different opportunities and risks that follow from trying to attend to its needs and interests. Let me offer a brief guide to the folks I think we've been trying to reach.

One audience, obviously, is our fellow historians. The number of major academic and literary prizes won by environmental historians over the past couple of decades is proof that our colleagues have been paying attention and are at least a little intrigued by what we've been up to. With this audience, we have an opportunity to make the case that "nature" is a fundamental

category of historical analysis, no less important than—indeed, deeply entangled with—class, race, and gender. Moreover, our project of exploring the human past as part of a web of systemic relationships within the natural world offers exciting opportunities for seeing things whole at a time when the historical profession seems desperately in need of such synthesis. More than most of the other "new" histories, environmental history erodes the boundaries among traditional historical subfields, be they national or thematic, and suggests valuable new ways of building bridges among them. The risk here is much like that of every other academic field: as a discipline matures, it tends to become ever more self-referential, less accessible to a wider audience, so that its practitioners increasingly talk only to each other. Valuable as it may be for us to demonstrate that our approach constitutes a significant contribution to academic history, we must also guard against focusing too narrowly on purely disciplinary imperatives that may distract us from larger and more important agendas.

Much the same thing can be said about our colleagues in other academic fields, from the humanities to the social and natural sciences. If the case is strong that environmental history offers an unusual opportunity for synthesis across historical subfields, it is even stronger for the many other disciplines that analyze environmental change. Environmental history has already demonstrated its ability to draw on the insights of radically different fields—ecology, geography, economics, anthropology, and many others—in its attempts to construct a more fully integrated synthesis.

Moreover, it has generally been far more successful than most allied disciplines in making these insights available to wider audiences, probably because of the narrative literary styles that remain much stronger in history than in other academic fields.[8] But the risks we face in speaking to our non-historical academic colleagues go beyond the usual danger of academic self-referentiality that I have already mentioned. Quite simply, it takes an awful lot of work to communicate with colleagues in other fields, and there are few institutional rewards for doing so. One does not generally get jobs, promotions, or tenure by teaching the basics of one's own discipline to people on the other side of campus who haven't thought about history since high school. Scientists often react to our eclecticism and our contextualized, narrative styles of explanation with more than a little suspicion that we lack rigour; in trying to defend ourselves against such suspicions, we may drift unconsciously toward seeking alien forms of rigour that our field can never attain. At an even more basic level, to speak to such folks in the first place, one has to spend considerable energy just learning their vocabulary—a vocabulary for which most of our fellow historians have little use and less patience. And so the risk we run, especially if we are young scholars trying to get established in our own discipline, is to inhabit an intellectual space so liminal that no one will adequately recognize the merits of our work. In trying to absorb and respond to the complicated agendas of other disciplines, we run the risk of not adequately serving our own.[9]

But the "usefulness" of environmental history is surely not limited to our fellow academics. If our histories are to help change the world, they must reach beyond the walls of the academy to affect the views of people who do more than just study the past. Under this heading fall many different groups. One is the policy makers,

who represent an especially seductive opportunity. By challenging us to focus our research on very concrete modern problems, they tempt us to believe that the insights we contribute may actually influence the course of events in the real world. By speaking to power, so the story goes, we may capture a little of that power for ourselves. And yet there is considerable risk here too. By taking as our starting point only the questions that policy makers ask, we may misspecify the terms of our own analysis, treating as givens the very categories we should be subjecting to criticism and thereby ignoring structural causes that may not be so malleable to current policy or management tools. Worse, the prospect of wielding power may tempt us to see reality through the eyes of power. This in turn leads us away from critiques that locate the roots of environmental problems in the very power we are seeking to influence or wield.

Comparable opportunities and risks attend our efforts to write history that speaks to environmental activists who may lack formal power but may be no less involved in the policy process. Since many environmental historians are uncomfortable with power and can more readily imagine protesting a policy than implementing it, they often see fellow environmentalists as their natural audience—and indeed have much to offer that audience. When, for instance, we write about the successes and failures of past organizing efforts, it's nice to think that our work might empower contemporary movements, helping them avoid past mistakes by focusing on efforts and initiatives that seem most likely to produce positive environmental change. But much like the policy makers, activists usually care more about effective strategies and usable stories than they do about good history.

Both groups share an instrumental view of the past that entails a search for "what works." Just as in the policy arena, this intensely practical focus may discourage analyses that explain environmental problems in relation to deep structural forces that may not be responsive to grassroots organizing. Furthermore, activists often seek provocative stories that can serve as inspiring moral fables with clear heroes and villains. Neither of these impulses may be conducive to good history, since they tempt us toward what might be called environmentalist realism—a genre no more aesthetically pleasing or intellectually compelling than socialist realism, and in the long run no more effective.[10]

If policy makers and activists both constitute dangerously narrow audiences, one might think we would do well to go off in search of that holy grail of crossover academic writing, the "general public." This is high on my own list of priorities, since I believe environmental history can profoundly inform public understanding of contemporary environmental issues by placing those issues in a broader historical context. Doing so increases people's understanding not just of the environment, but of history itself: the very eccentricity of our field makes it a highly attractive way to reinvigorate public interest in history and demonstrate the relevance of the past to the present. But we all know this is an uphill battle, given the low level of American public awareness of history in general. The mistaken assumptions and romantic myths that many people bring not just to history but to nature create endless distortions and misreadings that can defeat even our best-intentioned efforts at education. Moreover, the public fascination for "newness" (itself a consequence of short memories and weak historical consciousness) tempts

the historian into bold overstatement and provocative storytelling that potentially obscure one of the most important qualities of the past: its twinned strangeness and familiarity, its frequent tendency to pair the most ordinary causes with the most extraordinary effects and vice versa. In the end, our efforts to provoke the public with "new" stories may ultimately prove self-defeating once those stories too begin to seem "old."

But there is perhaps one other, more ultimate audience whose needs we seek to articulate and whose standards we hope to meet: non-human nature, the Earth itself. This will no doubt seem an odd, even mystical, item to include on my list, since nature neither speaks our language nor reads our books and so can't really be an "audience" for our work in any meaningful sense. And yet I'm sure that many environmental historians measure the "usefulness" of what they do in precisely this way: by whether or not it contributes to the health and integrity of natural systems.[11] In this sense, one of the richest and most exciting challenges of our field is the chance to enlist historical scholarship in the service of improving human relationships with nature. Simply put, we are trying to write histories that speak as much for the Earth and the rest of creation as they do for the human past. And yet inevitably, here too there are deep problems. In trying to speak on behalf of this non-human audience that can never talk back in the language we ourselves use, we can never finally be completely sure that we've gotten the story right, or that our own definition of "usefulness"—a peculiarly human concept if ever there was one— matches the conditions that drive natural systems.[12] Given the anthropocentrism that governs utilitarianism and narrative alike, any search for the "uses" to which nature itself might put our environmental histories is fraught with uncertainty—if not absurdity.

Our conclusions about the problem of audience must thus be ambiguous. We cannot escape the dilemma it poses, for if we fail to consider just whom we are addressing, our work won't even be read, let alone be useful. On the other hand, the competing needs of our different audiences can either tempt us to become so narrowly academic that we forget what it means to be useful, or encourage us to become so pragmatic, polemical, or present-minded that we forget what it means to do good history. In trying to discover the "uses" of environmental history, we perennially find ourselves between the Scylla of our disciplinary commitment to the autonomy of the past, and the Charybdis of our concern about modern problems seemingly so prodigious that they threaten to overwhelm all our traditional ways of understanding the ties that link past, present, and future. The difficulty of navigating between the rock of history and the whirlpool of prophecy in a world where we supposedly face both the death of nature and the end of history is no small reason why so many of our audiences despair after hearing our stories.[13]

WHAT WE'VE LEARNED

All of this no doubt seems pretty vague and abstract, so let me offer a more concrete description of the useful lessons that environmental historians have thus far taught us in their work. There are two ways of doing this. I can either tally up a long list of practical lessons that have important implications for very specific environmental phenomena, or I can make a few much more general observations about the peculiar benefits that flow from thinking historically as we consider human relationships with nature. My own preference is for the latter task, if only because the former is potentially so endless. But before moving on, let me at least suggest the kinds of practical lessons I think can be drawn from our work. Here are just a few of my personal favourites:

- When people buy and sell things in a market, they link together ecosystems and encourage change, rarely understanding the full ecological implications of what they are doing. Along with many others, this has been a central concern of my own work, and I can restate it with one of a favourite metaphor: the more complicated the paths in and out of town, the more obscure they become and the easier it is to forget them.[14]

- Tools and technology are immensely important in shaping natural environments, but their effects are powerfully mediated by the cultures in which they are embedded.[15]

- When people migrate from one ecosystem to another, they carry with them other organisms—plants, animals, microbes—whose success or failure in the new location is often crucial in determining the success or failure of the migration.[16]

- Having learned to enjoy the spectacular effects of an oxidizing environment, people the world over have long been inordinately fond of fire, thereby reshaping the world around them in the service of their pyromania.[17]

- Men and women often experience the world in very different ways, so that one cannot hope to understand the way a culture relates to an environment without examining the ways it engenders the natural world.[18]

- "Ideas of nature ... are the projected ideas of men."[19]

Such lessons as these are still quite general, but I can list others that are much more focused:

- Early conservationists were obsessed with questions of economically efficient production, while later environmentalists have been equally obsessed with questions of ecologically responsible consumption.[20]
- A capitalist ethos, in combination with an economic cycle of boom and bust and an unusually long drought, was the principal cause of the environmental disaster known as the Dust Bowl—and, by extension, of other disasters as well.[21]
- People mismanage fish (and any other common property resource) when they misunderstand the dynamics of ecosystems and apply to them too rigid a definition of sustainable production.[22]
- In American history, the horse was not simply a European invader, but a complex cultural entity that became attached to different human communities in very different ways: an English colonial horse was very different from a Spanish conquistador horse was very different from a Comanche trading horse was very different from a Sioux raiding horse was very different from a Pawnee herding horse.[23]
- If you want to understand people's environmental values, watch what they throw away and how they do the throwing—and take a look

at what they do with plastic pink flamingos as well.[24]

- Beware of bugs that come from afar.[25]

I could go on indefinitely with these lists, piling up the many lessons, large and small, that have made environmental history such an exciting field for the past quarter century, but I trust I have made my point. Arguments such as these are the meat of our subject, the news we have to share with the rest of the world, and I think we can be rightly proud of the contributions we've made and are continuing to make. These insights—when situated in a particular place and time—are the concrete goals of our historical practice, for history ceases to be history when it cuts itself loose from concrete particularities. And yet I think we also have deeper lessons that are equally valuable, lessons that have less to do with our actual findings than with the ways we've done the finding.

One reason I emphasize the importance of our historical practice is that there are impulses within environmentalism that are quite strongly ahistorical or even anti-historical, placing environmental history in some considerable but little noticed tension with the larger political movement that helped spawn it. This tension is fascinating in its own right, and it significantly complicates the already difficult task that environmental historians face in trying to make themselves "useful" to their fellow environmentalists. One of the long-standing impulses that environmentalism shares with its great ancestor, romanticism, has been to see human societies, especially those affected by capitalist urban-industrialism and the cultural forces of modernity, in opposition to nature. Ironically, environmentalism often commits itself to a fundamentally dualistic vision even as it appeals for holism. According to the standard terms of this dualism, nature is assumed to be stable, balanced, homeostatic, self-healing, purifying, and benign, while modern humanity, in contrast, is assumed to be environmentally unstable, unbalanced, disequilibrating, self-wounding, corrupting, and malign.

Implicit in this opposition is the belief that ideal nature is essentially without history as we know it, save on the very long time-scales that affect plate tectonics, biological evolution, and climatic change. Another way of putting this is to say that natural time is cyclical time, while the time of modern humanity is linear. Time's cycle is the proof of nature's self-healing homeostasis and equilibrium, while time's arrow is the proof of humanity's self-corrupting instability and disequilibrium. Humanity's arrow is the fall, while nature's cycle is salvation.[26] These metaphorical dualisms are among the most powerful in our culture, with roots that stretch back literally to biblical times, and by stating them in this way I do not intend to critique one or the other half of their implied dialectic. As with most dualisms, both poles of the opposition reveal important truths even as they work to disguise their mutual interdependence. I simply want to note that the environmentalist affection for natural equilibrium and cyclical time as the Archimidean foundation from which to judge the human drama as it unfolds in linear time necessarily implies a not-so-disguised flight from history. The natural or primitive Utopia that serves as counterpoint for so many environmentalist critiques of modern society posits a rupture between past and future so radical as to imply what Francis Fukuyama would call an "end of history."[27]

However one may feel about this Utopian environmentalist vision—and it has many attractive features—it collides at numerous points with the intellectual agenda that environmental historians have set for themselves. Our task, after all, far from trying to escape from history into nature, is to pull nature itself into the stream of human history. Whatever affection we may feel for the attractions of cyclical time and natural equilibrium, our chief stock in trade is linear time and disequilibrium: we study change. Perhaps one might argue that this is a temporary phenomenon. Maybe, for instance, we tell linear narratives of environmental degradation as moral fables whose purpose is to transform people's consciousness and behaviour in ways that will ultimately mean an end to linear time, heralding the coming millennium when cyclical time will reign once again over a stable equilibrium that applies as much to humanity as to nature. But I'm frankly dubious that many of us really believe this: most historians have pretty powerful negative reactions to pronouncements like Fukuyama's about "the end of history"—and not just because we have a professional vested interest in linear time!

The assumptions of our discipline more or less commit us to the task of historicizing everything we study, whether it be human cultures or natural systems. We know all too well that modern Americans have attitudes toward the natural world profoundly different from those of the Native peoples who first inhabited this continent, just as we know that the plants and animals that share the American landscape with us have been significantly affected by those different attitudes. The more we study the history of cultural and environmental systems, the more difficult it is not to be impressed by how dramatically those systems have changed over time. Even our ideas of nature as a repository for sacred and eternal values—values that are among the bedrock foundations for environmental ethics that many of us would embrace—are themselves products of very specific cultural histories. We can trace their stories back through romanticism to earlier cultural vocabularies in which words like the sublime, the picturesque, the pastoral, and the beautiful served as the trail markers for a complex convergence of beliefs drawn from antiquity, from Judeo-Christian traditions, and from the newly emerging philosophies of the Enlightenment. Just as the historicizing impulse of the nineteenth century helped erode the traditional biblical authority of received religion (a movement that in the guise of Unitarianism and Ralph Waldo Emerson's Transcendentalism also supplied some of the roots of American romantic values about nature), so too does the historicizing impulse of environmental history potentially challenge some of the more unreflective assumptions on which environmentalism tries to ground its own authority.

Is this a bad thing? I think not. If the grounding assumptions of modern environmentalism are susceptible to criticism for being historically naive, then surely they deserve to be criticized. We shouldn't evade that task for fear that it will weaken the larger political movement, since any movement worth defending—as environmentalism surely is—can only be strengthened by fostering rigorous critical analysis and debate. In a very different context, Eugene Genovese once wrote of socialist historians that "[w]e are so convinced we are right that we believe we have nothing whatever to fear from the truth about anything.... Our pretensions, therefore, lead us to the fantastic idea that all good (true, valid, competent) history serves our interest and that all poor (false, invalid, incompetent) history

serves the interest of our enemies—or at least of someone other than ourselves."[28] Although I've never been able to muster quite this level of self-assurance about my own political beliefs, I share Genovese's conviction that it is always best to look at the world with clear eyes. Indeed, I believe that historical habits of thought are profoundly valuable, offering our best antidote to naive assumptions, decontextualized arguments, excessive generalizations, and plain old-fashioned wishful thinking—all of which pose problems for contemporary environmentalism. It is here, I think, that we will discover the most important uses of environmental history.

THINKING LIKE A HISTORIAN

Let me move toward a close by offering what seem to me to be some of the core lessons that make environmental history useful not just in its specific claims but in its habits of thought. I'll state these as a general set of very broad, very simple morals for the stories we've been telling. They are among the deepest articles of faith for at least this environmental historian, articles of faith that I suspect many of my colleagues share.

1. All Human History Has a Natural Context

This is so obvious to most environmental historians that it is almost a truism of our subfield, and yet it is also the claim that seems to come as the greatest surprise to our colleagues. History since the 1930s has had a powerful bias toward cultural determinism, spawned in part as a reaction against the extreme environmental determinism that characterized some fields of history and geography in the pre–Second World War era when racialist theories held sway. The chief defenders of materialist history in the intervening period were Marxists who had their own reasons for de-emphasizing the natural context of human history. Their critics in turn used the attack on Marxism as a reason to reject all determinisms as inherently destructive to human freedom. One important contribution of environmental history, then, has been to reintroduce materialist styles of analysis to the study of past human–environment interactions while trying to finesse a full-blown determinism. Our strategy has been to argue for a dialogue between humanity and nature in which cultural and environmental systems powerfully interact, shaping and influencing each other, without either side wholly determining the outcome. One can restate this prescriptively as follows: *In studying environmental change, it is best to assume that most human activities have environmental consequences, and that change in natural systems (whether induced by humans or by nature itself) almost inevitably affects human beings.* As a corollary, most environmental historians would add that human beings are not the only actors who make history. Other creatures do too, as do large natural processes, and any history that ignores their effects is likely to be woefully incomplete.

2. Neither Nature nor Culture Is Static
This is the historicist argument I've already mentioned. Any vision of a past human place in nature that posits an ideal relationship of permanent stability or balance must defend itself against almost overwhelming evidence to the contrary. Descriptions of historical eras in which human populations were supposedly in eternal equilibrium with equally stable natural systems are almost surely golden-age myths. A comparable rejection of stasis

has occurred within the modern science of ecology, where the notion of a permanent climax community as postulated by Frederic Clements and his followers now seems thoroughly discredited. In its stead, we have a newly dynamic, even stochastic or chaotic ecology in which history plays a crucial role in shaping the pattern and process of ecosystems whether or not people are involved.

Recognizing the dynamism of natural and cultural systems does not, of course, mean that all change is good or that there are no benchmarks for comparing one kind of change with another. Most past societies, for instance, have not altered the natural world at anything like the rate or scale that has typified the modern era. To argue otherwise would be to engage in a different form of myth-making, in which the values and behaviours of different cultures toward nature are assumed to be everywhere and always the same— "economic man" being undoubtedly the most familiar subspecies of the genre. The insights of environmental history tend to be powerfully anti-essentialist, lying in the middle ground between the golden-age myth of permanent equilibrium and the economistic myth of a reductively universal human nature. Our work suggests that nature and culture change all the time, but that the *rate* and *scale* of such change can vary enormously. Perhaps this is why we feel some kinship with a Braudelian vision of history in which the different time scales of *la longue durée, la vie materiélle*, and *l'histoire événementielle* weave together to form the tapestry of the past.[29] Although our general bias is often toward the *longue durée*, we understand that the interactions of environment, economy, political institutions, social norms, cultural values, and natural processes are endlessly complex. Any simple formula for understanding their interactions is almost sure to be wrong. Restated prescriptively, this suggests that *the relationship between nature and culture should always be viewed as a problem in comparative dynamics, not statics.* Naive assumptions about the stability of natural systems can produce behaviour that is as environmentally destructive as it is culturally inappropriate.[30] As a corollary, essentialist arguments about past cultures and environments are almost always historically suspect.

3. All Environmental Knowledge Is Culturally Constructed and Historically Contingent—Including Our Own

On the surface, this will probably seem the most radical challenge that environmental history has to offer environmentalists who regard nature as a source of absolute authority for their vision of how people ought to behave in the world. Here again we encounter the problem of sacred versus historical time. If one is inclined to regard nature as an eternal realm of absolute facts, stable processes, and permanent values, it is not at all reassuring to discover that such beliefs have clear historical roots and that people in other times and other places and other cultures have held very different views. Much of what they took to be permanent and absolute has since changed, and the same will likely happen to many of our own most cherished beliefs as well. The historicist impulse seems to undermine sacred knowledge and replace it with a relativist world in which nature is apparently no more than what we think it is, with literally everything up for grabs. If static nature is our moral compass, then historicism threatens to set us adrift on an unfamiliar sea with no way of taking our bearings.

But one must be careful here, for this lesson can be pushed much too far. It must somehow be paired, however paradoxically, with the implied realism of my first lesson. Most environmental historians take it as a strong article of faith that the natural world exists quite apart from what we believe about it, that it powerfully affects the course of human history, and that if our beliefs diverge too far from its realities, we will eventually suffer at least as much as it will. Recognizing the culturally constructed character of our own knowledge is thus quite different from a claim that the world does not exist, or that people invent it merely as an idea in their heads. Rather, it acknowledges the chastening fact that we can never know nature at first hand. Instead we encounter it only through the many lenses of our own beliefs, cultural institutions, and structures of knowledge, all of which can only hope to approximate natural reality in a mimetic or metaphorical fashion, never actually replicate it. Rather than interpret this argument as a defence of human arrogance—asserting that we can do whatever we like because nature is whatever we wish it to be and will do whatever we want—I prefer to see the constructedness of human knowledge as proof of our own fallibility. The moral I find in this story, in other words, points us toward humility, tolerance, and self-criticism.

This lesson has several corollaries that are well worth noting. However unsettling it may be to become more aware of the historical origins of one's own beliefs, it is also liberating because it encourages us to explore different ways of thinking about the human relationship to nature that our own dogmatic blinders might have prevented us from seeing. Conversely, once one begins to understand the origins of one's own ways of thinking about nature, one may be better able to avoid falling into familiar ruts. One may, for instance, more easily recognize the romantic impulses that sometimes afflict environmentalist thinking, and more easily remember that scientific knowledge is rarely so absolute as its devotees sometimes pretend. One way of understanding our task is to think of trying to synthesize in historical perspective the divergent but complementary approaches of ecology and ethnoecology. Despite their apparent opposition, they are in fact equally valuable and the tension between them can be immensely fruitful.

Let me sum up this third lesson more prescriptively: *Recognizing the historical contingency of all knowledge helps us guard against the dangers of absolute, decontextualized "laws" or "truths," which can all too easily obscure the diversity and subtlety of environments and cultures alike.* An historical, social-constructionist perspective takes seemingly transparent, absolute environmental "facts" and places them in cultural contexts that render them at once more problematic, more interesting, and more instructive. Paradoxically, by making reality more contingent, the historicist approach to knowledge lends greater realism to our understanding of nature and culture alike.

My final lesson may seem oddly put, but seems to me the core of what sets environmental history apart from most other fields that seek to understand and influence the way we relate to the natural world. It describes a peculiar quality that characterizes most historical writing and sets it apart from the social and natural sciences. It is simply this:

4. Historical Wisdom Usually Comes in the Form of Parables, Not Policy Recommendations or Certainties

The significance of this point is hardly intuitive for anyone who is not a historian.

Whenever I lecture to the general public or to scholars in the social or natural sciences, I'm invariably asked afterwards for my predictions about the future course of environmental change. Just as invariably, I explain that historians usually make reluctant prophets, despite the teleological similarities between the stories we tell about the past and the prophecies that others may wish us to make about what will happen in the future. The power of our history derives from the fact that, when speaking about the past, we can at least pretend that we know the end of the story. Doing so enables us to make our arguments and narratives point toward the present and hence seem to explain it, if only for the brief period in which that supposed "ending" continues to hold good. This sense of narrative closure is never available to us for the future, the very contingency of which is what prophecy seeks to contain and resist. Because historians cannot help but respect the awesome, terrifying complexity of past cause and effect, and because we recognize the dangers of ideology even as we embrace it as a necessary consequence of the narrative form, most of us—unlike many of our colleagues in the sciences—are reluctant to predict the future course of events.

This is not to say that we are silent about the future, or that we regard our histories as irrelevant to present concerns. Instead we adopt a much older, albeit less seductively scientific, rhetorical strategy. Rather than make *predictions* about what *will* happen, we offer *parables* about how to interpret what *may* happen. Strange as it may sound, I believe this may be the most important contribution we environmental historians can make in a world where expert knowledge has for the most part forgotten the peculiar form of wisdom that the parable represents.

Santayana was probably wrong in implying that those who study the past can avoid repeating it, because in fact the past never repeats (and yet always repeats) itself. Instead, any series of past events can seem to resemble almost any other series of events, past or present, while at the same time differing in ways that seem no less important. In struggling to compare past and present so as to draw lessons for the future, we inevitably turn to analogy as one of our chief analytical tools. Analogy, alas, is never clean, is always subject to criticism, can often have diametrically contradictory implications, and is one of the reasons we historians rarely aspire to certainty in the parallels and differences we draw between past and present. But these problems with analogical reasoning are also one of its chief strengths: It continually reminds us that we are engaged in an interpretive, hermeneutic enterprise, not a quest for absolute knowledge, and that competing interpretations about the meaning of the past for the present are not only possible but inevitable. Analogy is the logical foundation for metaphor and parable alike, all three of which are near the heart of our scholarly practice. The job of historical scholarship is to provide the richest possible contextual field within which to frame and discipline our analogies, not because we expect historical insight to give absolute answers—it won't—but because it is the best source we have for *questions* whose subtlety and complexity can mirror that of the world we wish to understand. It is our own best route to mimesis, self-knowledge, and—to repeat again that old-fashioned word—wisdom.

Hence the affection we historians feel for the parable: by seeing the past as a story to be told rather than as a problem to be solved, we leave ourselves open to analogies, metaphors, resonances, and interpretive contexts that would probably be obscured by a more rigidly rule-bound analytical approach. In their book *Thinking in Time: The Uses of History for Decision-Makers*, Richard E. Neustadt and Ernest R. May label this approach "Goldberg's Rule" and, appropriately enough, tell a story to explain the label. After describing to a class of corporate executives the historian's habit of explaining past events by telling stories about them, one of their students, Avram Goldberg, responded by exclaiming, "Exactly right! When a manager comes to me, I don't ask him, 'What's the problem?' I say, 'Tell me the story.' That way, I find out what the problem *really* is."[31] What distinguishes environmental historians from environmental scientists and policy experts is our tendency to frame our work around one common question: "What's the story?" Moreover, like most modern historians, we have a special fondness for stories that convey a sense of irony, because irony best expresses our sense of the multivalent complexity of the world. It reflects one of the central insights our field explores, which is that whenever people act to change the natural world, the ensuing story has unexpected endings, because our actions seem always to have unexpected consequences. This in turn suggests a deeper moral still about the incompleteness of our knowledge of the world and the unexamined assumptions we have made about it.

To repeat: *Environmental history is at least as important for the way it asks and answers questions—by analogy, metaphor, and parable and the search to discover their meanings—than for any specific problems it may actually solve.* As such, it is a powerful and indispensable antidote to scientific and analytical approaches that aspire to greater and more unitary certainty in their search for knowledge.

GROUND FOR HOPE

Is telling parables about nature and the human past a useful thing to do? Yes. I believe so in my bones, which is what I told my students when they expressed despair about the seemingly hopeless lessons they thought they had learned from our course in environmental history. Let me close by returning for a moment to my secular pulpit to repeat some of the articles of faith I shared with those students.

The answers we environmental historians give to the question "What's the story?" have the great virtue that they remind people of the immense human power to alter and find meaning in the natural world — and the even more immense power of nature to respond. At the same time, they remind us that whatever we do in nature, we can never know in advance all the consequences of our actions. This need not necessarily point toward despair or cynicism, but rather toward a healthy respect for the complexity and unpredictability of history, which is much akin to the complexity and unpredictability of nature itself. The proper lesson of such complexity, I believe, should be to teach us humility. It should make us more critical of our own certainty and self-righteousness, and deepen our respect for the subtlety and mystery of the lives we lead on this planet, entangled as we are in the warp and woof of linear and cyclical, secular and sacred time.

Humility and constant attentiveness to that which we do not know seem to me essential to what we might call honesty in our relationships with each other and with the world around us. We can't not act if we are to remain alive — we have to use nature, we have to participate in the earthly webs of killing and consumption that sustain every creature on this planet —

but we must also act carefully — *act with care* — being as attentive as we can be to the consequences of what we do. The chief moral of my own version of environmental history is the one I tried to embed in the title of my book *Changes in the Land*. To live as human beings on this planet is to change the world around us. That much is inescapable. Environmental history tries to reconstruct the endless layers of change that we and the Earth have traced upon each other. It is the history recorded in Aldo Leopold's tree rings, the history recorded by the marks of his saw upon the good oak as he cut it down, the history recorded by the memories in the hatted head with its shadow on the stump: All of these are inextricably bound together.[32] There cannot be people outside of nature; there can only be people *thinking* they are outside of nature. By the same token, in the world in which we now live, there cannot be a nature separate from humanity. We are in this together as the *Whole Earth Catalog* once declared: "we are as Gods, and might as well get good at it."[33]

Tracing patterns on the landscape is something all living creatures do, and people are the furthest thing possible from an exception to this rule. The lines and shapes we draw on the land reflect the lines and shapes we carry inside our own heads, and we cannot understand either without understanding both at the same time. This means that the material history of environmental change is simultaneously a spiritual history of human consciousness and a political-economic history of human society. They can never finally be separated from each other, and it would be foolish even to try. I find a mysterious sort of wonder and beauty in that fact. Even our most abstract, grid-like shapes upon the land are also statements about our different visions of community: amongst ourselves and other people, ourselves and other living creatures, ourselves and the Earth itself. The struggle to live rightly in relation to the Earth and its creatures does not end, and the problems it poses are never solved. In seeking to tame the Earth, we have taken upon ourselves the burden of tending and caring for the garden we have sought to make of it. We have become responsible for the Earth, and must now accept the moral consequences of that fact. In caring for the Earth and its creatures we must also learn to care for ourselves, because taming nature with respect and love means taming ourselves as well.

These are the moral dilemmas to which the parables of environmental history must always return. In the particularism of its storytelling—its focus on particular people at particular times in particular landscapes—environmental history reminds us of the endlessly diverse human ways of using and living in nature. I personally take considerable solace in this diversity and particularism, for they remind us that—all appearances to the contrary, even in an era of "Global Change"—there is not One Big Problem called "The Environment." There is rather a near infinitude of smaller problems, each expressing a different relationship of use and meaning between people and the world around them. Although we will never solve the One Big Problem that does not in fact exist, we can never stop solving those smaller environmental problems, which together come very close to defining what it means to be alive. All of us change the world around us, and yet different people choose to confront their problems and make their changes in strikingly different ways. The diversity of their experiences, past and present, can serve almost as a laboratory for exploring the multitude of choices we ourselves face. Stories about the past lives of such people teach us how difficult it is to act in ways that benefit humanity and nature both—and yet how crucial it is to try. By telling parables that trace the often obscure connections between human history and ecological change, environmental history suggests where we ought to go looking if we wish to reflect on the ethical implications of our own lives.

And that, on reflection, seems quite a useful thing to do.

NOTES

1. My title and central question are borrowed, of course, from Herbert Joseph Muller, *The Uses of the Past: Profiles of Past Societies* (New York: Oxford University Press, 1957).
2. Roderick Nash, *Wilderness and the American Mind* (New Haven: Yale University Press, 1967, 1973, 1982).
3. Samuel P. Hays, *Conservation and the Gospel of Efficiency: The Progressive Conservation Movement, 1890–1920* (Cambridge: Harvard University Press, 1959); Samuel P. Hays, *Beauty, Health and Permanence: Environmental Politics in the United States, 1955–1985* (New York: Cambridge University Press, 1987).
4. Donald Worster, *Nature's Economy: The Roots of Ecology* (San Francisco: Sierra Club Books, 1977); Donald Worster, *Dust Bowl: The Southern Plains in the 1930s* (New York: Oxford University Press, 1979); Donald Worster, *Rivers of Empire: Water, Aridity, and the Growth of the American West* (New York: Pantheon, 1985).
5. Carolyn Merchant, *The Death of Nature: Women, Ecology, and the Scientific Revolution* (New York: Harper & Row, 1960); Carolyn Merchant, *Radical Zoology: The Search for a Livable World* (New York: Routledge, 1992).

6. Joel Tarr's published output is prodigious, but representative examples include "The Search for the Ultimate Sink: Urban Air, Land, and Water Pollution in Historical Perspective," *Records of the Columbia Historical Society of Washington, D.C.* 51 (1984): 1–29; Joel Tarr and K. Koons, "Railroad Smoke Control: A Case Study in the Regulation of a Mobile Pollution Source," in Mark Rose and George Daniels, eds., *Energy and Transport: Historical Perspectives on Policy Issues* (Beverly Hills: Sage, 1982), 71–92; and *Retrospective Assessment of Waste Water Technology in the United States: 1800–1972*, A Report to the National Science Foundation/RANN, October 1977, with F.C. McMichael et al. Pyne's classic work in American history is Stephen J. Pyne, *Fire in America: A Cultural History of Wildland and Rural Fire* (Princeton: Princeton University Press, 1982).

7. Stephen J. Pyne, *Introduction to Wildland Fire: Fire Management in the United States* (New York: John Wiley, 1984).

8. For this reason, I also believe that environmental history is an almost ideal subject for bridging the deep chasm that separates the natural sciences from the rest of the modern university, thereby offering a potentially crucial way of defending a coherent vision of liberal education in institutions that sometimes seem to have forgotten the meaning of that phrase. Although I will not elaborate this argument explicitly in the pages that follow, it is implicit in everything I say.

9. A subtler intellectual risk of an interdisciplinary field like environmental history is that its less skillful practitioners, as well as students just beginning their studies, may sail out into the waters of several disciplines before they have quite mastered one. Too often we forget that by becoming steeped in a single discipline—an act we often criticize as "narrowing"—we gain a crucial experience in rigour. How to retain this sense of rigour and make it serve as our intellectual compass as we venture out across disciplinary boundaries is perhaps the greatest single challenge of graduate training in environmental history.

10. Much the same thing can be said about laudable recent efforts to broaden environmental history (and one hopes environmentalism as well) to include groups other than the well-to-do White folks (many of them male) who have for the most part dominated environmental politics. Among those whose stories can only contribute to the diversity and richness of environmental history are women, multicultural people of colour, poor people, and workers. But again there's a temptation toward white-hat-black-hat narratives in which oppressors and victims conduct their struggles in degraded landscapes that simply mirror the terms of social oppression in too mechanically predictable a way. Moreover, the recent history of multiculturalism suggests that there are special dangers here of essentialist styles of reasoning that can be quite ahistorical.

11. At this and several other points in this essay, I trust that readers will hear my echoes of Aldo Leopold's *Sand County Almanac* (New York: Oxford University Press, 1949), 224–225.

12. This suggests one important way in which environmental history differs from the other "new" histories of post-1960s historiography. Whereas fields like women's history and African-American history have sought to recover the "lost" voices of "ordinary people" by letting their subjects "speak for themselves," we can never hope to discover quite so certain or autonomous a voice for the natural actors that participate in our own narratives. Their silence must remain deeper and more profound, and their stories more genuinely alien from our own.

13. The reference to the death of nature echoes Merchant's *Death of Nature* and Bill McKibben, *The End of Nature* (New York: Random House, 1989); the reference to the end of history is to Francis Fukuyama, *The End of History and the Last Man* (New York: Free Press, 1992).

14. William Cronon, *Changes in the Land: Indians, Colonists, and the Ecology of New England* (New York: Hill & Wang, 1983); William Cronon, *Nature's Metropolis: Chicago and the Great West* (New York: W.W. Norton, 1991); and William Cronon, "Kennecott Journey: The Paths out of Town," in William Cronon, George Miles, and Jay Gitlin, eds., *Under an Open Sky: Rethinking America's Western Past* (New York: W.W. Norton & Co., 1992).

15. See, for instance, Richard White, *Land Use, Environment, and Social Change: The Shaping of Island County, Washington* (Seattle: University of Washington Press, 1980); Richard White, *The Roots of Dependency: Subsistence, Environment, and Social Change among the Choctaws, Pawnees, and Navajos* (Lincoln: University of Nebraska Press, 1983); and Calvin Martin, *Keepers of the Game: Indian-Animal Relationships and the Fur Trade* (Berkeley: University of California Press, 1978).

16. Alfred W. Crosby, Jr., *The Columbian Exchange: Biological and Cultural Consequences of 1492* (Westport: Greenwood, 1972); Alfred W. Crosby, Jr., *Ecological Imperialism: The Biological Expansion of Europe, 900–1900* (New York: Cambridge University Press, 1986).

17. Pyne, *Fire in America*.

18. Merchant, *Death of Nature*; and Carolyn Merchant, *Ecological Revolutions: Nature, Gender, and Science in New England* (Chapel Hill: University of North Carolina Press, 1989).

19. This quotation is one of the wisest and most profound statements in a most wise and profound essay: Raymond Williams, "Ideas of Nature," in Raymond Williams, *Problems in Materialism and Culture* (London: Verso, 1980), 82.

20. Hays, *Conservation and the Gospel of Efficiency and Beauty, Health, and Permanence*.

21. Worster, *Dust Bowl*.

22. Arthur F. McEvoy, *The Fisherman's Problem: Ecology and Law in the California Fisheries, 1850–1980* (New York: Cambridge University Press, 1986).

23. White, *Roots of Dependency*.

24. Here I refer to Martin V. Melosi, *Garbage in the Cities: Refuse, Reform, and the Environment, 1880–1980* (College Station: Texas A&M University Press, 1981); and to the uncompleted Yale doctoral dissertation of Jennifer Price (working title: "Right Maps: Imaginative Encounters with Birds in Modern America"), about American attitudes toward nature as reflected in certain key species of birds.

25. Crosby, *Columbian Exchange* and *Ecological Imperialism*.

26. Among the most accessible discussions of this distinction between time's arrow and time's cycle are Mircea Eliade, *That Myth of the Eternal Return*, translated by Willard R. Trask (New York: Pantheon, 1954); and Stephen Jay Gould, *Time's Arrow, Time's Cycle: Myth and Metaphor in the Discovery of Geological Time* (Cambridge: Harvard University Press, 1987).

27. Francis Fukuyama, *End of History*. Examples of radical or deep ecological critiques of linear time include Bill Devall and George Sessions, *Deep Ecology: Living as if Nature Mattered* (Salt Lake City: Gibbs Smith, 1985); Jeremy Rifkin, *Time Wars: The Primary Conflict in Human History* (New York: Henry Holt, 1987); and Calvin Luther Martin, *In the Spirit of the Earth: Rethinking History and Time* (Baltimore: Johns Hopkins University Press, 1992). But I should note in passing that environmentalists can also tell linear narratives about heroic environmentalism: Nash's whig history of wilderness consciousness and the rights of nature is probably the most obvious case in point.

28. Eugene D. Genovese, *In Red and Black: Marxian Explorations in Southern and Afro-American History* (New York: Pantheon, 1971), 4.

29. This tripartite division occurs in all of Braudel's work, but was most famously articulated in his classic *The Mediterranean and the Mediterranean World in the Age of Phillip II*, translated by Sian Reynolds (New York: Harper & Row, 1972).

30. An excellent example of the dangerous consequences of naive assumptions about natural equilibria can be found in Arthur McEvoy, *Fisherman's Problem*.

31. Richard E. Neustadt and Ernest R. May, *Thinking in Time: The Uses of History for Decision-Makers* (New York: Free Press, 1986), 106.

32. The echo here is that of Leopold's "Good Oak," *Sand County Almanac*, 6–18.

33. *Whole Earth Catalog: Access to Tools* (Menlo Park: Portola Institute, Spring, 1969), inside front cover.

Chapter Three
Eve
Nature and Narrative

Carolyn Merchant

A Penobscot Indian story from northern New England explains the origin of maize. A great famine had deprived people of food and water. A beautiful Indian maiden appeared and married one of the young men of the tribe, but soon succumbed to another lover, a snake. On discovery she promised to alleviate her husband's sorrow if he would plant a blade of green grass clinging to her ankle. First he must kill her with his ax, then drag her body through the forest clearing until all her flesh had been stripped, and finally bury her bones in the centre of the clearing. She then appeared to him in a dream and taught him how to tend, harvest, and cook corn and smoke tobacco.[1]

This agricultural origin story taught Indians how to plant their corn in forest clearings and also that the Earth would continue to regenerate the human body through the corn plant. It features a woman (the corn maiden) and a male lover as central actors. It begins with the state of nature as drought and famine. Nature is a desert, a poor place for human existence. The plot features a woman as saviour. Through a willing sacrifice in which

her body is returned to the Earth, she introduces agriculture to her husband and to the women who subsequently plant the corn, beans, and squash that provide the bulk of the food sustaining the life of the tribe. The result is an agroecological system based on the planting of interdependent polycultures in forest gardens. The story type is ascensionist and progressive. Women transform nature from a desert into a garden. From a tragic situation to despair and death, a comic, happy, and optimistic situation of continued life results. In this story, the valence of women as corn mothers is good; they bring bountiful gifts. The valence of nature ends as a good. The Earth is an agent of regeneration. Death is transformed into life through a reunification of the corn mother's body with the Earth. Even death results in a higher good.[2]

Into this bountiful world of corn mothers enter the Puritan fathers bringing their own agricultural origin story of Adam and Eve. The biblical myth begins where the Indian story ends—with an ecological system of polycultures in the Garden of Eden. A woman, Eve, shows "the man," Adam, how to pick fruit from the Tree of the Knowledge of Good and Evil and harvest the fruits of the garden. Instead of attaining a resultant good, the couple is cast out of the garden into a desert; instead of moving from desert to garden, as in the Indian story, the biblical story moves from garden to desert. The Fall from paradise is caused by a woman. Men must labour in the Earth by the sweat of their brow to produce food. Here a woman is also the central actress and like the Indian story it contains elements of violence toward women. But the plot is declensionist and tragic, not progressive and comic as in the Indian story. The end result is a poorer state of nature than in the beginning. The valence of woman is bad. The end valence of nature is bad. Here men become the agents of transformation. They become saviours, who through their own agricultural labour have the capacity to recreate the lost garden on Earth.[3]

According to Benjamin Franklin, Indians quickly perceived the difference between the two accounts. Franklin satirically writes that when the Indians were apprised of the "historical facts on which our [own] religion is founded; such as the fall of our first parents by eating an apple, … an Indian orator stood up, …" to thank the Europeans for their story. "What you have told us … is all very good. It is, indeed, bad to eat apples. It is much better to make them all into cider. We are much obliged by your kindness in coming so far to tell us these things which you have heard from your mothers; in return I will tell you some of those which we have heard from ours."[4]

Historical events reversed the plots of the Indian and the European origin stories. The Indians' comic happy ending changed to a story of decline and conquest, while Euramericans were largely successful in creating a New World garden. Indeed,

the story of Western civilization since the seventeenth century and its advent on the American continent can be conceptualized as a grand narrative of fall and recovery. The concept of recovery, as it emerged in the seventeenth century, not only meant a recovery from the Fall, but also entailed restoration of health, reclamation of land, and recovery of property.[5]

The recovery plot is the long, slow process of returning humans to the Garden of Eden through labour in the Earth. Three subplots organize its argument: Christian religion, modern science, and capitalism. The Genesis story of the Fall provides the beginning; science and capitalism the middle; recovery of the garden the end. The initial lapsarian moment (the lapse from innocence) is the decline from garden to desert as the first couple is cast from the light of an ordered paradise into a dark, disorderly wasteland.

The Bible, however, offered two versions of the origin story that led to the Fall. In the Genesis 1 version, God created the land, sea, grass, herbs, and fruit; the stars, sun, and moon; and the birds, whales, cattle, and beasts—after which he made "man in his own image … ; male and female created he them." Adam and Eve were instructed, "be fruitful and multiply, and replenish the earth, and subdue it" and were given "dominion over the fish of the sea, and over the fowl of the air, arid over every living thing that rnoveth upon the earth." In the Genesis 2 version, thought to have derived from a different and earlier tradition, God first created the plants and herbs, next "man" from dust, and then the garden of Eden with its trees for food (including the Tree of Life and the Tree of the Knowledge of Good and Evil in the centre) and four rivers flowing out of it. He then put "the man" in the garden "to dress and keep it," formed the beasts and fowls from dust, and brought them to Adam to name. Only then did he create Eve from Adam's rib. Genesis 3 narrates the Fall from the garden, beginning with Eve's temptation by the serpent, the consumption of the fruit from the Tree of the Knowledge of Good and Evil (which in the Renaissance becomes an apple), the expulsion of Adam and Eve from the garden "to till the ground from which he was taken," and finally God's placement of the cherubims and flaming sword at the entrance of the garden to guard the Tree of Life.[6]

During the Renaissance, artists illustrated the Garden of Eden story through woodcuts and paintings, one of the most famous of which is Lucas Cranach's 1526 painting of Eve offering the apple to Adam, after having been enticed by the snake coiled around the Tree of the Knowledge of Good and Evil. Writers from Dante to Milton depicted the Fall and subsequent quest for paradise, while explorers searched for the garden first in the Old and then in the New Worlds. Although settlers endowed new lands and peoples with Eden-like qualities, a major effort to recreate the Garden of Eden on Earth ultimately ensued. Seventeenth-century botanical gardens and zoos marked early efforts to reassemble the parts of the garden dispersed throughout the world after the Fall and the Flood.[7]

But beginning in the seventeenth century and proceeding to the present, New World colonists have undertaken a massive effort to reinvent the whole Earth in the image of the Garden of Eden. Aided by the Christian doctrine of redemption and the inventions of science, technology, and capitalism ("arte and Industrie"), the long-term goal of the recovery project has been to turn the Earth itself into a vast cultivated garden. The strong interventionist version in Genesis 1 legitimates recovery through

domination, while the softer Genesis 2 version advocates dressing and keeping the garden through human management (stewardship). Human labour would redeem the souls of men and women, while cultivation and domestication would redeem the earthly wilderness. The End Drama envisions a reunification of the Earth with God (the Parousia), in which the redeemed earthly garden merges into a higher heavenly paradise. The Second Coming of Christ was to occur either at the outset of the thousand-year period of his reign on Earth (the millennium) or at the Last Judgment when the faithful were reunited with God at the resurrection.[8]

Greek philosophy offered the intellectual framework for the modern version of the recovery project. Parmenidean oneness represents the unchanging natural law that has lapsed into the appearances of the Platonic world. This fallen phenomenal world is incomplete, corrupt, and inconstant. Only by recollecting of the pure unchanging forms can the fallen partake of the original units. Recovered and Christianized in the Renaissance, Platonism provided paradigmatic ideals (such as that of the Garden of Eden) through which to interpret the earthly signs and signatures leading to the recovery.[9]

Modern Europeans added two components to the Christian recovery project—mechanistic science and *laissez-faire* capitalism to create a grand master narrative of Enlightenment. Mechanistic science supplies the instrumental knowledge for reinventing the garden on Earth. The Baconian-Cartesian-Newtonian project is premised on the power of technology to subdue and dominate nature, on the certainty of mathematical law, and on the unification of natural law into a single framework of explanation. Just as the alchemists tried to speed up nature's labour through human intervention in the transformation of base metals into gold, so science and technology hastened the recovery project by inventing the tools and knowledge that could be used to dominate nature. Francis Bacon saw science and technology as the way to control nature and hence recover the right to the garden given to the first parents. "Man by the Fall, fell at the same time from his state of innocency and from his dominion over creation. Both of these losses can in this life be in some part repaired: the former by religion and faith; the latter by arts and science." Humans, he asserted, could "recover that right over nature which belongs to it by divine bequest," and should endeavour "to establish and extend the power and dominion of the human race itself over the [entire] universe."[10]

The origin story of capitalism is a movement from desert back to garden through the transformation of undeveloped nature into a state of civility and order.[11] Natural resources—"the ore in the mine, the stone unquarried [and] the timber unfelled"—are converted by human labour into commodities to be exchanged on the market. The Good State makes capitalist production possible by imposing order on the fallen worlds of nature and human nature. Thomas Hobbes's nation-state was the end result of a social contract created for the purpose of controlling people in the violent and unruly state of nature. John Locke's political theory rested on the improvement of undeveloped nature by mixing human labour with the soil and subduing the Earth through human dominion. Simultaneously, Protestantism helped to speed the recovery by sanctioning increased human labour just as science and technology accelerated nature's labour.[12]

Crucial to the structure of the recovery narrative is the role of gender encoded into the story. In the Christian religious story, the original oneness is male and the Fall is caused by a female, Eve, with Adam, the innocent bystander, being forced to pay the consequences as his sons are pushed into developing both pastoralism and farming.[13] While fallen Adam becomes the inventor of the tools and technologies that will restore the garden, fallen Eve becomes the Nature that must be tamed into submission. In the Western tradition, fallen Nature is opposed by male science and technology. The Good State that keeps unruly nature in check is invented, engineered, and operated by men. The Good Economy that organizes the labour needed to restore the garden is likewise a male-directed project.

Nature, in the Edenic recovery story, appears in three forms. As original Eve, nature is virgin, pure, and light—land that is pristine or barren, but having the potential for development. As fallen Eve, nature is disorderly and chaotic; a wilderness, wasteland, or desert requiring improvement: dark and witchlike, the victim and mouthpiece of Satan as serpent. As mother Eve, nature is an improved garden; a nurturing Earth bearing fruit; a ripened ovary; maturity. Original Adam is the image of God as creator, initial agent, activity. Fallen Adam appears as the agent of earthly transformation, the hero who redeems the fallen land. Father Adam is the image of God as patriarch, law, and rule—the model for the kingdom and state. These meanings of nature as female and agency as male are encoded as symbols and myths into American lands as having the potential for development, but needing the male hero, Adam. Such symbols are not essences because they do not represent characteristics necessary or essential to being female or male. Rather, they are historically constructed and derive from the origin stories of European settlers and European cultural and economic practices transported to and developed in the American New World. That they may appear to be essences is a result of their historical construction in Western history, not their immutable characteristics.

The Enlightenment idea of progress is rooted in the recovery of the garden lost in the Fall—the bringing of light to the dark world of inchoate nature. The lapsarian origin story is thus reversed by the grand narrative of Enlightenment that lies at the very heart of modernism. The controlling image of Enlightenment is the transformation from desert wilderness to cultivated garden. This complex of Christian, Greco-Roman, and Enlightenment components touched and reinforced each other at critical nodal points. As a powerful narrative, the idea of recovery functioned as ideology and legitimation for settlement of the New World, while capitalism, science, and technology provided the means of transforming the material world.

GRECO-ROMAN ROOTS OF THE RECOVERY NARRATIVE

In creating a recovery narrative that reversed the lapsarian moment of the Fall, Europeans reinforced the Christian image of the precipitous Fall from the Garden of Eden with pagan images of a gradual decline from the golden age. Hesiod (eighth century BC) told of the time of immortal men who lived on Olympus where all was "of gold" and "the grain-giving soil bore its fruits of its own accord in unstinted plenty, while they at their leisure harvested their fields in contentment amid abundance."[14] Ovid, in the *Metamorphoses* (AD 7), pictured the golden age as a time when a bountiful (unplowed) Mother Earth brought forth grains, fruits, honey, and nectar and

people were peaceful, "unaggressive, and unanxious." Only in the decline of the subsequent silver, bronze, and iron ages did strife, violence, swindling, and war set in.[15]

Whereas Hesiod and Ovid offered elements that reinforced the Fall, Virgil and Lucretius introduced components of a recovery story that moved from "savagery" to "civilization." Nature was a principle of development, deriving from the Latin word *nascere*, "to be born." Each stage of development was inherent in the preceding stage, an actualization of a prior potential. The word "nation" derived from the same word, hence the state was born from the state of nature.[16] Virgil (70–19 BC) depicted a narrative of development from nature to nation that moved through four stages mimicking the human life cycle: (1) death and chaos, a world filled with pre-social "wild" peoples (winter); (2) birth and the pastoral, in which people grazed sheep on pastured lands (spring); (3) youth or farming by plowing and planting gardens (summer); (4) maturity, or the city (Rome) in the Garden (fall). For Virgil these four stages were followed by a return to death and chaos, whereas in the Christian myth the recovery was followed by redemption and a return to the original garden. Yet within each of Virgil's stages lies the potential to lapse back prematurely into the earlier chaotic or "savage" state.

The second or pastoral stage is like the Christian Garden of Eden—its loss is mourned and its innocence yearned for—but in the Roman story, it passes "naturally" to the third, or agricultural, stage.

Virgil's *Georgics* narrates the agricultural period in which humans actively labour in the Earth to cultivate it and themselves. Both society's potential and the Earth's potential are actualized and perfected.

When farmers till the ground and tend their crops, nature's bounty brings forth fruits: "Father Air with fruitful rains" descended on the "bosom of his smiling bride" to feed her "teeming womb."[17] The *Aeneid* reveals the fourth stage—the emergence of Rome as a city of culture and civilization within the pastoral and agricultural landscapes—*urbs in horto*—the city in the garden. The four developmental phases of nature and nation exist both temporally as stages and spatially as zones. The city is an actualization of movement from a chaotic "wild" periphery to a pastoral outer zone, a cultivated inner zone, and a "civilized" central place. Because nature is viewed as a cyclical development, the decline and fall of Rome is preordained in the final return to winter and chaos. Yet out of chaos comes a second golden age as "the great line of the ages is born anew." The "virgin" (Justice) returns and a "newborn boy" appears "at whose coming the iron race shall first cease and a golden race will spring up in the whole world." At this point the Roman and Christian versions of a second return converge, offering Europeans and Americans the possibility of the recovery of an Edenic golden age.[18]

Lucretius provides the elements for Thomas Hobbes's origin story of capitalism and the Good State as an emergence from the "state of nature." Lucretius's *De Rerum Naturum (Of the Nature of Things)* closely prefigures Hobbes's *Leviathan*. For both Lucretius and Hobbes the early state of human nature is disorderly, lawless, and chaotic. According to Lucretius, before the discovery of plow agriculture, wild beasts consumed humans and starvation was rampant.[19] But early civilization, nurtured by the taming of fire and the cooking of food, foundered on the discovery of gold, as human greed spawned violent wars. Just as Hobbes saw individual

men in the state of nature as unruly and warlike, so Lucretius lamented that "things down to the vilest lees of brawling mobs succumbed, whilst each man sought unto himself dominion and supremacy." Just as Hobbes argued that people voluntarily gave up their ability to kill each other in the state of nature and entered into a civil contract enforced by the state, so Lucretius held that people out of their own free will submitted to laws and codes. The creation of civil law thus imposes order on disorderly humans, offering the possibility of recovery from the state of nature.[20]

Yet Lucretius's poem, as it came down to the Renaissance, ended not in recovery, but in death, as plague and pestilence overcame Athens. The poem breaks off on a note of extreme pessimism and utter terror; piles of dead bodies burn on funeral pyres and all hope is forsaken. Like Lucretius, Hobbes (who was also deemed an atheist) offered a profoundly pessimistic view of nature, human nature, and divinity. Humans who are basically competitive and warlike contest with each other on the commons and in the marketplace in the creation of a capitalist economy.[21]

Like civilization, nature for Lucretius ends in death and a return to the chaos of winter. As did humans, the Earth "whose name was mother" went through stages of life and death. She brought forth birds, beasts, and humans. The fields were like wombs, and the Earth's pores gave forth milk like a mother's breasts. Yet when the Earth had aged, she was like a worn-out old woman.[22]

In the seventeenth century, the Greek cyclical stories of nature and human society that ended in death and destruction were converted to the Christian redemption story during the battle between ancients and moderns. The declensionist narrative depicting a slide downward from golden age to iron age, from original wisdom to ignorance, from human giants to midgets, was transformed by the hope of recovery. Both nature and human nature were capable of redemption. Science and technology offered the means of transforming nature, labour in the Earth the means of saving human souls. The Earth could be plowed, cultivated, and improved as human beings mixed their labour with the soil. (For Locke, as opposed to Hobbes, the state of nature is good and labour has a positive valence.) Thus both the cultivated Earth and cultivated humans would be prepared for the final moment of redemption, or Parousia, when Earth would merge with heaven, recreating the original oneness. With the discovery of the New World, a New Earth could be reconstructed with the image of the original garden as paradigmatic ideal.

THE AMERICAN HEROIC RECOVERY NARRATIVE

In America, the recovery narrative propelled settlement and "improvement" of the American continent by Europeans. Euramerican men acted to reverse the decline initiated by Eve by turning it into an ascent back to the garden. Using science, technology, and biblical imagery, they changed first the eastern wilderness and then the western deserts into cultivated gardens. Sanctioned by the Genesis origin story, they subdued the "wilderness," replenished the Earth, and appropriated Indian homelands as free lands for settlement. Mercantile capitalism cast America as the site of natural resources, Africa as the source of enslaved human resources, and Europe as the locale of resource management. Timber, barrel staves, animal hides, herbal medicines, tobacco, sugar, and cotton

were extracted from nature in the great project of "improving" the land. Men, as fallen Adam, became the heroic agents who transformed and redeemed fallen Nature.[23]

In New England, European settlers converted a "hideous and desolate wilderness" into "a second England for fertileness" in the space of a few decades. The Pilgrim migration, as recorded in the text of William Bradford, conforms to the six elements of the mythic heroic narrative identified by Russian folklorist Vladimer Propp: (1) the hero's initial absence; (2) his transference from one place to another; (3) the combat between hero and villain; (4) the hero's receipt of a gift; (5) the victory; (6) the final repair of the hero's initial absence.[24] In this case the hero, Bradford, leads his people through trials and tests in the struggle to recreate the garden in the New World.

In the first phase of the New England recovery story, the land is absent of the hero. Indian corn fields are abandoned and the Indians, victims of disease. As John Cotton put it: "When the Lord chooses to transplant his people, he first makes a country ... void in that place where they reside."[25] In the second, or transference phase, the hero, William Bradford, is transported from Old England to New England by ship. A spatial translocation takes place between two kingdoms, that of the Antichrist (the fleshpots of Old England) and the New Canaan, or promised land of New England. In the third, or combative phase, the hero is tested through struggle with the villain—the devil acting through nature. The mythic struggle between hero and villain is played out as a struggle between Bradford and the wilderness—the tempestuous ocean and the desolate forest, a land filled with "wild beasts and wild men." Bradford's faith in God and his leadership of his people are continually called on, as storms wreak havoc with the small ship, the *Mayflower*, and the little band of settlers struggles to survive the grim winter on the shores of an unforgiving land. In the fourth phase the hero receives a gift from a helper, in this case "a special instrument sent [from] God" through the Indian Squanto, who not only speaks the Pilgrims' own language, but shows them how to "set their corn, where to take fish, and to procure other commodities." The fifth phase is the victory of the hero, as the corn is harvested, cabins and stockade are built, and the struggling band survives its first year. Nature, as wilderness, has been defeated. In the sixth and climactic phase the hero's initial absence has been repaired, the misfortunes are liquidated, and the Pilgrims are reborn. They celebrate their triumph over wilderness by their first harvest, achieved through the miracle of the recreated garden. By tilling and replenishing the land, the recovery of the garden in the New World has been launched and the American recovery myth created.[26]

Pilgrim victory was followed by Puritan victory when the Massachusetts Bay Colony added thousands of additional settlers to the new land, repeating the heroic journey across the Atlantic to advance the Edenic recovery. As the *Arabella* left England for the New World in 1629, Puritan refugees listened to John Winthrop quoting Genesis 1:28, "Be fruitful and multiply, and replenish the earth and subdue it." Boston pastor Charles Morton adhered to both the Genesis origin story and the Baconian ideal when he wrote in 1728 that, because of the sin of the first parents, agriculture and husbandry must be used to combat weeds and soil sterility through fencing, tilling, manuring, and draining the land. Almanac maker Nathaniel Ames in 1754 helped to

justify the mechanistic science of the body in Edenic terms when he informed his readers that the Divine artificer initially had made the body of man "a machine capable of endless duration," but that, after Eve's ingestion of the forbidden apple, the living principle within had fallen into disharmony with the body, disrupting the smooth functioning of its parts.[27]

In the Chesapeake region, by the early eighteenth century, tobacco planters converted an "unjustly neglected" and "abused" Virginia into a ravishing garden of pleasure. Robert Beverley predicated Virginia's potential as a "Garden of the World," akin to Canaan, Syria, and Persia, on his countrymen's ability to overcome an "unpardonable laziness."[28] Tobacco cultivation became the means of participating in the European market, while simultaneously improving the land through labour. But the recovery was ever in danger from new lapsarian moments if people allowed themselves to indulge in laziness, narcotics, or alcohol. During the eighteenth and nineteenth centuries, migrants from the original colonies and immigrants from Europe explored, settled, and "improved" the uplands west of the Atlantic Coast, the intervales of the Appalachian Mountains, and the lowlands of the Mississippi Valley.

In the late 1820s and 1830s, Thomas Cole of the Hudson River school of painters depicted the American recovery narrative and the dangers of both the original and subsequent lapsarian moments. His *Expulsion from the Garden of Eden* (1827–1828) contrasts the tranquil, original garden on the right with the bleak, chaotic desert on the left, while in the centre God expels Adam and Eve through a gate. The garden features a meandering stream and luxuriant vegetation, while the desert comprises barren rock, hot winds, a wild cataract, an erupting volcano, and a wolf attacking a deer. *The Oxbow* (1836) portrays the possibility of recovery through recreating the Garden on Earth. The painting moves from dark wilderness on the left to an enlightened, tranquil, cultivated landscape on the right, bordering the curve of the peaceful Connecticut River. In the background, cutover scars in the forest on the hill apparently spell the Hebrew letters Noah, which, when viewed upside down from a God's eye view, form the word *shaddai*, meaning, "the Almighty." God's presence in the landscape recognizes God's covenant with Noah and anticipates the final reunion of God and the Earth at the Parousia. Humans can therefore redeem the land itself as garden, even as they redeem themselves through labouring in the Earth.[29]

In a series of paintings from the 1830s, Cole depicted the movement from "savagery" to "civilization" and the problem of lapsing back into the darkness of wilderness. Of an 1831 painting, *A Wild Scene*, he wrote, "The first picture must be a savage wilderness ... the figures must be savage—clothed in skins & occupied in the Chase— ... as though nature was just waking from chaos."[30] A subsequent series, *The Course of Empire*, followed Virgil's stages of emergence from "savagery"—*The Savage State, The Pastoral State, Consummation of Empire, Destruction of Empire*, and *Desolation*—to warn of lapsarian dangers that thwart progress and end in the ruin of civilization.

Ralph Waldo Emerson eulogized the recovered garden achieved through human dominion over nature in glowing rhetoric: "This great savage country should be furrowed by the plough, and combed by the harrow; these rough Alleganies should know their master; these foaming torrents should be bestridden by proud arches of stone; these wild prairies should be loaded with wheat; the swamps with rice; the hill-tops should pasture innumerable sheep and cattle.... How much better when the whole land is a garden, and the people have grown up in the bowers of a paradise."[31] Only after intensive development of the eastern seaboard did a small number of nineteenth-century urban artists, writers, scientists, and explorers begin to deplore the effects of the "machine in the garden."[32]

Similarly, Euramericans acted out the recovery narrative in transforming the Western deserts during the second half of the nineteenth century. The elements of the story again conform to the elements of Propp's heroic narrative. The land is absent of the heroes—the migrants themselves. They are transferred across inhospitable desert lands; engage in combat with hostile Indians, diseases, and starvation; receive gifts from God in the form of gold and free land; emerge victorious over nature and Indian; and liquidate the initial absence of the hero by filling and replenishing the land. In filling the land through settlement, the migrants heeded John Quincy Adams's 1846 call for expansion into Oregon: "to make the wilderness blossom as the rose, to establish laws, to increase, multiply, and subdue the earth, which we are commanded to do by the first behest of the God Almighty." They likewise heard Thomas Hart Benton's call to manifest destiny that the White race had "alone received the divine command to subdue and replenish the Earth: for it: is the only race that ... hunts out new and distant lands, and even a New World, to subdue and replenish."[33]

With Reverend Dwinell, they commemorated the 1869 joining of the Central Pacific and Union Pacific railroads, using the Bible to sanction human alteration of the landscape. "Prepare ye the way of the Lord, make straight in the desert a highway before our God. Every valley shall be exalted, and every mountain and hill shall be made low and the crooked shall be made straight and the rough places plain." And in settling, ranching, and plowing the Great Plains, they reversed the biblical Fall from Eden by turning the "Great American Desert" into yet another "Garden of the World." The reclamation of arid lands west of the hundredth meridian through the technologies of irrigation fulfilled the biblical mandate of making the desert blossom as the rose, while making the land productive for capitalist agriculture.[34]

At the end of the nineteenth century, Frederick Jackson Turner's essay on the closing of the frontier in American history epitomized the heroic recovery narrative. The six phases of the heroic victory are again present in Turner's narrative, although it warns of impending declension as the frontier closes. (1) The frontier is defined by the absence of settlement and civilization. "Up to and including 1880, the country had a frontier of settlement, but at present the unsettled area has been ... broken." (2) Europeans are transferred across space as the succession of frontier lines moves west, and they "adapt ... to changes involved in crossing the continent." Stand at Cumberland gap and watch the procession—the buffalo following the trail to the salt lick, the trapper, the miner, the rancher, and the farmer follow each other in succession; stand at South Pass a century later and watch the same succession again. (3) The individual hero is in combat with the villain—again the wilderness, Indians, and wild beasts. "The wilderness masters the colonist." The encounter with wilderness "strips off the garments" of European civilization and "puts him in the log cabin of the Cherokee

and Iroquois." (4) The heroes receive the gift of free land. But "never again," Turner warns, "will such gifts of free land offer themselves." (5) The encounter with the frontier transforms hero into victor. "Little by little he transforms the wilderness, but the outcome is not the old Europe ... here is a new product that is American." (6) Democracy and American civilization "in a perennial rebirth" fill the land, liquidating the initial absence. "Democracy is born of free land."[35] With frontier expansion, temporal recovery through science and capitalism merges with spatial recovery through acquisition of private property.

INDIANS IN THE RECOVERY NARRATIVE

The heroic recovery narrative that guided settlement is notable for its treatment of Indians. Wilderness is the absence of civilization. Although many Euramericans apparently perceived Indians as the functional equivalent of wild animals, they nevertheless believed the Indian survivors possessed the potential to be "civilized" and hence to participate in the recovery as settled farmers. American officials changed the Indians' own origin stories to make them descendants of Adam and Eve; hence they were not even indigenous to America. Thomas L. McHenry, who formulated Indian policy in the 1840s, said that the whole "family of man" came from "one original and common stock" of which the Indian was one branch. "Man ... was put by his creator in the garden, which was eastward in Eden, whence flowed the river which parted, and became into four heads; and that from his fruitfulness his [the Indian] species were propagated." The commissioner of Indian affairs in 1868 deemed them "capable of civilization and christianization." A successor in

1892 argued that since Indian children were "made in the image of God, being the likeness of their Creator," they had the "same possibilities of growth and development" as other children. An Indian baby could become "a cultivated refined Christian gentleman or lovely woman."[36]

Euramericans attempted to transform Indians from hunters into settled farmers by first removing them to lands west of the Mississippi, then to reservations, and later by allotting them 160-acre plots of private property. Thomas Jefferson saw them as capable of participating in the recovery narrative when he told a delegation in 1802 that he would be pleased to see them "cultivate the earth, to raise herds of useful animals and to spin and weave."[37] With Indians largely vanquished and moved to reservations by the 1890s, twentieth-century conservationists turned "recovered" Indian homelands into parks, set aside wilderness areas as people-free reserves where "man himself is a visitor who does not remain," and managed forests for maximum yield and efficiency. With the taming of wilderness, desert, and "wild men," the recovery story reached an apparently happy ending.[38]

But Indians, for the most part, rejected the new narrative. With some exceptions, they resisted the roles into which they were cast and the lines they were forced to speak. They objected to characterizations of their lands as wilderness or desert, calling them simply home. As Chief Luther Standing Bear put it, "We did not think of the great open plains, the beautiful rolling hills, and winding streams with tangled growth, as 'wild.' Only to the white man was nature a wilderness and only to him was the land 'infested' with 'wild' animals and 'savage' people. To us it was tame. Earth was bountiful...."[39]

While adopting the Christian religion, Indians often emphasized those aspects compatible with traditional beliefs and participated in the ceremonial and celebratory aspects with greater enthusiasm than in the more austere, otherworldly practices.[40] Although taught to read and cipher, they often rejected White society's science and technology as useless for living. As Franklin satirized the colonists' effort, the Indians, when offered the opportunity to attend the College of William and Mary in Virginia, politely considered the matter before refusing:

> Several of our young people were formerly brought up at the colleges of the northern provinces; they were instructed in all your sciences; but when they came back to us they were bad runners; ignorant of every means of living in the woods; unable to bear either cold or hunger; knew neither how to build a cabin, take a deer, or kill an enemy; spoke our language imperfectly, and were therefore neither fit for hunters, warriors, or counsellors; they were totally good for nothing. We are, however, none the less obliged by your kind offer, tho' we decline accepting it; and to show our grateful sense of it, if the gentlemen of Virginia will send us a dozen of their sons, we will take great care of their education, instruct them in all we know, and make men of them.[41]

FEMALE NATURE IN THE RECOVERY NARRATIVE

An account of the history of American settlement as a lapsarian and recovery narrative must also consider the crucial role of nature conceptualized as female in the very structure of the plot. The rhetoric of American settlement is filled with language that casts nature as female object to be transformed and men as the agents of change. Allusions to Eve as virgin land to be subdued, as fallen nature to be redeemed through reclamation, and as fruitful garden to be harvested and enjoyed are central to the particular ways in which American lands were developed. The extraction of resources from "nature's bosom," the penetration of "her womb" by science and technology, and the "seduction" of female land by male agriculture reinforced capitalist expansion.[42]

Images of nature as female are deeply encoded into the texts of American history, art, and literature and function as ideologies for settlement. Thus Thomas Morton, in praising New England as a new Canaan, likened its potential for development by "art and industry" to a "faire virgin longing to be sped and meet her lover in a Nuptiall bed." Now, however, "her fruitfull wombe, not being enjoyed is like a glorious tombe."[43] Male agriculturalists saw in plow technology a way to compel female nature to produce. Calling Bacon "the grand master of philosophy" in 1833, Massachusetts agricultural improver Henry Colman promoted Bacon's approach to recovering the garden through agriculture. "The effort to extend the dominion of man over nature," he wrote, "is the most healthy and most noble of all ambitions." He characterized the Earth as a female whose productivity could help to advance the progress of the human race. "Here man exercises dominion over nature ... commands the earth on which he treads to waken her mysterious energies ... compels the inanimate earth to teem with life; and to impart sustenance and power, health and happiness to the countless multitudes who hang on her breast and are dependent on her bounty."[44]

A graphic example of female nature succumbing to the male plow is provided by Frank Norris in his 1901 novel *The Octopus*, a story of the transformation of

California by the railroad. Here the Earth is female, sexual, and alive. Norris writes,

> The great brown earth turned a huge flank to [the sky], exhaling the moisture of the early dew.... One could not take a dozen steps upon the ranches without the brusque sensation that underfoot the land was alive, ... palpitating with the desire of reproduction. Deep down there in the recesses of the soil, the great heart throbbed once more, thrilling with passion, vibrating with desire, offering itself to the caress of the plough, insistent, eager, imperious. Dimly one felt the deep-seated trouble of the earth, the uneasy agitation of its members, the hidden tumult of its womb, demanding to be made fruitful, to reproduce, to disengage the eternal renascent germ of Life that stirred and struggled in its loins....[45]

In Norris's novel, the seduction of the female Earth was carried out on a massive scale by thousands of men operating their plows in unison on a given day in the spring. "Everywhere throughout the great San Joaquin," he wrote, "unseen and unheard, a thousand ploughs up-stirred the land, tens of thousands of shears clutched deep into the warm, moist soil."[46] And Norris leaves no doubt that the men's technology, the plow, is also male and that the seduction becomes violent rape:

> It was the long stroking caress, vigorous, male, powerful, for which the Earth seemed panting. The heroic embrace of a multitude of iron hands, gripping deep into the brown, warm flesh of the land that quivered responsive and passionate under this rude advance, so robust as to be almost an assault, so violent as to be veritably brutal. There, under the sun and under the speckless sheen of the sky, the wooing of the Titan began, the vast primal passion, the two world-forces,

the elemental Male and Female, locked in a colossal embrace, at grapples in the throes of an infinite desire, at once terrible and divine, knowing no law, untamed, savage, natural, sublime.[47]

The narrative of frontier expansion is a story of male energy subduing female nature, taming the wild, plowing the land, recreating the garden lost by Eve. American males lived the frontier myth in their everyday lives, making the land safe for capitalism and commodity production. Once tamed by men, the land was safe for women. To civilize was to bring the land out of a state of savagery and barbarism into a state of refinement and enlightenment. This state of domestication, of civility, is symbolized by woman and "womanlike" man. "The man of training, the civilizee," reported *Scribner's Monthly* in November 1880, "is less manly than the rough, the pioneer."[48]

But the taming of external nature was intimately linked to the taming of internal nature, the exploitation of non-human nature to the exploitation of human nature. The civilization process not only removed wild beasts from the pastoral lands of the garden, it suppressed the wild animal in men. Crévecoeur, in 1782, noted that on the frontier, "men appear to be no better than carnivorous animals ... living on the flesh of wild animals." Those who farmed the middle settlements, on the other hand, were "like plants," purified by the "simple cultivation of the earth," becoming civilized through reading and political discourse.[49] Or, as Richard Burton put it in 1861, "The civilizee shudders at the idea of eating wolf."[50] Just as the Earth is female to the farmer who subdues it with the plow, so wilderness is female to the male explorer, frontiersman, and pioneer who tame it with the brute strength of the ax, the trap, and the gun. Its valence, however, changes from the negative satanic forest of William Bradford and the untamed wilderness of the pioneer (fallen Eve) to the positive pristine Eden and Mother Earth of John Muir (original and Mother Eve) and the parks of Frederick Law Olmsted. As wilderness vanishes before advancing civilization, its remnants must be preserved as test zones for men (epitomized by Theodore Roosevelt) to hone male strength and skills.[51]

Civilization is the final end, the telos, toward which "wild" Nature is destined. The progressive narrative undoes the declension of the Fall. The "end of nature" is civilization. Civilization is thus nature natured. *Natura naturata*—the natural order, or nature ordered and tamed. It is no longer nature naturing, *Natura naturans*—nature as creative force. Nature passes from inchoate matter endowed with a formative power to a reflection of the civilized natural order designed by God. The unruly energy of wild female nature is suppressed and pacified. The final happy state of nature natured is female and civilized—the restored garden of the world.[52]

THE CITY IN THE GARDEN

The city represents the next stage of the recovery narrative—the creation of the City in the Garden (Virgil's *urbs in horto*) by means of the capitalist market. The city epitomizes the transformation of female nature into female civilization through the mutually reinforcing powers of male energy and interest-earning capital. Frank Norris, in his second novel, *The Pit* (1903), reveals the connections.[53] In writing of Chicago and the wheat pit at the Board of Trade (a story brilliantly told in Willian Cronon's *Nature's Metropolis*, inspired in part by Norris's book), Norris depicts the city as female.[54] The city is the locus of power that operates in the natural world, sweeping everything toward its centre. It is the bridge between civilized female form and the raw matter surrounding hinterlands, drawing that matter toward it, as natural resources are transformed into capitalist commodities. Chicago, writes Norris,

> the Great Grey City, brooking no rival, imposed its dominion upon a reach of country larger than many a kingdom of the Old World. For thousands of miles beyond its confines was the influence felt. Out, far out, far away in the snow and shadow of Northern Wisconsin forests, axes and saws bit at the bark of century old trees, stimulated by this city's energy. Just as far to the southward pick and drill leaped to the assault of veins of anthracite moved by her central power. Her force turned the wheels of harvester and seeder a thousand miles distant in Iowa and Kansas. Her force spun the screws and propellers of innumerable squadrons of lake steamers crowding the Sault Sainte Marie. For her and because of her all the Central States, all the Great Northwest roared with traffic and industry; sawmills screamed; factories, their smoke blackening the sky, slashed and flamed; wheels turned, pistons leaped in their cylinders; cog gripped cog; beltings clasped the drums of mammoth wheels; and converters of forges belched into the clouded air their tempest breath of molten steel.[55]

The city transforms the matter of nature in the very act of pulling it inward. Like Plato's female soul of the world, turning herself within herself, the city provides the source of motion that permeates and energizes the world around it, the bridge

between raw changing matter and final civilized form. In Norris's novel, men at first seem subordinate to the city's higher force, acting merely as agents in the preordained purpose of transforming nature into civilization. They facilitate the change from *Natura naturans* into *Natura naturata*, from natural resource into fabricated product. Operating the steam engines, sawmills, factories, lumber barges, grain elevators, trains, and switches that make Chicago an industrial city, workers shout and signal as trains daily debouch businessmen, bringing with them trade from country to city. This process of "civilization in the making," says Norris, is like a "great tidal wave," an "elemental," "primordial" force, "the first verses of Genesis." It "subdu[es] the wilderness in a single generation," through the "resistless subjugation of ... the lakes and prairies."[56]

Yet behind the scenes other men, the capitalist speculators of the Chicago Board of Trade, attempt to manipulate the very forces of nature, pushing the transformation faster and faster. Capitalism mystifies by converting living nature into dead matter and by changing inert metals into living money.[57] To the capitalist puppeteers, nature is a doll-like puppet controlled by the strings of the wheat trade that changes money into interest-earning capital. Male minds calculate the motions that control the inert matter below.

To Norris's capitalist, Curtis Jadwin, nature is dead. Only money is alive, growing and swelling through the daily trade of the wheat pit. With the bulls and bears of the marketplace the only apparent living things he encounters, Jadwin utterly fails to account for the Earth and the wheat as alive. Yet as Jadwin, the bull trader, corners the market to obtain complete control over the bears, driving the price higher and higher, the living wheat planted by hundreds of farmers throughout the heartland rises from the soil as a gigantic irrepressible force. The capitalist's manipulation of apparently dead nature has immense environmental consequences. Jadwin, Norris writes, had "laid his puny human grasp upon Creation and the very earth herself." The "great mother ... had stirred at last in her sleep and sent her omnipotence moving through the grooves of the world, to find and crush the disturber of her appointed courses."[58]

But in the late nineteenth century, as the frontier closes, forests disappear, and the land is made safe for civilization, American men begin to lament the loss of wild nature. There is an apparent need to retain wilderness as a place for men to test maleness, strength, and virility and an apparent association of men with nature.[59] Similarly, women are symbolized as the moral model that suppresses internal sexual libido. But nature as wilderness does not *become* male, nor does civilization *become* female in a reversal of the so-called universal association of female to nature and male to culture identified by Sherry Ortner.[60] There is no real reversal of male/female valences in the closing chapters of the story of frontier expansion. In the story of American progress, males continue to be the transforming agents between active female nature and civilized female form, making the land safe for women and men alike, suppressing both unpredictable external nature and unruly internal nature.

Nor are nature and culture, women and men, binary opposites with universal or essential meanings. Nature, wilderness, and civilization are socially constructed concepts that change over time and serve as stage settings in the progressive narrative. So too are the concepts of male and female and the roles that men and women act

out on the stage of history. The authors of such powerful narratives as *laissez-faire* capitalism, mechanistic science, manifest destiny, and the frontier story are usually privileged elites with access to power and patronage. Their words are read by persons of power who add the new stories to the older biblical story. As such, the books become the library of Western Culture. The library, in turn, functions as ideology when ordinary people read, listen to, internalize, and act out the stories told by their elders—the ministers, entrepreneurs, newspaper editors, and professors who teach and socialize the young.

The most recent chapter of the book of the recovery narrative is the transformation of nature through biotechnology. From genetically engineered apples to Flavr-Savr tomatoes, the fruits of the original (evolved) garden are being redesigned so that the salinated irrigated desert can continue to blossom as the rose. In the recovered Garden of Eden, fruits ripen faster, have fewer seeds, need less water, require fewer pesticides, contain less saturated fat, and have longer shelf lives. The human temptation to engineer nature is reaching too close to the powers of God, warn the Jeremiahs, who depict the snake coiled around the Tree of the Knowledge of Good and Evil as the DNA spiral. But the progressive engineers who design the technologies that allow the recovery to accelerate see only hope in the new fabrications.

The twentieth-century Garden of Eden is the enclosed shopping mall decorated with trees, flowers, and fountains in which people can shop for nature at the Nature Company, purchase "natural" clothing at Esprit, sample organic foods and rainforest crunch in kitchen gardens, buy twenty-first-century products at Sharper Image, and play virtual reality games in which SimEve is reinvented in Cyberspace. This Garden in the City recreates the pleasures and temptations of the original garden and the golden age where people can peacefully harvest the fruits of Earth with gold grown by the market. The mall, enclosed by the desert of the parking lots surrounding it, is covered by glass domes reaching to heaven, accessed by spiral staircases and escalators affording a vista over the whole garden of shops. The "river that went out of Eden to water the garden" is reclaimed in meandering streams lined with palm trees and filled with bright orange carp. Today's malls feature stone grottos, trellises decorated with flowers, life-sized trees, statues, birds, animals, and even indoor beaches that simulate paradigmatic nature as a cultivated, benign garden. With their engineered spaces and commodity fetishes, they epitomize consumer capitalism's vision of the recovery from the Fall.[61]

CRITIQUES OF THE RECOVERY NARRATIVE

The modern version of the recovery narrative, however, has been subjected to scathing criticism. Postmodern thinkers contest its Enlightenment assumptions, while cultural feminists and environmentalists reverse its plot, depicting a slow decline from a prior golden age, not a progressive ascent to a new garden on Earth. The critics' plot does not move from the tragedy of the Fall to the comedy of an earthly paradise, but descends from an original state of oneness with nature to the tragedy of nature's destruction. Nevertheless, they too hope for a recovery, one rapid enough to save the Earth and society by the mid-twenty-first century. The meta-narrative of recovery does not change, but the declensionist

plot, into which they have cast prior history, must be radically reversed. The postmodern critique of modernism is both a deconstruction of Enlightenment thought and a set of reconstructive proposals for the creation of a better world.

The identification of modernism as a problem rather than as progress was sharply formulated by Max Horkheimer and Theodor Adorno in the opening sentences of their 1944 *Dialectic of Enlightenment*: "The fully enlightened earth radiates disaster triumphant. The program of the enlightenment was the disenchantment of the world; the dissolution of myths and the substitution of knowledge for fancy." They criticize both Francis Bacon's concept of the domination of nature and Karl Marx and Friedrich Engels's optimism that the control of nature would lead to advancement. They faulted the reduction of nature to mere number by mechanistic science and capitalism: "Number becomes the canon of the Enlightenment. The same equations dominate bourgeois justice and commodity exchange.... Myth turns into enlightenment and nature into mere objectivity."[62]

Among the critics of modernism are many feminists and environmentalists who propose a reversal that will initiate a new millennium in the twenty-first century. Cultural feminists and cultural ecofeminists see the original oneness as female, the *Terra Mater* of the neolithic era, from which emerged the consciousness of differences between humans and animals, male and female, people and nature, leading to dominance and submission. The advent of patriarchy initiates a long decline in the status of women and nature. Men's plow agriculture took over women's gathering and horticultural activities, horse-mounted warriors injected violence into a largely peaceful Old European culture, and male gods replaced female Earth deities in origin stories. In the proposed recovery, Eve is revisioned as the first scientist, Sophia as ultimate wisdom, and the goddess as symbol of female power and creativity. Feminist religious history redirects inquiry into the gendered nature of the original oneness as both male and female. The recovery would therefore be a feminist or an egalitarian world.[63]

Feminist science sees the original mind as having no sex, and hence accessible to male and female minds alike. It has been men, many feminists would argue, who invented the science and technology and organized the market economies that made nature the victim in the ascent of "man." For such feminists, the new narrative entails reclaiming women's roles in the history of science and asserting female power in contemporary science and technology. Hence both sexes can participate in the recovery.[64]

Environmentalism, like feminism, reverses the plot of the recovery narrative, seeing history as a slow decline, not a progressive movement that made the desert blossom as the rose. The recovery story is a false story; an original garden became a degraded desert. Pristine nature, not innocent man, has fallen. The decline from Eden was slow, rather than a precipitous lapsarian moment as in the Adam and Eve origin story. Over the millennia from the paleolithic to the present, nature has been the victim of both human hubris and social changes that overcome "the necessities of nature" through domestication, cultivation, and commodification of every aspect of an original, evolved, pre-human garden. So-called advances in science, technology, and economy actually accelerate the decline.[65]

As the twentieth century drew to a close and the second great millennium since the birth of Christ reached its end, the environmental decline approached a crisis. The greenhouse effect, the population explosion, the destruction of the ozone layer, the extinction of species, and the end of wilderness are all subplots in a grand narrative of environmental endism. Predictions of crisis, such as those of Paul Ehrlich in "Ecocatastrophe" (1969), the Club of Rome in *Limits to Growth* (1972) and of Bill McKibben in *The End of Nature* (1989), abound, as first (evolved, pre-human) nature is totally subsumed by humans and the human artifacts of second (commodified) nature.[66]

Like feminists, environmentalists want to rewrite the modern progressive story. Viewing the plot as declensionist rather than progressive, they nevertheless opt for a recovery that must be put in place by the mid-twenty-first century. "Sustainability" is a new vision of the recovered garden, one in which humanity will live in a relationship of balance and harmony with the natural world. Environmentalists who press for sustainable development see the recovery as achievable through the spread of non-degrading forms of agriculture and industry. Preservationists and deep ecologists strive to save pristine nature as wilderness before it can be destroyed by development. Restoration ecologists wish to marshal human labour to restore an already degraded nature to an earlier, pristine state. Social ecologists and green parties devise new economic and political structures that overcome the domination of human beings and non-human nature. Women and nature, minorities and nature, other animals and nature, will be fully included in the recovery. The regeneration of nature and people will be achieved through social and environmental justice. The End Drama envisions a post-patriarchal, socially just ecotopia for the post-millennial world of the twenty-first century.[67]

CHAOS THEORY AND PARTNERSHIP ETHICS

Seeing Western history as a recovery narrative, with feminism and environmentalism as reversals of the plot, brings up the question of the character of the plot itself. The declensionist and progressive plots that underlie the meta-narrative of recovery both gain power from their linearity. Linearity is not only conceptually easy to grasp, but it is also a property of modernity itself. Mechanistic science, progress, and capitalism all draw power from the linear functions of mathematical equations—the upward and downward slopes of straight lines and curves. To the extent that these linear slopes intersect with a real material world, they refer to a limited domain only. Chaos theory and complexity theory suggest that only the unusual domain of mechanistic science can be described by linear differential equations. The usual—that is, the domain of everyday occurrences, such as the weather, turbulence, the shapes of coastlines, the arrhythmic fibrillations of the human heart—cannot be so easily described. The world is more complex than we know or indeed can ever know. The comfortable predictability of the linear slips away into the uncertainty of the indeterminate—into discordant harmonies and disorderly order.

The appearance of chaos as an actor in science and history in the late twentieth century is not only symptomatic of the breakdown of modernism, mechanism, and, potentially, capitalism, but suggests the possibility of a new birth, a new world, a new millennium—the order out of chaos narrative of Ilya Prigogine and Isabelle Stengers. But chaos theory also fundamentally destabilizes the very concept of nature as a standard or referent. It disrupts the idea of the "balance of nature," of nature as resilient actor or mother who will repair the errors of human actors and continue as fecund garden (Eve as mother). It questions the possibility that humans as agents can control and master nature through science and technology, undermining the myth of nature as virgin female to be developed (Eve as virgin). Chaos is the re-emergence of nature as power over humans, nature as active, dark, wild, turbulent, and uncontrollable (fallen Eve). Ecologists characterize "Mother Nature" as a "strange attractor" while turbulence is seen to be encoded with gendered images of masculine channels and feminine flows.[68] Moreover, in the chaotic narrative, humans lose the hubris of fallen Adam that the garden can be recreated on Earth. The world is not created by a patriarchal God *ex nihilo*, but emerges out of chaos. Thus the very possibility of the recovery of a stable original garden—the plot of the recovery meta-narrative—is itself challenged.

Recognition of history as a meta-narrative raises the further question of the relativity of the histories through which we are educated and of our own lives as participants in the plots they tell. Like our nineteenth-century counterparts, we live our lives as characters in the grand narrative into which we have been socialized as children and conform as adults. That narrative is the story told to itself by the dominant society of which we are a part. We internalize narrative as ideology. Ideology is a story told by people in power. Once we identify ideology as a story—powerful and compelling, but still only a story—we realize that by rewriting the story, we can challenge the structures of power. We recognize that all stories can and should be challenged.

But can we actually step outside the story into which we are cast as characters and enter into a story with a different plot? More important, can we change the plot of the grand master narrative of modernism? Where do I as author of this text stand in relationship to it? As a product of modernism, mechanism, and capitalism, I have internalized the values of the recovery narrative I have sought to identify. I participate in the progressive recovery narrative in my daily work, my wages for intellectual labour, my aspirations for a better material life, and my enjoyment of the profits my individual achievements have wrought. Yet I also believe, despite the relativism of environmental endism, that the environmental crisis is real—that the vanishing frogs, fish, and songbirds are telling us a truth. I am also a product of linear thinking and set up this recovery narrative to reflect the very linearity of progressive history. This is history seen from a particular point of view, the view I have identified as the dominant ideology of modernism. I also believe my recovery narrative reflects a fundamental insight into how nature has been historically constructed as a gendered object.

Yet both history and nature are extremely complex, complicated, and non-linear. What would a chaotic, non-linear, non-gendered history with a different plot look like? Would it be as compelling as the linear version, even if that linear version were extremely nuanced and complicated? A postmodern history might posit characteristics other than those identified with modernism, such as many authorial voices; a multiplicity of real actors; acausal, non-sequential events; non-essentialized symbols and meanings; dialectical action and process rather than the imposed logos of form; situated and contextualized, rather than universal knowledge. It would be a story (or multiplicity of stories) that perhaps can only be acted and lived, not written at all.

I too yearn for a recovery from environmental declension—for my own vision of a post-patriarchal, socially just ecotopia for the third millennium. My vision entails a partnership ethic between humans (whether male or female), and between humans and non-human nature. For most of human history, non-human nature has had power over humans. People accepted fate while propitiating nature with gifts, sacrifices, and prayer (often within hierarchical human relationships). Since the seventeenth century, however, some groups of people have increasingly gained great power over nature and other human groups through the interlinked forces of science, technology, capitalism (and state socialism), politics, and religion.

A partnership ethic would bring humans and non-human nature into a dynamically balanced, more nearly equal relationship. Humans, as the bearers of ethics, would acknowledge non-human nature as an autonomous actor that cannot be predicted or controlled except in very limited domains. We would also acknowledge that we have the potential to destroy life as we currently know it through nuclear power, pesticides, toxic chemicals, and unrestrained economic development, and exercise specific restraints on that ability. We would cease to create profit for the few at the expense of the many. We would instead organize our economic and political forces to fulfill people's basic needs for food, clothing, shelter, and energy, and to provide security for health, jobs, education, children, and old age. Such forms of security would rapidly reduce population growth rates since a major means of providing security would not depend on having large numbers of children, especially boys. A partnership ethic would be a relationship between a human community and a non-human community in a particular place, a place that recognizes its connections to the larger world through economic and ecological exchanges. It would be an ethic in which humans act to fulfill both human needs and nature's needs by restraining human hubris. Guided by a partnership ethic, people would select technologies that sustained the natural environment by becoming co-workers and partners with non-human nature, not dominators over it. [...]

A partnership ethic implies a remythicizing of the Edenic recovery narrative or the writing of a new narrative altogether. The new myth would not accept the patriarchal sequence of creation, or even the milder phrase "male and female, created he them," but might instead emphasize simultaneous creation, co-operative male/female evolution, or even an emergence out of chaos or the Earth. It would not accept the idea of subduing the Earth, or even dressing and keeping the garden, since both entail total domestication and control by human beings. Instead, each earthly place would be a home, or community, to be shared with other living and non-living things. The needs of both humans and non-humans would be dynamically balanced. If such a story can be rewritten or experienced, it would be the product of many new voices and would have a complex plot and a different ending. As in the corn mother origin story, women and the Earth, along with men, would be active agents. The new ending, however, will not come about if we simply read and reread the story into which we were born. The new story can be rewritten only through action.

NOTES

1. Roland Nelson, Penobscot, as recorded by Frank Speck, "Penobscot Tales and Religious Beliefs," *Journal of American Folklore* 48, no. 187 (January–March 1935): 1–107, on 75. This corn mother origin story is a variant on a number of eastern United States and Canadian transformative accounts, recorded from oral traditions, that attribute the origins of corn to a mythical corn mother, who produces corn from her body, grows old, and then instructs her lover or son how to plant and tend corn. The killing of the corn mother in most of the origin stories may symbolize a transition from gathering/hunting to active corn cultivation. The snake lover may be an influence from the Christian tradition or a more universal symbol of the renewal of life (snakes shed their skins) and/or the male sexual organ. On corn mother origin stories, see John Witthoft, *Green Corn Ceremonialism in the Eastern Woodlands* (Ann Arbor: University of Michigan Press, 1949), 77–85; Joe Nicholas, Malechite, Tobique Point, Canada, August 1910, as recorded by W.H. Mechling, *Malechite Tales* (Ottawa: Government Printing Bureau, 1914), 87–88; for the Passamaquoddy variant, see *Journal of American Folklore* 3 (1890): 214; for Creek and Natchez variants, see J.R. Swanton, "Myths and Tales of the Southeastern Indians," *Bulletin of the Bureau of American Ethnology* no. 88 (1929): 9–17; on Iroquois variants, see Jesse Cornplanter, *Legends of the Longhouse* (Philadelphia: J.B. Lippincott, 1938) and Arthur Parker, "Iroquois Use of Maize and Other Food Plants," *New York State Museum Bulletin* no. 144 (November 1, 1910): 36–39; Gudmund Hart, "The Corn Mother in America and Indonesia," *Anthropos* 46 (1951): 853–914. Examples of corn mother origin stories from the Southwest include the Pueblo emergence from the dark interior of the Earth into the light of the fourth world where Corn Mother plants Thought Woman's gift of corn. See Ramon Gutierrez, *When Jesus Came the Corn Mothers Went Away* (Stanford: Stanford University Press, 1991). For a discussion of the relationship of the corn mother to Mother Earth, see Sam Gill, *Mother Earth: An American Story* (Chicago: University of Chicago Press, 1987), 4, 125.

2. On Great Plains environmental histories as progressive and declensionist plots, see William Cronon, "A Place for Stories: Nature, History and Narrative," *Journal of American History* 78 (March 1992): 1347–1376. The Indian and European origin stories can be interpreted from a variety of standpoints other than the declensionist and progressive narrative formats I have emphasized here (such as romance and satire). Additionally, the concepts of desert, wilderness, and garden are nuanced and elaborate motifs that change valences over time in ways I have not tried to deal with here.

3. *Holy Bible*, King James version, Genesis, Book 1. On the comic and tragic visions of the human, animal, vegetable, mineral, and unformed worlds, see Northrop Frye, *Fables of Identity* (New York: Harcourt Brace, 1963), 19–20. In the comic state, or vision, the human world is a community, the animal world consists of domesticated flocks and birds of peace, the vegetable world is a garden or park with trees, the mineral world is a city or temple with precious stones and starlit domes, and the unformed world is a river. In the tragic state or vision, the human world is an anarchy of individuals; the animal world is filled with beasts and birds of prey (such as wolves, vultures, and serpents); the vegetable world is a wilderness, desert, or sinister forest; the mineral world is filled with rocks and ruins; and the unformed world is a sea or flood. The plot of the tragedy moves from a better or comic state to a worse or tragic state; the comedy from an initial tragic state to a comic outcome. On history as narrative, see Hayden White, *Metahistory: The Historical Imagination in Nineteenth-Century Europe* (Baltimore: Johns Hopkins University Press, 1973); Hayden White, *Tropics of Discourse: Essays in Cultural Criticism* (Baltimore: Johns Hopkins University Press, 1978); Hayden White, *The Content of the Form: Narrative Discourse and Historical Representation* (Baltimore: Johns Hopkins University Press, 1987).

4. Benjamin Franklin, "Remarks Concerning the Savages of North America," in *Franklin's Wit and Folly: The Bagatelles*, edited by Richard E. Amacher (New Brunswick: Rutgers University Press, 1953), 89–98. Franklin's story is probably satirical rather than literal.

5. The concept of a recovery from the original Fall appears in the early modern period. See the *Oxford English Dictionary*, compact edition, vol. 2, 2447: The act of recovering oneself from a mishap, mistake, fall, etc. See Bishop Edward Stillingfleet, *Origines Sacrae* (London, 1662), II, i, sec. 1: "The conditions on which fallen man may expect a recovery." William Cowper, *Retirement* (1781), 138: "To ... search the themes, important above all Ourselves, and our recovery from our fall." See also Richard Eden, *The Decades of the Newe Worlde or West India* (1555), 168: "The recoverie of the kyngedome of Granata." The term "recovery" also embraced the idea of regaining a "natural" position after falling and a return to health after sickness. It acquired a legal meaning in the sense of gaining possession of property by a verdict or judgment of the court. In common recovery, an estate was transferred from one party to another. John Cowell, *The Interpreter* (1607), s.v. "recoverie": "A true recoverie is an actuall or reall recoverie of anything, or the value thereof by Judgement." Another meaning was the restoration of a person or thing to a healthy or normal condition, or a return to a higher or better state, including the reclamation of land. Anonymous, *Captives Bound in Chains ... the Misery of Graceless Sinners, and the Hope of Their Recovery by Christ* (1674); Bishop Joseph Buder, *The Analogy of Religion Natural and Revealed* (1736), II, 295: "Indeed neither Reason nor Analogy would lead us to think ... that the Interposition of Christ ... would be of that Efficacy for Recovery of the World, which Scripture teaches us it was." Joseph Gilbert, *The Christian Atonement* (1836), i, 24: "A modified system, which shall include the provision of means for recovery from a lapsed state." James Martineau, *Essays, Reviews, and Addresses* (1880–1891), II, 310: "He is fitting to be among the prophets of recovery, who may prepare for us a more wholesome future." John Henry Newman, *Historical Sketches* (1872–1873) II, 1, iii, 121: "The special work of his reign was the recovery of the soil."

6. On the Genesis 1, or priestly version (Genesis P), composed in the fifth century BC versus the Genesis 2, or Yahwist version (Genesis J), composed in the ninth or tenth century BC and their relationships to the environmental movement, see J. Baird Callicott, "Genesis Revisited: Muirian Musings on the Lynn White, Jr. Debate," *Environmental Review* 14, nos. 1–2 (Spring/Summer 1990): 65–92. Callicott argues that Lynn White, Jr. mixed the two versions in his famous article "The Historical Roots of Our Ecologic Crisis," *Science* 155 (1967): 1203. On the historical traditions behind the Genesis stories, see Artur Weiser, *The Old Testament: Its Formation and Development*, translated by Dorthea M. Barton (New York: Association Press, 1961).

7. John Prest, *The Garden of Eden: The Botanic Garden and the Recreation of Paradise* (New Haven: Yale University Press, 1981), 1–37; J.A. Phillips, *Eve: The History of an Idea* (San Francisco: Harper & Row, 1984); Francis Russell, *The World of Dürer* (New York: Time, 1967), 83, 109.

8. "Paradise" derives from the old Persian word for "enclosure" and in Greek and Latin takes on the meaning of "garden." Its meanings include heaven, a state of bliss, an enclosed garden or park, and the Garden of Eden. "Parousia" derives from the Latin *parere*, meaning to produce or bring forth. The Parousia is the idea of the End of the World, expressed as the hope set forth in the New Testament that "he shall come again to judge both the quick and the dead." See A.L. Moore, *The Parousia in the New Testament* (Leiden: E.J. Brill, 1966). I thank Anthony Chennells for bringing this concept to my attention. Capitalism and Protestantism were initially mutually reinforcing in their common hope of a future golden age. But as capitalism became more materialistic and worldly, it began to undercut the Church's parousia hope. Communism retained the idea of a future golden age in its concern for community and future direction (pp. 2–3). The parousia hope was a driving force behind the Church's missionary work in its early development and in the New World (p. 5). The age of glory was a gift of God; an acknowledgment of the future inbreaking of God (JHWH) into history (pp. 16, 17). "The scene of the future consumation is a radically transformed earth. The coming of this Kingdom was conceptualized as a sudden catastrophic moment, or as preceded by the Messianic kingdom, during which it was anticipated that progressive work would take place" (p. 20). "Concerning the central figure in the awaited End-drama there is considerable variation. In some visions the figure of Messiah is entirely absent. In such cases 'the kingdom was always represented as under the immediate sovereignty of God'" (p. 21). "The divine intervention in history was the manifestation of the Kingdom of God.... [T]his would involve a total transformation of the present situation, hence the picture of world renewal enhanced sometimes by the idea of an entirely supernatural realm" (pp. 25–26). "The fourth *Eclogue* of Virgil presents the hope of a 'golden age' but in fundamental contrast to apocalyptic expectation; although it is on a cosmic scale, it is the hope of revolution from within rather than of intervention from without" (p. 28).

9. Max Oelschlaeger, *The Idea of Wilderness: From Prehistory to the Age of Ecology* (New Haven: Yale University Press, 1991), 49–60.

10. Francis Bacon, "Novum Organum," in *Works*, 14 vols., edited by James Spedding, Robert Leslie Ellis, and Douglas Devon Heath (London: Longmans Green, 1870), vol. 4, 114–115, 247–248. See also Bacon's statement, "I mean (according to the practice in civil causes) in this great plea or suit granted by the divine favor and providence (whereby the human race seeks to recover its right over nature) to examine nature herself and the arts upon interrogatories." Bacon, "Preparative towards a Natural and Experimental History," *Works*, vol. 4, 263. William Leiss, *The Domination of Nature* (New York: George Braziller, 1972), 48–52; Merchant, *The Death of Nature: Women, Ecology, and the Scientific Revolution*, 185–186; Charles Whitney, *Francis Bacon and Modernity* (New Haven: Yale University Press, 1986), 25.

11. Marshall Sahlins, *Culture and Practical Reason* (Chicago: University of Chicago Press, 1976), 53: "The development from a Hobbesian state of nature is the origin myth of Western capitalism."

12. On the definition of natural resources, see John Yeats, *Natural History of Commerce* (London, 1870); 2; Thomas Hobbes, "Leviathan" (1651), in *English Works*, 11 vols. (reprint edition, Aslen: Scientia, 1966), vol. 3, 145, 158; John Locke, *Two Treatises of Government* (1690), edited by Peter Laslett (Cambridge: Cambridge University Press, 1960), Second Treatise, Chapter 5, secs. 28, 32, 35, 37, 46, 48.

13. The Fall from Eden may be interpreted (as can the corn mother origin story; see note 1) as representing a transition from gathering/hunting to agriculture. In the Garden of Eden, Adam and Eve pick the fruits of the trees without having to labour in the earth (Genesis 1: 29–30; Genesis 2: 9). After the Fall they had to till the ground "in the sweat of thy face" and eat "the herb of the field" (Genesis 3: 18, 19, 23). In Genesis 4, Abel, "keeper of sheep," is the pastoralist, while Cain, "tiller of the ground," is the farmer. Although God accepted Abel's lamb as a first fruit, he did not accept Cain's offering. Cain's killing of Abel may represent the ascendancy of farming over pastoralism. Agriculture requires more intensive labour than either pastoralism or gathering. See Oelschlaeger, *The Idea of Wilderness*; Callicott, "Genesis Revisited," 81.

14. Victor Rotenberg, "The Lapsarian Moment," mss. Hesiod, "Works and Days," in *Theogony and Works and Days*, translated by M.L. West (Oxford: Oxford University Press, 1988), 40.

15. Publius Ovid, *Metamorphoses* (written AD 7), translated by Rolfe Humphries (Bloomington: Indiana University Press, 1955), Book 1, p. 6, lines 100–111.

16. On the meanings of nature and nation and the following interpretation of Virgil, see Kenneth Olwig, *Nature's Ideological Landscape* (London: George Allen and Unwin, 1984), 3–9. In the *Eclogues*, Virgil characterized the pastoral landscape as the grazing of tame animals on grassy hillsides. Human labour domesticated animals, transformed the forest into meadows, and dammed springs to form pools for watering livestock. But the shepherd was relatively passive, watching flocks while reclining in the shade of a remnant forest tree.

17. Olwig, *Nature's Ideological Landscape*, 6. Agriculture is initiated by Jove, who "endowed that cursed thing the snake with venom and the wolf with thirst for blood." "Toil ... taught men the use and method of the plough." Agricultural instruments were hammered out by the use of fire, becoming "weapons hardy rustics need ere they can plow or sow the crop to come." Virgil, *Georgics*, I, 151–152, as quoted in Olwig, *Nature's Ideological Landscape*, 6.

18. Olwig, *Nature's Ideological Landscape*, 3–9; Virgil, *Georgics*, II (1946), 106–107; Virgil, *Eclogues*, IV, 4–34. Virgil's temporal and spatial stages prefigure Frederick Jackson Turner's frontier stages and Johann Heinrich von Thünen's rings, discussed by William Cronon in the conversion of hinterland resources (first nature) into commodities (second nature) in Chicago. See William Cronon, *Nature's Metropolis: Chicago and the Great West* (New York: Norton, 1991), 46–54.

19. Lucretius, *Of the Nature of Things*, translated by William Ellery Leonard (New York: E.P. Dutton, 1950), Book V, lines 922–1008. Lucretius's image of "the state of nature was strikingly similar to that of Thomas Hobbes in *Leviathan*. Lucretius wrote that in the early days, "men led a life after the roving habit of wild beasts." They chased and ate wild animals and were in turn hunted and devoured by them. In the state of nature, they "huddled in groves, and mountaincaves, and woods" without any regard for "the general good" and did not "know to use in common any customs, any laws." Just as Hobbes characterized life before civil law as "nasty, brutish, and short," so Lucretius wrote that "the clans of savage beasts" would make "sleep-time horrible for those poor wretches." Men were "snatched upon and gulped by fangs," while those who escaped "with bone and body bitten, shrieked," as the "writhing pangs took them from life." In a time before agricultural plenty, starvation was rampant as "lack of food gave o'er men's fainting limbs to dissolution." Procreation, for Lucretius, was likewise beastlike and brutal. Men took women "with impetuous fury and insatiate lust" or bribed them with berries and fruit. When finally women moved "into one dwelling place" with men, "the human race began to soften," as they saw "an offspring born from out themselves." Neighbours intervened on behalf of women and children and urged compassion for the weak.

20. Lucretius, *Of the Nature of Things*, Book V, lines 1135–1185: "So next some wiser heads instructed men to found the magisterial office, and did frame codes that they might consent to follow laws." ... "For humankind, o'er wearied with a life fostered by force ... of its own free will yielded to laws and strictest codes." Because "each hand made ready in its wrath to take a vengeance fiercer than by man's fair laws," people voluntarily submitted to "fear of punishment."

21. Lucretius, *Of the Nature of Things*, Book VI, lines 1136–1284. "For now no longer men did mightily esteem the old Divine, the worship of the gods: the woe at hand did overmaster."

22. Lucretius, *Of the Nature of Things*, Book V, lines 811–870.

23. On Edenic imagery in American history, see R.W.B. Lewis, *The American Adam: Innocence, Tragedy, and Tradition in the Nineteenth Century* (Chicago: University of Chicago Press, 1955); David Noble, *The Eternal Adam and the New World Garden: The Central Myth in the American Novel since 1830* (New York: George Braziller, 1968); David Watt, *The Fall into Eden: Landscape and Imagination in California* (New York: Cambridge University Press, 1986); Cecelia Tichi, *New World, New Earth: Environmental Reform in American Literature from the Puritans through Whitman* (New Haven: Yale University Press, 1979).

24. Vladimir Propp, "Morphology of the Folktale," *International Journal of American Linguistics* 24, no. 4 (October 1958): 46–48. Roland Barthes, "The Struggle with the Angel," *Image, Music, Text*, translated by Stephen Heath (New York: Noonday Press, 1977), 139–141.

25. Quoted in Peter N. Carroll, *Puritanism and the Wilderness, 1629–1700* (New York: Columbia University Press, 1969), 13–14.

26. William Bradford, *History of Plymouth Plantation, 1620–1647*, 2 vols., edited by Worthington C. Ford (Boston: Published for the Massachusetts Historical Society by Houghton Mifflin Co., 1912).

27. Charles Morton, *Compendium Physicae*, from the 1697 manuscript copy (Boston: Colonial Society of Massachusetts Publications, 1940), vol. 33, xi, xxix, xxiii, xxxi; Nathaniel Ames, *An Astronomical Diary or Almanac* (Boston: J. Draper, 1758), endpapers.

28. Robert Beverley, *The History and Present State of Virginia* (London: R. Parker, 1705), 246–248.

29. Matthew Baigell, *Thomas Cole* (New York: Watson Guptill, 1981), plates 15, 16. On Cole's use of Eden as metaphor, see Henry Adams, "The American Land Inspired Cole's Prescient Visions," *Smithsonian* 25, no. 2 (May 1994): 99–107.

30. Baigell, *Thomas Cole*, plates 10, 15.

31. Ralph Waldo Emerson, "The Young American," *The Dial* 4 (April 1844): 484–507, quotation on 489, 491.

32. Leo Marx, *The Machine in the Garden: Technology and the Pastoral Ideal in America* (New York: Oxford University Press, 1964).

33. John Winthrop, "Winthrop's Conclusions for the Plantation in New England," *Old South Leaflets* no. 50 (1629) (Boston: Directors of the Old South Work, 1897), no. 50: 4–5; John Quincy Adams, *Congressional Globe* 29, no. 1 (1846): 339–342; Thomas Hart Benton, ibid., 917–918.

34. Reverend Dwinell, quoted in John Todd, *The Sunset Land, or the Great Pacific Slope* (Boston: Lee and Shepard, 1870), 252; Henry Nash Smith, *Virgin Land: The American West as Symbol and Myth* (Cambridge: Harvard University Press, 1950); Marx, *The Machine in the Garden*.

35. Frederick Jackson Turner, "The Significance of the Frontier in American History," American Historical Association, Annual Report for the Year 1893 (Washington: 1894), 199–227.

36. Francis Paul Prucha, *The Indians in American Society* (Berkeley: University of California Press, 1985), quotations on 7, 10.

37. Prucha, *Indians in American Society*, quotation on 12.

38. Prucha, *Indians in American Society*, 14–20; Lloyd Burton, *American Indian Water Rights and the Limits of the Law* (Lawrence: University of Kansas Press, 1991), 6–34; Carolyn Merchant, *Ecological Revolutions: Nature, Gender, and Science in New England* (Chapel Hill: University of

North Carolina Press, 1989), chapters 2 and 3; William Cronon, *Changes in the Land: Indians, Colonists, and the Ecology of New England* (New York: Hill and Wang, 1983); Richard White, *The Roots of Dependency: Subsistence, Environment, and Social Change among the Choctaws, Pawnees, and Navajos* (Lincoln: University of Nebraska Press, 1983); "Wilderness Act," 1964.

39. Chief Luther Standing Bear, *Land of the Spotted Eagle* (Boston: Houghton Mifflin, 1933), xix. On the ethnocentricity of wilderness values, see J. Baird Callicott, "The Wilderness Idea Revisited: The Sustainable Development Alternative," *The Environmental Professional* 13 (1991): 236–245.

40. Neal Salisbury, "Red Puritans: The 'Praying Indians' of Massachusetts Bay and John Eliot," *William and Mary Quarterly*, 3rd ser. 31, no. 1 (1974): 27–54; William Simmons, "Conversion from Indian to Puritan," *New England Quarterly* 52, no. 2 (1979): 197–218.

41. Franklin, "Remarks Concerning the Savages of North America," 91.

42. On images and metaphors of nature as female in American history, see Annette Kolodny, *The Lay of the Land: Metaphor as Experience and History in American Life and Letters* (Chapel Hill: University of North Carolina Press, 1975); Annette Kolodny, *The Land before Her: Fantasy and Experience of the American Frontier, 1630–1860* (Chapel Hill: University of North Carolina Press, 1984); Vera Norwood and Janice Monk, eds., *The Desert Is No Lady: Southwestern Landscapes in Women's Writing and Art* (New Haven: Yale University Press, 1987); Vera Norwood, *Made from This Earth: American Women and Nature* (Chapel Hill: University of North Carolina Press, 1993); Sam Gill, *Mother Earth* (Chicago: University of Chicago Press, 1987).

43. Thomas Morton, *New English Canaan*, in *Tracts and Other Papers ...*, edited by Peter Force (Washington: 1838), vol. 2, 10.

44. Henry Colman, "Address before the Hampshire, Franklin and Hampden Agricultural Society Delivered in Greenfield, Oct. 23, 1833" (Greenfield: Phelps and Ingersoll, 1833), 5–6, 15, 27.

45. Frank Norris, *The Octopus, a Story of California* (New York: Penguin Books, 1986; originally published 1901), 127. I thank David Igler for bringing these passages to my attention.

46. Norris, *Octopus*, 127.

47. Norris, *Octopus*, 130–131.

48. *Scribner's Monthly* 21, no. 1 (November 1880): 61. On the association of women with civilization and culture in nineteenth-century America, see Christopher Lasch, *The New Radicalism in America, 1889–1963* (New York: W.W. Norton, 1965), 65; Nancy Woloch, *Women and the American Experience* (New York: Knopf, 1984), Chapter 6; Merchant, *Ecological Revolutions*, Chapter 7.

49. J. Hector St. John de Crévecoeur, "What is an American?" *Letters from an American Farmer* (New York: E.P. Dutton, 1957; originally published 1782), 39–43.

50. Richard F. Burton, *The City of the Saints and across the Rocky Mountains to California* (1861) (New York: Knopf, 1963), 72.

51. See also Roderick Nash, *Wilderness and the American Mind*, 3rd ed. (New Haven: Yale University Press, 1982); Richard Slotkin, *Regeneration through Violence: The Mythology of the Frontier* (Middletown: Wesleyan University Press, 1973); Richard Slotkin, *Gunfighter Nation: The Myth of the Frontier in Twentieth-Century America* (New York: Atheneum, 1992).

52. On the Greek and Renaissance distinction between *Natura naturans* and *Natura naturata*, see Eustace M.W. Tillyard, *The Elizabethan World Picture* (New York: Random House Vintage, 1959), 42: "This giving a soul to nature—nature, that is, in the sense of *natura naturans*, the creative force, not of *natura naturata*, the natural creation—was a mildy unorthodox addition to the spiritual or intellectual beings.... Hooker, orthodox as usual, is explicit on this matter. [Nature] cannot be allowed a will of her own.... She is not even an agent ... [but] is the direct and involuntary tool of God himself." See also Whitney, *Bacon and Modernity*, 123: "[T]he extreme dehumanization of [nature by] the Baconian scientist ... is linked not simply to a complementary dehumanization of the feminine object of study, but to a somewhat

anachronistic return to a more robust feminine image of nature as *natura naturans*." Spinoza also used the two terms, but with rather different meanings than implied here. See Baruch Spinoza, *Spinoza Selections*, edited by John Wild (New York: Charles Scribner's Sons, 1930), 80–82; Harry A. Wolfson, *The Philosophy of Spinoza*, 2 vols. (New York: Meridian, 1958), vol. I, 253–255.

53. Frank Norris, *The Pit, a Story of Chicago* (1903) (New York: Grove Press, 1956).

54. William Cronon, *Nature's Metropolis: Chicago and the Great West* (New York: W.W. Norton, 1991). Cronon quotes the passage below from *The Pit* on the page preceding his "Prologue."

55. Norris, *The Pit*, 62.

56. Norris, *The Pit*, 60–63.

57. On Marx's concept of the endowment of money with organic, living properties and its application among the Indians of the Cauca Valley in Colombia, see Michael Taussig, "The Genesis of Capitalism amongst a South American Peasantry: Devil's Labor and the Baptism of Money," *Comparative Studies in Society and History* 19 (April 1977): 130–153.

58. Norris, *The Pit*, 374.

59. Nash, *Wilderness and the American Mind*.

60. Carol P. MacCormack, "Nature, Culture, and Gender," in *Nature, Culture, and Gender*, edited by Carol MacCormack and Marilyn Strathern (Cambridge: Cambridge University Press, 1980), 6–7; Sherry Ortner, "Is Female to Male as Nature Is to Culture?" in *Woman, Culture, and Society*, edited by Michelle Rosaldo and Louise Lamphere (Stanford: Stanford University Press, 1974), 67–87.

61. Philip Elmer-Dewitt, "Fried Gene Tomatoes," *Time* (May 30, 1994): 54–55; Richard Keller Simon, "The Formal Garden in the Age of Consumer Culture: A Reading of the Twentieth-Century Shopping Mall," in *Mapping American Culture*, edited by Wayne Franklin and Michael Steiner (Iowa City: University of Iowa Press, 1992), 231–250.

62. Max Horkheimer and Theodor Adorno, *Dialectic of Enlightenment* (1944) (New York: Continuum, 1993), quotations on 3, 7, 9.

63. Maria Gimbutas, *The Goddesses and Gods of Old Europe, 6500–3500 BC* (Berkeley: University of California Press, 1982); Merlin Stone, *When God Was a Woman* (New York: Harcourt Brace Jovanovich, 1976); Riane Eisler, *The Chalice and the Blade* (San Francisco: Harper & Row, 1988); Elinor Gadon, *The Once and Future Goddess* (San Francisco: Harper & Row, 1989); Monica Sjöö and Barbara Mor, *The Great Cosmic Mother: Rediscovering the Religion of the Earth* (San Francisco: Harper & Row, 1987); Pamela Berger, *The Goddess Obscured: The Transformation of the Grain Protectress from Goddess to Saint* (Boston: Beacon Press, 1985). On cultural ecofeminism, see some of the essays in *Reweaving the World: The Emergence of Ecofeminism*, edited by Irene Diamond and Gloria Orenstein (San Francisco: Sierra Club Books, 1990).

64. Londa Schiebinger, *The Mind Has No Sex? Women in the Origins of Modern Science* (Cambridge: Harvard University Press, 1989); Evelyn Fox Keller, *Reflections on Gender and Science* (New Haven: Yale University Press, 1985).

65. Examples include Oelschlaeger, *The Idea of Wilderness*; Donald Worster, *The Wealth of Nature* (New York: Oxford University Press, 1993); Barry Commoner, *The Closing Circle: Nature, Man, and Technology* (New York: Knopf, 1971).

66. Bill McKibben, *The End of Nature* (New York: Random House, 1989). For a critique, see Tom Athanasiou, "U.S. Politics and Global Warming" (Westfield: Open Magazine Pamphlet Series, 1991).

67. Carolyn Merchant, *Radical Ecology: The Search for a Livable World* (New York: Routledge, 1992).

68. Alan Hastings, Carole L. Hin, Stephen Ellner, Peter Turchin, and H. Charles J. Godfray, "Chaos in Ecology: Is Mother Nature a Strange Attractor?" *American Review of Ecological Systems* 24, no. 1 (1993): 1–33; N. Katherine Hayles, "Gender Encoding in Fluid Mechanics: Masculine Channels

and Feminine Flows," *Differences: A Journal of Feminist Cultural Studies* 4, no. 2 (1992): 16–44; N. Katherine Hayles, *Chaos Bound: Orderly Disorder in Contemporary Literature and Science* (Ithaca: Cornell University Press, 1990); N. Katherine Hayles, ed., *Chaos and Order: Complex Dynamics in Literature and Science* (Chicago: University of Chicago Press, 1991); Daniel Botkin, *Discordant Harmonies: A New Ecology for the Twenty-First Century* (New York: Oxford University Press, 1990); James Gleick, *Chaos: The Making of a New Science* (New York: Viking, 1987); Edward Lorenz, *The Essence of Chaos* (Seattle: University of Washington Press, 1993).

Chapter Four

A Death-Defying Attempt to Articulate a Coherent Definition of Environmental History

Douglas R. Weiner

In a recent article in the *Journal of Historical Geography*, J.M. Powell began with a well-meaning attempt at humour: *"Question.* Why is environmental history like Belgium?" *Answer.* "Because it was entirely the product of a resident collective imagination."[1] There is no denying that this is true. I would only add that it is also the product of a resident collective toleration of a good deal of intellectual uncertainty, diversity, and even incoherence.

Most recently, this uncertainty was reflected in the short essays on the essence and future directions of the field that constituted the heart of the January 2005 issue of *Environmental History*. Harriet Ritvo characterized the field as "an unevenly spreading blob," while others questioned the utility of the very term "environment."[2]

Human ecologists—when there still were such academic beings—long nourished the hope that close study could reveal correlations between types of environments and the kind of adaptations that humans make to them.[3] Fifty years ago, in June 1955, Princeton University hosted the international symposium "Man's Role in Changing the Face of the Earth." Setting the tone were the disciplinary approaches of historical geography, cultural anthropology, and human ecology, exemplified by the title of Lewis Mumford's talk, "The Natural History of Urbanization." Experts still believed then that humans could be understood largely as biological subjects whose behaviour could be explained by importing models from zoology, ecology, and other fields. Scholars still held out hope then that science could identify healthy natural norms and that such knowledge would aid us in identifying and checking modern, self-destructive, human-engineered "pathologies."[4]

Whether they were primitive geographical determinists like the earlier Ellsworth Huntington and Ellen Semple or more sophisticated researchers like Julian Steward, the picture these scholars drew was predicated on stable, essentialized understandings of nature and culture.

Since the appearance of John Cole and Eric Wolf's 1974 study comparing the lifeways and household organization of neighbouring Italian and German villages in the Trentino, no one can seriously adopt a strong environmental determinist position.[5] Nature's role, it turned out, is much more complicated and subtle, defying attempts to establish one-to-one correspondences with social and cultural forms.

Nor are we much clearer about how particular societies have affected their environments. As Richard White has suggested, we can make believable statements about some effects by local actors on environments of local scale, but beyond that we get into a domain better addressed by chaos theory.[6]

Even if the environment is not determinative in human affairs, and an understanding of our effects on the environment still is viewed through a glass darkly, then surely at least the term must evoke some agreed upon image? Nothing of the kind, argues Anna Tsing; at least no discernable *positive* figuration has emerged. The "environment" is what and where we variously want it to be. "On the one hand," she writes, "environmental rhetoric is widely used and accepted. On the other, no one agrees about what this rhetoric should do for humans and nature, and the struggle is on to bend environmental rhetoric to particular, contradictory purposes— Wise Use or preservation; privatization, national heritage, or tribal autonomy; international restructuring or democratic internationalism; and much more."[7]

From another angle, if the environment is everything from the microparticle to the universe, then all history, it may be argued, is collapsible to environmental history, which in turn ceases to be distinguishable from history as such. Epistemologically

radical geographers and cultural ecologists such as David Demeritt, while taking pains to emphasize that they believe in the existence of a material world out there, argue that it is impossible to know that world in-itself. "[Its] apparent reality is never pre-given," writes Demeritt; "it is an emergent property," Demerritt continues, quoting Joseph Rouse, that "'depends on the configuration of practices within which [it] becomes manifest.'"[8] In plain English, our cognitive maps of the world are continually being produced and revised, and their production is closely tied up with our systems of politics and economics and the practices associated with those systems. So what we study when we study the human–nature relationship is a set of shadows and distorted images—a moving target. The objects of our study, social actors, are armed with their own socially constructed cognitive maps, which we, armed with our own maps and tools, try to understand.

SCIENTIFIC ECOLOGY

One such tool, scientific ecology, already has bumped up against the limits of its own abilities to produce a predictive, positivist picture of nature. As early as the mid-1930s Arthur Tansley, in an often-overlooked conceptual breakthrough, cautioned that while ecosystems were, to his mind, ontologically real interlocking networks of the organic and inorganic world ranging in scale "from the universe as a whole down to the atom," they could be studied only through their "artificial" abstraction out of the skein of life. Researchers could select the scale of the truncated "system" under study for reasons of convenience and relevance to the purposes of the study. "The whole method of science," he wrote, crediting Hyman Levy's Marxist-inflected formulation, "is to isolate systems mentally for the purposes of study, so that the series of *isolates* we make become the actual objects of our study, whether the isolate be a solar system, a planet, a climatic region, a plant or animal community, an individual organism, an organic molecule, or an atom. Actually, the systems we isolate mentally are not only included as parts of larger ones, but they also overlap, interlock and interact with one another. The isolation is partly artificial, but is the only possible way in which we can proceed."[9]

In the headlong rush to create a positivist predictive science, as Donald Worster has brilliantly described in *Nature's Economy*, ecology imported (and re-exported) models and metaphors based on energy flow, ideas of homeostasis in physiology, and attempts to correlate land mass dimensions with species richness.[10] In all of this, Tansley's reflective warning got lost, and ecologists mistook their imagined figurations of "nature" for actual empirical objects of study, ironically without ever developing the capacity to wield the predictive power they desired.

By the 1990s, science studies, cultural ecology, and environmental history had caught up with the ecologists, who, with rare exceptions like Norm Christensen, had additionally continued to make the increasingly dubious assumption that the "nature" that they studied was "pure," untainted by the complications of human "culture."[11] Far from being the result of a conscious application of Tansley's idea that we are condemned to abstract "artificial" units of study from the web of reality, it was, on the contrary, the expression of the field's understanding of true, "healthy," untainted nature as one unsullied by human presence. One root of the problem, as cultural ecologist Bruce Winterhalder noted, was that ecology, suffering from

physics-envy, had tried to deny history. By the mid-1970s, Richard Lewontin, in Winterhalder's paraphrase, already recognized that "the full set of functional relationships that would enable us to predict the phenotype for each genotype in all possible environments (technically, the norm of reaction) is effectively 'hidden' to analytical view." That is, the path of evolution is littered with non-repeatable events such as the "origin of life" that preclude a definitive cause-and-effect-based reconstruction because knowledge of both the spectrum of factors at the time and the particular ways in which they interacted with each other will forever remain obscure. Using the same logic, Winterhalder argued that it is similarly impossible to predict how collections of organisms and environments—ecosystems, if you will—will evolve, or to explain exhaustively their past histories.[12] A genuine acceptance of Darwin's vision of an emergent evolution precludes the dream of a predictive ecology. As many of us already knew, ecology is no more—or less—scientific than history itself. Tansley's advice remains as solid as it ever was. Our science, our history, our maps of the world can never be certified as eternally and absolutely true—or complete. They are only as good as the degree to which they allow us to attain certain kinds of real-world objectives, not only in the domain of technology but also the production of systems of science, political economy, and social relations.

Now theorists are pushing us to consider the ultimate implications of this. Kristin Asdal, for example, is but one of a host of thinkers who are asking whether there is a "nature" out there to be represented at all, let alone whether "ecology" can in fact represent it.[13] She writes: "Environmental history appears today as a sub-discipline within history, and attempts to bring nature in as a co-creator of histories.... However, in all of this, environmental history has apparently been oblivious to more recent philosophies and histories of science. Feminist philosophy of science, for example, has posed fundamental questions about what we at any given time perceive as science—and nature. The question is thus *which* nature and *which* conceptions of science should be brought in. One might also ask, which ethic?"[14]

Obviously, to restate Demeritt's point, no one is denying the existence of a real world out there. Neither Demeritt nor Asdal, nor even Donna Haraway, I assume, are about to walk out of Alan Sokal's twenty-first floor apartment window because they believe that modern physics, with its laws of gravity, is socially produced. It is, but the science also works on a pragmatic level, which is why we accept its provisional truths, even while appreciating, as Bruno Latour and Stephen Woolgar have noted, that those truths are artifacts of the metaphors, instrumentation, and world views of their scientific creators.[15] Ecologists must remember that when they go into the field and take measurements with instruments, when they assign a given life form to a given taxon, they are producing in fact the very nature that they study. We as environmental historians, when we borrow from the ecologists, need to be aware of that as well. If we turn our backs on the notion that we socially construct science and nature, we risk buying into a delusory, uncritical, received picture of the "real world." If we fall silent in response to the challenge of continually destroying and producing new pictures of "reality," then we abandon the field to those who are not as squeamish or self-conscious. As awful as it seems, we cannot return to the imagined Eden of empiricism. To a great

extent, of course, we may lay the blame for this on the irruption of post-structuralism and science studies, but they are merely the messengers bearing the bad news.

A VERY SHORT HISTORY OF THE FIELD

Once upon a time, Environmental History (that is, as an institutional form) and environmental history (as an intellectual project) existed approximately in such a prelapsarian state. The origin story of the field was that it came together in the early to mid-1970s, influenced by activism of a certain type, linked to a particular understanding of scientific ecology, and built on a dichotomized picture of nature and culture facing each other across a formidable ontological divide. Let me say strongly that there was never any kind of "party line" at all. An early issue of the *Environmental History Newsletter* stressed that "one need not be a card-carrying conservationist to teach in American environmental studies. On the contrary, we need a variety of views and ideologies."[16] And one of the first issues of *Environmental Review*, which was subtitled "an interdisciplinary journal," boasted that its "historical framework allows many different viewpoints to come together." "Far too often," it complained, "our understanding of man's relation to nature has been distorted by limited technological, economic, and political questions and answers." True enough. But there was the lurking implication that a "true" picture could somehow emerge from the discussions in the review's dispassionate pages.[17]

To account for the rise of Environmental History institutionally, it is also important to recall that the intellectual concerns of its founding scholars—wilderness and the American mindset, pollution, forest history, the histories of conservation, preservation, and irrigation, social and intellectual understandings of nature, the relationship between social systems and environmental change, and, conversely, the effects of the environment on societies—could not find an honoured place at the table of the so-called mainstream subfields in history, including the history of science and the history of technology. Eventually, historians of agriculture, landscape gardening, and cities, and historically minded geographers, anthropologists, sociologists, and others joined our society, our meetings, and our discursive network. Intellectual concord, however, could only be sustained as long as there was general agreement about the objects of our study. Things were fine as long as nature was in its corner, culture was in its corner, and both behaved properly, as neo-Platonic entities are supposed to.

By the mid-1980s, however, the reflux of the science wars and of post-structuralism had finally washed over our field too. In the pages of the *Journal of American History*, at ASEH meetings, and in other publications we debated the ontological viability and utility of concepts such as wilderness and the balance of nature.

After two decades of tortured language and undecipherable jargon, it is fair to say that, like it or not, we are all post-structuralists now. Every speaker, we recognize, is situated—even the deconstructionist. Although the physical scientists are still bringing up the rear, almost all of us recognize that a posture of epistemological invulnerability, of godlike omniscience, is beyond us mortals. We must give up on the attainment of a perfectly transcendent position, an *aperçu* that Immanuel Kant had already arrived at in the 1780s.[18]

Instead, we must content ourselves with an understanding of the world that is an eternally dynamic set of cognitive frameworks in which the brass ring of absolute truth is never grasped. In fact, the material reality that we seek to understand constitutes an evolving, eternally moving target requiring ever newer cognitive maps. We cannot study the emergence of the Asian bird flu today with mental pictures of the world drawn from the epidemiological, geographical, and social understandings of 1919. This situation alone has made our jobs exponentially harder.

Another annoying insight of those naughty French (and other) theorists also refuses to go away. Knowledge also is relative because it is always linked in some way to power. What we produce as narrative accounts and historical understandings always bears some relation to our social aims, our visions of desired and abjured social relationships, notions of justice and political organization, and, most fundamentally, notions of selfhood and psychologies, all considered as an integrated posture toward the world. Our accounts, however carefully and dispassionately written, are always at the service of *some* understanding of common sense, some reading of "human nature" and social dynamics. Our accounts therefore also help to constitute environments—and social relations. We are unavoidably always on one or another side of power. Pierre Bourdieu once called sociology a "contact sport" and we can certainly say the same for history.[19]

That our scholarship cannot avoid taking sides is not a pleasant realization for folks like us who, for the most part, landed in academia to pursue a quiet, contemplative life. Yet, this appreciation of the saturation of history and knowledge with questions about power give us the most powerful tool to reconstitute the "environment" as a central and prepossessing category of analysis. To respond to Anna Tsing's complaint, the reason that the idea of a generic environment in environmental history appears to be, and is, incoherent—and this applies equally to the idea of generic "humans" in history—is that at every invocation, "environment" is a figure and a metonym.[20] What we are doing when we speak about "the environment" is always instantiating it in some particular fashion: as a polluted river, as wilderness, as wildlife habitat, or as "virgin lands." These images of the "environment" seem to call out for some kind of action: cleanup and restoration, preservation, management, or exploitation. Yet at the same time, these images seek to conceal and suppress alternative imaginings of a place and its uses. They also hide the fact that behind each figuration of "the environment" is a social agenda. Because "the environment" is a term that has a universal ring to it and pretends to embrace the general good, it serves as an excellent mask for particular interests.

Consequently, every "environmental" struggle is, at its foundations, a struggle among interests about power. In fact, I would go further. Every environmental story is a story about power. Who will control access to resources and amenities? What are the trade-offs? Who is expected to bear the risks and encumbered costs? Who believes that they will benefit? Who controls the range of choices that are made available to publics as they decide major questions? What are the conditions of those choices? Who is actually doing the choosing? What outcomes are being projected by "experts"? How are the expected risks and benefits being represented? How will social structures and relationships change as a result of particular choices?

We may either write a celebration of power or exhume the dark side of power. It matters little whether it is a story about Panama Disease and bananas, a park in Rio de Janeiro, a famine in late nineteenth-century India, exobiology, landscape change in southeastern Spain, the rise of suburbia, Chernobyl, environmental non-governmental organizations (NGOs) and Amazonia, burial practices, a donkey massacre in Bophuthatswanaland, *alleged* deforestation in Ethiopia or Guinea, or logging and forest management in the Pacific Northwest. Nor may we content ourselves with producing stories with cardboard cutout characters, with black hats and white hats, although it's always nice to take a swat at venality. The dark side of power is never that obvious. Let me provide an example from my own research.

By the 1930s the Soviet regime had embarked on a course of deep transformation (or production) of society and nature, aided by an "industrial strength" repressive apparatus. Practically the lone public opponents of these death-dealing megalomaniacal and tributary schemes were the activists and members of the U.S.S.R.'s nature-protection movement, most prominently a group of botanists and zoologists. They miraculously survived the bloody 1930s and even succeeded in nurturing a new generation of heterodox activists. As I developed the story in the early 1990s, I thought to myself, "What incredible luck! I've stumbled on the story of the origins of Soviet and Russian civil society—this is the motherlode!" It would have been easy, and temporarily beneficial—from the standpoint of my career—to tell the story that way, but I knew it was wrong. I realized with a certain regret that I could not tell the story that Western public opinion and Soviet and Russian Studies wanted to hear. That realization came not only because I was a trained—that is, disciplined, historian—who honours the duty to account for all of the facts as I have [co-]produced them through my investigation, but also because I had begun to take seriously Jacques Derrida's admonition that we should not content ourselves with our previous excavations but continually make new incisions. I needed to reveal that these heroic conservation activists were themselves elitists, not democrats, and were serving their own unarticulated professional interests that often were at odds with those of farmers, pastoralists, and other groups. Not only were they not the nucleus of a future Russian civil society, the activists, with their guild mentality, were in fact evidence precisely of the inability of such a civil society to cohere across group lines.[21] Ultimately, the activists remained a part of the problem of a society continually unable to resist the depredations of tributary elites, a story that still requires further excavation.

Every historical figuration of the environment has contained within itself social trade-offs, and, at the extreme endpoint of its logic, dispossession and even genocide. Most environmental historians already are familiar with the research of Kenneth Olwig and Karen Fog Olwig, Mark Spence, Jane Carruthers, John MacKenzie, Ted Catton, John Sandlos, and others, who have shown that the idea of "wilderness" was predicated on the mental and physical dispossession of locals, be they Scottish Highlanders, Arawaks and Caribs, Native Americans, South African Dantus, Maasai, Inuit, or others.[22]

Many also are aware of the growing scholarship on nature protection and landscape planning in Nazi Germany.[23] Here, too, a naive reading of this history—

"Well, the Nazis accomplished at least something positive"—can no longer be indulged as stemming from a lack of information.[24] The celebration of an idealized, well-tended nature created by German *Kultur* presupposed a contrasting, debased nature of the Poles, Jews, and Russians that "cried out" for its own obliteration. This figuration legitimated the "cleansing" and improvement of the eastern lands.[25] Heinrich Wiepking-Jorgensmann, a professor of landscape gardening who was tapped by Heinrich Himmler to help in the redesign of the landscapes of the "Annexed East Lands," wrote: "The landscape is always a form, an expression ... of the people living within it. It can be the gentle countenance of its soul and spirit, just as it can be the grimace of its soullessness or of human and spiritual depravity. In any case, it is the infallible, distinctive mark of that which a race feels, thinks, creates and does.... Therefore the Germans' landscapes differ in all ways from those of the Poles and the Russians— just as do the peoples themselves. The murders and cruelties of the Eastern races are engraved, razor-sharp, into the grimaces of their native landscapes."[26] The *Basics of Planning for the Building-up of the Eastern Areas* took for granted the mass elimination of the existing population and its replacement by German settlers.[27]

Ruling elites have attempted to redesign entire landscapes, not just with warped racial figurations but also under the aegis of "science and progress," perpetrating horrendous crimes in the process. The Bolsheviks had come to the consensus early that the landscape of "Communism" would be a gleaming industrial one. To pay for and build that hoped-for landscape of progress, Josef Stalin needed to create an actual landscape of renewed serfdom, confining peasants in the new collective farms where their labour and productive power was completely captured. Against the backdrop of peasant resistance and severe drought, Stalin's regime continued to requisition grain in Ukraine and the North Caucasus, condemning perhaps 5 million or more to horrible deaths from starvation.[28] An additional 1.5 million—a full 25 percent of the population of the Kazakh ethnos—perished as "backward" nomadic herding was classed as incompatible with the Soviet vision of a "progressive" agrarian landscape of state farms. Thirty-three years later, in a colossal act of historical retouching, Nikita Khrushchev could call those regions purged of herders "virgin lands," awaiting their wheat planters from Ukraine, Russia, and from among the political deportees already in the east. Also taking a cue from Stalin, China's Mao Zedong, trying to industrialize in record time, precipitated a famine unparalleled in human history: estimates of the famine dead range from between 20 to 50 million people.[29]

And the inexpressible tragedy of the Holocaust cannot occlude the painful realization that from the first *aliyah*, or wave of Jewish immigration to Palestine from Europe in the 1880s, Zionists armed themselves with a European-derived colonialist figuration of "progress." That vision involved afforestation and efforts to turn the Jewish state into a productive garden ("making the desert bloom") at the expense of lands confiscated from Arab use. This view, in large measure, was partly based on the presumption that Arab traditional agriculture and pastoralism were somehow more primitive, and therefore less justifiable, uses of the land.[30] In the 1910s, no less a figure than David Ben Gurion, the future prime minister, described Zionist settlers such as himself as "a company of conquistadors," likening

their settlement to that of the New World with its "fierce fights" with "wild nature and wilder Redskins."[31] Selig Soskin, closest aide to Otto Warburg, even pointed to "Aryan" German colonization in Africa as the model for the development of agricultural technology in Palestine.[32] Vladimir Jabotinsky, founder of Revisionist Zionism, admitted that: "Any native people ... will not voluntarily allow, not only a new master, but a new partner. And so it is for the Arabs ... Culturally they are 500 years behind us, spiritually they do not have our endurance and strength of will, but ... [t]hey look upon Palestine with the same instinctive love and true fervor that any Aztec looked upon his Mexico or any Sioux looked upon his prairie ... Zionist colonization, even the most restricted, must either be terminated or carried out in defiance of the will of the native population."[33] Regrettably, it must be conceded that these candid early sentiments—reworked for public international consumption as the slogan "A land without people for a people without land"—prefigured the later dispossession of Arab Palestinians as they were overtaken by their own Catastrophe (*Nakba*).[34] Here, too, further exhumation is necessary, for much of the afforestation campaign developed by the Jewish National Fund (first in 1912 as the Olive Tree Fund) was of "marginal" economic value, instead serving as a symbol of the settlers' "struggle with nature and the conquest of the land."[35] More than that—like the Kiev ravine of Babi Yar that Nikita Khrushchev filled in and planted over, rendering invisible the actual site of the Nazi slaughter of 33,000 Jews—smiling forests grow over the spectres of razed Arab villages: true landscapes of green oblivion.[36]

Not just Israeli, but all state forestry projects—whether American, Indian, or

Russian—derive generally from the same source of cameral absolutism in Europe. Ravi Rajan has provided an important genealogy of colonial forestry practices showing that the rational, optimistic, science-based scenario of sustained-yield forest management was inescapably based on a denigration of allegedly "inefficient" folk uses of the forest, whether by German peasants or Indian farmers and tribals, to justify enclosures and removals.[37] Through force and hegemony, the European cognitive map has carried the day and become the global standard.

The African-American experience is another ocean of examples of how power inflects environmental histories. In an incisive study, Sylvia Hood Washington has shown that with emancipation, the tar and turpentine industry in North Carolina and Georgia collapsed, as former slaves refused to perform such environmentally hazardous work. In response, Southern states enacted laws permitting the states to lease convict labour to private interests. By 1890, almost 30,000 prisoners were leased out, and the tar and turpentine industries were again assured of a labour supply. We can only wonder how many of those almost entirely African-American prisoner-lessees were arrested in order to provide for the recovery of the industry.[38] Similarly, James Giesen has shown how it is impossible to write about the boll weevil infestation in the Mississippi Delta (or anywhere else in the United States) without also talking about the racialized nature of the labour force. Should we really be surprised that, as Giesen tells us, a successful experiment by African Americans in renting and managing a cotton plantation ended after only a couple of years, because the experiment promised to overturn the public shibboleth that African Americans could not work without White supervision?[39] Or that the town of Rosewood, Florida, had to be destroyed because it gave lie to the myth that African Americans could not run their own farms successfully?

These examples, it bears repeating, attest to the truism that part of the production of "knowledge" is the production of "ignorance." I use ignorance here in two senses. On the one hand, the bearers of "superior knowledge" counterpose their "enlightenment" to the beliefs and practices of those who still are in the dark, producing a picture of ignorance. On the other, the new knowledge makes its own adepts ignorant by its arrogant dismissal of whole realms of experience, whole universes of possibilities—or by destroying them when they threaten to subvert the legitimacy of the "superior knowledge."

Just as "ignorance" of African Americans' real capabilities continually had to be enforced, so too did ignorance of Native practices in the Canadian North, which, as John Sandlos has shown, also involved a large dose of misrepresentation of "the Other."[40] Moreover, as Sandlos takes pains to point out, "the misrepresentation of tradition is more than a mere misunderstanding. By extending a degree of intellectual ownership over traditional Native sources of legitimacy and by appealing to a romantic ideal of the Native hunter whose only legitimacy lies in a dimly remembered past, [Canadian] federal bureaucrats could now define the northern landscape completely in accordance with their own political, social, and economic goals. The federal government's efforts to preserve a wilderness sublime for the last remnants of the large buffalo herds necessitated a parallel effort to legislate Native people, their culture, and their values out of the wild northern landscape."

Such a "wilderness sublime"—as figured by White bureaucrats and zoologists—was promoted as embodying the general good of the nation, which was imperilled by alleged overhunting by Natives corrupted by modernity.[41] Similar to the story told by Ted Catton about Alaska, Sandlos's story is about a set of framings that put marginalized Native peoples in a double-bind: remain "unspoiled" and thereby, under the new conditions, unviable, or try to adapt, which stimulated charges that they were betraying their heritage and therefore did not deserve continued "rights" to their territory.[42] The ignorance of the colonizers, among other things, consisted precisely in ignoring that their beliefs and practices necessarily had to lead to this double-bind, to the condemnation of Natives to a status of non-viability.

Perhaps the account that most powerfully reveals the high stakes associated with our environmental maps and the practices based upon them is Mike Davis's controversial history of the millennial-scale El Niño events toward the end of the nineteenth century.[43] *Late Victorian Holocausts* makes the case that the environment that the British imperialists saw in India was a landscape of potential profits—whether from opium, cotton, indigo, wheat, or taxation. In the process of realizing that landscape of profit, Davis shows, the British created a landscape of unprecedented mass death, as they did in Ireland. With the delusory and self-deluding logics of "progress" and "improvement of the Natives," the British sought at once to naturalize and legitimate their rule. With the rhetorics of "free trade" and Social Darwinism they sought to bury the fact that the famines and epidemics of 1876–1879 and 1896–1902, which took between 12 and 29 million lives, were not the mere consequences of El Niño, not principally

acts of nature, not the responsibility of a "lazy," "improvident," "inefficient," and Malthusian rural underclass who deserved to go under (as Lytton, Lord Cromer, and an 1881 official report put it), but the products of state terror and coerced dispossession. They were the ultimate products of the landscape of cotton, opium, wheat, and indigo tribute.

Davis's book, like no other, highlights the trade-offs between the "primitive" yet locally adaptive millet-sorghum-pulse subsistence economy, with reasonably sustainable well-water-based irrigation, coppicing, and grazing in communal forests and pastures, supplemented by community-based provisions for famine relief, on the one hand, and on the other, the "progressive" colonial economy, with its commercially driven irrigation that ruined the village wells, waterlogged the soils, created malaria epidemics, and drove millions of people to horrible starvation deaths, but made plenty of money for British banks, commercial houses, and the state. In the Earl of Lytton's and Lord Elgin's India, as in Stalin's Soviet Union, and Mao's China, environmental history must reveal how the social costs of the production of environments of "progress" are paid for with the bodies of millions.

But Americans are also paying. While Gerald Markowitz and David Rosner in *Deceit and Denial* make the case that the production by corporations of ignorance about the risks of industrial pollution was wilful, Christine Rosen's research points to an even more nuanced, and hence more dangerous, situation.[44] Because the citizenry, in the pre-Environmental Protection Agency era, did not possess the proper tools and knowledge to recognize what is truly health-threatening, the law and business practices allowed the production and discharge of harmful substances as long as they were generally "unrecognized" as such and unchallenged in the courts. As long as the onus was on claimants, the law permitted businesses to claim plausible deniability because it placed the prevention of social risk below the production of commodities.

WHAT IS ENVIRONMENTAL HISTORY?

Because of the diffuse nature of its original suite of concerns, Environmental History has become over the past 30 years a very big tent. That is attractive. But that has also created an institutional "boundary problem" of cetacean proportions. There are no ways to license "environmental historians," no collectively accepted criteria to distinguish "true" environmental historians from "impostors." Anyone who can claim to be an environmental historian and get away with it becomes one. This is exactly what stimulated Powell's "collective imagination" quip, which opens up onto a much more serious question, first raised a few years ago by James Secord—namely: "Who should be credentialed to interpret the world and our place in it?"[45]

These two questions—what is environmental history and who can interpret the world—must be resolved in tandem. While there is, of course, plenty of room for all kinds of research under the big tent of Environmental History, I would propose that the most important work that we can do and the work that can unite all of our various efforts is to bring to the light of day the unexamined metonymic "shadows" of environmental figurations and practices. Every figuration of the "environment"—by distributing different opportunities for environmental

access and decision-making power to different "types" and groups—potentially encodes exclusion, dispossession, or even genocide.

We must persist with investigations in order to uncover how, at the end of the day, actors and Others in our stories were left with deficient choices. Why were they asked to choose between, or among, dispossessions? It is the history of the narrowing of choices that ultimately will be the most useful for us if we are to turn around our nihilistic, fatalistic century and begin to build a world of increasingly rich choices. One clarification must be made. This should not be read to mean that only the history of power and conflicts may be written. That would be silly as well as exclusivist. What I have tried to underscore, simply, is that all of our histories must reflect an awareness that their subjects, and our writing of them, are inflected by power.

Others will object that the framework I propose fails to take seriously the agency and power of the non-human world. After all, Mike Davis's book would be unthinkable without the current scientific discourse about the ENSO (El Niño/La Niña) phenomenon, but, more importantly, without the unprecedented extreme weather events of the late nineteenth century. That is a given. I would add only that what is of interest to us is precisely the reactions of specific societies to these events. Nature is a part of the dialectic, but its impact is always refracted through the social structures and pathways of power of our societies, which was the point that Davis tried to underscore in his study.

I am not bothered by the absence of a coherent, essentializable object of study—"the environment"—and continue to proudly define myself as a member of the field of Environmental History, understood as a locus for exploration and intellectual adventure rather than as a capturable, bounded entity in the way that, for example, Medieval British History or United States History, 1877 to the present, seem to be and claim to be—although we know better! As an aside, I cannot resist noting that a decade and a half ago, Leo Marx similarly suggested, in a death-defying review essay that appeared in *Technology and Culture*, that his own discipline, the history of technology, had no coherent object of study.[46] The only problem with his incendiary essay was that he didn't extend his logic!

Seriously, those fields with seemingly well-delineated boundaries, which have little difficulty justifying their academic existence, easily slide into comfortable ruts. They continue to operate with yesterday's commonsensical categories of analysis, even as the world has moved on. We continue to teach national histories even as there is an "erosion of the nation state, national economies and natural cultural identities" before our very eyes.[47] On the other hand, those fields (like our own) with visibly imperfectly delineated, or worse, highly permeable, migratory, or even barely discernible boundaries provide unexcelled opportunities for intellectual conversations and explorations of a scope, variety, and daring rarely encountered in more fixed fields.

We in environmental history are aware that we are not producing a "correspondence" of the world in a primitive scientistic sense, but rather a situated, inspired understanding of the world—a story that could be useful as a guide to refashioning ourselves. We also are providing a model of the process whereby people may arrive at a disciplined, reflexive knowledge of themselves, open to revision. These are not simply professional aims. They are models for democratic citizenship. We are, regrettably, past the point where we can blithely believe we are generating a positivist picture of empirical reality. The alternatives are actively to serve power, to become an antiquarian (which is the same thing), or to develop cognitive models of the world that will help us create a future where the exercise of domination will become just a little bit harder.

By virtue of our academic positions we have somehow ended up as socially "credentialed" interpreters of the world. However, our official credentials mean nothing if we do not produce interpretations of the world that are increasingly shared, not hoarded, and that create in our publics a toleration for, if not comfort with, uncertainty, reflexivity, ideas of an evolving selfhood in an evolving world, and the public's responsibility for the conscious production of that world. Unlike the Wizard of Oz, if we are to help build a demystified and democratic social order, we must reveal our feet of clay. We must tear that scientistic curtain that protects our "exalted" status within the social hierarchy to reveal ourselves as seekers who are only a couple of steps ahead of the rest of the population in learning how to be reflexive interpreters and creators of our lives. Along with de-essentializing our field, we need a renewed turn, as I once called for, to an attention to process, to means as opposed to ends.[48] As practitioners of a fearlessly reflexive scholarly venture, Environmental Historians can set a model for a new open praxis that needs to emerge if human civilization on Planet Earth is to have a chance. And if anyone wants to know what "environmental history" is, you just send them to me!

NOTES

With sincere thanks to my colleague, Hermann Rebel, for imposing clarity on what was an incoherent "unevenly spreading blob."

1. J.M. Powell, "Historical Geography and Environmental History: An Australian Interface," *Journal of Historical Geography* 22 (1996): 253.
2. Harriet Ritvo, "Discipline and Indiscipline," *Environmental History* 10 (January 2005): 75.
3. Sec Eugene Cittadino, "The Failed Promise of Human Ecology," in *Science and Nature: Essays in the History of the Environmental Sciences*, edited by Michael Shortland (London: British History of Science Society, 1993), 251–283.
4. William L. Thomas, Jr., ed., *Man's Role in Changing the Face of the Earth* (Chicago: University of Chicago Press, 1956).
5. John W. Cole and Eric R. Wolf, *The Hidden Frontier: Ecology and Ethnicity in an Alpine Village* (New York: Academic Press, 1974).
6. Richard White, "The Nationalization of Nature," *Journal of American History* 86 (December 1999): 976–986.
7. Anna L. Tsing, "Nature in the Making," in *New Directions in Anthropology and Environment*, edited by Carole L. Crumley (Walnut Creek: Altamira Press, 2001), 3.
8. David Demeritt, "Science, Social Construction, and Nature," in *Making Reality: Nature at the Millennium*, edited by Bruce Braun and Noel Castree (London: Routledge, 1998), 178—citing Joseph Rouse, *Knowledge and Power: Toward a Political Philosophy of Science* (Ithaca: Cornell University Press, 1987), 160–161.
9. Arthur G. Tansley, "The Use and Abuse of Vegetational Concepts and Terms," *Ecology* 16 (1935): 299–300. Hyman Levy's work was *The Universe of Science* (London: Watts, 1932).
10. Donald Worster, *Nature's Economy: A History of Ecological Ideas* (San Francisco: Sierra Club Books, 1977).
11. Norman Christensen, "Landscape History and Ecological Change," *Journal of Forest History* 33 (1989): 116–125.
12. Bruce P. Winterhalder, "Concepts in Historical Ecology: The View from Evolutionary Ecology," in *Historical Ecology: Cultural Knowledge and Changing Landscapes*, edited by Carole L. Crumley (Santa Fe: School of American Research Press, 1994), 20.
13. Kristin Asdal, "The Problematic Nature of Nature: The Post-constructivist Challenge to Environmental History," *History and Theory* 42 (December 2003): 60–74.
14. Ibid., 61.
15. Bruno Latour and Stephen Woolgar, *Laboratory Life: The Social Construction of Scientific Facts* (London: Sage Publications, 1979).
16. Donald Worster, "American Environmental Studies," *Environmental History Newsletter* 3 (March 1976): 6.
17. *Environmental Review*, issue P/76. From the subscription-information page.
18. Immanuel Kant, *The Critique of Judgement* (1790; reprint, New York: Hafner Press, 1951).
19. See Pierre Bourdieu say it on film in Veronique Fregosi, Annie Gonzalez, and Pierre Carles, directors, *La sociologie est un sport de combat*. New York: Icarus Films, 2001 [C-P Productions, France].
20. On the use of "figures," see especially Erich Auerbach, *Mimesis: The Representation of Reality in Western Literature*, translated by Willard R. Trask (Princeton: Princeton University Press, 1968), 73.
21. Douglas R. Weiner, *A Little Corner of Freedom: Russian Nature Protection from Stalin to Gorbachev* (Berkeley and Los Angeles: University of California Press, 1999).

22. Karen Fog Olwig and Kenneth Olwig, "Underdevelopment and the Development of 'Natural' Park Ideology," *Antipode* 11 (1979): 16–25; Mark Spence, *Dispossessing the Wilderness: Indian Removal and the Making of the National Parks* (New York: Oxford University Press, 1999); John MacKenzie, *The Empire of Nature: Hunting, Conservation and British Imperialism* (Manchester: University of Manchester Press, 1988); Jane Carruthers, *The Kruger National Park: A Social and Political History* (Pietermaritzburg: University of Natal Press, 1995); and Karl Jacoby, *Crimes against Nature: Squatters, Poachers, Thieves, and the Hidden History of American Conservation* (Berkeley and Los Angeles: University of California Press, 2001).

23. Joachim Radkau and Frank Uekbitter, eds., *Naturschutz und Nationalsozialismus* (Frankfurt am Main and New York: Campus Verlag, 2003).

24. For such a naive reading, see Anna Bramwell, *Ecology in the Twentieth Century: A History* (New Haven: Yale University Press, 1989).

25. Gerd Groening and Joachim Wolshcke-Bulmahn, "Politics, Planning, and the Protection of Nature: Political Abuse of Early Ecological Ideas in Germany, 1933–1945," *Planning Perspectives* 2 (1987): 127–148.

26. Ibid., 139.

27. Ibid., 139–140.

28. Estimates of the victims of the Soviet terror-famine of 1932–1933, whose epicentre was Ukraine and the North Caucasus, range from 5 to 8 million. See Robert Conquest, *Harvest of Sorrow* (Oxford: Oxford University Press, 1986).

29. On the Chinese terror-famine, which took over 30 million lives, see Judith Shapiro's gripping *Mao's War against Nature: Politics and the Environment in Revolutionary China* (New York: Cambridge University Press, 2001).

30. See Bernard Avishai, *The Tragedy of Zionism* (New York: Helios Press, 2002); and Ilan Pappé, *A History of Modern Palestine: One Land, Two Peoples* (Cambridge: Cambridge University Press, 2004); but also Gershon Shafir, *Land, Labor and the Origins of the Israeli-Palestinian Conflict, 1882–1914* (Cambridge and New York: Cambridge University Press, 1989); Derek J. Penslar, *Zionism and Technocracy: The Engineering of Jewish Settlement in Palestine, 1870–1918* (Bloomington: Indiana University Press, 1991); Shaul Ephraim Cohen, *The Politics of Planting: Jewish-Palestinian Competition for Control of Land in the Jerusalem Periphery* (Chicago: University of Chicago Press, 1993); Baruch Kimmerling, *Zionism and Territory: The Socio-Territorial Dimension of Zionist Politics* (Berkeley: University of California Institute of International Studies, 1983).

31. Quoted in Abdelwahab M. Elmessiri, *The Land of Promise: A Critique of Political Zionism* (New Brunswick: North American, 1977), 113.

32. Ibid.

33. Quoted in Benjamin Beit-Hallahmi, *Original Sins: Reflections on the History of Zionism and Israel* (London: Pluto Press, 1992), 99–100.

34. Ibid., 72.

35. Kimmerling, *Zionism and Territory*, 203–204.

36. A.B. Yehoshua, "Facing the Forests," in *Three Days and a Child*, translated by M. Arad (New York: Doubleday, 1970).

37. S. Ravi Rajan, "Imperial Environmentalism or Environmental Imperialism? European Forestry, Colonial Foresters, and the Agendas of Forest Management in British India, 1800–1900," in *Nature and the Orient: The Environmental History of South and Southeast Asia*, edited by Richard Grove, Vinita Darmodaran, and Satpal Sangwan (Delhi: Oxford University Press India, 1995), 324–371.

38. Sylvia Hood Washington, "Tar Babies and Turpentine: African Americans and the 'Fuel of the Future,' 1830–1930," American Society for Environmental History annual meeting poster session, Houston, March 19, 2005.

39. James Giesen, "'The Boll Weevil Works While the Darkies Play': Labor, Race, and the Cotton Boll Weevil," American Society for Environmental History annual meeting, Houston, March 17, 2005.

40. John Sandlos, "From the Outside Looking in: Aesthetics, Politics, and Wildlife Conservation in the Canadian North," *Environmental History* 6 (January 2001): 6–31.

41. Ibid., 13.

42. Theodore Catton, *Inhabited Wilderness: Indians, Eskimos, and National Parks in Alaska* (Albuquerque: University of New Mexico Press, 1997).

43. Mike Davis, *Late Victorian Holocausts: El Niño Famines and the Making of the Third World* (London: Verso, 2001).

44. Christine Meisner Rosen, "'Knowing' Industrial Pollution: Nuisance Law and the Power of Tradition in a Time of Rapid Economic Change, 1840–1864," *Environmental History* 8 (October 2003): 565–597.

45. James Secord, remarks at the conference "Nature's Histories" at the Max-Planck Gesellschaft, Berlin, August 1997.

46. Leo Marx, review of *In Context: History and the History of Technology: Essays in Honor of Melvin Kranzberg, Technology and Culture* 32 (April 1991): 394–396.

47. Allan Pred, "The Nature of Denaturalized Consumption and Everyday Life," in Braun and Castree, *Remaking,* 150, (quoting Stuart Hall, "The Local and the Global: Globalization and Ethnicity," in *Culture, Globalization and the World-System: Contemporary Conditions for the Representation of Identity,* edited by Anthony D. King (Binghamton: Department of Art and Art History, 1991), 25.

48. Douglas R. Weiner, "Demythologizing Environmentalism," *Journal of the History of Biology* 25 (Fall 1992): 385–411.

Critical Thinking Questions

1. Worster argues that environmental historians should be conversant with other disciplines, especially from within the sciences. How can environmental historians ensure that "history" remains as their methodological focus when working with these other, highly varied, disciplines?
2. What, according to Worster, is the role of culture in human relationships with the environment through time?
3. Worster argues that environmental history as a field first developed in the 1970s. Is that necessarily true, in your opinion? What factors converged to create the new field?
4. Was Cronon right to attempt an upbeat end to his environmental history course?
5. Why is Cronon's identification of "the problem of audience" so crucial to successful environmental history? What are the audiences that he identifies as important for the environmental historian?
6. In what ways does Cronon see parables as useful forms of communicating historical wisdom? Do you agree with his perspective?
7. What is the "recovery narrative" as outlined by Merchant? Is it a persuasive explanation for the actions of New World colonists beginning in the 1600s?
8. In what ways is the human–environment relationship a gendered one, in Merchant's view?
9. Weiner suggests that ecology possessed in-built limitations because it focused on nature as a discrete, definable phenomenon delineated by space and time. Why was this a problem, in Weiner's view, and how do environmental historians seek to solve the problem of seeing nature as an "object of study" instead of as a "process"?
10. In what ways, according to Weiner, do political ideologies produce definable environmental impacts? Does this process continue today, in your view?
11. Each of the four readings in this section possesses a political component, either expressly or implicitly. In what ways does the political, even prescriptive, viewpoint affect the theory of environmental history? Does this strengthen or weaken the field?

FURTHER READING

Donald Worster, "The Vulnerable Earth: Toward a Planetary History," from *The Ends of the Earth: Perspectives on Modern Environmental History*, edited by Donald Worster (Cambridge: Cambridge University Press, 1988): 3–20.

Worster's introduction to the volume from which the reading "Doing Environmental History" in this reader is drawn discusses the impact of so-called planetary history, and the perspective it generates, on the origination of the field of environmental history. Written with an eye to the upcoming quincentenary of Columbus's arrival in the Americas, Worster explicitly links the fields of world history and environmental history to produce a broad-sweep understanding of historical forces over the last 500 years.

Alfred W. Crosby, "The Past and Present of Environmental History," *American Historical Review* 100, no. 4 (October 1995): 1177–1189.

Crosby's article here is a useful overview of the development of what would become called "environmental history." He examines the environment as an object of historical study throughout the twentieth century, and thus provides a strong sense of the longer-term currents in historiographical and philosophical thought that led to the understanding of the environmental impact on history.

Mart A. Stewart, "Environmental History: Profile of a Developing Field," *The History Teacher* 31, no. 3 (May 1998): 351–368.

Stewart complement's Crosby's piece above by focusing more narrowly on the development of environmental history since 1980 or so. He describes and analyzes the remarkable expansion of the field and some of the tensions that the growth has produced.

Journal of American History 76, no. 4 (March 1990): 1087–1147.

The editors of this journal devoted considerable space to the publication of seven essays by Donald Worster, Alfred W. Crosby, Carolyn Merchant, William Cronon, Richard White, and Stephen J. Pyne on the topic of new directions and definitions in environmental history. The points of contact—and the sharp disagreements—are important indicators of the strengths, and the strains, evident in what was then a rapidly expanding field that was still finding its feet.

RELEVANT WEB SITES

American Society for Environmental History
http://www.h-net.org/~environ/ASEH/home.html

This Web site gives a brief explanation of the ASEH, its activities, and membership requirements. The site serves as an excellent portal to a wide variety of materials (book reviews, on-line course syllabi, etc.) as well as access to the Environmental History Bibliography database.

European Society for Environmental History
http://adm-websrv3b.sdu.dk/

A brief history and the mission of the ESEH are outlined on this Web site. It serves as a clearinghouse for information on past and upcoming conferences and ESEH membership. It also provides information on publishers, new publications, hosts the European Environmental History bibliography and ESEH publications, and links to other environmental history Web sites.

Pre-contact Environmental History

The environmental history of the Americas prior to the arrival of Europeans remains a contentious and challenging area of study, a fact that is reflected in the two readings presented here. In the first, William Denevan challenges most, if not all, of the European explorers' and settlers' long-held assumptions of the New World. These assumptions, as Merchant outlined in the previous section, were a crucial component of the heroic narrative of dominance and subjugation created by Europeans as justification for their actions. To Europeans of the Enlightenment, and for their heirs, the idea of an "empty America" composed of wilderness and unruliness, within which only a few Native peoples ghosted through the landscape, hardly altering it by their activities or passage, presented a powerful foil against which they could measure their progressive impact on the environment of the Americas.

But Denevan, in this controversial essay, argues that the Native populations of the Americas were huge in number at the time of the European contact in the late fifteenth and sixteenth centuries. Not only that: he advances the idea that Natives throughout the Americas, from the Andes and the Amazon to the Arctic, substantially altered and managed their environments. This seems to do away with the idea of a pristine and untouched landscape upon which the adventure of subjugation and settlement was played, with Europeans as leading actors. Not only does this undermine the myth of a uniquely Euroamerican-driven process of environmental change, it also significantly affects the equally deep myth of Native peoples invariably living harmoniously with nature prior to European contact.

This has significant ramifications. If ecosystems from the Amazonian rain forest to the prairie grasslands were the result of Native activity in the Americas, then we must reorient our perceptions of Native peoples as environmental actors, seeing them as far more active agents of environmental change than we have done so thus far. On the basis of the substantially larger population figures reported by Denevan, we must also recast the magnitude of the disaster visited on Native peoples as a consequence of European contact. It is true that for decades scholars have recognized the impact as being tremendous. But it is now clear that it represented perhaps the greatest demographic catastrophe in human history.

In pre-industrial times there is no doubt that fire was the most important environmentally transformative tool available to humans. Shepard Krech discusses the myriad uses that Native Americans made of fire, and reinforces Denevan's claim that the Americas were characterized by anthropogenic environments (that is, those created or heavily influenced by human activity) on a far wider scale than allowed by the pristine myth developed by Europeans. After Europeans arrived and began to engage Natives in fur trapping and other industries, the latter continued to employ fire to different ends. It could be used as an aid to maximize game concentrations, or it could be employed to drive animals away from competitors' hunting and trapping grounds. Fire was also used as a weapon, both against other Native peoples and against White encroachment. In most cases the environmental impact was significant.

Both Denevan's and Krech's essays are important correctives to the idea of Native peoples as invisible, or near-invisible, inhabitants of a vast and unaffected landscape, one that was entirely pristine when Europeans arrived and settlers began to push westward across North America. Instead it is clear that Native peoples were active agents, changing environments to suit their economic, social, cultural, and even military needs. That Natives were such active agents partially undermines the late twentieth-century notion that only through reference to traditional environmental knowledge (TEK) could Euroamerican environmental degradation be reversed. Unfortunately, this idea that it might be possible to ameliorate environmental damage and to return as far as possible to some kind of wilderness ideal seems to have been informed by the pristine myth outlined by Merchant in the previous section and questioned by Denevan and Krech here. This is not to say that TEK does not retain significant importance as a consideration in conservation strategies in Canada and the United States today. As Krech demonstrates, Native pyrotechnology indicated a tremendously varied and deep understanding of fire's role in ecosystem maintenance and renewal. This in turn supports the idea that Native peoples possessed an intimate relationship with their environment. But that environment was what geographers refer to as a "built environment," one that exists, and only persists, as a consequence of human actions.

This raises the question: What should be the core ethic of conservation strategies in the United States and Canada if it cannot be a return to pristine wilderness? And what are the roles of Native peoples in conservation and preservation policy, and in new economic activities such as ecotourism, when it is clear that they practised environmentally exploitative activities on a wide scale as well?

Chapter Five
The Pristine Myth
The Landscape of the Americas in 1492

William M. Denevan

"This is the forest primeval ..."
—*Evangeline: A Tale of Acadie* (Longfellow, 1847)

What was the New World like at the time of Columbus?—"Geography as it was," in the words of Carl Sauer (1971: x).[1] The Admiral himself spoke of a "Terrestrial Paradise," beautiful and green and fertile, teeming with birds, with naked people whom he called "Indians" living there. But was the landscape encountered in the sixteenth century primarily pristine, virgin, a wilderness, nearly empty of people, or was it a humanized landscape, with the imprint of Native Americans being dramatic and persistent? The former still seems to be the more common view, but the latter may be more accurate.

The pristine view is to a large extent an invention of nineteenth-century romanticist and primitivist writers such as W.H. Hudson, Cooper, Thoreau, Longfellow, and Parkman, and painters such as Catlin and Church.[2] The wilderness image has since become part of the American heritage, associated "with a heroic pioneer past in need of preservation" (Pyne 1982: 17; also see Bowden 1992: 22). The pristine view was restated clearly in 1950 by John Bakeless in his book *The Eyes of Discovery*:

There were not really very many of these redmen ... the land seemed empty to invaders who came from settled Europe ... that ancient, primeval, undisturbed wilderness ... the streams simply boiled with fish ... so much game ... that one hunter counted a thousand animals near a single salt lick ... the virgin wilderness of Kentucky ... the forested glory of primitive America. (Bakeless 1950: 13, 201, 223, 314, 407)

But then he mentions that Indian "prairie fires ... cause the often-mentioned oak openings ... Great fields of corn spread in all directions ... the Barrens ... without forest," and that "Early Ohio settlers found that they could drive about through the forests with sleds and horses" (Bakeless 1950: 31, 304, 308, 314). A contradiction?

In the ensuing 40 years, scholarship has shown that Indian populations in the Americas were substantial, that the forests had indeed been altered, that landscape change was commonplace. This message, however, seems not to have reached the public through texts, essays, or talks by both academics and popularizers who have a responsibility to know better.[3]

Kirkpatrick Sale in 1990, in his widely reported *Conquest of Paradise*, maintains that it was the Europeans who transformed nature, following a pattern set by Columbus. Although Sale's book has some merit and he is aware of large Indian numbers and their impacts, he nonetheless champions the widely held dichotomy of the benign Indian landscape and the devastated Colonial landscape. He overstates both.

Similarly, *Seeds of Change: Christopher Columbus and the Columbian Legacy*, the popular book published by the Smithsonian Institution, continues the litany of Native American passivity:

> [P]re-Columbian America was still the First Eden, a pristine natural kingdom. The native people were transparent in the landscape, living as natural elements of the ecosphere. Their world, the New World of Columbus, was a world of barely perceptible human disturbance. (Shetler 1991: 226)

To the contrary, the Indian impact was neither benign nor localized and ephemeral, nor were resources always used in a sound ecological way. The concern here is with the form and magnitude of environmental modification rather than with whether or not Indians lived in harmony with nature with sustainable systems of resource management. Sometimes they did; sometimes they didn't. What they did was to change their landscape nearly everywhere, not to the extent of post-Colonial Europeans but in important ways that merit attention.

The evidence is convincing. By 1492 Indian activity throughout the Americas had modified forest extent and composition, created and expanded grasslands, and rearranged microrelief via countless

artificial earthworks. Agricultural fields were common, as were houses and towns and roads and trails. All of these had local impacts on soil, microclimate, hydrology, and wildlife. This is a large topic, for which this essay offers but an introduction to the issues, misconceptions, and residual problems. The evidence, pieced together from vague ethnohistorical accounts, field surveys, and archaeology, supports the hypothesis that the Indian landscape of 1492 had largely vanished by the mid-eighteenth century, not through a European superimposition, but because of the demise of the Native population. The landscape of 1750 was more "pristine" (less humanized) than that of 1492.

INDIAN NUMBERS

The size of the Native population at contact is critical to our argument. The prevailing position, a recent one, is that the Americas were well populated rather than relatively empty lands in 1492. In the words of the sixteenth-century Spanish priest, Bartolomé de las Casas, who knew the Indies well:

> All that has been discovered up to the year forty-nine [1549] is full of people, like a hive of bees, so that it seems as though God had placed all, or the greater part of the entire human race in these countries. (Las Casas, in MacNutt 1909: 314)

Las Casas believed that more than 40 million Indians had died by the year 1560. Did he exaggerate? In the 1930s and 1940s, Alfred Kroeber, Angel Rosenblat, and Julian Steward believed that he had. The best counts then available indicated a population of between 8–15 million Indians in the Americas. Subsequently, Carl Sauer, Woodrow Borah, Sherburne F. Cook, Henry

Dobyns, George Lovell, N. David Cook, myself, and others have argued for larger estimates. Many scholars now believe that there were between 40–100 million Indians in the hemisphere (Denevan 1992). This conclusion is primarily based on evidence of rapid early declines from epidemic disease prior to the first population counts [...].

I have recently suggested a New World total of 53.9 million (Denevan 1992 [1972]: xxvii). This divides into 3.8 million for North America, 17.2 million for Mexico, 5.6 million for Central America, 3.0 million for the Caribbean, 15.7 million for the Andes, and 8.6 million for lowland South America. These figures are based on my judgment as to the most reasonable recent tribal and regional estimates. Accepting a margin of error of about 20 percent, the New World population would lie between 43–65 million. Future regional revisions are likely to maintain the hemispheric total within this range. Other recent estimates, none based on totalling regional figures, include 43 million by Whitmore (1991: 483), 40 million by Lord and Burke (1991), 40–50 million by Cowley (1991), and 80 million for just Latin America by Schwerin (1991: 40). In any event, a population between 40–80 million is sufficient to dispel any notion of "empty lands." Moreover, the Native impact on the landscape of 1492 reflected not only the population then but the cumulative effects of a growing population over the previous 15,000 years or more.

European entry into the New World abruptly reversed this trend. The decline of Native American populations was rapid and severe, probably the greatest demographic disaster ever [...]. Old World diseases were the primary killer. In many regions, particularly the tropical lowlands, populations fell by 90 percent or more in the first century after contact. Indian populations (estimated) declined in Hispaniola from 1 million in 1492 to a few hundred 50 years later, or by more than 99 percent; in Peru from 9 million in 1520 to 670,000 in 1620 (92 percent); in the Basin of Mexico from 1.6 million in 1519 to 180,000 in 1607 (89 percent); and in North America from 3.8 million in 1492 to 1 million in 1800 (74 percent). An overall drop from 53.9 million in 1492 to 5.6 million in 1650 amounts to an 89 percent reduction (Denevan 1992: xvii–xxix). The human landscape was affected accordingly, although there is not always a direct relationship between population density and human impact (Whitmore et al. 1990: 37).

The replacement of Indians by Europeans and Africans was initially a slow process. By 1638 there were only about 30,000 English in North America (Sale 1990: 388), and by 1750 there were only 1.3 million Europeans and slaves (Meinig 1986: 247). For Latin America in 1750, Sánchez-Albornoz (1974: 7) gives a total (including Indians) of 12 million. For the hemisphere in 1750, the *Atlas of World Population History* reports 16 million (McEvedy and Jones 1978: 270). Thus the overall hemispheric population in 1750 was about 30 percent of what it may have been in 1492. The 1750 population, however, was very unevenly distributed, mainly located in certain coastal and highland areas with little Europeanization elsewhere. In North America in 1750, there were only small pockets of settlement beyond the coastal belt, stretching from New England to northern Florida (see maps in Meinig 1986: 209, 245). Elsewhere, combined Indian and European populations were sparse, and environmental impact was relatively minor.

Indigenous imprints on landscapes at the time of initial European contact varied regionally in form and intensity. The following are examples for vegetation and wildlife, agriculture, and the built landscape.

VEGETATION

The Eastern Forests

The forests of New England, the Midwest, and the Southeast had been disturbed to varying degrees by Indian activity prior to European occupation. Agricultural clearing and burning had converted much of the forest into successional (fallow) growth and into semi-permanent grassy openings (meadows, barrens, plains, glades, savannas, prairies), often of considerable size.[4] Much of the mature forest was characterized by an open, herbaceous understory, reflecting frequent ground fires. "The de Soto expedition, consisting of many people, a large horse herd, and many swine, passed through ten states without difficulty of movement" (Sauer 1971: 283). The situation has been described in detail by Michael Williams in his recent history of American forests: "Much of the 'natural' forest remained, but the forest was not the vast, silent, unbroken, impenetrable and dense tangle of trees beloved by many writers in their romantic accounts of the forest wilderness" (1989: 33).[5] "The result was a forest of large, widely spaced trees, few shrubs, and much grass and herbage.... Selective Indian burning thus promoted the mosaic quality of New England ecosystems, creating forests in many different states of ecological succession" (Cronon 1983: 49–51).

The extent, frequency, and impact of Indian burning is not without controversy. Raup (1937) argued that climatic change

rather than Indian burning could account for certain vegetation changes. Emily Russell (1983: 86), assessing pre-1700 information for the Northeast, concluded that "[t]here is no strong evidence that Indians purposely burned large areas," but Indians did "increase the frequency of fires above the low numbers caused by lightning," creating an open forest. But then Russell adds: "In most areas climate and soil probably played the major role in determining the precolonial forests." She regards Indian fires as mainly accidental and "merely" augmental to natural fires, and she discounts the reliability of many early accounts of burning.

Forman and Russell (1983: 5) expand the argument to North America in general: "regular and widespread Indian burning (Day 1953) [is] an unlikely hypothesis that regretfully has been accepted in the popular literature and consciousness." This conclusion, I believe, is unwarranted given reports of the extent of prehistoric human burning in North America and Australia (Lewis 1982), and Europe (Patterson and Sassaman 1988: 130), and by my own and other observations on current Indian and peasant burning in Central America and South America; when unrestrained, people burn frequently and for many reasons. For the Northeast, Patterson and Sassaman (1988: 129) found that sedimentary charcoal accumulations were greatest where Indian populations were greatest.

Elsewhere in North America, the Southeast is much more fire prone than is the Northeast, with human ignitions being especially important in winter (Taylor 1981). The Berkeley geographer and Indianist Erhard Rostlund (1957: 1960) argued that Indian clearing and burning created many grasslands within mostly open forest in the so-called "prairie belt" of Alabama. As improbable as it may seem, Lewis (1982) found Indian burning in the subarctic, and Dobyns (1981) in the Sonoran desert. The characteristics and impacts of fires set by Indians varied regionally and locally with demography, resource management techniques, and environment, but such fires clearly had different vegetation impacts than did natural fires owing to differences in frequency, regularity, and seasonality.

Forest Composition

In North America, burning not only maintained open forest and small meadows but also encouraged fire-tolerant and sun-loving species. "Fire created conditions favorable to strawberries, blackberries, raspberries, and other gatherable foods" (Cronon 1983: 51). Other useful plants were saved, protected, planted, and transplanted, such as American chestnut, Canada plum, Kentucky coffee tree, groundnut, and leek (Day 1953: 339–340). Gilmore (1931) described the dispersal of several native plants by Indians. Mixed stands were converted to single species dominants, including various pines and oaks, sequoia, Douglas fir, spruce, and aspen (M. Williams 1989: 47–48). The longleaf, slash pine, and scrub oak forests of the Southeast are almost certainly an anthropogenic subclimax created originally by Indian burning, replaced in early Colonial times by mixed hardwoods, and maintained in part by fires set by subsequent farmers and woodlot owners (Garren 1943). Lightning fires can account for some fire-climax vegetation, but Indian burning would have extended and maintained such vegetation (Silver 1990: 17–19, 59–64).

Even in the humid tropics, where natural fires are rare, human fires can dramatically influence forest composition. A good example is the pine forests of Nicaragua (Denevan 1961). Open pine stands occur both in the northern highlands (below 5,000 feet) and in the eastern (Miskito) lowlands, where warm temperatures and heavy rainfall generally favour mixed tropical montane forest or rain forest. The extensive pine forests of Guatemala and Mexico primarily grow in cooler and drier, higher elevations, where they are in large part natural and pre-human (Watts and Bradbury 1982: 59). Pine forests were definitely present in Nicaragua when Europeans arrived. They were found in areas where Indian settlement was substantial, but not in the eastern mountains where Indian densities were sparse. The eastern boundary of the highland pines seems to have moved with an eastern settlement frontier that has fluctuated back and forth since prehistory. The pines occur today where there has been clearing followed by regular burning and the same is likely in the past. The Nicaraguan pines are fire tolerant once mature, and large numbers of seedlings survive to maturity if they can escape fire during their first three to seven years (Denevan 1961: 280). Where settlement has been abandoned and fire ceases, mixed hardwoods gradually replace pines. This succession is likely similar where pines occur elsewhere at low elevations in tropical Central America, the Caribbean, and Mexico.

Midwest Prairies and Tropical Savannas

Sauer (1950, 1958, 1975) argued early and often that the great grasslands and savannas of the New World were of anthropogenic rather than climatic origin, that rainfall was generally sufficient to support trees. Even non-agricultural Indians expanded what may have been pockets of natural, edaphic grasslands at the expense of forest. A fire burning to the edge of a grass/forest boundary will penetrate the drier forest margin and push back the edge, even if the forest itself is not consumed (Mueller-Dombois 1981: 164). Grassland can therefore advance significantly in the wake of hundreds of years of annual fires. Lightning-set fires can have a similar impact, but more slowly if less frequent than human fires, as in the wet tropics.

The thesis of prairies as fire induced, primarily by Indians, has its critics (Borchert 1950; Wedel 1957), but the recent review of the topic by Anderson (1990: 14), a biologist, concludes that most ecologists now believe that the eastern prairies "would have mostly disappeared if it had not been for the nearly annual burning of these grasslands by the North American Indians" during the last 5,000 years. A case in point is the nineteenth-century invasion of many grasslands by forests after fire had been suppressed in Wisconsin, Illinois, Kansas, Nebraska, and elsewhere (M. Williams 1989: 46).

The large savannas of South America are also controversial as to origin. Much, if not most of the open vegetation of the Orinoco Llanos, the Llanos de Mojos of Bolivia, the Pantanal of Mato Grosso, the Bolívar savannas of Colombia, the Guayas savannas of coastal Ecuador, the *campo cerrado* of central Brazil, and the coastal savannas north of the Amazon, is of natural origin. The vast *campos cerrados* occupy extremely senile, often toxic oxisols. The seasonally inundated savannas of Bolivia, Brazil, Guayas, and the Orinoco owe their existence to the intolerance of woody species to the extreme alternation of lengthy flooding or waterlogging and severe desiccation during a long dry season. These savannas, however, were and are burned by Indians and ranchers, and such fires have expanded the savannas

into the forests to an unknown extent. It is now very difficult to determine where a natural forest/savanna boundary once was located (Hills and Randall 1968; Medina 1980).

Other small savannas have been cut out of the rain forest by Indian farmers and then maintained by burning. An example is the Gran Pajonal in the Andean foothills in east-central Peru, where dozens of small grasslands (*pajonales*) have been created by Campa Indians—a process clearly documented by air photos (Scott 1978). *Pajonales* were in existence when the region was first penetrated by Franciscan missionary explorers in 1733.

The impact of human activity is nicely illustrated by vegetational changes in the basins of the San Jorge, Cauca, and Sinú rivers of northern Colombia. The southern sector, which was mainly savanna when first observed in the sixteenth century, had reverted to rain forest by about 1750 following Indian decline, and had been reconverted to savanna for pasture by 1950 (Gordon 1957: 69, map). Sauer (1966: 285–288; 1975: 8) and Bennett (1968: 53–55) cite early descriptions of numerous savannas in Panama in the sixteenth century. Balboa's first view of the Pacific was from a "treeless ridge," now probably forested. Indian settlement and agricultural fields were common at the time, and with their decline the rain forest returned.

Anthropogenic Tropical Rain Forest

The tropical rain forest has long had a reputation for being pristine, whether in 1492 or 1992. There is, however, increasing evidence that the forests of Amazonia and elsewhere are largely anthropogenic in form and composition. Sauer (1958: 105) said as much at the Ninth Pacific Science Congress in 1957 when he challenged the statement of tropical botanist Paul

Richards that, until recently, the tropical forests have been largely uninhabited, and that prehistoric people had "no more influence on the vegetation than any of the other animal inhabitants." Sauer countered that Indian burning, swiddens, and manipulation of composition had extensively modified the tropical forest.

"Indeed, in much of Amazonia, it is difficult to find soils that are not studded with charcoal" (Uhl et al. 1990: 30). The question is, to what extent does this evidence reflect Indian burning in contrast to natural (lightning) fires, and when did these fires occur? The role of fire in tropical forest ecosystems has received considerable attention in recent years, partly as result of major wild fires in East Kalimantan in 1982–1983 and small forest fires in the Venezuelan Amazon in 1980–1984 (Goldammer 1990). Lightning fires, though rare in moist tropical forest, do occur in drier tropical woodlands (Mueller-Dombois 1981: 149). Thunderstorms with lightning are much more common in the Amazon compared to North America, but in the tropics lightning is usually associated with heavy rain and non-combustible, verdant vegetation. Hence Indian fires undoubtedly account for most fires in prehistory, with their impact varying with the degree of aridity.

In the Río Negro region of the Colombian-Venezuelan Amazon, soil charcoal is very common in upland forests. C-14 dates range from 6260–6250 B.P., well within human times (Saldarriaga and West 1986). Most of the charcoal probably reflects local swidden burns; however, there are some indications of forest fires at intervals of several hundred years, most likely ignited by swidden fires. Recent wild fires in the upper Río Negro region were in a normally moist tropical forest (3,530 mm annual rainfall) that had experienced several years of severe drought. Such infrequent wild fires in prehistory, along with the more frequent ground fires, could have had significant impacts on forest succession, structure, and composition. Examples are the pine forests of Nicaragua, mentioned above, the oak forests of Central America, and the babassu palm forests of eastern Brazil. Widespread and frequent burning may have brought about the extinction of some endemic species.

The Amazon forest is a mosaic of different ages, structure, and composition resulting from local habitat conditions and disturbance dynamics (Haffer 1991). Natural disturbances (tree falls, landslides, river activity) have been considerably augmented by human activity, particularly by shifting cultivation. Even a small number of swidden farmers can have a widespread impact in a relatively short period of time. In the Río Negro region, species-diversity recovery takes 60–80 years and biomass recovery 140–200 years (Saldarriaga and Uhl 1991: 312). Brown and Lugo (1990: 4) estimate that today about 40 percent of the tropical forest in Latin America is secondary as a result of human clearing and that most of the remainder has had some modification despite current low population densities. The species composition of early stages of swidden fallows differs from that of natural gaps and may "alter the species composition of the mature forest on a long-term scale" (Walschburger and von Hildebrand 1991: 262). While human environmental destruction in Amazonia currently is concentrated along roads, in prehistoric times Indian activity in the upland (interfluve) forests was much less intense but more widespread (Denevan 1992).

Indian modification of tropical forests is not limited to clearing and burning.

Large expanses of Latin American forests are humanized forests in which the kinds, numbers, and distributions of useful species are managed by human populations. Doubtless this applies to the past as well. One important mechanism in forest management is manipulation of swidden fallows (sequential agroforestry) to increase useful species. The planting, transplanting, sparing, and protection of useful wild fallow plants eliminate clear distinctions between field and fallow (Denevan and Padoch 1988). Abandonment is a slow process, not an event. Gordon (1982: 79–98) describes managed regrowth vegetation in eastern Panama, which he believes extended from Yucatán to northern Colombia in pre-European times. The Huastec of eastern Mexico and the Yucatec Maya have similar forms of forest gardens or forest management (Alcorn 1981; Gómez-Pompa 1987). The Kayapó of the Brazilian Amazon introduce and/or protect useful plants in activity areas ("nomadic agriculture") adjacent to villages or campsites, in foraging areas, along trails, near fields, and in artificial forest-mounds in savanna (Posey 1985). In managed forests, both annuals and perennials are planted or transplanted, while wild fruit trees are particularly common in early successional growth. Weeding by hand was potentially more selective than indiscriminate weeding by machete (Gordon 1982: 57–61). Much dispersal of edible plant seeds is unintentional via defecation and spitting out.

The economic botanist William Balée (1987, 1989) speaks of "cultural" or "anthropogenic" forests in Amazonia in which species have been manipulated, often without a reduction in natural diversity. These include specialized forests (babassu, Brazil nuts, lianas, palms, bamboo), which currently make up at least 11.8 percent (measured) of the total upland forest in the Brazilian Amazon (Balée 1989: 14). Clear indications of past disturbance are the extensive zones of *terra preta* (black earth), which occur along the edges of the large floodplains as well as in the uplands (Balée 1989: 10–12; Smith 1980). These soils, with depths to 50 cm or more, contain charcoal and cultural waste from prehistoric burning and settlement. Given high carbon, nitrogen, calcium, and phosphorus content, *terra preta* soils have a distinctive vegetation and are attractive to farmers. Balée (1989: 14) concludes that "large portions of Amazonian forests appear to exhibit the continuing effects of past human interference." The same argument has been made for the Maya lowlands (Gómez-Pompa et al. 1987) and Panama (Gordon 1982). There are no virgin tropical forests today, nor were there in 1492.

WILDLIFE

The indigenous impact on wildlife is equivocal. The thesis that "overkill" hunting caused the extinction of some large mammals in North America during the late Pleistocene, as well as subsequent local and regional depletions (Martin 1978: 167–172), remains controversial. By the time of the arrival of Cortez in 1519, the dense populations of Central Mexico apparently had greatly reduced the number of large game, given reports that "they eat any living thing" (Cook and Borah 1971–1979: (3), 135, 140). In Amazonia, local game depletion apparently increases with village size and duration (Good 1987). Hunting procedures in many regions seem, however, to have allowed for recovery because of the "resting" of hunting zones intentionally or as a result of shifting of village sites.

On the other hand, forest disturbance increased herbaceous forage and edge effect, and hence the numbers of some animals (Thompson and Smith 1970: 261–264). "Indians created ideal habitats for a host of wildlife species ... exactly those species whose abundance so impressed English colonists: elk, deer, beaver, hare, porcupine, turkey, quail, ruffed grouse, and so on" (Cronon 1983: 51). White-tailed deer, peccary, birds, and other game increases in swiddens and fallows in Yucatan and Panama (Greenberg 1991; Gordon 1982: 96–112; Bennett 1968). Rostlund (1960: 407) believed that the creation of grassy openings east of the Mississippi extended the range of the bison, whose numbers increased with Indian depopulation and reduced hunting pressure between 1540–1700, and subsequently declined under White pressure.

AGRICULTURE

Fields and Associated Features

To observers in the sixteenth century, the most visible manifestation of the Native American landscape must have been the cultivated fields, which were concentrated around villages and houses. Most fields are ephemeral, their presence quickly erased when farmers migrate or die, but there are many eyewitness accounts of the great extent of Indian fields. On Hispaniola, Las Casas and Oviedo reported individual fields with thousands of *montones* (Sturtevant 1961: 73). These were manioc and sweet potato mounds 3–4 m in circumference, of which apparently none have survived. In the Llanos de Mojos in Bolivia, the first explorers mentioned *percheles,* or corn cribs on pilings, numbering up to 700 in a single field, each holding 30–45 bushels of food (Denevan 1966: 98). In northern Florida in 1539, Hernando de Soto's army passed through numerous fields of maize,

beans, and squash, their main source of provisions; in one sector, "great fields ... were spread out as far as the eye could see across two leagues of the plain" (Garcilaso de la Vega 1980 [1605]: (2), 182; also see Dobyns 1983: 135–146).

It is difficult to obtain a reliable overview from such descriptions. Aside from possible exaggeration, Europeans tended not to write about field size, production, or technology. More useful are various forms of relict fields and field features that persist for centuries and can still be recognized, measured, and excavated today. These extant features, including terraces, irrigation works, raised fields, sunken fields, drainage ditches, dams, reservoirs, diversion walls, and field borders number in the millions and are distributed throughout the Americas (Denevan 1980 [...]). For example, about 500,000 ha of abandoned raised fields survive in the San Jorge Basin of northern Colombia (Plazas and Falchetti 1987: 485), and at least 600,000 ha of terracing, mostly of prehistoric origin, occur in the Peruvian Andes (Denevan 1988: 20). There are 19,000 ha of visible raised fields in just the sustaining area of Tiwanaku at Lake Titicaca (Kolata 1991: 109) and there were about 12,000 ha of *chinampas* (raised fields) around the Aztec capital of Tenochtitlan (Sanders et al. 1979: 390). Complex canal systems on the north coast of Peru and in the Salt River Valley in Arizona irrigated more land in prehistory than is cultivated today. About 175 sites of Indian garden beds, up to several hundred acres each, have been reported in Wisconsin (Gartner 1992). These various remnant fields probably represent less than 25 percent of what once existed, most being buried under sediment or destroyed by erosion, urbanization, plowing, and bulldozing. On the other hand, an inadequate effort has been made to search for ancient fields.

Erosion

The size of Native populations, associated deforestation, and prolonged intensive agriculture led to severe land degradation in some regions. Such a landscape was that of Central Mexico, where by 1519 food production pressures may have brought the Aztec civilization to the verge of collapse even without Spanish intervention (Cook and Borah 1971–79: (3), 129–176).[6] There is good evidence that severe soil erosion was already widespread, rather than just the result of subsequent European plowing, livestock, and deforestation. Cook examined the association between erosional severity (gullies, barrancas, sand and silt deposits, and sheet erosion) and pre-Spanish population density or proximity to prehistoric Indian towns. He concluded that "an important cycle of erosion and deposition therefore accompanied intensive land use by huge primitive populations in central Mexico, and had gone far toward the devastation of the country before the white man arrived" (Cook 1949: 86).

Barbara Williams (1972: 618) describes widespread *tepetate*, an indurated substrate formation exposed by sheet erosion resulting from prehistoric agriculture, as "one of the dominant surface materials in the Valley of Mexico." On the other hand, anthropologist Melville (1990: 27) argues that soil erosion in the Valle de Mezquital, just north of the Valley of Mexico, was the result of overgrazing by Spanish livestock starting before 1600: "there is an almost total lack of evidence of environmental degradation before the last three decades of the sixteenth century." The Butzers, however, in an examination of Spanish land grants, grazing patterns, and soil and vegetation ecology, found that there was only light intrusion of Spanish livestock (sheep and cattle were moved frequently)

into the southeastern Bajfo near Mezquital until after 1590 and that any degradation in 1590 was "as much a matter of long-term Indian land use as it was of Spanish intrusion" (Butzer and Butzer 1994). The relative roles of Indian and early Spanish impacts in Mexico still need resolution; both were clearly significant but varied in time and place. Under the Spaniards, however, even with a greatly reduced population, the landscape in Mexico generally did not recover due to accelerating impacts from introduced sheep and cattle.[7]

THE BUILT LANDSCAPE

Settlement

The Spaniards and other Europeans were impressed by large flourishing Indian cities such as Tenochtitlán, Quito, and Cuzco, and they took note of the extensive ruins of older, abandoned cities such as Cahokia, Teotihuacán, Tikal, Chan Chan, and Tiwanaku (Hardoy 1968). Most of these cities contained more than 50,000 people. Less notable, or possibly more taken for granted, was rural settlement—small villages of a few thousand or a few hundred people, hamlets of a few families, and dispersed farmsteads. The numbers and locations of much of this settlement will never be known. With the rapid decline of Native populations, the abandonment of houses and entire villages and the decay of perishable materials quickly obscured sites, especially in the tropical lowlands.

We do have some early listings of villages, especially for Mexico and Peru. Elsewhere, archaeology is telling us more than ethnohistory. After initially focusing on large temple and administrative centres, archaeologists are now examining rural sustaining areas, with remarkable results. See, for example, Sanders et al. (1979) on the Basin of Mexico, Culbert and Rice (1990) on the Maya lowlands, and Fowler (1989) on Cahokia in Illinois. Evidence of human occupation for the artistic Santarém Culture phase (Tapajós chiefdom) on the lower Amazon extends over thousands of square kilometres, with large nucleated settlements (Roosevelt 1991: 101–102).

Much of the rural pre-contact settlement was semi-dispersed (*rancherías*), particularly in densely populated regions of Mexico and the Andes, probably reflecting poor food transport efficiency. Houses were both single-family and communal (pueblos, Huron long houses, Amazon malocas). Construction was of stone, earth, adobe, daub and wattle, grass, hides, brush, and bark. Much of the dispersed settlement not destroyed by depopulation was concentrated by the Spaniards into compact grid/plaza style new towns (*congregaciones, reducciones*) for administrative purposes.

Mounds

James Parsons (1985: 161) has suggested that "[a]n apparent mania for earth moving, landscape engineering on a grand scale runs as a thread through much of New World prehistory." Large quantities of both earth and stone were transferred to create various raised and sunken features, such as agricultural landforms, settlement and ritual mounds, and causeways.

Mounds of different shapes and sizes were constructed throughout the Americas for temples, burials, settlement, and as effigies. The stone pyramids of Mexico and the Andes are well known, but equal monuments of earth were built in the Amazon, the Midwest U.S., and elsewhere. The Mississippian period complex of 104 mounds at Cahokia near East St. Louis supported 30,000 people; the largest, Monk's Mound, is currently 30.5 m high

and covers 6.9 ha (Fowler 1989: 90, 192). Cahokia was the largest settlement north of the Rio Grande until surpassed by New York City in 1775. An early survey estimated "at least 20,000 conical, linear, and effigy mounds" in Wisconsin (Stout 1911: 24). Overall, there must have been several hundred thousand artificial mounds in the Midwest and South. De Soto described such features still in use in 1539 (Silverberg 1968: 7). Thousands of settlement and other mounds dot the savanna landscape of Mojos in Bolivia (Denevan 1966). At the mouth of the Amazon on Marajó Island, one complex of 40 habitation mounds contained more than 10,000 people; one of these mounds is 20 m high while another is 90 ha in area (Roosevelt 1991: 31, 38).

Not all of the various earthworks scattered over the Americas were in use in 1492. Many had been long abandoned, but they constituted a conspicuous element of the landscape of 1492 and some are still prominent. Doubtless, many remain to be discovered, and others remain unrecognized as human or prehistoric features.

Roads, Causeways, and Trails

Large numbers of people and settlements necessitated extensive systems of overland travel routes to facilitate administration, trade, warfare, and social interaction (Hyslop 1984; Trombold 1991). Only hints of their former prominence survive. Many were simple traces across deserts or narrow paths cut into forests. A suggestion as to the importance of Amazon forest trails is the existence of more than 500 km of trail maintained by a single Kayapó village today (Posey 1985: 149). Some prehistoric footpaths were so intensively used for so long that they were incised into the ground and are still detectable, as has recently been described in Costa Rica (Sheets and Sever 1991).

Improved roads, at times stone-lined and drained, were constructed over great distances in the realms of the high civilizations. The Inca road network is estimated to have measured about 40,000 km, extending from southern Colombia to central Chile (Hyslop 1984: 224). Prehistoric causeways (raised roads) were built in the tropical lowlands (Denevan 1991); one Maya causeway is 100 km long, and there are more than 1,600 km of causeways in the Llanos de Mojos. Humboldt reported large prehistoric causeways in the Orinoco Llanos. Ferdinand Columbus described roads on Puerto Rico in 1493. Caspar de Carvajal, travelling down the Amazon with Orellana in 1541, reported "highways" penetrating the forest from river bank villages. Joseph de Acosta (1880 [1590]: (1), 171) in 1590 said that between Peru and Brazil, there were "waies as much beaten as those betwixt Salamanca and Valladolid." Prehistoric roads in Chaco Canyon, New Mexico, are described in Trombold (1991). Some routes were so well established and located that they have remained roads to this day.

RECOVERY

A strong case can be made for significant environmental recovery and reduction of cultural features by the late eighteenth century as a result of Indian population decline. Henry Thoreau (1949: 132–137) believed, based on his reading of William Wood, that the New England forests of 1633 were more open, more park-like, with more berries and more wildlife, than Thoreau observed in 1855. Cronon (1983: 108), Pyne (1982: 51), Silver (1990: 104), Martin (1978: 181–182), and Williams (1989: 49) all maintain that the eastern

forests recovered and filled in as a result of Indian depopulation, field abandonment, and reduction in burning. While probably correct, these writers give few specific examples, so further research is needed. The sixteenth-century fields and savannas of Colombia and Central America also had reverted to forest within 150 years after abandonment (Parsons 1975: 30–31; Bennett 1968: 54). On his fourth voyage in 1502–1503, Columbus sailed along the north coast of Panama (Veragua). His son Ferdinand described lands that were well peopled, full of houses, with many fields, and open with few trees. In contrast, in 1681 Lionel Wafer found most of the Caribbean coast of Panama forest covered and unpopulated. On the Pacific side in the eighteenth century, savannas were seldom mentioned; the main economic activity was the logging of tropical cedar, a tree that grows on the sites of abandoned fields and other disturbances (Sauer 1966: 132–133, 287–288). An earlier oscillation from forest destruction to recovery in the Yucatan is instructive. Whitmore et al. (1990: 35) estimate that the Maya had modified 75 percent of the environment by AD 800, and that following the Mayan collapse, forest recovery in the central lowlands was nearly complete when the Spaniards arrived.

The pace of forest regeneration, however, varied across the New World. Much of the southeastern U.S. remained treeless in the 1750s according to Rostlund (1957: 408, 409). He notes that the tangled brush that ensnarled the "Wilderness Campaign of 1864 in Virginia occupied the same land as did Captain John Smith's 'open groves with much good ground between without any shrubs'" in 1624; vegetation had only partially recovered over 240 years. The Kentucky barrens in contrast were largely reforested by the early nineteenth century

(Sauer 1963 [1927]: 30). The Alabama Black Belt vegetation was described by William Bartram in the 1770s as a mixture of forest and grassy plains, but by the nineteenth century, there was only 10 percent prairie and even less in some counties (Rostlund 1957: 393, 401–403). Sections of coastal forests never recovered, given colonist pressures, but Sale's (1990: 291) claim that "the English were well along in the process of eliminating the ancient Eastern woodlands from Maine to the Mississippi" in the first 100 years is an exaggeration.

Wildlife also partially recovered in eastern North America with reduced hunting pressure from Indians; however, this is also a story yet to be worked out. The white-tailed deer apparently declined in numbers, probably reflecting reforestation plus competition from livestock. Commercial hunting was a factor on the coast, with 80,000 deerskins being shipped out yearly from Charleston by 1730 (Silver 1990: 92). Massachusetts enacted a closed season on deer as early as 1694, and in 1718 there was a three-year moratorium on deer hunting (Cronon 1983: 100). Sale (1990: 290) believes that beaver were depleted in the Northeast by 1640. Other fur-bearers, game birds, elk, buffalo, and carnivores were also targeted by White hunters, but much game probably was in the process of recovery in many eastern areas until a general reversal after 1700–1750.

As agricultural fields changed to scrub and forest, earthworks were grown over. All the raised fields in Yucatán and South America were abandoned. A large portion of the agricultural terraces in the Americas were abandoned in the early colonial period (Donkin 1979: 35–38). In the Colca Valley of Peru, measurement on air photos indicates 61 percent terrace abandonment (Denevan 1988: 28). Societies vanished or declined everywhere and whole villages

with them. The degree to which settlement features were swallowed up by vegetation, sediment, and erosion is indicated by the difficulty of finding them today. Machu Picchu, a late prehistoric site, was not rediscovered until 1911.

The renewal of human impact also varied regionally, coming with the Revolutionary War in North America, with the rubber boom in Amazonia, and with the expansion of coffee in southern Brazil (1840–1930). The swamplands of Gulf Coast Mexico and the Guayas Basin of Ecuador remained hostile environments to Europeans until well into the nineteenth century or later (Siemens 1990; Mathewson 1987). On the other hand, Highland Mexico–Guatemala and the Andes, with greater Indian survival and with the establishment of haciendas and intensive mining, show less evidence of environmental recovery. Similarly, Indian fields in the Caribbean were rapidly replaced by European livestock and sugar plantation systems, inhibiting any sufficient recovery. The same is true of the sugar zone of coastal Brazil.

CONCLUSIONS

By 1492, Indian activity had modified vegetation and wildlife, caused erosion, and created earthworks, roads, and settlements throughout the Americas. This may be obvious, but the human imprint was much more ubiquitous and enduring than is usually realized. The historical evidence is ample, as are data from surviving earthworks and archaeology. And much can be inferred from present human impacts. The weight of evidence suggests that Indian populations were large, not only in Mexico and the Andes, but also in seemingly unattractive habitats such as the rain forests of Amazonia, the swamps of Mojos, and the deserts of Arizona.

Clearly, the most humanized landscapes of the Americas existed in those highland regions where people were the most numerous. Here were the large states, characterized by urban centres, road systems, intensive agriculture, a dispersed but relatively dense rural settlement pattern of hamlets and farmsteads, and widespread vegetation and soil modification and wildlife depletion. There were other, smaller regions that shared some of these characteristics, such as the Pueblo lands in the southwestern U.S., the Sabana de Bogotá in highland Colombia, and the central Amazon floodplain, where built landscapes were locally dramatic and are still observable. Finally, there were the immense grasslands, deserts, mountains, and forests elsewhere, with populations that were sparse or moderate, with landscape impacts that mostly were ephemeral or not obvious but nevertheless significant, particularly for vegetation and wildlife, as in Amazonia and the northeastern U.S. In addition, landscapes from the more distant past survived to 1492 and even to 1992, such as those of the irrigation states of the north coast of Peru, the Classic Maya, the Mississippian mound builders, and the Tiwanaku Empire of Lake Titicaca.

*****

The pristine myth cannot be laid at the feet of Columbus. While he spoke of "Paradise," his was clearly a humanized paradise. He described Hispaniola and Tortuga as densely populated and "completely cultivated like the countryside around Cordoba" (Colón 1976: 165). He also noted that "the islands are not so thickly wooded as to be impassable," suggesting openings from clearing and burning (Columbus 1961: 5).

The roots of the pristine myth lie in part with early observers unaware of human impacts that may be obvious to scholars today, particularly for vegetation and wildlife.[8] But even many earthworks such as raised fields have only recently been discovered (Denevan 1966, 1980). Equally important, most of our eyewitness descriptions of wilderness and empty lands come from a later time, particularly 1750–1850 when interior lands began to be explored and occupied by Europeans. By 1650, Indian populations in the hemisphere had been reduced by about 90 percent, while by 1750 European numbers were not yet substantial and settlement had only begun to expand. As a result, fields had been abandoned, while settlements vanished, forests recovered, and savannas retreated. The landscape did appear to be a sparsely populated wilderness. This is the image conveyed by Parkman in the nineteenth century, Bakeless in 1950, and Shetler as recently as 1991. There was some European impact, of course, but it was localized. After 1750 and especially after 1850, populations greatly expanded, resources were more intensively exploited, and European modification of the environment accelerated, continuing to the present.

It is possible to conclude not only that "the virgin forest was not encountered in the sixteenth and seventeenth centuries; [but that] it was invented in the late eighteenth and early nineteenth centuries" (Pyne 1982: 46). However, "paradoxical as it may seem, there was undoubtedly much more 'forest primeval' in 1850 than in 1650" (Rostlund 1957: 409). Thus the "invention" of an earlier wilderness is in part understandable and is not simply a deliberate creation that ennobled the American enterprise, as suggested by Bowden (1992: 20–23). In any event, while pre-European landscape alteration has been demonstrated previously, including by several geographers, the case has mainly been made for vegetation and mainly for eastern North America. As shown here, the argument is also applicable to most of the rest of the New World, including the humid tropics, and involves much more than vegetation.

The human impact on environment is not simply a process of increasing change or degradation in response to linear population growth and economic expansion. It is instead interrupted by periods of reversal and ecological rehabilitation as cultures collapse, populations decline, wars occur, and habitats are abandoned. Impacts may be constructive, benign, or degenerative (all subjective concepts), but change is continual at variable rates and in different directions. Even mild impacts and slow changes are cumulative, and the long-term effects can be dramatic. Is it possible that the thousands of years of human activity before Columbus created more change in the visible landscape than has occurred subsequently with European

settlement and resource exploitation? The answer is probably yes for most regions for the next 250 years or so, and for some regions right up to the present time. American flora, fauna, and landscape were slowly Europeanized after 1492, but before that they had already been Indianized. "It is upon this imprint that the more familiar Euro-American landscape was grafted, rather than created anew" (Butzer 1990: 28). What does all this mean for protectionist tendencies today? Much of what is protected or proposed to be protected from human disturbance had Native peoples present, and environmental modification occurred accordingly and in part is still detectable.

The pristine image of 1492 seems to be a myth, then, an image more applicable to 1750, following Indian decline, although recovery had only been partial by that date. There is some substance to this argument, and it should hold up under the scrutiny of further investigation of the considerable evidence available, both written and in the ground.

NOTES

1. Sauer had a life-long interest in this topic (1963 [1927], 1966, 1971, 1980).
2. See Nash (1967) on the "romantic wilderness" of America; Bowden (1992: 9–12) on the "invented tradition" of the "primeval forest" of New England; and Manthorne (1989: 10–21) on artists' images of the tropical "Eden" of South America. Day (1953: 329) provides numerous quotations from Parkman on "wilderness" and "vast," "virgin," and "continuous" forest.
3. For example, a 1991 advertisement for a Time-Life video refers to "the unspoiled beaches, forests, and mountains of an earlier America" and "the pristine shores of Chesapeake Bay in 1607."
4. On the other hand, the ability of Indians to clear large trees with inefficient stone axes, assisted by girdling and deadening by fire, may have been overestimated (Denevan 1992). Silver (1990: 51) notes that the upland forests of Carolina were largely uninhabited for this reason.
5. Similar conclusions were reached by foresters Maxwell (1910) and Day (1953); by geographers Sauer (1963), Brown (1948: 11–19), Rostlund (1957), and Bowden (1992); and by environmental historians Pyne (1982: 45–51), Cronon (1983: 49–51), and Silver (1990: 59–66).
6. B. Williams (1989: 730) finds strong evidence of rural overpopulation (66 percent in poor crop years, 11 percent in average years) in the Basin of Mexico village of Asunción, ca. AD 1540,

which was probably "not unique but a widespread phenomenon." For a contrary conclusion, that the Aztecs did not exceed carrying capacity, see Ortiz de Montellano (1990: 119).

7. Highland Guatemala provides another prehistoric example of "severe human disturbance" involving deforestation and "massive" soil erosion (slopes) and deposition (valleys) (Murdy 1990: 186). For the central Andes there is some evidence that much of the *puna* zone (3200–4500 m), now grass and scrub, was deforested in prehistoric times (White 1985).

8. The English colonists in part justified their occupation of Indian land on the basis that such land had not been "subdued" and therefore was "land free to be taken" (Wilson 1992: 16).

REFERENCES

Acosta, Joseph [Jose] de. 1880 [1590]. *The Natural and Moral History of the Indies*, vols. 60, 61. Translated by E. Gimston. London: Hakluyt Society.

Alcorn, J.B. 1981. "Huastec Noncrop Resource Management: Implications for Prehistoric Rain Forest Management." *Human Ecology* 9: 395–417.

Anderson, R.C. 1990. "The Historic Role of Fire in the North American Grassland." In *Fire in North American Tallgrass Prairies*, edited by S.L. Collins and L.L. Wallace, 8–18. Norman: University of Oklahoma Press.

Bakeless, J. 1950. *The Eyes of Discovery: The Pageant of North America as Seen by the First Explorers*. New York: J.B. Lippincott.

Balée, W. 1987. "Cultural Forests of the Amazon." *Garden* 11: 12–14, 32.

_____. 1989. "The Culture of Amazonian Forests," in *Advances in Economic Botany*, edited by Darrell A. Posey and W. Balée, vol. 7, 1–21. New York: New York Botanical Garden.

Bennett, C.F. 1968. *Human Influences on the Zoogeography of Panama*. Ibero-Americana 51. Berkeley: University of California Press.

Borchert, J. 1950. "Climate of the Central North American Grassland." *Annals of the Association of American Geographers* 40: 1–39.

Bowden, M.J. 1992. "The Invention of American Tradition." *Journal of Historical Geography* 18: 3–26.

Brown, R.H. 1948. *Historical Geography of the United States*. New York: Harcourt, Brace.

Brown, S., and A. Lugo. 1990. "Tropical Secondary Forests." *Journal of Tropical Ecology* 6: 1–32.

Butzer, K.W. 1990. "The Indian Legacy in the American Landscape." In *The Making of the American Landscape*, edited by M.P. Conzen, 27–50. Boston: Unwin Hyman.

_____, and E.K. Butzer. 1994. "The Sixteenth-Century Environment of the Central Mexican Bajío: Archival Reconstruction from Spanish Land Grants." In *Culture, Form, and Place*, edited by K. Mathewson, 89–124. Baton Rouge: Geoscience and Man.

Colón, C. 1976. *Diario del descubrimiento*, vol. 1, edited by M. Alvar. Madrid: Editorial La Muralla.

Columbus, C. 1961. *Four Voyages to the New World: Letters and Selected Documents*, edited by R.H. Major. New York: Corinth Books.

Cook, S.F. 1949. *Soil Erosion and Population in Central Mexico*. Ibero-Americana 34. Berkeley: University of California Press.

_____, and W. Borah. 1971–1979. *Essays in Population History*, 3 vols. Berkeley: University of California Press.

Cowley, G. 1991. "The Great Disease Migration." In *1492–1992, When Worlds Collide: How Columbus's Voyages Transformed Both East and West. Newsweek*, Special Issue (Fall/Winter): 54–56.

Cronon, W. 1983. *Changes in the Land: Indians, Colonists, and the Ecology of New England*. New York: Hill and Wang.

Culbert, T.P., and D.S. Rice, eds. 1990. *Pre-Columbian Population History in the Maya Lowlands.* Albuquerque: University of New Mexico Press.

Day, G.M. 1953. "The Indian as an Ecological Factor in the Northeastern Forest." *Ecology* 34: 329–346.

Denevan, W.M. 1961. "The Upland Pine Forests of Nicaragua." *University of California Publications in Geography* 12: 251–320.

_____. 1966. *The Aboriginal Cultural Geography of the Llanos de Mojos of Bolivia.* Ibero-Americana 48. Berkeley: University of California Press.

_____. 1980. "Tipología de configuraciones agrícolas prehispánicas." *America Indigena* 40: 619–652.

_____. 1988. "Measurement of Abandoned Terracing from Air Photos: Colca Valley, Peru." *Yearbook, Conference of Latin Americanist Geographers* 14: 20–30.

_____. 1991. "Prehistoric Roads and Causeways of Lowland Tropical America." In *Ancient Road Networks and Settlement Hierarchies in the New World,* edited by C.D. Trombold, 230–242. Cambridge: Cambridge University Press.

_____, ed. 1992 [1976]. *The Native Population of the Americas in 1492,* 2nd ed. Madison: University of Wisconsin Press.

_____. 1992. "Stone vs. Metal Axes: The Ambiguity of Shifting Cultivation in Prehistoric Amazonia." *Journal of the Steward Anthropological Society* 20: 153-165.

_____, and C. Padoch, eds. 1988. *Swidden-Fallow Agroforestry in the Peruvian Amazon. Advances in Economic Botany,* vol. 5. New York: New York Botanical Garden.

Dobyns, H.F. 1981. *From Fire to Flood: Historic Human Destruction of Sonoran Desert Riverine Oases.* Socorro: Ballena Press.

_____. 1983. *Their Number Become Thinned: Native American Population Dynamics in Eastern North America.* Knoxville: University of Tennessee Press.

Donkin, R.A. 1979. *Agricultural Terracing in the Aboriginal New World.* Viking Fund Publications in Anthropology 56. Tucson: University of Arizona Press.

Doolittle, W.E. 1990. *Canal Irrigation in Prehistoric Mexico: The Sequence of Technological Change.* Austin: University of Texas Press.

Forman, R.T.T., and E.W.B. Russell. 1983. "Evaluation of Historical Data in Ecology." *Bulletin of the Ecological Society of America* 64: 5–7.

Fowler, M. 1989. *The Cahokia Atlas: A Historical Atlas of Cahokia Archaeology.* Studies in Illinois Archaeology 6. Springfield: Illinois Historic Preservation Agency.

Garcilaso de la Vega, The Inca. 1980 [1605]. *The Florida of the Inca: A History of the Adelantado, Hernando de Soto,* 2 vols., translated and edited by J.C. Varner and J.J. Varner. Austin: University of Texas Press.

Garren, K.H. 1943. "Effects of Fire on Vegetation of the Southeastern United States." *The Botanical Review* 9: 617–654.

Gartner, W.G. 1992. "The Hulbert Creek Ridged Fields: Pre-Columbian Agriculture near the Dells, Wisconsin." Master's thesis, Department of Geography, University of Wisconsin, Madison.

Gilmore, M.R. 1931. "Dispersal by Indians a Factor in the Extension of Discontinuous Distribution of Certain Species of Native Plants." *Papers of the Michigan Academy of Science, Arts and Letters* 13: 89–94.

Goldammer, J.C, ed. 1990. *Fire in the Tropical Biota: Ecosystem Processes and Global Challenges.* Ecological Studies 84. Berlin: Springer-Verlag.

Gómez-Pompa, A. 1987. "On Maya Silviculture." *Mexican Studies* 3: 1–17.

_____, J. Salvador Flores, and V. Sosa. 1987. "The "Pet Kot": A Man-Made Forest of the Maya." *Interciencia* 12: 10–15.

Good, K.R. 1987. "Limiting Factors in Amazonian Ecology." In *Food and Evolution: Toward a Theory of Human Food Habitats*, edited by M. Harris and E.B. Ross, 407–421. Philadelphia: Temple University Press.

Gordon, B.L. 1957. *Human Geography and Ecology in the Sinú Country of Colombia*. Ibero-Americana 39. Berkeley: University of California Press.

_____. 1982. *A Panama Forest and Shore: Natural History and Amerindian Culture in Bocas del Torn*. Pacific Grove: Boxwood Press.

Greenberg, L.S.C. 1991. "Garden-Hunting among the Yucatec Maya." *Etnoecológica* 1: 30–36.

Haffer, J. 1991. "Mosaic Distribution Patterns of Neotropical Forest Birds and Underlying Cyclic Disturbance Processes." In *The Mosaic-Cycle Concept of Ecosystems*, edited by H. Remmert, 83–105. Ecological Studies, vol. 85. Berlin: Springer-Verlag.

Hardoy, J. 1968. *Urban Planning in Pre-Columbian America*. New York: George Braziler.

Hills, T.L., and R.E. Randall, eds. 1968. *The Ecology of the Forest/Savanna Boundary*. Savanna Research Series 13. Montreal: McGill University.

Hyslop, J. 1984. *The Inka Road System*. New York: Academic Press.

Kolata, A.L. 1991. "The Technology and Organization of Agricultural Production in the Tiwanaku State." *Latin American Antiquity* 2: 99–125.

Lewis, H.T. 1982. "Fire Technology and Resource Management in Aboriginal North America and Australia." In *Resource Managers: North American and Australian Hunter-Gatherers*, edited by N.M. Williams and E.S. Hunn, 45–67. AAAS Selected Symposia 67. Boulder: Westview Press.

Lord, L., and S. Burke. 1991. "America before Columbus." *U.S. News and World Report* (July 8): 22–37.

McEvedy, C., and R. Jones. 1978. *Atlas of World Population History*. New York: Penguin Books.

MacNutt, F.A. 1909. *Bartholomew de las Casas: His Life, His Apostolate, and His Writings*. New York: Putnam's.

Manthorne, K.E. 1989. *Tropical Renaissance: North American Artists Exploring Latin America, 1839–1879*. Washington: Smithsonian Institution Press.

Martin, C. 1978. *Keepers of the Game: Indian-Animal Relationships and the Fur Trade*. Berkeley: University of California Press.

Mathewson, K. 1987. "Landscape Change and Cultural Persistence in the Guayas Wetlands, Ecuador." Ph.D. dissertation, Department of Geography, University of Wisconsin, Madison.

Maxwell, H. 1910. "The Use and Abuse of Forests by the Virginia Indians." *William and Mary College Quarterly Historical Magazine* 19: 73–103.

Medina, E. 1980. "Ecology of Tropical American Savannas: An Ecophysiological Approach." In *Human Ecology in Savanna Environments*, edited by D.R. Harris, 297–319. London: Academic Press.

Meinig, D.W. 1986. *The Shaping of America: A Geographical Perspective on 500 Years of History*, vol. 1, *Atlantic America, 1492–1800*. New Haven: Yale University Press.

Melville, E.G.K. 1990. "Environmental and Social Change in the Valle del Mezquital, Mexico, 1521–1600." *Comparative Studies in Society and History* 32: 24–53.

Mueller-Dombois, D. 1981. "Fire in Tropical Ecosystems." In *Fire Regimes and Ecosystem Properties: Proceedings of the Conference*, Honolulu, 1978, 137–176. General Technical Report WO-26. Washington: U.S. Forest Service.

Murdy, C.N. 1990. "Prehispanic Agriculture and Its Effects in the Valley of Guatemala." *Forest and Conservation History* 34: 179–190.

Nash, R. 1967. *Wilderness and the American Mind*. New Haven: Yale University Press.

Ortiz de Montellano, B.R. 1990. *Aztec Medicine, Health, and Nutrition*. New Brunswick: Rutgers University Press.

Parsons, J.J. 1975. "The Changing Nature of New World Tropical Forests since European Colonization." In *The Use of Ecological Guidelines for Development in the American Humid Tropics*: 28–38. International Union for Conservation of Nature and Natural Resources Publications, n.s., 31. Morges.

———. 1985. "Raised Field Farmers as Pre-Columbian Landscape Engineers: Looking North from the San Jorge (Colombia)." In *Prehistoric Intensive Agriculture in the Tropics*, edited by I.S. Farrington, 149–165. International Series 232. Oxford: British Archaeological Reports.

———, and W.M. Denevan. 1967. "Pre-Columbian Ridged Fields." *Scientific American* 217, no. 1: 92–100.

Patterson, W.A., III, and K.E. Sassaman. 1988. "Indian Fires in the Prehistory of New England." In *Holocene Human Ecology in Northeastern North America*, edited by G.P. Nicholas, 107–135. New York: Plenum.

Plazas, C., and A.M. Falchetti. 1987. "Poblamiento y adecuación hidráulica en el bajo Río San Jorge, Costa Atlantica, Colombia." In *Prehistoric Agricultural Fields in the Andean Region*, edited by W.M. Denevan, K. Mathewson, and G. Knapp, 483–503. International Series 359. Oxford: British Archaeological Reports.

Posey, D.A. 1985. "Indigenous Management of Tropical Forest Ecosystems: The Case of the Kayapó Indians of the Brazilian Amazon." *Agroforestry Systems* 3: 139–158.

Pyne, S.J. 1982. *Fire in America: A Cultural History of Wildland and Rural Fire*. Princeton: Princeton University Press.

Raup, H.M. 1937. "Recent Changes in Climate and Vegetation in Southern New England and Adjacent New York." *Journal of the Arnold Arboretum* 18: 79–117.

Roosevelt, A.C. 1991. *Moundbuilders of the Amazon: Geophysical Archaeology on Marajo Island, Brazil*. San Diego: Academic Press.

Rostlund, E. 1957. "The Myth of a Natural Prairie Belt in Alabama: An Interpretation of Historical Records." *Annals of the Association of American Geographers* 47: 392–411.

———. 1960. "The Geographic Range of the Historic Bison in the Southeast." *Annals of the Association of American Geographers* 50: 395–407.

Russell, E.W.B. 1983. "Indian-Set Fires in the Forests of the Northeastern United States." *Ecology* 64: 78–88.

Saldarriaga, J.C., and C. Uhl. 1991. "Recovery of Forest Vegetation Following Slash-and-Burn Agriculture in the Upper Río Negro." In *Rainforest Regeneration and Management*, edited by A. Gómez-Pompa, T.C. Whitmore, and M. Hadley, 303–312. Paris: UNESCO.

———, and D.C. West. 1986. "Holocene Fires in the Northern Amazon Basin." *Quaternary Research* 26: 358–366.

Sale, K. 1990. *The Conquest of Paradise: Christopher Columbus and the Columbian Legacy*. New York: Alfred A. Knopf.

Sánchez-Albornoz, N. 1974. *The Population of Latin America: A History*. Berkeley: University of California Press.

Sanders, W.T., J.R. Parsons, and R.S. Santley. 1979. *The Basin of Mexico: Ecological Processes in the Evolution of a Civilization*. New York: Academic Press.

Sauer, C.O. 1950. "Grassland Climax, Fire, and Man." *Journal of Range Management* 3: 16–21.

———. 1958. "Man in the Ecology of Tropical America." *Proceedings of the Ninth Pacific Science Congress, 1957* 20: 104–110.

———. 1963 [1927]. "The Barrens of Kentucky." In *Land and Life: A Selection from the Writings of Carl Ortwin Sauer*, edited by J. Leighly, 23–31. Berkeley: University of California Press.

———. 1966. *The Early Spanish Main*. Berkeley: University of California Press.

———. 1971. *Sixteenth-Century North America: The Land and the People as Seen by the Europeans*. Berkeley: University of California Press.

_____. 1975. "Man's Dominance by Use of Fire." *Geoscience and Man* 10: 1–13.

_____. 1980. *Seventeenth-Century North America.* Berkeley: Turtle Island Press.

Schwerin, K.H. 1991. "The Indian Populations of Latin America." In *Latin America, Its Problems and Its Promise: A Multidisciplinary Introduction*, 2nd ed., edited by J.K. Black, 39–53. Boulder: Westview Press.

Scott, G.A.J. 1978. *Grassland Development in the Gran Pajonal of Eastern Peru.* Hawaii Monographs in Geography 1. Honolulu: University of Hawaii.

Sheets, P., and T.L. Sever. 1991. "Prehistoric Footpaths in Costa Rica: Transportation and Communication in a Tropical Rainforest." In *Ancient Road Networks and Settlement Hierarchies in the New World*, edited by C.D. Trombold, 53–65. Cambridge: Cambridge University Press.

Shetler, S. 1991. "Three Faces of Eden." In *Seeds of Change: A Quincentennial Commemoration*, edited by H.J. Viola and C. Margolis, 225–247. Washington: Smithsonian Institution Press.

Siemens, A.H. 1990. *Between the Summit and the Sea: Central Veracruz in the Nineteenth Century.* Vancouver: University of British Columbia Press.

Silver, T. 1990. *A New Face on the Countryside: Indians, Colonists, and Slaves in South Atlantic Forests, 1500–1800.* Cambridge: Cambridge University Press.

Silverberg, R. 1968. *Mound Builders of Ancient America: The Archaeology of a Myth.* Greenwich: New York Graphic Society.

Smith, N.J.H. 1980. "Anthrosols and Human Carrying Capacity in Amazonia." *Annals of the Association of American Geographers* 70: 553–566.

Stout, A.B. 1911. "Prehistoric Earthworks in Wisconsin." *Ohio Archaeological and Historical Publications* 20: 1–31.

Sturtevant, W.C. 1961. "Taino Agriculture." In *The Evolution of Horticultural Systems in Native South America, Causes and Consequences: A Symposium*, edited by J. Wilbert, 69–82. Caracas: Sociedad de Ciencias Naturales La Salle.

Taylor, D.L. 1981. *Fire History and Fire Records for Everglades National Park.* Everglades National Park Report T-619. Washington: National Park Service, U.S. Department of the Interior.

Thompson, D.Q., and R.H. Smith. 1970. "The Forest Primeval in the Northeast—a Great Myth?" *Proceedings, Tall Timbers Fire Ecology Conference* 10: 255–265.

Thoreau, H.D. 1949. *The Journal of Henry D. Thoreau*, vol. 7, *September 1, 1854–October 30, 1855*, edited by B. Torrey and F.H. Allen. Boston: Houghton Mifflin.

Trombold, C.D., ed. 1991. *Ancient Road Networks and Settlement Hierarchies in the New World.* Cambridge: Cambridge University Press.

Uhl, C., D. Nepstad, R. Buschbacher, K. Clark, B. Kauffman, and S. Subler. 1990. "Studies of Ecosystem Response to Natural and Anthropogenic Disturbances Provide Guidelines for Designing Sustainable Land-Use Systems in Amazonia." In *Alternatives to Deforestation: Steps toward Sustainable Use of the Amazon Rain Forest*, edited by A.B. Anderson, 24–42. New York: Columbia University Press.

Walschburger, T., and P. von Hildebrand. 1991. "The First 26 Years of Forest Regeneration in Natural and Man-Made Gaps in the Colombian Amazon." In *Rain Forest Regeneration and Management*, edited by A. Gómez-Pompa, T.C. Whitmore, and M. Hadley, 257–263. Paris: UNESCO.

Watts, W.A., and J.P. Bradbury. "Paleoecological Studies at Lake Patzcuaro on the West-Central Mexican Plateau and at Chalco in the Basin of Mexico." *Quaternary Research* 17: 56–70.

Wedel, W.R. 1957. "The Central North American Grassland: Man-Made or Natural?" *Social Science Monographs* 3: 39–69. Washington: Pan American Union.

White, S. 1985. "Relations of Subsistence to the Vegetation Mosaic of Vilcabamba, Southern Peruvian Andes." *Yearbook, Conference of Latin Americanist Geographers* 11: 3–10.

Whitmore, T.M. 1991. "A Simulation of the Sixteenth-Century Population Collapse in the Basin of Mexico." *Annals of the Association of American Geographers* 81: 464–487.

_____, B.L. Turner, II, D.L. Johnson, R.W. Kates, and T.R. Gottschang. 1990. "Long-Term Population Change." In *The Earth as Transformed by Human Action*, edited by B.L. Turner II et al., 25–39. Cambridge: Cambridge University Press.

Williams, B.J. 1972. "Tepetate in the Valley of Mexico." *Annals of the Association of American Geographers* 62: 618–626.

_____. 1989. "Contact Period Rural Overpopulation in the Basin of Mexico: Carrying-Capacity Models Tested with Documentary Data." *American Antiquity* 54: 715–732.

Williams, M. 1989. *Americans and Their Forests: A Historical Geography*. Cambridge: Cambridge University Press.

Wilson, S.M. 1992. "That Unmanned Wild Countrey": Native Americans Both Conserved and Transformed New World Environments." *Natural History* (May): 16–17.

Wood, W. 1977 [1635]. *New England's Prospect*, edited by A.T. Vaughan. Amherst: University of Massachusetts Press.

Chapter Six
Fire

Shepard Krech

In 1632, somewhere off the mid-Atlantic Coast, a Dutch mariner and merchant named David Pietersz. de Vries wrote about land "smelt before it is seen." An entire continent to the west lay beyond sensation except for the smell of smoke in the air. On arrival, Europeans would find thick clouds of hazy smoke enveloping the land, grasslands reduced to charred stubble, and park-like forests clear of undergrowth. Fire had clearly modified this landscape, and sometimes scarred it deeply—not lightning-caused fire (although some was), but fire ignited by Indians.

Yet when Europeans depicted America in literary texts or on canvas, they often imagined a pristine, primeval land in which an unbroken, vast, tangled, and impenetrable forest figured prominently. Predisposed to find wilderness, as well as savagery romanticized or brutalized, they discovered both. Over centuries, the image of the forest has been consistent, from "infinite thick woods" in the seventeenth century, to the Hudson River School's canvases of deep twisted woods in the nineteenth, to "the shadows and the gloom of mighty forests" in the early twentieth. None immortalized the forest more lastingly than Henry Wadsworth Longfellow in *The Song of Hiawatha*, but as he implored all who "Love the shadow of the forest" to listen to the wild traditions of the Native Hiawatha, his imagery was overwhelmingly of an interminable, tangled, solitary, wailing, darksome, moaning, gloomy forest primeval.[1]

This contradiction between the evidence of fires set by man and the literary and artistic image of a primeval forest left the role of Indians misunderstood.[2] Through the years, Indians have been damned either because they could not set fires with significant consequences or because they could and did. In the second half of the nineteenth century, when evolutionary schemes were everywhere on the air, many argued that Indian fires—the fires of "primitive" or "savage" men—could not possibly have had much impact compared to the fires of civilized men. They castigated Indians for technological incompetence. But by the early twentieth century, when historical evidence for Indians' having used fire on a continental scale had emerged (and for European settlers following suit until controlled by local jurisdictions concerned for the safety of agriculture and settlement), critics developed a new appreciation for the power of Indian fires. At this very time, however, forest ecologists branded fires as always destructive and for virtually three-quarters of a century made every effort to halt them in national forests and parks. Even John Muir, the great preservationist, regarded fire as "the great master-scourge of forests." In this context, Indians were damned again, this time for being careless and profligate and more "destructive" than White men. Forest ecologists concerned with forest conservation defined by regulated timber harvest said that Indians could not possibly have cared about "the fall of some score millions of feet of prime timber in a forest conflagration." They thought that Indians were inherently incendiaries, uncontrollable pyromaniacs, and arsonists, and they likened their burning to sin.

The view that Indian burning was benign never, however, completely disappeared. It re-emerged in the 1940s when Indians were depicted as conservationists whose campfires never spread. According to the U.S. Department of Agriculture, there was no proof that Indians regularly set large-scale fires; lacking matches, they set "small campfires that they tended carefully." Another forest resident, Smokey the Bear, was beginning his long reign as icon of fire prevention and forest protection. Smokey's slogan, "Remember, only you can prevent forest fires," was soon on everyone's lips, and real live Smokeys in Washington's National Zoo warned generations of children and their parents about the incompatibility of fire and forest, and the importance of conservation. Some Indians adopted Smokey's philosophy—one going so far as to say that fighting fires is "the one thing the White man does that makes sense." These ideas later resonated with

the unequivocal post-1970 positioning of Indians as Noble Indians at one with nature. Not only were Indians regarded as benign burners, but also they were seen as ecologically prescient: their fires were based on ecological (systemic) knowledge, and they were conservation-minded.[3]

Forest ecologists, chastened dramatically by the Yellowstone fires of 1988, now tell us that fire prevention often does not make much ecological sense. But what about Indians and fire? How extensively did they burn? Why did they burn? Did they set small and innocuous fires or large and destructive ones? Did they manage carefully the size and intensity of their fires or were their burns uncontrolled? Did they understand the ecological consequences of fire? In fact, North American Indians burned often and for myriad ends, many of which were practical, like keeping voracious mosquitoes and flies at bay. The most important related to subsistence, aggression, communication, and travel.[4]

*****

Indians used fire to improve subsistence more than for any other end. Across the continent, they deployed fire to improve their access to animals, to improve or eliminate forage for the animals they depended on for food, and to drive and encircle animals. In early seventeenth-century Massachusetts Bay Colony, William Wood reported that Indians burned each November "when the grass is withered and leaves dried." Fire, he said, "consumes all the underwood and rubbish which otherwise would overgrow the country, making it unpassable, and spoil their much affected hunting." Others agreed, like the Puritan Edward Johnson who said that Indians in Massachusetts burned the woods so that "they may not be hindered in hunting Venson, and Bears, in the winter season." Wood saw understory growth only "in swamps and low grounds that are wet," and on land that Indians "hath not ... burned" because they had died from epidemic disease. As in Massachusetts, this occurred in much of the East (and elsewhere on the continent), where Indians burned forest in order to eliminate understory and promote the growth of grasses and other forage favoured by large animals. The practice was not universal, but many burned the woods once a year or twice annually, often in spring and fall. It was the "frequent fiering of the woods," as Johnson remarked, that led so many Europeans to remark on the regular spacing and open grassy ground between oak, walnut, beech, and other trees: they were "thin of Timber in many places, like our Parkes in England."

Others linked burning specifically to hunting large animals like deer—by using fire to create favourable ecological niches. In New England, Virginia, and elsewhere in the East, burning undeniably created extensive meadows bordering open forest, with "edge" habitats ideal for the growth of berry bushes and grasses, where deer browsed and other animals fed. The fire history of landscapes could be seen in "oak openings," hundreds or thousands of acres of cleared meadowlands with oaks, poplars, and other trees scattered throughout, and with strawberry vines growing among charred stumps.[5]

Throughout the continent, especially in the fall, Indians hunted animals by surrounding them with fire. In California, the Shasta encircled mule deer; in the East, the Delaware surrounded white-tailed deer; and on the Plains, many did the same with the buffalo. The hunts demanded co-operation—in the East, dozens and perhaps hundreds of hunters co-operated in lighting fires around a herd of deer

and then killing all that were trapped. At times the fire surround was deliberately broken in places, and at times it was totally enclosed. The seventeenth-century Illinois used both techniques. Sometimes Indians set grasslands on fire around a buffalo herd, except for several places through which animals were encouraged to pass to escape the flames, where they would wait and kill some "six score in a day." Sometimes, as Nicolas Perrot described for the seventeenth-century Illinois as well as the Iowa, Pawnee, and perhaps Omaha, people of "an entire village" enclosed buffalo inside a summer grass fire. Hundreds of men surrounded a herd, evidently at night, and once day broke they set fire to the dry grass on all sides, and a circle of flames and smoke gradually took form. Men ran at animals attempting to escape, keeping them inside the flames. "This produces the same effect to the sight as four ranks of palisades, in which the buffalo are enclosed. When the savages see that the animals are trying to get outside of it, in order to escape the fires which surround them on all sides ... they run at them and compel them to re-enter the enclosure; and they avail themselves of this method to kill all the beasts." On some hunts they evidently killed as many as 1,500 buffalo. The Sioux, according to Henry R. Schoolcraft, used an identical technique, first encircling animals and then setting grass on fire. Buffalo, "having a great dread of fire, retire towards the centre of the grasslands as they see it approach, and here being pressed together in great numbers, many are trampled under foot, and the Indians rushing in with their arrows and musketry, slaughter immense numbers in a short period."[6]

Indians often set fires designed to move animals from one place to another. "Hundreds of *Indians*," Mark Catesby, the naturalist, wrote in the early eighteenth-century Southeast, "[spread] themselves in Length thro' a great Extent of Country," then "set the Woods on Fire" in order to drive and hunt deer in October. On the Plains they used fire to drive animals toward pounds. On the northern Plains in the fall of 1808, Alexander Henry described how three men set grass or dung ablaze upwind of a herd of several hundred buffalo, and slowly drove the herd before them, until in sight of a man covered with a robe, who, disguised as a buffalo, piqued the animals' curiosity and led them onward into a pound to their deaths. At times the objective was not a pound but woods or grasslands—wherever animals could more easily be killed. They burned grasslands to force animals from them into river bottoms below, and they fired thick underbrush along river bottoms to drive them out. In the Southwest, Indians used fire to drive deer from ponderosa pine forests into canyons.[7]

Indians routinely burned lands so that animals could not use them. On the Plains, Indians set fires to ruin forage and force animals to find grasses in areas where they were more easily hunted. This was common in late fall, when Indians evidently burned large sections of grasslands and subsequently easily found buffalo, elk, and deer in timbered river bottoms where grasses remained unburned. When buffalo were scarce, several "sagacious" men once set grass afire "so as to denude the grasslands, except within an area of fifteen or twenty miles contiguous to the camp," which "reduc[ed] greatly," as stated, "the labor of the hunt." Fire may also have attracted some animals: in California, hunters evidently set fires in meadows, then waited to ambush the deer lured to the flames, it was said, from curiosity.[8]

Indians also burned regularly to set the stage for new plant growth and for the return of animals and men when conditions were right. After burning, animals left until another season when they might return for new palatable growth. Again, this was common on the Plains. In the spring or fall, Indians often set fire to grasslands to improve grazing conditions. They ignited spring fires to improve the forage for buffalo in the summer, and fall fires for better pasturage the following spring. The fires, which were sometimes hundreds of miles across, worked. In September 1804, Meriwether Lewis saw recently burnt grassland with 3-inch-high grass being consumed by "vast herds of buffalo, deer, elk, and antelope ... feeding in every direction as far as the eye of the observer could reach."[9] In California, Indians regularly burned small grasslands or meadows of grasses and ferns embedded in forests, so that seed-producing grasses would be productive and enticing to deer, elk, and bears.[10]

Thus, almost everywhere Indians burned the land to surround, drive, frighten, or scorch the animals and reptiles they sought to eat, and to create proper foraging conditions, either in the same or in the following year, for small and large mammals and the predators (including themselves) that sought them.

Indians throughout North America also used fire to increase the production of berries, seeds, nuts, and other gathered foods. Whether or not this end was deliberate or a by-product of others for which fire was deployed, Indians who set fires in eastern woods created not just foraging ground for deer and other large animals and woods free of understory, but niches on the fringes of meadows, and between and under trees free of competition for strawberries, blackberries, and raspberries, all of which flourished, often in profusion. "In its season," one seventeenth-century observer near the Potomac River remarked, "your foot can hardly direct itself where it will not be dyed in the blood of large and delicious strawberries."[11] In the West, a large number of Indians, including three dozen groups in California alone, burned to increase seed and grass yields. The Yurok and Karok burned to improve growth, production, or harvest of huckleberry bushes, hazel sticks (for basket making), "bear grass" (a lily), "wild rice" (grass seed) plants, acorns, and tobacco. Indians burned Oregon's Willamette Valley in late fall prior to collecting "wild wheat," which was probably tarweed. Through regular fires, Indians in western Washington encouraged camas and bracken, which were important sources of carbohydrates, as well as nettles, to grow in profusion, and their fires perpetuated the grasslands where these plants grew.[12]

Indians also used fire in herding and farming. As William Clark remarked in 1805 of upper Missouri River Indians who burned near their villages every spring "for the benefit of their horses," Indians on the Plains and elsewhere set fires regularly to improve pasture for horse herds.[13] Where they farmed maize, beans, squash, melons, pumpkins, and other crops, Indians used fire extensively to clear land, destroy plants competing with crops, and deposit ash on soil. Throughout the East, Indians cleared hundreds and, at times, thousands of acres of ground for their crops. From New England came many reports of fields several hundred acres in extent. In Virginia, mention was made of thousands of acres of cleared land under cultivation. When Europeans arrived on the East Coast, they discovered massive quantities of maize, calculated later in hundreds of thousands of bushels, thousands of

barrels, or of sufficient quantity to fill the holds of several ships or to keep people alive for months. East and west, fields and meadows in which burning ceased when they were abandoned by Indians dead from disease or driven away by their enemies again soon became forest.[14]

*****

Indians across North America may have used fire most often in contexts relating to subsistence, but they also collectively employed fire as a weapon, as a means of communication, and to improve travel. As men and women of European descent soon discovered, many Indians unhesitatingly used fire as an offensive and defensive weapon. They employed tactical fires effectively to drive unwanted strangers or enemies from cover or away altogether. Fires lit for these purposes were especially common on the eighteenth- and nineteenth-century Plains where, for example, the Sioux lit the grassy plains to the windward of traders such as David Thompson, hoping to drive him and his men into low-lying woods and a waiting ambush. The Blackfeet ignited thickets in which Kit Carson and other trappers had taken refuge. The Blood took similar action against other trappers. One nineteenth-century governmental agent, finding fires lit to his windward several nights in succession, countered the threat by burning all the grass surrounding his camp. Indians used fire against each other as well as against Europeans: A Sioux war party set fire to grass to the windward of two Omaha men with a Sioux captive that they had come across hidden in a ravine, hoping to drive them from cover. Indians set the woods and grasslands ablaze so often in Iowa in the 1830s for these and other reasons, said C.A. Murray, that "the whole country around" was "completely burnt up and devastated" and he and his party had to go elsewhere to find forage and game.

Economic and political motives often propelled Indians to use fire aggressively, as against traders and trappers such as David Thompson and Kit Carson, who trespassed and poached, threatening the new Indian economic livelihood. The Cree set fire to their own territory against trespassing Assiniboine hunters, hoping to drive them back to their own territory. Indians near Fort Garry, angry at the implications of a merger of fur trading companies for the favourable exchange rates they had been receiving in a period of competition, burned grasslands to keep buffalo at a distance and starve the traders. In other instances, Indians burned other Native peoples' hunting territory to force them farther away, or burned near a post to prevent hunting and enhance the value of their own provisions. The strategy sometimes backfired: in 1781, the Cree and Assiniboine set fires so that they alone would have provisions to trade, but drove buffalo so far away that they were compelled to beg for food. Plains Indians set fire to grasslands to destroy the forage of the horses and mules belonging to their enemies and pursuers, who consequently were more vulnerable or compelled to give up the chase. In addition, Indians burned for the smoke that masked offensive movements or obscured retreat from danger—or that agitated horses and concealed men bent on stampeding and stealing them. Throughout North America, American Indians burned the land pragmatically to confuse, hinder, maim, or kill their enemies, Indian or White, to drive them from or into cover, or to mask their own actions.[15]

Indians also ignited fires to signal each other in a variety of habitats—even subarctic boreal forest and taiga, where Northern Athapaskan people announced their presence or communicated prearranged messages with moss and lichen fires or by burning an entire spruce. But these types of fires were especially prevalent on the prairies and Plains during the summer season. Explorers, trappers, missionaries, soldiers, and others, some no doubt apprehensive about what the flames meant, attributed as many as four of every ten fires to signalling. "Firing the prairie" many called it. The blazes were quick and effective, propelling messages from one group to the next, and could be "read" as far as 100 miles from their source in as short a time as one-half hour. Indians sometimes arranged signals in advance, lighting a grass fire when they decamped, for example. They ignited fires to alert others to the presence of buffalo and call them together for hunting. Or, as reported by a prisoner of the Sioux in the late seventeenth century and by many others on the Plains in the centuries that followed, they lit blazes to announce the return of war parties.

Communication was evidently the predominant cause of summer burns in the central portions of the Plains. Here (and elsewhere) Indians used fire to communicate with each other almost endless possible matters including success in war, arrival of White people, sightings of buffalo, and so on. Often large fires, ignited either on ridges (perhaps diminishing the damage they might do) or on the open plain or grasslands, could and did rage out of control. The soldier and author Bacqueville de la Potherie, who was on the upper Mississippi River in the late seventeenth century, remarked that Indians lit fires in the spring and fall so that different bands knew where others were, but that any single fire became "so strong, especially when the wind rises, and when the nights are dark, that it is visible forty leagues away." Still others spoke of the loss of hundreds of acres from the most "trivial signal"; Captain John Palliser, who was on the Canadian prairies in the late 1850s, remarked that Indians "frequently fire the prairie for the most trivial reasons," especially "signals to telegraph to one another concerning a successful horse-stealing exploit, or in order to proclaim the safe return of a war party." He thought them "very careless about the consequences of such an occurrence," and spoke of "disastrous effects" including "denuding the land of all useful trees" as well as "cut[ting] off the buffalo sometimes from a whole district of country," thereby causing "great privation and distress." There are numerous similar examples. After the Flathead lit one fire in the 1830s to announce their arrival with trappers, "the flames ran over the neighboring hills with great violence, sweeping all before them, above the surface of the ground." Indians also lit fires to warn members of their own tribe of danger. Some Sioux, for example, attacked by the Ojibwa in July 1840, lit three fires to alert other Sioux nearby, and they in turn did the same until all had "set the grasslands on fire in various directions."[16]

In some parts of North America, Indians lit fires to ease travel. In the East, Europeans often remarked that Indians burned the woods to make travel through them easier. According to Thomas Morton, if Indians in seventeenth-century Massachusetts did not set fire to the forest in the spring and fall, then it "would otherwise be so overgrown with underweed, that it would be all a coppice wood, and the people would not be able in any wise to pass through the country out of a beaten path." Fires burned until extinguished by rain in this "custom

of firing the country," but despite any ill effects the fires provided "the means to make it passable"—as well as "very beautiful, and commodious." Elsewhere, John Smith spoke of being able to "gallop a horse among these woods any way except when the creeks or rivers shall hinder," and other Europeans, of driving carts and coaches through eastern woods.[17]

Some have argued that Indian burning "was almost universal." It may indeed have been the most prevalent tool employed by Indians to manipulate their environment, but there is no need for hyperbole for a practice that, while widespread, varied from one tribe to the next—as did the ecological effects from one vegetation zone to the next.[18]

Anecdotal historical evidence often suggests that while most Indians used fire, some did not; they burned some areas annually but others less frequently and a few, like California redwood forests and tundra, not at all. Moreover, even those ecosystems burned frequently, like ponderosa pine and chaparral, were not torched totally or with absolute regularity, for Europeans were occasionally stymied by the brush or undergrowth in them. Unfortunately, information on the extent, intensity, and duration of the fires is often poorly known, and while observers might note grasslands or forests burning in every direction, they do not know whether they were set alight by man or lightning.

It is clear that when lit at optimum times of the year, fires had a positive impact on the growth of grasses and animal forage, but in their pragmatism, Indians were not always concerned with how far, fast, or hot each and every fire burned. Objectives such as delivering signals, or killing, discomforting, or hindering one's enemies (the most commonly reported uses of fire in the Plains) were not always compatible with control. And accounts of campsite fires burning thousands of acres are legion. Whether this was "careless" behaviour, as many disapprovingly labelled it, depends on what, precisely, must be taken care of and in what way. All this does not mean that Indians were not ecologically or systemically aware, only that they did not always think of the ecological consequences of all the fires they lit. The fires used aggressively or to communicate were not kindled with identical considerations in mind as the fires lit to enhance the productivity of econiches.

To bring tree-ring, historical, and ethnographic evidence to bear on specific burns is seldom possible. In some regions where Indians set fires frequently, like the eastern prairies, there are few or no trees to provide corroborating evidence from fire scars. And extrapolating from current fires to ones in the past is inadvisable: today's burns, like the 1988 Yellowstone conflagration, can be exacerbated by decades or a century of fuel buildup and are poor analogs for Indian fires in forests burned with regularity. Ethnographic information must also be used with caution, in part because some Indians have clearly adopted the opinion, common until recently among forest ecologists, that fires were uniformly destructive, and that they did not set destructive fires; and in part because by the twentieth century some Indians have forgotten the details of Aboriginal burning. The latter is certainly the case with the coastal southern California Chumash, who have no memory of burning. Yet written historical sources make clear that they once burned grasslands extensively in the summer, probably following seed collecting, in order to promote new seed growth and hunt rabbits, and that their burning practices were changed first

by friars who gathered the people in mission compounds and later, in the late eighteenth century, by regulations suppressing burning. One hundred years after the regulations, the Chumash had cultural amnesia with respect to fire.[19]

For these reasons, determining the ecological consequences of fire, and the precise Indian role, is a more daunting task than unearthing the widespread anecdotal evidence for burning. Yet even where the relationship between fire and forest is relatively inaccessible, there is much about which one can speculate. The East is one such region that is difficult to approach, in large measure because at a very early date the Indians succumbed to epidemics and the Europeans altered landscapes, and the earliest written historical sources are frequently anecdotal or ambiguous. It is nevertheless clear that Indians in this region used fire to drive or surround white-tailed deer, clear the forest of underbrush, produce forage that deer found "alluring" (as one nineteenth-century writer put it), improve travel, and prepare ground for crops and trees for fuel.

Yet it is difficult to discern how often Indians actually ignited fires for subsistence and other ends. In fact there was most likely considerable variation linked to population size and density, with the greatest frequency of fires (and greatest ecological impact) found where the population was highest and the land most densely settled. Except for its northernmost portions, the East was not one solid forest but a mosaic of forest and grasslands, the latter ranging in size from small meadows to the vast 1,000 square-mile Shenandoah Valley. The grasslands especially were produced and maintained by fire. They were manipulated and managed habitats whose history could be read in their scattered charred stumps and stunted trees. Their high vegetation attracted buffalo, turkey, and many other mammals and birds. Deer browsed them and their edges. Indians planted corn in some meadows. So important was fire to the maintenance of grasslands that after Indians died from disease or abandoned them, these clearings quickly reverted to forest. Animals that had been at home there went elsewhere or their populations dwindled. One vanished entirely: the heath hen became extinct when fire was eliminated from eastern scrub oak grasslands. Eliminate fire and the change from a mixed grassland-forest habitat to forest alone could be rapid: after only two decades in Virginia, the results included trees with "very good board timber," as Robert Beverley, the early eighteenth-century historian, put it.[20]

*****

Fire (or its absence) affects plants and animals in ecosystems in predictable ways. Without fire, jack pine cones are kept closed by a resinous material and cannot release their seeds; only the heat of fire destroys the volatile resin and opens the cones for propagation. Black spruce is similar. The impact on animals is often beneficial. More often than not, fires burn unevenly and animals small and large survive in safe areas, or what are known as *refugia*. In many cases birds and mammals repopulate burn areas within one or two breeding seasons, and plants increase in vigour due to the ash. While some birds and mammals like the spruce grouse, marten, caribou, wolverine, grizzly bear, and fisher fare poorly during fires, most species notably hold their numbers or increase following the initial aftereffects of fire. Fire also predictably affects succession in ecosystems. In the absence of fire, mesquite, jumping cholla, prickly pear, and other shrubs and trees take over grasslands in the Southwest; following fire,

trembling aspen and paper birch become far more abundant in many regions.[21]

Prior to the suppression of fires in the nineteenth century, many of North America's forest and grassland ecosystems were fire-succession ecosystems; that is, fires produced and maintained them. Forest and fire ecologists appreciate the association between regular fire and ecological types and successions in ponderosa pine, chaparral, longleaf pine, and grassland habitats. Native peoples, keen observers of the environment, surely understood the associations long before. Not only were these ecosystems pyrogenic (produced by fire) but also they were anthropogenic (produced by man) to the degree that the fires that ran through them were also. Through their fires, North American Indians probably played some role in the creation of, and more certainly maintained, a number of fire-succession ecosystems.

Ponderosa pine forest, widespread in the West, is a classic example. From the seventeenth through the nineteenth centuries, either lightning or man-caused fires periodically burned forests of ponderosa pine and other conifers in the California Sierras, Sierra Madre, and northern Coast Ranges, and on the eastern slope of Oregon's Cascades. Fires eliminated dense competing understory plants and trees, kept the fire hazard low, and maintained the forest as ponderosa pine. Farther south, in Arizona's 12,000 square miles of ponderosa pine, fires produced forests that elicited mid-nineteenth-century praise for "gigantic pines, intersected frequently with open glades, sprinkled all over with mountains, meadows, and wide savannas, and covered with the richest grasses." The burns were spaced at intervals of 2 to 20 years in California, 11 to 17 years in Oregon's Cascades, and 5 to 12 years in Arizona.

But most White people misunderstood how those forests came into being. John Wesley Powell, for example, the explorer, geologist, and ethnologist, remarked from the Southwest in 1879 that forest protection "is reduced to one single problem—Can the forests be saved from fire?" He could not have been more misguided. After European settlers arrived, the grasses underneath and between the pines, which had fed deer, produced tons of hay for domesticated animals. But once sheep, increasing numbers of which belonged to Navajo herders, cropped the piney slopes, and people no longer lit fires, ponderosa pine forests changed radically. Then and now, in the absence of fire, ponderosa pine trees grow so thickly that the forest stagnates or (as has happened farther north in Oregon) Douglas fir and other shade-tolerant species quickly develop in dense clumps. In other words, ponderosa pine forests "saved" from fire are forests transformed—and because of the accumulation of surplus fuel, they become even more combustible and susceptible to explosive, destructive fire.[22]

Like ponderosa pine, chaparral, a scrub- or brushland community widespread in California, is also fire-induced. Chamise, manzanitas, oaks, and ceanothus are prominent species in chaparral. Chamise, a remarkable bush that is simultaneously flammable and fire-tolerant, dominates this ecological community. Its dead branches and resinous leaves readily catch fire, but following fire, shrubs sprout and grow rapidly on blackened chamise stumps. Chamise seeds are also very resistant to fire. They lie dormant in soil until after fire has passed, when seedlings erupt in profusion within a matter of weeks or months. Most species in chaparral produce seeds within five years after a fire, but if chaparral is burned again too soon, many

seedlings that had sprouted and been browsed can die, with an increased chance of erosion from rains.

Many Indians, it seems, actively managed chaparral with fire, for reasons that are familiar by now: to produce better forage for deer, increase yields of berries, ease the collection of seeds and bulbs, and suppress the destructive consequences of lightning-caused fires. They placed a priority especially on maintaining optimum forage for deer in the winter and spring. Most burns took place in the fall and favoured the growth of species browsed by deer in the spring, which was a difficult season for subsistence. Indians lit some spring fires evidently to reduce brushiness and induce sprouting species. Within a matter of months after a fire, mule deer, jackrabbits, quail, other birds, and mice and voles recover their numbers in burned-over, resprouted chaparral. Sometimes they become more numerous. Deer are healthier. Overall, chaparral endures as a robust ecological community only if fires persist.[23]

Like chaparral and ponderosa pine, longleaf pine forests of the Southeast are also intimately associated with fire. After germination, each longleaf pine seedling concentrates energy in its root system, which grows prodigiously, and is protected from fire by thick, resistant bark. In its next growth phase, which sometimes comes on the heels of fire, it shoots up rapidly above ground so that its needles are above the level of the next fire. In this way longleaf pines survive fire, but that is only half the story: Regular fires actually perpetuate longleaf pine forests. Fires remove competing trees and shrubs, and destroy a fungal disease that attacks needles. In the absence of fire, longleaf pines fail to reproduce or survive, and within one to seven decades, depending on the soil, drainage, and location, forests of oak, gum, beech-maple, and other pines succeed longleaf pine forests.[24]

*****

Some assert that with their fires, Indians were responsible for the formation of the vast grassland ecosystem of the Great Plains; others, that they did not form it but probably helped maintain it; and still others, that they did neither because their technology could not possibly have played such a formative role in an ecosystem so large. Whatever the influence of Indian fires, there are strong climatic and environmental reasons for doubting that fires were the only or even the major formative one. In the central and western Plains, compared to eastern portions, there is less moisture from rain and snow, lower humidity, higher winds, and more periodic drought. Singly or in combination, these conditions prevent forest formation and growth and would lead to extensive grasslands without help from fires. Yet the increased precipitation found as one moves east makes tree growth far more a reality in the eastern and northern high-grass portions of the Plains, where fire played a greater role maintaining grasslands: for centuries observers remarked on the charred stumps or extensive root systems of mature trees ravaged by regular fires, their remnants enhancing deep grassland soils. In the east and north, fires—some lightning-caused, others anthropogenic—were important in checking the natural succession of grassland by forest. When fires were checked, aspens, oaks, and willows proliferated. In the north, aspen groves expanded, and in the east, oak openings closed as groves of trees broke up the grasslands and in places, forest eventually consumed them.[25]

The effect of grassland fires depends on the same factors as elsewhere: the season of the burn, time of the last burn, heat of the fire, wind, temperature, terrain, soil, moisture, and so on. Grassland fires move with extraordinary speed when grasses are dry and the wind is up, but they also move irregularly over uneven terrain, sometimes skip over areas, and rarely consume plants so completely that their roots are burned. After the fires pass, burned areas cool quickly. Productivity often increases following grassland fires because surface litter is removed. Tall-grass prairie needs at least three years to return to its preburn state, though grazing animals like buffalo return immediately to tender young plants growing after the burn. But not all grassland fires are benign and restorative. When they are too frequent or hot, when moisture is low, or when heavy rains follow fires and cause erosion, plants may not easily recover.[26]

*****

In Oregon's Willamette Valley, as on the Great Plains, Indian fires shocked many White men. Settlers were horrified by the "long lines of fire and smoke," and by "ravages" of fire, "dense volumes" of smoke, and "sheets of flame." They were unable to make the connection between the valley's extensive grasslands and regular fires. And they rarely saw the use of fire as indicative of ecosystemic knowledge.

White settlers quickly discovered that the Kalapuya Indians torched Willamette Valley grasslands regularly each year between July and November, preferably in the fall. They burned for several reasons related to hunting white-tailed deer, their most important source of animal protein, using fire both to encircle deer and to induce them to browse in the unburned thickets or in the fresh grass that appeared in weeks. These Indians also used fire to harvest tarweed seeds, a staple they stored for winter and later consumed as meal. At maturity, tarweed is covered by a sticky resin that makes harvest exceedingly awkward. Fires both burned off this substance along with any interfering grasses, and dried the pods in which seeds would be collected. Through burning, the Kalapuya also killed grasshoppers, which they gathered up and ate with relish, and safely harvested wild honey. They also probably lit fires to clear underbrush from beneath oak trees, easing acorn collection, enhancing the growth of berry and hazel bushes, both of which grow following fire, and favouring the growth of camas, lupine, and other wild roots.

The Kalapuya maintained the Willamette Valley as an ecosystem consisting, when people of European descent arrived, of grasslands broken by "orchard"-like groves of oaks regularly spaced on hillsides. But when horrified settlers stopped the burning, which they regarded as incompatible with their settlements, they doomed the grasslands. Trees invaded, oak openings became dense woods, and Douglas firs descended from the Coast Range and from the Cascades to overrun everything before them.[27]

As with the Kalapuya, the burning practices of the Cree, Denesóline, Dene Th'a, and Dunne-za of northern Alberta lend support to the idea that Indians drew on a vast storehouse of knowledge about ecosystems and fire ecology before kindling a fire. These Indians understood that if controlled, burns promoted diversity, mosaic-like habitats, and renewal of growth. They knew that through fire they could create "edge" or "ecotonal" habitats where resources were concentrated and alluring to game. They burned at times that made sense for the production of grasses, berries, bulbs, shrubs, and other early fire-succession species attractive to small and large animals—including the moose,

bears, snowshoe hares, muskrats, beavers, lynx, and foxes they sought especially. They recognized that berry bushes would grow in burned areas and attract bears, and that moose came to roll in the ashes of a fire. They understood that fire favoured the growth of certain grasses and the mice that grazed on them, and that foxes and martens would come to prey on the mice. They knew that muskrats needed the fresh growth of reeds that fire produces, and that when aspens and poplars returned some four or five years following a burn, so also would beavers that fed on and used them in their construction projects.

In spring, these people burned sloughs and streamsides, meadows and large grasslands, and deadfall areas that were unproductive for game and represented a dangerous accumulation of fuel. They were knowledgeable of the ways that fires could escape their control, and before they burned they weighed their objectives against the season of year, time of day, slope of land, wind and other weather conditions, and the presence or absence of natural firebreaks. Sometimes, in the containing wet of spring, they left fires to smoulder in logs; sometimes, when they left their trapping territories until the next year, they ignited fires behind them.

The knowledge of these peoples was "ecological"—that is, systemic, relational, and interactional. One-half century after fire-suppression policies put an end to these and other practices, a Dunne-za woman reflected on the knowledge of those who burned, which she had known only as a child: "They must be very wise, eh? Those people? That time?"[28]

*****

Most people of European descent failed to appreciate not only the Indian ecosystemic knowledge affecting the deployment of fire but especially the myriad indigenous understandings of the natural world. As a rule, Indians animated trees and animals affected by burning and considered fire as a powerful force or being. This affected how Indians interacted with plants and animals. For example, for Indians in the Plateau and elsewhere, conifers and other trees were of symbolic importance and at times were placed or burned at the centre of ritual enclosures. Before picking the harvest from berry bushes in anticipation of burning the slopes in order to enhance production the next season, a Northwest Coast Kwakwaka'wakw woman might say to berries, as Franz Boas reported, "I have come, Supernatural Ones, you, Long-Life-Makers, that I may take you, for that is the reason why you have come, brought by your creator, that you may come and satisfy me; you, Supernatural Ones; and this, that you do not blame me for what I do to you when I set fire to you the way it is done by my root (ancestor) who set fire to you in his manner when you get old on the ground that you may bear much fruit. Look! I come now dressed with my large basket and my small basket that you may go into it, Healing Woman; you, Supernatural Ones. I mean this, that you may not be evilly disposed towards me, friends. That you may only treat me well."[29]

Of course, depicting the Kalapuya or the people of northern Alberta as ecologically aware burners, or the Kwakwaka'wakw as respectful toward the berries whose habitats they burn is not the complete story. Their ecological awareness or respect did not stop them or other people from lighting fires at times that, while convenient for itinerant burners, were inopportune from other standpoints. It did not stop them from converting plants and animals into various useful products (indeed they may have equated the health

of the environment, and their relationship with it, as contingent on such conversion). Moreover, drawing on the earlier mention of the Chumash, a note of caution must be introduced to our understanding of the burning practices of the Native peoples in northern Alberta. These people had stopped burning some 50 to 75 years before reconstructing fire use for late twentieth-century ethnographers. Had they forgotten those long-ago practices that might have had adverse results? Were memories compromised by the contemporary ideology that Indians were ecologically minded conservationists? After all, some fires lit by these people unintentionally escaped their control to ravage marten and fisher habitat. The anthropologist Henry Lewis remarked that one man "set a fire along his trapline because he happened to be there at that particular time (which wasn't the best time to do it) but he understood what would follow in the days, weeks, and months" to come. From the nineteenth and twentieth centuries comes evidence that subarctic Indians lit, but did not always extinguish, smoky fires against voracious insects when they rested during summer travels. These, together with fires lit for purposes of communication and subsistence, may account for the extensive stretches of burned ground that traders and others often encountered on their travels through the subarctic.[30]

A similar case involves the Flathead, Pend d'Oreille, and Kootenai of western Montana, whose territory embraced different ecological zones, including ponderosa pine and Douglas fir forests. Today, these Native peoples deny that their forebears set fires during summers or droughts when they might rage destructively. In arguing that the fires lit by their ancestors did not burn out of control, they sound like the Indians of northern Alberta. They emphasize that they set fires annually to protect the forest against disease and conflagration, to improve hunting through better browse for mule deer and elk, to destroy browse and force deer to places convenient for hunting, to increase the production of berries and licorice-root (a medicine), to improve grazing for horses (after 1730), to clear campsites, and to communicate.

Written historical sources confirm that these Indians burned lowland forests and grasslands where they spent most of their time, and support the idea that they ignited fires for subsistence reasons and understood their ecosystemic consequences. Documentary and fire-scar data suggest that Indians burned with enough regularity to maintain econiches in different stages of succession and ages. They kindled fairly low- or moderate-intensity fires in the spring or fall, which in theory prevented catastrophic hot "crown" fires, promoted the growth of grasses under trees, and produced an open landscape.

But the written sources also suggest, contrary to oral sources, that these Indians did not control fires systematically, that they left fires to burn themselves out, that they did not kindle fires in the same place year after year (which would have greatly damaged browse), that burning varied between and within groups, and that some people did not burn at all. Finally, here as elsewhere—and again in contrast to generally silent oral testimony—when Indians ignited fires for purposes such as communication (such as the one lit in August 1833 to announce the arrival of trappers that roared beyond control into a vast, violent conflagration), or neglected or did not extinguish campfires, these Indians sometimes produced hot summer fires with the clear potential to run away.

Most of what the Flathead, Pend d'Oreille, and Kootenai say their ancestors did is confirmed by historical and fire-scar data, but not all is. Nor does oral testimony reveal the entire story of burning in the past. In their remarks on past practices, people in general tend to emphasize the beneficial results of their actions and to elide destructive consequences. Perhaps, like the Chumash, the Flathead and others simply lack a full memory of former burning practices.[31]

*****

But the harmful consequences of fire simply cannot be ignored. The evidence that Indians lit fires that then were allowed to burn destructively and without regard to ecological consequences is abundant. In 1796, David Thompson remarked that Indians were "frequently very careless" when it came to extinguishing fires set in northern forests. These fires, he said, could take off in flammable coniferous trees and burned "until stopped by some large swamp or lake." They killed spruce grouse and other animals and left vast areas "unsightly." Thompson speculated that "this devastation is nothing to the Indian," for the simple reason that "his country is large."

Others confirmed Thompson's impressions, detailing fires that Indians ignited inadvertently, became far larger than intended, burned until rains quenched them, and consumed tens and hundreds of thousands of acres in grasslands and forest. Farther north, Thompson's contemporary Andrew Graham spoke of extensive fires lit "every summer" and "not a track of a living thing" the following winter. In eastern forests and on grasslands, many Indian-lit fires raged until extinguished by rain. Peter Fidler, for example, a Hudson's Bay Company surveyor who was on the northern Plains in the 1790s, spoke of a fire left unextinguished by the Piegan: "They did not put out their fire when they left it, which spread amongst the dry grass & ran with great velocity & burnt with very great fury, which enlightened the night like day, and appeared awfully grand." Some Piegans disapproved of that fire because it might alert hostile Indians to their presence at a season when fires that appeared were what "Inds. make accidentally." Six days later, the fire still raged "very furiously" and grasslands continued to burn for another four days.

These examples could be multiplied. Observers depicted many Indians, including the Ojibwa, Cree, Mandan, Arapahoe, Gros Ventres, Shoshone, Blackfeet, Assiniboine, and various Northern Athapaskan people, as "careless" burners, by which they meant that Indian fires accidentally blew up into vast conflagrations or burned until rain fell (which may have been what the Indians intended).[32]

At times, as in Saskatchewan in 1789 when a blaze was so hot that the following year "scarcely a blade was seen," fires destroyed plant life. When they roared into huge, hot infernos, burning several hundred thousand square miles and lasting for weeks, they "destroyed," as the trader Daniel Harmon remarked, "great numbers of buffaloes" in the late eighteenth and early nineteenth centuries. Other traders concurred. Charles McKenzie saw "whole herds of Buffaloes with their hair singed—some were blind; and half roasted carcasses strewed our way." Alexander Henry remarked on "blind buffalo ... seen every moment wandering off" following late-fall grassland fires on the northern Plains. "The poor beasts," Henry said, "have all the hair singed off; even skin in many places is shrivelled up and terribly burned, and their eyes are swollen and closed fast. It was really pitiful to see them

staggering about, sometimes running afoul of a large stone, at other times tumbling downhill and falling into creeks not yet frozen over. In one spot we found a whole herd lying dead."

Others spoke of deer, elk, buffalo, and wolves dead from fires, of herds of up to 1,000 animals killed, and of thousands of beavers immolated. Little wonder that smoke made animals nervous and sometimes drove them away. Fires also sometimes destroyed horses and other property and even occasionally torched men and women—as in Saskatchewan in 1812, a very dry year, when "dreadful" fires burned 11 Blackfeet Indians to death.[33]

*****

In recent years, the debate over Indian fire has continued in the context of discussions of the Wilderness Act of 1964. It is basically a dispute over whether or not Indian fires were a "natural" form of fire management and, if they were not, whether fires should be set in wilderness areas. Wildlife biologists, foresters, fire ecologists, and others in the Forest Service, Bureau of Land Management, National Park Service, Fish and Wildlife Service, and other agencies are involved. Beginning in the 1960s, fire-suppression policies in place for over a half-century began to change, and today forest managers burn with greater regularity than at any previous time. Their prescribed and controlled burns implicitly acknowledge the role that fires once played in ecosystems. But the Wilderness Act defines wilderness as "untrammeled by man," as "primeval" in character, and as "affected primarily by the forces of nature, with the imprint of man's work substantially unnoticed." The debate reveals the persistent gap between those who argue that Indians (unlike humans in general) lived in harmony with nature, that their fires had an impact equal to or less than lightning fires, and that they had set fires for hundreds if not thousands of years and therefore qualified as a natural force; and those who disagree, arguing in contrast that Indian fires were not all benign, that thousands of years is not significant in ecosystem evolution, and that "natural" means non-human.

The tension is between those who think that Indians were somehow non-technological or pre-technological, had no impact on the environment, and were therefore "natural," and those who disagree. It recalls the earlier day when many questioned the ability of Indians to light fires of evolutionary consequence. But wilderness, as others have emphasized, is an artifact of a time and place—the twentieth-century United States—and untrammelled wilderness "is a state of mind." By the time Europeans arrived, North America was a manipulated continent. Indians had long since altered the landscape by burning or clearing woodland for farming and fuel. Despite European images of an untouched Eden, this nature was cultural not virgin, anthropogenic not primeval, and nowhere is this more evident than in the Indian uses of fire.[34]

NOTES

1. Stephen J. Pyne's *Fire in America: A Cultural History of Wildland and Rural Fire* (Princeton: Princeton University Press, 1982), 72–83 and passim, and Michael Williams's *Americans & Their Forests: A Historical Geography* (Cambridge: Cambridge University Press, 1989), 22–49 and passim, are valuable works on the subject of fire. Stephen J. Pyne, "Indian Fires," *Natural History* 92 (February 1983): 6–11; David Pietersz. de Vries, *Voyages from Holland to America A.D. 1632–1644*, translated by Henry C. Murphy (New York, 1853), 31 ("smelt before it is seen"); Julian Harris Salomon, "Indians That Set the Woods on Fire," *Conservationist* 38 (March–April 1984): 34–39; Timothy Silver, *New Face on the Countryside: Indians, Colonists and Slaves in South Atlantic Forests, 1500–1800* (Cambridge: Cambridge University Press, 1990), 59 (in 1524 off Carolina, Verrazzano smelled "the sweet fragrance [of smoke] a hundred leagues away" and saw Indians burning); Henry Wadsworth Longfellow, *The Song of Hiawatha* (New York: Grosset and Dunlap, n.d.), 9, 41, 131, 139, 140, 194, 255, 274, 275 ("interminable ... gloomy" are his); Francis Parkman, *France and England in North America* (Boston: Little, Brown and Company, 1909), 179 ("shadows and gloom"); John Josselyn, *New England Rarities Discovered* (London: Printed for G. Widdowes, 1672), 4 ("infinite thick woods"); Gordon M. Day, "The Indian as an Ecological Factor in the Northeastern Forest," *Ecology* 34 (1953): 329–346; Roderick Nash, "Sorry, Bambi, but Man Must Enter the Forest: Perspectives on the Old Wilderness and the New," in *Proceedings—Symposium and Workshop on Wilderness Fire, Missoula, Montana, November 15–18, 1983*, General Technical Report INT-182 (Ogden: Intermountain Forest and Range Experiment Station, Forest Service, U.S. Department of Agriculture, 1985), 264–268.
2. Omer Stewart was an early proponent of a significant Indian role in burning: Omer C. Stewart, "Burning and Natural Vegetation in the United States," *Geographical Review* 41 (1951): 317–320; "Forest Fires with a Purpose," *Southwestern Lore* 20 (1954): 42–46; "Why Were the Prairies Treeless?," *Southwestern Lore* 20 (1954): 59–64; "The Forgotten Side of Ethnogeography," in *Method and Perspective in Anthropology: Papers in Honor of Wilson D. Wallis*, edited by Robert F. Spencer (Minneapolis: University of Minnesota Press, 1954), 221–311; "Forest and Grass Burning in the Mountain West," *Southwestern Lore* 21 (June 1955): 5–9; "Fire as the First Great Force Employed by Man," in *Man's Role in Changing the Face of the Earth*, edited by William L. Thomas, Jr. (Chicago: University of Chicago Press, 1956), 115–133; "Barriers to Understanding

the Influence of Use of Fire by Aborigines on Vegetation," in *Proceedings of the 2nd Annual Tall Timbers Fire Ecology Conference—March 14–15, 1963* (Tallahassee: Tall Timbers Research Station, 1963), 117–126.

3. Hu Maxwell, "The Use and Abuse of Forests by the Virginia Indians," *William and Mary Quarterly* 19 (1910): 73–103, 86, 88 ("destructive," "savage"); Day, "The Indian as an Ecological Factor"; Stewart, "Forest Fires with a Purpose," 42–43 (Indians as conservationists); Clarence K. Collins, "Indian Firefighters of the Southwest," *Journal of Forestry* 60 (February 1962): 87–91 ("the one thing ..."); Stewart, "Barriers to Understanding the Influence of Use of Fire by Aborigines on Vegetation," 119 ("small campfires"); Henry T. Lewis, *Patterns of Indian Burning in California: Ecology and Ethnohistory*, Ballena Press Anthropological Papers No. 1 (Ramona: Ballena Press, 1973); L.T. Burcham, "Fire and Chaparral before European Settlement," in *Symposium on Living with the Chaparral Proceedings*, edited by Murray Rosenthal (San Francisco: Sierra Club, 1974), 101–120, 115 ("the great master-scourge"); Henry T. Lewis, "Maskuta: The Ecology of Indian Fires in Northern Alberta," *Western Canadian Journal of Anthropology* 7, no. 1 (1977): 15–52; Pyne, *Fire in America*, 72–83, 176–180 (Smokey); Stephen J. Pyne, "Vestal Fires and Virgin Lands: A Historical Perspective on Fire and Wilderness," in *Proceedings—Symposium and Workshop on Wilderness Fire*, 254–262; Nash, "Sorry, Bambi, but Man Must Enter the Forest."

4. Early sixteenth-century southeastern Texas Indians and many subarctic Indians deployed fire to keep insects away (Stewart, "Burning and Natural Vegetation"; Harold J. Lutz, *Aboriginal Man and White Man as Historical Causes of Fires in the Boreal Forest, with Particular Reference to Alaska*, Yale University, School of Forestry Bulletin no. 65 [New Haven: Yale University, 1959], 18–20).

5. William Wood, *New England's Prospect*, edited by Alden T. Vaughan (Amherst: University of Massachusetts Press, 1993), 38; Edward Johnson, *Johnson's Wonder-Working Providence 1628–1651*, edited by J. Franklin Jameson (New York: Charles Scribner's Sons, 1910), 85; Day, "The Indian as an Ecological Factor," 335; Stewart, "Fire as the First Great Force Employed by Man," 121; Daniel Q. Thompson and Ralph H. Smith, "The Forest Primeval in the Northeast—A Great Myth," in *Proceedings of the Tall Timbers Fire Ecology Conference* 10 (1970): 255–265; Calvin Martin, "Fire and Forest Structure in the Aboriginal Eastern Forest," *Indian Historian* 6, no. 4 (1973): 38–42, 54; Conrad Taylor Moore, "Man and Fire in the Central North American Grassland 1535–1890: A Documentary Historical Geography," Ph.D. dissertation, University of California, Los Angeles, 1972, 75–76.

6. Maxwell, "The Use and Abuse of Forests," 89; Emma Helen Blair, *The Indian Tribes of the Upper Mississippi Valley and Region of the Great Lakes, as Described by Nicolas Perrot ...*, vols. I–II (Cleveland: Arthur H. Clark Company, 1911), vol. I, 120, 122; Donald J. Lehmer, "The Plains Bison Hunt—Prehistoric and Historic," *Plains Anthropologist* 8, no. 22 (1963): 211–217 (Perrot); Henry R. Schoolcraft, *A Narrative Journal of Travel through the Northwestern Regions of the United States ... in the Year 1820*, edited by Mentor L. Williams (East Lansing: Michigan State College Press, 1953), 185; Moore, *Man and Fire*, 20–21 (Schoolcraft); Lewis, *Patterns of Indian Burning in California*, 55; Emily W.B. Russell, "Indian-Set Fires in the Forests of the Northeastern United States," *Ecology* 64 (1983): 78–88.

7. Mark Catesby, *The Natural History of Carolina, Florida, and the Bahama Islands*, vols. I–II (London, 1771), vol. II, xii; Moore, *Man and Fire*, 22, 67–69; Charles F. Cooper, "Changes in Vegetation, Structure, and Growth of Southwestern Pine Forests since White Settlement," *Ecological Monographs* 30, no. 2 (1960): 129–164, 138; J.G. Nelson and R.E. England, "Some Comments on the Causes and Effects of Fire in the Northern Grasslands Area of Canada and the Nearby United States, 1750–1900," in *Proceedings of the 1977 Rangeland Management and Fire Symposium*, prepared by Connie M. Bourassa and Arthur P. Brackebusch (Missoula: University of Montana School of Forestry, 1978), 39–47.

8. Richard Irving Dodge, *The Plains of the Great West and Their Inhabitants Being a Description of the Plains, Game, Indians, &c. of the Great North American Desert* (New York: G.P. Putnam's Sons, 1877), 29 (several "sagacious" men); Moore, *Man and Fire*, 22–23; Lewis, *Patterns of Indian Burning in California*, 73.

9. Frank Bergon, ed., *The Journals of Lewis and Clark* (New York: Penguin Books, 1989), 95; [John Palliser, James Hector, and others], *The Journals, Detailed Reports, and Observations Relative to the Exploration, by Captain Palliser, of That Portion of British America ... during the Years 1857, 1858, 1859, and 1860* (London: Eyre and Spottiswoode, 1863), 54; Joel A. Allen, *The American Bisons, Living and Extinct* (New York: Arno Press, 1974 [1876]), 202; Clark Wissler, "Material Culture of the Blackfoot Indians," *Anthropological Papers of the American Museum of Natural History*, vol. V, Part 1 (New York: American Museum of Natural History, 1910), 50; Joe Ben Wheat, "The Olsen-Chubbuck Site: A Paleo-Indian Bison Kill," *American Antiquity* 37 (1972): 1–169, 93; Moore, *Man and Fire*, 24–25, 74–75, 91; David A. Dary, *The Buffalo Book: The Full Saga of the American Animal* (Chicago: The Swallow Press, [1974]), 39–40; Nelson and England, "Some Comments on the Causes and Effects of Fire"; Eleanor Verbicky-Todd, *Communal Buffalo Hunting among the Plains Indians: An Ethnographic and Historic Review* (Alberta: Archaeological Survey of Alberta, 1984), 156.

10. Lewis, *Patterns of Indian Burning in California*, 66–69.

11. Maxwell, "The Use and Abuse of Forests."

12. Stewart, "Fire as the First Great Force," 120; Stewart, "Forest and Grass Burning in the Mountain West"; Lewis, *Patterns of Indian Burning in California*, 41, 50–53, 61–62; Harold Weaver, "Effects of Fire on Temperate Forests: Western United States," in *Fire and Ecosystems*, edited by T.T. Kozlowski and C.E. Ahlgren (New York: Academic Press, 1974), 279–319; Richard White, "Indian Land Use and Environmental Change: Island County, Washington, a Case Study," *Arizona and the West* 17, no. 4 (1975): 327–338; Helen H. Norton, "The Association between Anthropogenic Prairies and Important Food Plants in Western Washington," *Northwest Anthropological Research Notes* 13 (1979): 175–200. Tarweed is *Madia sativa*.

13. Moore, *Man and Fire*, 24, 75.

14. Day, "The Indian as an Ecological Factor"; Maxwell, "The Use and Abuse of Forests"; Moore, *Man and Fire*, 56.

15. Moore, *Man and Fire*, 18 (Murray), 16–96 passim; Mavis A. Loscheider, "Use of Fire in Interethnic and Intraethnic Relations on the Northern Plains," *Western Canadian Journal of Anthropology* 7, no. 4 (1977): 82–96; Gregory Thomas, "Fire and the Fur Trade: The Saskatchewan District 1790–1840," *Beaver* 308 (Autumn 1977): 34–39.

16. [Palliser, Hector, and others], *The Journals, Detailed Reports, and Observations*, 57, 89; Blair, *The Indian Tribes of the Upper Mississippi Valley and Region*, 336ff. (Bacqueville de la Potherie); Moore, *Man and Fire*, 12–16, 53–54, 67, 70–72 (Flathead, Sioux); Loscheider, "Use of Fire in Interethnic and Intraethnic Relations," 89; George E. Gruell, "Indian Fires in the Interior West: A Widespread Influence," in *Proceedings—Symposium and Workshop on Wilderness Fire*, 68–74; Lutz, *Aboriginal Man and White Man*, 4–11.

17. John P. Dempsey, "New English Canaan by Thomas Morton of 'Merrymount': A Critical Edition," Ph.D. dissertation, Department of English, Brown University, 1998, 49–50; Maxwell, "The Use and Abuse of Forests," 88–89 (Smith).

18. Stewart, "Burning and Natural Vegetation," 319 ("almost universal"); Pyne, *Fire in America*, 72–83; Stephen F. Arno, "Ecological Effects and Management Implications of Indian Fires," in *Proceedings—Symposium and Workshop on Wilderness Fire*, 81–86; James K. Agee, *Fire Ecology of Pacific Northwest Forests* (Washington: Island Press, 1993), 53–57.

19. Moore, *Man and Fire*, 111–117; Loscheider, "Use of Fire in Interethnic and Intraethnic Relations"; Lewis, *Patterns of Indian Burning in California*, 64–65 (redwood); Burcham, "Fire and Chaparral

before European Settlement"; Jan Timbrook, John R. Johnson, and David D. Earle, "Vegetation Burning by the Chumash," *Journal of California and Great Basin Anthropology* 4, no. 2 (1982): 163–186.

20. Day, "The Indian as an Ecological Factor"; Maxwell, "The Use and Abuse of Forests," 94 (Beverley); Silas Little, "Effects of Fires on Temperate Forests: Northeastern United States," in *Fire and Ecosystems*, 225–250; Thompson and Smith, "The Forest Primeval"; Julia E. Hammett, "The Shapes of Adaptation: Historical Ecology of Anthropogenic Landscapes in the Southeastern United States," *Landscape Ecology* 7 (1992): 121–135; Russell, "Indian-Set Fires"; William A. Patterson, III and Kenneth E. Sassaman, "Indian Fires in the Prehistory of New England," in *Holocene Human Ecology in Northeastern North America*, edited by George P. Nicholas (New York: Plenum Press, 1988), 107–135 (fire and population), 115 ("alluring"—Timothy Dwight, 1823).

21. C.E. Ahlgren, "Effects of Fires on Temperate Forests: North Central United States," in *Fire and Ecosystems*, 195–223; H.G. Reynolds and J.W. Bohning, "Effects of Burning on a Desert Grass-Shrub Range in Southern Arizona," *Ecology* 37 (1956): 769–777; J.F. Bendell, "Effects of Fire on Birds and Mammals," in *Fire and Ecosystems*, 73–138.

22. E.I. Kotok, "Fire, a Major Ecological Factor in the Pine Region of California," *Fifth Pacific Science Congress, Proceedings* 5 (1933): 4017–4022; H.H. Biswell, "Man and Fire in Ponderosa Pine in the Sierra Nevada of California," *Sierra Club Bulletin* 44, no. 7 (1959): 44–53; Roehajat Emon Soeriaatmadja, "Fire History of the Ponderosa Pine Forests of the Warm Springs Indian Reservation, Oregon," thesis, Oregon State University, 1966; Harold Weaver, "Ecological Changes in the Ponderosa Pine Forest of the Warm Springs Indian Reservation in Oregon," *Journal of Forestry* 57 (January 1959): 15–20; Cooper, "Changes in Vegetation, Structure, and Growth of Southwestern Pine Forests," 129, 137 (Powell); Lewis, *Patterns of Indian Burning in California*, 71–80, 74–80; Melissa Savage, "Structural Dynamics of a Southwestern Pine Forest under Chronic Human Influence," *Annals of the Association of American Geographers* 81 (1991): 271–289.

23. Harold H. Biswell, "Effects of Fire on Chaparral," in *Fire and Ecosystems*, 321–364; Lewis, *Patterns of Indian Burning in California*, 20–31, 50–59, 82–84; William O. Wirtz, II, "Chaparral Wildlife and Fire Ecology," in *Symposium on Living with the Chaparral Proceedings*, 7–18; Burcham, "Fire and Chaparral before European Settlement."

24. Kenneth H. Garren, "Effects of Fire on Vegetation of the Southeastern United States," *Botanical Review* 9 (1943): 617–654; E.V. Komarek, "Effects of Fire on Temperate Forests and Related Ecosystems: Southeastern United States," in *Fire and Ecosystems*, 251–277.

25. W. Raymond Wood and Thomas D. Thiessen, *Early Fur Trade on the Northern Plains: Canadian Traders among the Mandan and Hidatsa Indians, 1738–1818* (Norman: University of Oklahoma Press, 1985), 230; Carl O. Sauer, "Grassland Climax, Fire, and Man," *Journal of Range Management* 3 (1950): 16–21; Stewart, "Why Were the Prairies Treeless?"; Stewart, "The Forgotten Side of Ethnogeography"; Stewart, "Fire as the First Great Force Employed by Man," 125–129; Waldo R. Wedel, "The Central North American Grassland: Man-Made or Natural?," *Social Science Monographs 3* (Washington: Pan American Union, 1957), 39–69; Moore, *Man and Fire*, 34–38, 82–86, 119–122; Richard J. Vogl, "Effects of Fire on Grasslands," in *Fire and Ecosystems*, 139–194; Carl O. Sauer, "Man's Dominance by Use of Fire," *Geoscience and Man* 10 (April 20, 1975): 1–13; George W. Arthur, *An Introduction to the Ecology of Early Historic Communal Bison Hunting among the Northern Plains Indians* (Ottawa: National Museums of Canada, 1975), 10–30; George E. Gruell, "Fire on the Early Western Landscape: An Annotated Record of Wildland Fires 1776–1900," *Northwest Science* 59, no. 2 (1985): 97–107.

26. Vogl, "Effects of Fire on Grasslands," 152–172; Gregory Thomas, "Fire and the Fur Trade"; Nelson and England, "Some Comments on the Causes and Effects of Fire"; Gruell, "Indian Fires in the Interior West"; Kenneth F. Higgins, *Interpretation and Compendium of Historical Fire*

Accounts in the Northern Great Plains, Resource Publication no. 161 (Washington: U.S. Department of the Interior, Fish and Wildlife Service, 1986).

27. Carl L. Johannessen, William A. Davenport, Artimus Millet, and Steven McWilliams, "The Vegetation of the Willamette Valley," *Annals of the Association of American Geographers* 61 (1971): 286–302; Robert Boyd, "Strategies of Indian Burning in the Willamette Valley," *Canadian Journal of Anthropology/Revue Canadienne d'Anthropologie* 5, no. 1 (Fall 1986): 65–86; Agee, *Fire Ecology of Pacific Northwest Forests*, 554–558.

28. Henry T. Lewis, "Maskuta: The Ecology of Indian Fires in Northern Alberta," *Western Canadian Journal of Anthropology* 7, no. 1 (1977): 15–52; Henry T. Lewis, "Indian Fires of Spring," *Natural History* 89 (1980): 76–78, 82, 83; Henry T. Lewis, *A Time for Burning*, Occasional Publication no. 17 (Edmonton: The Boreal Institute for Northern Studies, University of Alberta, 1982), 50 (Dunne-za quote); Henry T. Lewis, "Fire Technology and Resource Management in Aboriginal North America and Australia," in *Resource Managers: North American and Australian Hunter-Gatherers*, edited by Nancy M. Williams and Eugene S. Hunn (Boulder: Westview Press, 1982), 45–67; Henry T. Lewis, "Why Indians Burned: Specific Versus General Reasons," in *Proceedings— Symposium and Workshop on Wilderness Fire*, 75–80; Henry T. Lewis and Theresa A. Ferguson, "Yards, Corridors, and Mosaics: How to Burn a Boreal Forest," *Human Ecology* 16 (1988): 57–77; Richard White, "Indian Land Use and Environmental Change: Island County, Washington, a Case Study," *Arizona and the West* 17, no. 4 (1975): 327–338; John Dennis and Roland H. Wauer, "Role of Indian Burning in Wilderness Fire Planning," in *Proceedings—Symposium and Workshop on Wilderness Fire*, 296–298.

29. Franz Boas, *The Religion of the Kwakiutl Indians*, parts I—II. Columbia University Contributions to Anthropology, vol. X (New York: Columbia University Press, 1930), Part II, 205 ("I have come ..."); Nancy J. Turner, "'Burning Mountain Sides for Better Crops': Aboriginal Landscape Burning in British Columbia," *Archaeology in Montana* 32, no. 2 (1991): 57–73; Carling Malouf, "The Coniferous Forests and Their Uses in the Northern Rocky Mountains through 9,000 Years of Prehistory," in *Coniferous Forests of the Northern Rocky Mountains: Proceedings of the 1968 Symposium, Center for Natural Resources, September 17–20, 1968*, edited by Richard D. Taber (Missoula: University of Montana Foundation, 1969), 271–290.

30. Henry Lewis, personal communication (e-mail), March 11, 1996; Lutz, *Aboriginal Man and White Man*, 18–20; K.G. Davies, ed., *Northern Quebec and Labrador Journals and Correspondence 1819–1835*, vol. 24 (London: The Hudson's Bay Record Society, 1963), 202–204, 223.

31. Stephen W. Barrett, "Indian Fires in the Pre-settlement Forests of Western Montana," in *Proceedings of the Fire History Workshop*, October 20–24, 1980, Tucson, Arizona, technical coordinators M.A. Stokes and J.H. Dietrich; General Technical Report RM-81, Rocky Mountain Forest and Range Experiment Station (Fort Collins: Forest Service, U.S. Department of Agriculture, 1981), 35–41; Stephen W. Barrett, "Indians & Fire," *Western Wildlands* 6, no. 3 (1980): 17–21; Stephen W. Barrett, "Relationship of Indian-Caused Fires to the Ecology of Western Montana Forests," M.S. thesis, University of Montana, 1981; Stephen W. Barrett and Stephen F. Axno, "Indian Fires as an Ecological Influence in the Northern Rockies," *Journal of Forestry* 80 (1982): 647–651.

32. "Careless" must be placed in a context in which White people construed the behaviour of Indians derogatorily and in opposition to their own. J.B. Tyrrell, *David Thompson's Narrative of His Explorations in Western America, 1783–1812* (Toronto: Champlain Society, 1916), 137; Glyndwr Williams, ed., *Andrew Graham's Observations on Hudson's Bay 1767–91* (London: The Hudson's Bay Record Society, 1969), 270–271; Peter Fidler, "Journal of a Journey over Land from Buckingham House to the Rocky Mountains in 1792 & 3." Accession no. 79.269/89. Provincial Archives of Alberta, Edmonton, Alberta, 1–87, 49–54; [Palliser, Hector, and others], *The Journals, Detailed Reports, and Observations*, 57; Moore, *Man and Fire*, 26–28, 54–55; Lutz, *Aboriginal Man*

and White Man, 1–4, 13–16; Julian Harris Salomon, "Indians That Set the Woods on Fire," *Conservationist* 38 (March–April 1984): 34–39; Conrad J. Bahre, "Wildfire in Southeastern Arizona between 1859 and 1890," *Desert Plants* 7, no. 4 (1985): 190–194.

33. Moore, *Man and Fire*, 38–42 (Alexander Henry, Daniel Harmon), 61–62; Wood and Thiessen, *Early Fur Trade on the Northern Plains*, 230 (Charles McKenzie); Nelson and England, "Some Comments on the Causes and Effects of Fire"; Dean A. Shinn, "Historical Perspectives on Range Burning in the Inland Pacific Northwest," *Journal of Range Management* 33, no. 6 (November 1980): 415–423 (beavers); Thomas, "Fire and the Fur Trade"; Gruell, "Indian Fires in the Interior West"; Higgins, *Interpretation and Compendium of Historical Fire Accounts in the Northern Great Plains*.

34. Along the Little Tennessee River, many species of plants consistent with disturbed, cleared lands induced or maintained by fire appeared some 1,500 years ago: Paul A. Delcourt, Hazel R. Delcourt, Patricia A. Cridlebaugh, and Jefferson Chapman, "Holocene Ethnobotanical and Paleoecological Record of Human Impact on Vegetation in the Little Tennessee River Valley, Tennessee," *Quaternary Research* 25 (1986): 330–349; Jefferson Chapman, Hazel R. Delcourt, and Paul A. Delcourt, "Strawberry Fields, Almost Forever," *Natural History* 98 (September 1989): 51–59). On the recent debate over wilderness and related matters, see Pyne, "Vestal Fires and Virgin Lands"; Nash, "Sorry, Bambi, but Man Must Enter the Forest"; Bruce M. Kilgore, "What Is 'Natural' in Wilderness Fire Management?," in *Proceedings—Symposium and Workshop on Wilderness Fire*, 57–67; Clinton B. Phillips, "The Relevance of Past Indian Fires to Current Fire Management Programs," in ibid., 87–91; James K. Agee, "The Historical Role of Fire in Pacific Northwest Forests," in *Natural and Prescribed Fire in Pacific Northwest Forests*, edited by John D. Walstad, Steven R. Radosevich, and David V. Sandberg (Corvallis: Oregon State University Press, 1990), 25–38; Conrad Smith, "Yellowstone Media Myths: Print and Television Coverage of the 1988 Fires," in *Fire and the Environment: Ecological and Cultural Perspectives, Proceedings of an International Symposium*, Knoxville, Tennessee, March 20–24, 1990, edited by Stephen C. Nodvin and Thomas A. Waldrop (Asheville: Southeastern Forest Experiment Station, 1991), 321–327; William Cronon, ed., *Uncommon Ground: Rethinking the Human Place in Nature* (New York: W.W. Norton & Company, 1995).

Critical Thinking Questions

1. As Denevan clearly suggests in his article, the question of pre-contact human population figures for the Americas is one that has been sharply divisive in the past. Why is this, in your view, and what are the historical, political, and cultural implications of a large pre-contact population in the Americas? What are the implications for a small population?
2. What are the origins of the "pristine myth" in Denevan's view? To what extent are those origins connected with the recovery narrative described by Carolyn Merchant in the preceding section?
3. In what ways did Native peoples employ fire as an agent of anthropogenic change, according to Krech? How does this perspective link with the discussion outlined by Denevan in his article?
4. While Krech happily endorses the notion that Native peoples had intimate and deep knowledge of their environment, it is clear to him that generally speaking, Natives were not conservationist in their outlook. Modern environmentalists, however, have often pointed to Native practices as models of ecological sustainability. If Krech is right, what are the implications of his argument for the *idea* of the "ecological Indian" today?

FURTHER READING

Tim Flannery, *The Eternal Frontier: An Ecological History of North America and Its Peoples* (New York: Atlantic Monthly Press, 2001), especially Part IV.

   This is a highly readable and *longue durée* analysis of North America's environmental history, beginning some 65 *million* years ago. Flannery's gift is to place things in perspective, and his description of the rise and spread of pre-Columbian civilization in North America is very useful as context for the debates outlined in Denevan's article in this section.

Charles C. Mann, *1491: New Revelations of the Americas before Columbus* (New York: Knopf,

2005).

Mann is a science writer and journalist who has reported on the population controversy for almost 20 years. This book is a synthesis of that work and it serves as an excellent overview of recent work in palaeontology, archaeology, paleolinguistics, and anthropology that, taken together, is revolutionizing our understanding of pre-contact American history, and especially pre-contact environmental history. It is also an outstanding account of the major forms of pre-contact Native civilization in the Americas, which was clearly far more complex, diverse, and productive than hitherto believed.

Irene M. Spry, "Aboriginal Resource Use in the Nineteenth Century in the Great Plains of Modern Canada," in Kerry Abel and Jean Friesen, eds., *Aboriginal Resource Use in Canada: Historical and Legal Aspects* (Winnipeg: University of Manitoba Press, 1991), 81–92.

Spry is more willing than Krech to give Plains Natives the benefit of the doubt on the question of conservation. She accepts that wastefulness on the prairies did occur but, by and large, Natives acted in a generally conservationist manner. She also emphasizes the changing relationship of Plains Natives to their environment after the arrival of European traders, analyzing the way in which Natives swapped their traditional seasonal food-quest cycle for a fur-trapping cycle geared to supplying the requirements of external actors.

E.C. Pielou, *After the Ice Age: The Return of Life to Glaciated North America* (Chicago: University of Chicago Press, 1992), especially chapters 12–14.

Although slightly dated, Pielou's book still represents an excellent and highly accessible overview of the natural and human history of North America over the last 20,000 years or so. Of particular importance is her discussion (in Chapter 12) of the great waves of extinctions that occurred among the large mammals and which have had significant repercussions for subsequent human history on the continent.

RELEVANT WEB SITES

Head-Smashed-in Buffalo Jump
http://www.head-smashed-in.com/

The official Web site of the Head-Smashed-in Buffalo Jump UNESCO world heritage site. It provides a virtual tour of parts of the exhibit and the cliff itself. A guide to events and the interpretive centre is provided. It also gives a brief history of the Blackfoot and the buffalo hunt, and links to other sites dealing with archaeology and Native history.

A History of the Native People of Canada
http://www.civilization.ca/archeo/hnpc/npint01e.html

This is a major Web presence developed by the Canadian Museum of Civilization. It is designed to provide readers with a broad and comprehensive pre-contact history of Native peoples in what is today called Canada. In printed form it would run to approximately 1,800 pages, and so may be considered an exhaustive and exceptionally valuable resource in examining the pre-contact environmental history of Canada.

Canada's First Nations

http://www.ucalgary.ca/applied_history/tutor/firstnations/home.html

This is a multimedia tutorial created by the Applied History Research Group at the University of Calgary and Red Deer College. It focuses on the pre-contact history of Canada's Native peoples together with an examination of the impact of European contact.

Biology and Imperialism in North American Environmental History

If the previous section undermined the pristine myth, then the authors in this section administer the *coup de grace*. The first of them, Alfred Crosby, is responsible for popularizing the idea of the "Columbian Exchange" of biological material between the Old World and the New, and for demonstrating that the Old World received by far the better part of the deal. Crosby also developed the concept of the "demographic takeover," which describes the process by which European demographic dominance was established only in temperate lands that resembled Europe climatically—Canada, the United States, Uruguay, Argentina, Australia, and New Zealand. Although European dominance was exerted across large swathes of the tropical world, that was never a demographic dominance, as was the case in the temperate zones. Crosby shows that demographic takeovers occurred in lands not only where Europeans were physically comfortable and resistant to local pathogens themselves, but where their associated agricultural activities—whether animal or plant-based—were easily transplantable. He also outlines perhaps the central reason for the demographic takeover of the Americas: the astonishing pathogenicity of the micro-organisms brought by Europeans to the Native peoples there. This, more than anything else, explains the decimation of the sophisticated and, as we saw in the previous section, demographically numerous Native populations found in North and South America.

Ramsay Cook's "Making a Garden out of a Wilderness" is a specific study of colonization patterns and the social, cultural, and economic imperatives that defined and drove them. The early explorers of Acadia were interested in describing and enumerating, as precisely as possible, the natural resources of the region in which they found themselves (and here "natural resources" included, for these enumerators, the Native inhabitants). Very soon, however, this descriptive impulse was replaced by a progressive one, as Cook suggests in the title of his essay, itself derived from the writings of an early seventeenth-century Catholic priest sent to Acadia. Cook has employed Crosby's theoretical framework of biological imperialism (What can be used? What can be replaced by organisms from Europe?) to explain the European desire to improve upon wild nature as they found it. He shows in specific terms the nature of biological (and cultural, and economic) imperialism on the ground, as it were. He demonstrates the corrosive influence of the progressive ideology that governed the settlers' outlook, corrosive because of its dramatic impact on the Micmac way of life. Cook also uncovers the reaction of the Micmac to the threat represented by the European colonists, their animals, their way of life, and their micro-organisms. It is one of sorrow and recognition, of a realization that their ancient ways are thoroughly imperilled by the new situation. That he is able to demonstrate this on the basis of contemporary European accounts suggests either a blissful ignorance on the part of the settlers or, worse, a monstrous blindness.

As we have seen, disease was a key factor of the European expansion across North and South America. Measles, influenza, whooping cough, and other maladies claimed untold numbers of Native lives. But it was smallpox that was by far the greatest killer. By the nineteenth century an emphasis on hygiene and disease prevention had become a key component of European imperialist philosophy: as the century matured, public health and sanitation projects were undertaken in India, Africa, and Southeast and East Asia. Doubtless these activities were motivated in part by a sincere desire to improve the lives of the imperial subjects who suffered from illness and disease, but colder political and economic calculations were at work too. In any colonial situation disease was highly disruptive politically and socially, and an epidemic, especially a major one, could interrupt economic activities for considerable periods of time. Control of disease, either through preventative measures such as vaccination or public hygiene projects, or through responsive efforts such as quarantine, demonstrated the superiority of European medicine and, by extension, European culture and society. Disease control therefore provided evidence of European fitness to rule. This dimension of imperial history has been well explored by historians in the context of tropical and subtropical imperial efforts, but it has not been explored as thoroughly in the context of Canada. Paul Hackett shows in his study of the Hudson's Bay Company's efforts to control smallpox among the Native population that the impulse to do so was overwhelmingly economic: the Natives, with whom the Company traded, indeed upon whom the Company was almost totally reliant for trading resources, were exceptionally susceptible to smallpox and other diseases. As Hackett discovered, the efforts of the Company to prevent smallpox among their Native trading partners easily surpassed those of the U.S. federal government in both scope and consequences.

Chapter Seven
Ecological Imperialism
The Overseas Migration of Western Europeans as a Biological Phenomenon

Alfred W. Crosby

> Industrial man may in many respects be considered an
> aggressive and successful weed strangling other species and
> even the weaker members of its own.
> —Stafford Lightman, "The Responsibilities of Intervention in
> Isolated Societies," *Health and Disease in Tribal Societies*

Europeans in North America, especially those with an interest in gardening and botany, are often stricken with fits of homesickness at the sight of certain plants, which, like themselves, have somehow strayed thousands of miles westward across the Atlantic. Vladimir Nabokov, the Russian exile, had such an experience on the mountain slopes of Oregon:

> Do you recognize that clover?
> Dandelions, *l'or du pauvre?*
> (Europe, nonetheless, is over.)

A century earlier the success of European weeds in America inspired Charles Darwin to goad the American botanist Asa Gray: "Does it not hurt your Yankee pride that we thrash you so confoundly? I am sure Mrs. Gray will stick up for your own weeds. Ask her whether they are not more honest, downright good sort of weeds."[1]

The common dandelion, *l'or du pauvre*, despite its ubiquity and its bright yellow flower, is not at all the most visible of the Old World immigrants in North America. Vladimir Nabokov was a prime example of the most visible kind: the *Homo sapiens* of European origin. Europeans and their descendants, who comprise the majority of human beings in North America and in a number of other lands outside of Europe, are the most spectacularly successful overseas migrants of all time. How strange it is to find Englishmen, Germans, Frenchmen, Italians, and Spaniards comfortably ensconced in places with names like Wollongong (Australia), Rotorua (New Zealand), and Saskatoon (Canada), where obviously other peoples should dominate, as they must have at one time.

None of the major genetic groupings of humankind is as oddly distributed about the world as European, especially Western European, Whites. Almost all the peoples we call Mongoloids live in the single contiguous land mass of Asia. Black Africans are divided between three continents—their homeland and North and South America—but most of them are concentrated in their original latitudes, the tropics, facing each other across one ocean. European Whites were all recently concentrated in Europe, but in the last few centuries have burst out, as energetically as if from a burning building, and have created vast settlements of their kind in the South Temperate Zone and North Temperate Zone (excepting Asia, a continent already thoroughly and irreversibly tenanted). In Canada and the United States together they amount to nearly 90 percent of the population; in Argentina and Uruguay together to over 95 percent; in Australia to 98 percent; and in New Zealand to 90 percent. The only nations in the Temperate Zones outside of Asia that do not have enormous majorities of European Whites are Chile, with a population of two-thirds mixed Spanish and Indian stock, and South Africa, where Blacks outnumber Whites six to one. How odd that these two, so many thousands of miles from Europe, should be exceptions in *not* being predominantly pure European.[2]

Europeans have conquered Canada, the United States, Argentina, Uruguay, Australia, and New Zealand not just militarily and economically and technologically—as they did India, Nigeria, Mexico, Peru, and other tropical lands, whose Native peoples have long since expelled or interbred with and even absorbed the invaders. In the Temperate Zone lands listed above Europeans conquered and triumphed demographically. These, for the sake of convenience, we will call the Lands of the Demographic Takeover. There is a long tradition of emphasizing the contrasts between Europeans and Americans—a

tradition honoured by such names as Henry James and Frederick Jackson Turner—but the vital question is really why Americans are so European. And why the Argentinians, the Uruguayans, the Australians, and the New Zealanders are so European in the obvious genetic sense. The reasons for the relative failure of the European demographic takeover in the tropics are clear. In tropical Africa, until recently, Europeans died in droves of the fevers; in tropical America they died almost as fast of the same diseases, plus a few Native American additions. Furthermore, in neither region did European agricultural techniques, crops, and animals prosper. Europeans did try to found colonies for settlement, rather than merely exploitation, but they failed or achieved only partial success in the hot lands. The Scots left their bones as monument to their short-lived colony at Darien at the turn of the eighteenth century. The English Puritans, who skipped Massachusetts Bay Colony to go to Providence Island in the Caribbean Sea, did not even achieve a permanent settlement, much less a Commonwealth of God. The Portuguese, who went to northeastern Brazil, created viable settlements, but only by perching themselves on top of first a population of Native Indian labourers and then, when these faded away, a population of labourers imported from Africa. They did achieve a demographic takeover, but only by interbreeding with their servants. The Portuguese in Angola, who helped supply those servants, never had a breath of a chance to achieve a demographic takeover.[3] There was much to repel and little to attract the mass of Europeans to the tropics, and so they stayed home or went to the lands where life was healthier, labour more rewarding, and where White immigrants, by their very number, encouraged more immigration.

In the cooler lands, the colonies of the Demographic Takeover, Europeans achieved very rapid population growth by means of immigration, by increased lifespan, and by maintaining very high birthrates. Rarely has population expanded more rapidly than it did in the eighteenth and nineteenth centuries in these lands. It is these lands, especially the United States, that enabled Europeans and their overseas offspring to expand from something like 18 percent of the human species in 1650 to well over 30 percent in 1900. Today 670 million Europeans live in Europe, and 250 million or so other Europeans—genetically as European as any left behind in the Old World—live in the Lands of the Demographic Takeover, an ocean or so from home.[4] What the Europeans have done with unprecedented success in the past few centuries can accurately be described by a term from apiculture: they have swarmed.

They swarmed to lands that were populated at the time of European arrival by peoples as physically capable of rapid increase as the Europeans, and yet who are now small minorities in their homelands and sometimes no more than relict populations. These population explosions among colonial Europeans of the past few centuries coincided with population crashes among the Aborigines. If overseas Europeans have historically been less fatalistic and grim than their relatives in Europe, it is because they have viewed the histories of their nations very selectively. When he returned from his world voyage on the *Beagle* in the 1830s, Charles Darwin, as a biologist rather than a historian, wrote, "Wherever the European has trod, death seems to pursue the aboriginal."[5]

Any respectable theory that attempts to explain the Europeans' demographic triumphs has to provide explanations for at least two phenomena. The first is the decimation and demoralization of the Aboriginal populations of Canada, the United States, Argentina, and others. The obliterating defeat of these populations was not simply due to European technological superiority. The Europeans who settled in temperate South Africa seemingly had the same advantages as those who settled in Virginia and New South Wales, and yet how different was their fate. The Bantu-speaking peoples, who now overwhelmingly outnumber the Whites in South Africa, were superior to their American, Australian, and New Zealand counterparts in that they possessed iron weapons, but how much more inferior to a musket or a rifle is a stone-pointed spear than an iron-pointed spear? The Bantu have prospered demographically not because of their numbers at the time of first contact with Whites, which were probably not greater per square mile than those of the Indians east of the Mississippi River. Rather, the Bantu have prospered because they survived military conquest, avoided the conquerors, or became their indispensable servants—and in the long run because they reproduced faster than the Whites. In contrast, why did so few of the Natives of the Lands of the Demographic Takeover survive?

Second, we must explain the stunning, even awesome success of European agriculture, that is, the European way of manipulating the environment in the Lands of the Demographic Takeover. The difficult progress of the European frontier in the Siberian *taiga* or the Brazilian *sertão* or the South African *veldt* contrasts sharply with its easy, almost fluid advance in North America. Of course, the pioneers of North America would never have characterized their progress as easy: their lives were filled with danger, deprivation, and unremitting labour, but as a group they always succeeded in taming whatever portion of North America they wanted within a few decades and usually a good deal less time. Many individuals among them failed—they were driven mad by blizzards and dust storms, lost their crops to locusts and their flocks to cougars and wolves, or lost their scalps to understandably inhospitable Indians—but as a group they always succeeded—and in terms of human generations, very quickly.

In attempting to explain these two phenomena, let us examine four categories of organisms deeply involved in European expansion: (1) human beings; (2) animals closely associated with human beings—both the desirable animals such as horses and cattle and undesirable varmints such as rats and mice; (3) pathogens or micro-organisms that cause disease in humans; and (4) weeds. Is there a pattern in the histories of these groups that suggests an overall explanation for the phenomenon of the Demographic Takeover or that at least suggests fresh paths of inquiry?

Europe has exported something in excess of 60 million people in the past few hundred years. Great Britain alone exported over 20 million. The great mass of these White emigrants went to the United States, Argentina, Canada, Australia, Uruguay, and New Zealand. (Other areas to absorb comparable quantities of Europeans were Brazil and Russia east of the Urals. These would qualify as Lands of the Demographic Takeover except that large fractions of their populations are non-European.)[6]

In stark contrast, very few Aborigines of the Americas, Australia, or New Zealand ever went to Europe. Those who did often died not long after arrival.[7] The fact that the flow of human migration was almost entirely from Europe to her colonies and not vice versa is not startling—or

very enlightening. Europeans controlled overseas migration, and Europe needed to export, not import, labour. But this pattern of one-way migration is significant in that it reappears in other connections.

The vast expanses of forests, savannas, and steppes in the Lands of the Demographic Takeover were inundated by animals from the Old World, chiefly from Europe. Horses, cattle, sheep, goats, and pigs have for hundreds of years been among the most numerous of the quadrupeds of these lands, which were completely lacking in these species at the time of first contact with the Europeans. By 1600 enormous feral herds of horses and cattle surged over the pampas of the Rio de la Plata (today's Argentina and Uruguay) and over the plains of northern Mexico. By the beginning of the seventeenth century, packs of Old World dogs gone wild were among the predators of these herds.[8]

In the forested country of British North America population explosions among imported animals were also spectacular, but only by European standards, not by those of Spanish America. In 1700 in Virginia feral hogs, said one witness, "swarm like vermaine upon the Earth," and young gentlemen were entertaining themselves by hunting wild horses of the inland counties. In Carolina the herds of cattle were "incredible, being from one to two thousand head in one Man's Possession." In the eighteenth and early nineteenth centuries the advancing European frontier from New England to the Gulf of Mexico was preceded into Indian territory by an avant-garde of semiwild herds of hogs and cattle tended, now and again, by semi-wild herdsmen, White and Black.[9]

The first English settlers landed in Botany Bay, Australia, in January of 1788 with livestock, most of it from the Cape of Good Hope. The pigs and poultry thrived; the cattle did well enough; the sheep, the future source of the colony's good fortune, died fast. Within a few months two bulls and four cows strayed away. By 1804 the wild herds they founded numbered from 3,000 to 5,000 head and were in possession of much of the best land between the settlements and the Blue Mountains. If they had ever found their way through the mountains to the grasslands beyond, the history of Australia in the first decades of the nineteenth century might have been one dominated by cattle rather than sheep. As it is, the colonial government wanted the land the wild bulls so ferociously defended, and considered the growing practice of convicts running away to live off the herds as a threat to the whole colony, so the adult cattle were shot and salted down and the calves captured and tamed. The English settlers imported woolly sheep from Europe and sought out the interior pastures for them. The animals multiplied rapidly, and when Darwin made his visit to New South Wales in 1836, there were about a million sheep there for him to see.[10]

The arrival of Old World livestock probably affected New Zealand more radically than any other of the Lands of the Demographic Takeover. Cattle, horses, goats, pigs, and—in this land of few or no large predators—even the usually timid sheep went wild. In New Zealand herds of feral farm animals were practising the ways of their remote ancestors as late as the 1940s and no doubt still run free. Most of the sheep, though, stayed under human control, and within a decade of Great Britain's annexation of New Zealand in 1840, her new acquisition was home to a quarter million sheep. In 1974 New Zealand had over 55 million sheep, about 20 times more sheep than people.[11]

In the Lands of the Demographic Takeover the European pioneers were accompanied and often preceded by their domesticated animals, walking sources of food, leather, fibre, power, and wealth, and these animals often adapted more rapidly to the new surroundings and reproduced much more rapidly than their masters. To a certain extent, the success of Europeans as colonists was automatic as soon as they put their tough, fast, fertile, and intelligent animals ashore. The latter were sources of capital that sought out their own sustenance, improvised their own protection against the weather, fought their own battles against predators, and, if their masters were smart enough to allow calves, colts, and lambs to accumulate, could and often did show the world the amazing possibilities of compound interest.

The honey bee is the one insect of worldwide importance that human beings have domesticated, if we may use the word in a broad sense. Many species of bees and other insects produce honey, but the one that does so in greatest quantity and that is easiest to control is a native of the Mediterranean area and the Middle East, the honey bee (*Apis mellifera*). The European has probably taken this sweet and short-tempered servant to every colony he ever established, from Arctic to Antarctic Circle, and the honey bee has always been one of the first immigrants to set off on its own. Sometimes the advance of the bee frontier could be very rapid: the first hive in Tasmania swarmed 16 times in the summer of 1832.[12]

Thomas Jefferson tells us that the Indians of North America called the honey bees "English flies," and St. John de Crévecoeur, his contemporary, wrote that "[t]he Indians look upon them with an evil eye, and consider their progress into the interior of the continent as an omen of the white man's approach: thus, as they discover the bees, the news of the event, passing from mouth to mouth, spreads sadness and consternation on all sides."[13]

Domesticated creatures that travelled from the Lands of the Demographic Takeover to Europe are few. Australian Aborigines and New Zealand Maoris had a few tame dogs, unimpressive by Old World standards and unwanted by the Whites. Europe happily accepted the American Indians' turkeys and guinea pigs, but had no need for their dogs, llamas, and alpacas. Again the explanation is simple: Europeans, who controlled the passage of large animals across the oceans, had no need to reverse the process.

It is interesting and perhaps significant, though, that the exchange was just as one-sided for varmints, the small mammals whose migrations Europeans often tried to stop. None of the American or Australian or New Zealand equivalents of rats have become established in Europe, but Old World varmints, especially rats, have colonized right alongside the Europeans in the Temperate Zones. Rats of assorted sizes, some of them almost surely European immigrants, were tormenting Spanish Americans by at least the end of the sixteenth century. European rats established a beachhead in Jamestown, Virginia, as early as 1609, when they almost starved out the colonists by eating their food stores. In Buenos Aires the increase in rats kept pace with that of cattle, according to an early nineteenth-century witness. European rats proved as aggressive as the Europeans in New Zealand, where they completely replaced the local rats in the North Island as early as the 1840s. Those poor creatures are probably completely extinct today or exist only in tiny relict populations.[14]

The European rabbits are not usually thought of as varmints, but where there are neither diseases nor predators to hold down their numbers they can become the worst of pests. In 1859 a few members of the species *Orytolagus cuniculus* (the

scientific name for the protagonists of all the Peter Rabbits of literature) were released in southeast Australia. Despite massive efforts to stop them, they reproduced—true to their reputation—and spread rapidly all the way across Australia's southern half to the Indian Ocean. In 1950 the rabbit population of Australia was estimated at 500 million, and they were outcompeting the nation's most important domesticated animals, sheep, for the grasses and herbs. They have been brought under control, but only by means of artificially fomenting an epidemic of myxomatosis, a lethal American rabbit disease. The story of rabbits and myxomatosis in New Zealand is similar.[15]

Europe, in return for her varmints, has received muskrats and grey squirrels and little else from America, and nothing at all of significance from Australia or New Zealand, and we might well wonder if muskrats and squirrels really qualify as varmints.[16] As with other classes of organisms, the exchange has been a one-way street.

None of Europe's emigrants were as immediately and colossally successful as its pathogens, the micro-organisms that make human beings ill, cripple them, and kill them. Whenever and wherever Europeans crossed the oceans and settled, the pathogens they carried created prodigious epidemics of smallpox, measles, tuberculosis, influenza, and a number of other diseases. It was this factor, more than any other, that Darwin had in mind as he wrote of the Europeans' deadly tread.

The pathogens transmitted by the Europeans, unlike the Europeans themselves or most of their domesticated animals, did at least as well in the tropics as in the temperate Lands of the Demographic Takeover. Epidemics devastated Mexico, Peru, Brazil, Hawaii, and Tahiti soon after the Europeans made the first contact with Aboriginal populations. Some of these populations were able to escape demographic defeat because their initial numbers were so large that a small fraction was still sufficient to maintain occupation of, if not title to, the land, and also because the mass of Europeans were never attracted to the tropical lands, not even if they were partially vacated. In the Lands of the Demographic Takeover the Aboriginal populations were too sparse to rebound from the onslaught of disease or were inundated by European immigrants before they could recover.

The First Strike Force of the White immigrants to the Lands of the Demographic Takeover were epidemics. A few examples from scores of possible examples follow. Smallpox first arrived in the Rio de la Plata region in 1558 or 1560 and killed, according to one chronicler possibly more interested in effect than accuracy, "more than a hundred thousand Indians" of the heavy riverine population there. An epidemic of plague or typhus decimated the Indians of the New England coast immediately before the founding of Plymouth. Smallpox or something similar struck the Aborigines of Australia's Botany Bay in 1789, killed half, and rolled on into the interior. Some unidentified disease or diseases spread through the Maori tribes of the North Island of New Zealand in the 1790s, killing so many in a number of villages that the survivors were not able to bury the dead.[17]

After a series of such lethal and rapidly moving epidemics, then came the slow, unspectacular but thorough cripplers and killers like venereal disease and tuberculosis. In conjunction with the large numbers of White settlers, these diseases were enough to smother Aboriginal chances of recovery. First the blitzkrieg, then the mopping up.

The greatest of the killers in these lands was probably smallpox. The exception is New Zealand, the last of these lands to attract permanent European settlers. They came to New Zealand after the spread of vaccination in Europe, and so were poor carriers. As of the 1850s smallpox still had not come ashore, and by that time two-thirds of the Maori had been vaccinated.[18] The tardy arrival of smallpox in these islands may have much to do with the fact that the Maori today comprise a larger percentage (9 percent) of their country's population than that of any other Aboriginal people in any European colony or former European colony in either Temperate Zone, save only South Africa.

American Indians bore the full brunt of smallpox, and its mark is on their history and folklore. The Kiowa of the southern Plains of the United States have a legend in which a Kiowa man meets Smallpox on the plain, riding a horse. The man asks, "Where do you come from and what do you do and why are you here?" Smallpox answers, "I am one with the White men—they are my people as the Kiowas are yours. Sometimes I travel ahead of them and sometimes behind. But I am always their companion and you will find me in their camps and their houses." "What can you do," the Kiowa asks. "I bring death," Smallpox replies. "My breath causes children to wither like young plants in spring snow. I bring destruction. No matter how beautiful a woman is, once she has looked at me she becomes as ugly as death. And to men I bring not death alone, but the destruction of their children and the blighting of their wives. The strongest of warriors go down before me. No people who have looked on me will ever be the same."[19]

In return for the barrage of diseases that Europeans directed overseas, they received little in return. Australia and New Zealand provided no new strains of pathogens to Europe—or none that attracted attention. And of America's native diseases, none had any real influence on the Old World—with the likely exception of venereal syphilis, which almost certainly existed in the New World before 1492 and probably did not occur in its present form in the Old World.[20]

Weeds are rarely history makers, for they are not as spectacular in their effects as pathogens. But they, too, influence our lives and migrate over the world despite human wishes. As such, like varmints and germs, they are better indicators of certain realities than human beings or domesticated animals.

The term "weed" in modern botanical usage refers to any type of plant that— because of especially large numbers of seeds produced per plant, or especially effective means of distributing those seeds, or especially tough roots and rhizomes from which new plants can grow, or especially tough seeds that survive the alimentary canals of animals to be planted with their droppings—spreads rapidly and outcompetes others on disturbed, bare soil. Weeds are plants that tempt the botanist to use such anthropomorphic words as "aggressive" and "opportunistic."

Many of the most successful weeds in the well-watered regions of the Lands of the Demographic Takeover are of European or Eurasian origin. French and Dutch and English farmers brought with them to North America their worst enemies, weeds, "to exhaust the land, hinder and damnify the Crop."[21] By the last third of the seventeenth century at least 20 different types were widespread enough in New England to attract the attention of the English visitor John Josselyn, who identified couch grass, dandelion, nettles, mallowes, knot grass, shepherd's purse,

sow thistle, and clot burr and others. One of the most aggressive was plantain, which the Indians called "English-Man's Foot."[22]

European weeds rolled west with the pioneers, in some cases spreading almost explosively. As of 1823 corn chamomile and maywood had spread up to but not across the Muskingum River in Ohio. Eight years later they were over the river.[23] The most prodigiously imperialistic of the weeds in the eastern half of the United States and Canada were probably Kentucky bluegrass and white clover. They spread so fast after the entrance of Europeans into a given area that there is some suspicion that they may have been present in pre-Colombian America, although the earliest European accounts do not mention them. Probably brought to the Appalachian area by the French, these two kinds of weeds preceded the English settlers there and kept up with the movement westward until reaching the plains across the Mississippi.[24]

Old World plants set up business on their own on the Pacific Coast of North America just as soon as the Spaniards and Russians did. The climate of coastal southern California is much the same as that of the Mediterranean, and the Spaniards who came to California in the eighteenth century brought their own Mediterranean weeds with them via Mexico: wild oats, fennel, wild radishes. These plants, plus those brought in later by the Forty-niners, muscled their way to dominance in the coastal grasslands. These immigrant weeds followed Old World horses, cattle, and sheep into California's interior prairies and took over there as well.[25]

The region of Argentina and Uruguay was almost as radically altered in its flora as in its fauna by the coming of the Europeans. The ancient Indian practice, taken up immediately by the Whites, of burning off the old grass of the pampa every year, as well as the trampling and cropping to the ground of indigenous grasses and forbs by the thousands of imported quadrupeds who also changed the nature of the soil with their droppings, opened the whole countryside to European plants. In the 1780s Félix de Azara observed that the pampa, already radically altered, was changing as he watched. European weeds sprang up around every cabin, grew up along roads, and pressed into the open steppe. Today only a quarter of the plants growing wild in the pampa are native, and in the well-watered eastern portions, the "natural" ground cover consists almost entirely of Old World grasses and clovers.[26]

The invaders were not, of course, always desirable. When Darwin visited Uruguay in 1832, he found large expanses, perhaps as much as hundreds of square miles, monopolized by the immigrant wild artichoke and transformed into a prickly wilderness fit neither for man nor his animals.[27]

The onslaught of foreign and specifically European plants on Australia began abruptly in 1778 because the first expedition that sailed from Britain to Botany Bay carried some livestock and considerable quantities of seed. By May of 1803 over 200 foreign plants, most of them European, had been purposely introduced and planted in New South Wales, undoubtedly along with a number of weeds.[28] Even today so-called clean seed characteristically contains some weed seeds, and this was much more so 200 years ago. By and large, Australia's north has been too tropical and her interior too hot and dry for European weeds and grasses, but much of her southern coasts and Tasmania have been hospitable indeed to Europe's wilful flora.

Thus, many — often a majority — of the most aggressive plants in the temperate humid regions of North America, South America, Australia, and New Zealand are of European origin. It may be true that in every broad expanse of the world today where there are dense populations, with Whites in the majority, there are also dense populations of European weeds. Thirty-five of 89 weeds listed in 1953 as common in the state of New York are European. Approximately 60 percent of Canada's worst weeds are introductions from Europe. Most of New Zealand's weeds are from the same source, as are many, perhaps most, of the weeds of southern Australia's well-watered coasts. Most of the European plants that Josselyn listed as naturalized in New England in the seventeenth century are growing wild today in Argentina and Uruguay, and are among the most widespread and troublesome of all weeds in those countries.[29]

In return for this largesse of pestiferous plants, the Lands of the Demographic Takeover have provided Europe with only a few equivalents. The Canadian water weed jammed Britain's nineteenth-century waterways, North America's horseweed and burnweed have spread in Europe's empty lots, and South America's flowered galinsoga has thrived in her gardens. But the migratory flow of a whole group of organisms between Europe and the Lands of the Demographic Takeover has been almost entirely in one direction.[30] Englishman's foot still marches in seven league jackboots across every European colony of settlement, but very few American or Australian or New Zealand invaders stride the waste lands and unkempt backyards of Europe.

European and Old World human beings, domesticated animals, varmints, pathogens, and weeds all accomplished demographic takeovers of their own in the temperate, well-watered regions of North and South America, Australia, and New Zealand.

They crossed oceans and Europeanized vast territories, often in informal co-operation with each other—the farmer and his animals destroying native plant cover, making way for imported grasses and forbs, many of which proved more nourishing to domesticated animals than the native equivalents; Old World pathogens, sometimes carried by Old World varmints, wiping out vast numbers of Aborigines, opening the way for the advance of the European frontier, exposing more and more Native peoples to more and more pathogens. The classic example of symbiosis between European colonists, their animals, and plants comes from New Zealand. Red clover, a good forage for sheep, could not seed itself and did not spread without being annually sown until the Europeans imported the bumblebee. Then the plant and insect spread widely, the first providing the second with food, the second carrying pollen from blossom to blossom for the first, and the sheep eating the clover and compensating the human beings for their effort with mutton and wool.[31]

There have been few such stories of the success in Europe of organisms from the Lands of the Demographic Takeover, despite the obvious fact that for every ship that went from Europe to those lands, another travelled in the opposite direction.

The demographic triumph of Europeans in the temperate colonies is one part of a biological and ecological takeover that could not have been accomplished by human beings alone, gunpowder notwithstanding. We must at least try to analyze the impact and success of all the immigrant organisms together—the European portmanteau of often mutually supportive plants, animals, and microlife, which in its entirety can be accurately described as aggressive and opportunistic, an ecosystem simplified by ocean crossings and honed by thousands of years of competition in the unique environment created by the Old World Neolithic Revolution.

The human invaders and their descendants have consulted their egos, rather than ecologists, for explanations of their triumphs. But the human victims, the Aborigines of the Lands of the Demographic Takeover, knew better, knew they were only one of many species being displaced and replaced; knew they were victims of something more irresistible and awesome than the spread of capitalism or Christianity. One Maori, at the nadir of the history of his race, knew these things when he said, "As the clover killed off the fern, and the European dog the Maori dog—as the Maori rat was destroyed by the Pakeha (European) rat—so our people, also, will be gradually supplanted and exterminated by the Europeans."[32] The future was not quite so grim as he prophesied, but we must admire his grasp of the complexity and magnitude of the threat looming over his people and over the ecosystem of which they were part.

NOTES

1. Page Stegner, ed., *The Portable Nabokov* (New York: Viking, 1968), 527; Francis Darwin, ed., *Life and Letters of Charles Darwin* (London: Murray, 1887), vol. 2, 391.
2. *The World Almanac and Book of Facts 1978* (New York: Newspaper Enterprise Association, 1978), passim.
3. Philip D. Curtin, "Epidemiology and the Slave Trade," *Political Science Quarterly* 83 (June 1968): 190–216 passim; John Prebble, *The Darien Disaster* (New York: Holt, Rinehart & Winston, 1968), 296, 300; Charles M. Andrews, *The Colonial Period of American History* (New Haven: Yale University Press, 1934), vol. 1, n. 497; Gilberto Freyre, *The Masters and*

the Slaves, translated by Samuel Putnam (New York: Knopf, 1946), passim; Donald L. Wiedner, *A History of Africa South of the Sahara* (New York: Vintage Books, 1964), 49–51; Stuart B. Schwartz, "Indian Labor and New World Plantations: European Demands and Indian Responses in Northeastern Brazil," *American Historical Review* 83 (February 1978): 43–79 passim.

4. Marcel R. Reinhard, *Histoire de la population modiale de 1700 à 1948* (n.p.: Editions Domat-Montchrestien, n.d.), 339–411, 428–431; G.F. McCleary, *Peopling the British Commonwealth* (London: Farber and Farber, n.d.), 83, 94, 109–110; R.R. Palmer and Joel Colton, *A History of the Modern World* (New York: Knopf, 1965), 560; *World Almanac 1978*, 34, 439, 497, 513, 590.

5. Charles Darwin, *The Voyage of the Beagle* (Garden City: Doubleday Anchor Books, 1962), 433–434.

6. William Woodruff, *Impact of Western Man* (New York: St. Martin's, 1967), 106–108.

7. Carolyn T. Foreman, *Indians Abroad* (Norman: University of Oklahoma Press, 1943), passim.

8. Alfred W. Crosby, *The Columbian Exchange* (Westport: Greenwood, 1972), 82–88; Alexander Gillespie, *Gleanings and Remarks Collected during Many Months of Residence at Buenos Aires* (Leeds: B. DeWhirst, 1818), 136; Oscar Schmieder, "Alteration of the Argentine Pampa in the Colonial Period," *University of California Publications in Geography* 2 (27 September 1927): n. 311.

9. Robert Beverley, *The History and Present State of Virginia* (Chapel Hill: University of North Carolina Press, 1947), 153, 312, 318; John Lawson, *A New Voyage to Carolina* (n.p.: Readex Microprint Corp., 1966), 4; Frank L. Owsley, "The Pattern of Migration and Settlement of the Southern Frontier," *Journal of Southern History* 11 (May 1945): 147–175.

10. Commonwealth of Australia, *Historical Records of Australia* (Sydney: Library Committee of the Commonwealth Parliament, 1914), ser. 1, vol. 1, 550; vol. 7, 379–380; vol. 8, 150–151; vol. 9, 349, 714, 831; vol. 20, 92, 280, 682; vol. 20, 839.

11. Andrew H. Clark, *The Invasion of New Zealand by People, Plants, and Animals* (New Brunswick: Rutgers University Press, 1949), 190; David Wallechinsky, Irving Wallace, and A. Wallace, *The Book of Lists* (New York: Bantam, 1978), 129–130.

12. Remy Chauvin, *Traité de biologie de Vabeille* (Paris: Masson et Cie, 1968), vol. 1, 38–39; James Backhouse, *A Narrative of a Visit to the Australian Colonies* (London: Hamilton, Adams and Co., 1834), 23.

13. Merrill D. Peterson, ed., *The Portable Thomas Jefferson* (New York: Viking, 1975), 111; Michel-Guillaume St. Jean de Crévecoeur, *Journey into Northern Pennsylvania and the State of New York*, translated by Clarissa S. Bostelmann (Ann Arbor: University of Michigan Press, 1964), 166.

14. Bernabé Cobo, *Obras* (Madrid: Atlas Ediciones, 1964), vol. 1, 350–351; Edward Arber, ed., *Travels and Works of Captain John Smith* (New York: Burt Franklin, n. d.), vol. 2, xcv; K.A. Wodzicki, *Introduced Mammals of New Zealand* (Wellington: Department of Scientific and Industrial Research, 1950), 89–92.

15. Frank Fenner and F.N. Ratcliffe, *Myxomatosis* (Cambridge: Cambridge University Press, 1965), 9, 11, 17, 22–23; Frank Fenner, "The Rabbit Plague," *Scientific American* 190 (February 1954): 30–35; Wodzicki, *Introduced Mammals*, 107–141.

16. Charles S. Elton, *The Ecology of Invasions* (Trowbridge and London: English Language Book Society, 1972), 24–25, 28, 73, 123.

17. Juan López de Velasco, *Geografia y descripción universal de las Indias* (Madrid: Establecimiento Topografico de Fortanet, 1894), 552; Oscar Schmieder, "The Pampa: A Natural and Culturally Induced Grassland?" *University of California, Publications in*

Geography (September 27, 1927): 266; Sherburne F. Cook, "The Significance of Disease in the Extinction of the New England Indians," *Human Biology* 14 (September 1975): 486–491; J.H.L. Cumpston, *The History of Smallpox in Australia, 1788–1908* (Melbourne: Albert J. Mullet, Government Printer, 1914), 147–149; Harrison M. Wright, *New Zealand, 1769–1840* (Cambridge: Harvard University Press, 1959), 62. For further discussion of this topic, see Crosby, *Columbian Exchange*, chapters 1 and 2, and Henry F. Dobyns, *Native American Historical Demography: A Critical Bibliography* (Bloomington: Indiana University Press/Newberry Library, 1976).

18. Arthur C. Thomson, *The Story of New Zealand* (London: Murray, 1859), vol. 1, 212.

19. Alice Marriott and Carol K. Rachlin, *American Indian Mythology* (New York: New American Library, 1968), 174–175.

20. Crosby, *Columbian Exchange*, 122–164, passim.

21. Jared Eliot, "The Tilling of the Land, 1760," in *Agriculture in the United States: A Documentary History*, edited by Wayne D. Rasmussen (New York: Random House, 1975), vol. 1, 192.

22. John Josselyn, *New Englands Rarities Discovered* (London: G. Widdowes at the Green Dragon in St. Paul's Churchyard, 1672), 85, 86; Edmund Berkeley and Dorothy S. Berkeley, eds., *The Reverend John Clayton* (Charlottesville: University of Virginia Press, 1965), 24.

23. Lewis D. de Schweinitz, "Remarks on the Plants of Europe Which Have Become Naturalized in a More or Less Degree, in the United States," *Annals Lyceum of Natural History of New York*, vol. 3 *(1832) 1828–1836*, 155.

24. Lyman Carrier and Katherine S. Bort, "The History of Kentucky Bluegrass and White Clover in the United States," *Journal of the American Society of Agronomy* 8 (1916): 256–266; Robert W. Schery, "The Migration of a Plant: Kentucky Bluegrass Followed Settlers to the New World," *Natural History* 74 (December 1965): 43–44; G.W. Dunbar, ed., "Henry Clay on Kentucky Bluegrass," *Agricultural History* 51 (July 1977): 522.

25. Edgar Anderson, *Plants, Man, and Life* (Berkeley and Los Angeles: University of California Press, 1967), 12–15; Elna S. Bakker, *An Island Called California* (Berkeley and Los Angeles: University of California Press, 1971), 150–152; R.W. Allard, "Genetic Systems Associated with Colonizing Ability in Predominantly Self-Pollinated Species," in *The Genetics of Colonizing Species*, edited by H.G. Baker and G. Ledyard Stebbins (New York: Academic Press, 1965), 50; M.W. Talbot, H.M. Biswell, and A.L. Hormay, "Fluctuations in the Annual Vegetation of California," *Ecology* 20 (July 1939): 396–397.

26. Félix de Azara, *Descripción é historia del Paraguay y del Rio de la Plata* (Madrid: Imprenta de Sanchez, 1847), vol. 1, 57–58; Schmieder, "Alteration of the Argentine Pampa," 310–311.

27. Darwin, *Voyage of the Beagle*, 119–120.

28. *Historical Records of Australia*, set. 1, vol. 4, 234–241.

29. Edward Salisbury, *Weeds and Aliens* (London: Collins, 1961), 87; Angel Julio Cabrera, *Manual de la flora de los alrededores de Buenos Aires* (Buenos Aires: Editorial Acme S.A., 1953), passim.

30. Elton, *Ecology of Invasions*, 115; Hugo Ilitis, "The Story of Wild Garlic," *Scientific Monthly* 68 (February 1949): 122–124.

31. Otto E. Plath, *Bumblebees and Their Ways* (New York: Macmillan, 1934), 115.

32. James Bonwick, *The Last of the Tasmanians* (New York: Johnson Reprint Co., 1970), 380.

Chapter Eight
Making a Garden out of a Wilderness

Ramsay Cook

"The Land is a Garden of Eden before them, and behind them a
desolate wilderness."
—Joel 2:3

"Is not America," Fernand Braudel asks in *The Perspective of the World*, the third volume of his magisterial *Civilization and Capitalism*, "perhaps the true explanation of Europe's greatness? Did not Europe discover or indeed 'invent' America, and has Europe not always celebrated Columbus's voyages as the greatest event in history 'since the creation'?" And then the great French historian concludes that "America was ... the achievement by which Europe most truly revealed her own nature."[1] Braudel wrote without any intended irony, but that final remark about Europe truly revealing "her own nature" is perhaps what is really at issue in the current reassessment of the implications of Columbus's landfall at Guanahari, or San Salvador, as he named it in his first act of semiotic imperialism. ("Each received a new name from me,"[2] he recorded.)

That "nature" is captured by Marc Lescarbot in a sentence from his remarkable *History of New France*, published in 1609,

which goes right to the heart of what we have apparently decided to call the "encounter" between the Old World and the New that Columbus symbolizes. Acadia, he wrote, "having two kinds of soil that God has given unto man as his possession, who can doubt that when it shall be cultivated it will be a land of promise?"[3] I hardly need to explain why I think that sentence is so revealing of the European "nature," but I will. It forthrightly articulates the Renaissance European's conviction that man was chosen by the Creator to possess and dominate the rest of creation. And it further assumes that, for the land to be fully possessed, it must be cultivated: tilled, improved, developed. The result: a promised land, a paradise, a garden of delights. Lescarbot's observations seemed so axiomatic then, and for nearly five centuries afterward, that almost no one questioned his vision of a promised land—at least almost no European. But that has begun to change. Contemporary Europe, as much in its

Western as in its Eastern portions, struggles to redefine itself. Consequently Europe overseas, as J.G.A. Pocock recently argued in a brilliant essay,[4] is being forced to look again at the meaning of what Gomara in his *General History of the Indies* (1552) called "the greatest event since the creation of the world"[5]—the meaning of the "discovery of America."

The general shape of that reassessment has been emerging for more than a decade. Indeed, the seminal work, Alfred W. Crosby's powerful study, *The Columbian Exchange: Biological and Cultural Consequences of 1492*, is two decades old. In that book Crosby argued that most of the histories of European expansion had missed the real point. It was not principalities and powers but rather organisms, seeds, and animals that wrought the most fundamental changes in post-Columbian America. "Pandemic disease and biological revolution," not European technology and Christian culture, allowed Europeans "to transform as much of the New World as possible into the Old World."[6]

If Crosby offered a startling new explanation for the ease with which Europeans conquered the Americas, it remained for a Bulgarian cultural critic, living in Paris, to dissect the language and ideology of Columbian imperialism. In his brilliant if sometimes infuriatingly undocumented *The Conquest of America*, Tzvetan Todorov argued that sixteenth-century Europeans adopted two equally destructive attitudes toward the inhabitants of the New World. On the one hand, Amerindians were viewed as "savages," radically different and inferior to Europeans. Consequently, they could be enslaved. On the other hand, the Native peoples were seen "not only as equal but as identical." Consequently, they could be assimilated. Whether "noble savage" or

just "savage," Amerindians were never accepted on their own terms—different but equal. "Difference," Todorov wrote, "is corrupted into inequality, equality into identity."[7] *The Conquest of America* is, in essence, a subtle questioning of Eurocentricity, an assertion that the conventional story of 1492 had for too long been a monologue in which only European voices and values had been heard. Together, Crosby and Todorov—not just them, though they have been essential—have argued that to understand the coming together of Europe and America the ecological and intellectual worlds of both sides of the encounter must be brought into dialogue. If Europe discovered America in the centuries following 1492, it is equally true that America discovered Europe—and each revealed its "own nature."

What began as a trickle with Crosby and Todorov has since become a near flood. Its most extreme and popularized form is found in Kirkpatrick Sale's recent *Conquest of Paradise*, a book that has received far more attention, even from serious reviewers, than it warranted. Sale is not just critical of the Admiral of the Ocean Sea, though he certainly is that, but his principal point is an indictment of "the essential unsuitability of European culture for the task on which it was embarking."[8] In Sale's view, pre-Columbian America was a continent whose people lived in such harmony with each other, and with nature, as to approximate paradise. The European quest for gold, inspired by greed and God, destroyed that Edenic life. Exaggerated as his claims are, and driven as they are by a kind of moral certainty unbecoming to an historian, Sale's book nevertheless does raise some matters that will increasingly be part of the historian's agenda in examining the early period of the European entry into the Americas.

And that brings me back to Marc Lescarbot and the issues that might be considered in an ecological approach to the early history of Acadia. The history of the environment and the people who lived in pre-contact and especially proto-historical Acadia is not a new subject. Though his name is rarely mentioned in books about Canadian historical writing, the pioneer in environmental history in the Maritime region was an extraordinary scholar named William Francis Ganong. A graduate of the University of New Brunswick, Harvard, and Munich, Ganong taught natural science—botany was his specialty—at Smith College, Massachusetts, throughout his scholarly life. But he devoted most of his research to the natural history, geography, and general history of Acadia. His maps of early explorations, his editions of Nicolas Denys's *Description and Natural History of Acadia* and Chrestien Le Clercq's *New Relation of Gaspesia*, and his hundreds of articles in a variety of journals represent a contribution to early Canadian history that has yet to be properly recognized—at least by historians. When I set out on my current research work on the natural and anthropological history of early Canada, one of my ambitions was to compile an historical bird watcher's guide—a chronology of the discovery of the birds of Canada. I quickly found that Ganong had been there well before me: in 1909 he published in the *Transactions of the Royal Society of Canada* "The Identity of Plants and Animals mentioned by the Early Voyages to Eastern Canada and Newfoundland."

The second Maritime scholar for whom the environment was a necessary component of history was Alfred G. Bailey. His work is more generally known because of the reissue in 1969 of his seminal *Conflict of European and Eastern Algonkian Cultures 1304–1700*, which had first appeared in a small edition in 1937 and then dropped from sight. Professor Bruce Trigger has argued persuasively that Bailey was the first practitioner of what has come to be known as ethnohistory, the synthesis of historical and anthropological techniques. It is also true that Bailey had a profound sense of that symbiosis of environment and culture that made North American societies what they were, and an understanding that, when European social, religious, and economic practices altered that environment, Native culture could hardly remain unchanged.

Finally, there is the well-known work of the historical geographer Andrew Hill Clark. His *Acadia*, published in 1968, is a model of environmental history, though his emphasis is upon the impact of the newcomers on "the face of the earth" and the modification of European culture in the face of the demands of a new environment. Obviously, then, the foundations of Acadian ecological history have been well laid. A new approach, based on a wider conception of ecology, is William Cronon's *Changes in the Land Indians: Colonists and the Ecology of New England*. This excellent book has, I think, demonstrated that the early history of North America can be profitably recast. Cronon looks at the manner in which the interaction between Natives and newcomers in New England led to alterations in the landscape, the introduction of new crops and diseases, the reduction of animal populations, the clearing of the forest, and the establishment of a "world of fields and fences" legally enforced by a system that established rights of private ownership. Moreover, Cronon makes the crucial point that in this encounter the contrast was not, as Europeans usually argued, between wilderness and civilization, but rather "between two human ways of living, two ways of belonging to an ecology."[9] That this approach might be adapted to the history of Acadia is immediately obvious, especially since Cronon himself often resorted to the evidence of those conscientious French record-keepers whose works are so familiar to historians of the French empire in North America.

The process whereby a way of life—the European one—triumphed over the Amerindian one was fairly rapid in New England, being virtually completed by 1800, as Cronon demonstrates. Since the colonization of Acadia, a vast area that included present-day Nova Scotia, New Brunswick, Prince Edward Island, part of Quebec (Gaspesia), and a portion of Maine, advanced at a slower pace, the two ways of life existed side by side, interacting with one another, for a longer period of time. Leslie Upton's study, *Micmacs and Colonists*, noted that at the beginning of his period, the Treaty of Utrecht of 1713, the number of Micmac people of Acadia still roughly equalled European settlers even though the Native population had declined drastically. In the seventeenth century Acadia had been the scene of imperial competition and war between France and Britain and that had hardly been conducive to extensive settlement. By 1650 there were some 50 households at Port Royal and Le Have; that population had grown to about 900 souls by the 1680s. So, too, population was spreading from Port Royal to the Minas Basin, to Beaubassin and scattered along the north shore of the Bay of Fundy. "When the British took control for the third and last time," Upton writes, "capturing Port Royal in 1710, there were just over 1,500 native born Acadians with roots going back from two to four generations. The Micmac population stood at about the same number having

declined from 3,000 or so at the beginning of the seventeenth century. In one hundred years the French had been able to establish a white population only half the size of the Micmacs at their first arrival."[10] It is the very slowness of the process, and the richness of the documentation for the seventeenth century, that makes the study of the ecology of contact in Acadia so fascinating.

Between 1604 and 1708 six major writers—of varied background—composed accounts of what Nicolas Denys called "the natural history" of Acadia (and this leaves out Jacques Cartier's sixteenth-century account). Now I think it is highly significant that Denys believed that "natural history" included not just geography, geology, climate, and the flora and fauna, but also the inhabitants of the new land. The model, of course, was Pliny's *Natural History*, a work that played such a large part in determining what was discovered in America that the Roman almost deserves to be ranked with Columbus.[11] Nor was Denys in any way unique: Champlain, Lescarbot, Biard, Le Clercq, and Dièreville, the authors of the other five major contemporary accounts of seventeenth-century Acadia, all followed a similar recipe, though varying the amounts of the ingredients somewhat. Each of these writers approached Acadia from an ecological perspective, setting people squarely in their environment and noting the contrast between European and Amerindian ways of living in, and belonging to, their environment. Each of these seventeenth-century writers devoted a substantial portion of his book to descriptions of the natural world. There was a certain awesomeness about this wilderness and its abundance of birds, beasts, fish, and flowers. In 1604, at Seal Island, Champlain's keen eye identified firs, pines, larches, and poplars, and more than a dozen species of birds, not all of which he recognized. At nearly every landing he made similar observations.[12] He identified the skimmer with its extended lower bill, while Denys described the majestic bald eagle carrying off a rabbit in its talons. Marc Lescarbot provided comparable descriptions. His famous poem "Adieu à la Nouvelle France" is a versified catalogue of the environment, including that indigenous marvel, "*un oiselet semblable au papillon*," the hummingbird. This ornithological marvel, sometimes called the Bird of Heaven, provoked the same sense of admiration in early visitors to North America that the flamboyant parrot produced in explorers of South America, beginning with Columbus. The parrot, captured and transported, quickly became the symbol of America in post-Columbian art. The hummingbird, impossible to rear in captivity, remained in the Americas. Among insects the firefly was highly appealing, the mosquito detested.[13]

Of the creatures of the animal world none attracted more curiosity than the beaver. It became an animal more marvellous in the Plinean imaginations of Le Clercq and Denys than in reality. It performed as architect, mason, carpenter, even hod carrier, walking upright with a load of mud piled on its broad tail. A natural rear end loader! "Its flesh is delicate," Father Le Clercq reported, "and very much like that of mutton. The kidneys are sought by apothecaries, and are used with effect in easing women in childbirth, and in mitigating hysterics."[14] The value of beaver pelts hardly needed comment.

Then there was the moose that cured itself of epilepsy by scratching its ear with its own cloven hoof. At least the flying squirrel was the real thing. Denys's description of the cod fishery, not surprisingly since he was a merchant, was accurate and detailed. On the southeast coast of St. Mary's Bay, when the tide dropped, Champlain found mussels, clams, and sea snails, while elsewhere oysters abounded. The swarms of fish that swam in the waterways filled these men with excitement. On the Miramichi, Denys claimed that he had been kept awake all night by the loud sounds of salmon splashing. The fertility and agricultural potential of the new land were naturally a constant preoccupation. "The entire country is covered with very dense forests," Champlain wrote of the site that would become Annapolis Royal, "... except a point a league and a half up the river, where there are some oaks which are very scattered and a number of wild vines. These could be easily cleared and the place brought under cultivation." Nor did Champlain miss the minerals and metals—silver at Mink Cove, iron further north on Digby Neck.[15]

Moreover, it is from these writers that at least a partial sketch of the lives and customs of the Northeastern Algonquian peoples can be reconstructed. Champlain recorded the practice of swidden agriculture among the Abenaki—corn, beans, and squash—and even noted the use of horseshoe crab shells, probably as fertilizer.[16] Hunting and fishing methods were remarked upon, though the lack of detail is somewhat surprising, particularly when contrasted with the lengthy accounts of religious beliefs—or supposed lack of them—of the various inhabitants of Acadia. "Jugglery," or shamanism, and also medical practices were of particular interest—indeed, Dièreville even convinced himself of the efficacy of some shamanistic cures. So, too, dress, hairstyles, courtship and marriage customs, and ceremonies surrounding childbirth and death were carefully recorded and sometimes compared to classical and contemporary European practices. Lescarbot, for example, concluded that the Jesuits were quite mistaken in attempting to force Christian monogamy on the Micmacs, arguing that indigenous marriage customs would best be "left in the state in which they were found." In contrast to some Jesuit writers—and Brian Moore—Lescarbot judged the Aboriginal peoples very modest in sexual matters. This he attributed partly to their familiarity with nakedness, but chiefly "to their keeping bare the head, where lies the fountain of the spirits which excite to procreation, partly to the lack of salt, of hot spices, of wine, of meats which provoke desire, and partly to their frequent use of tobacco, the smoke of which dulls the senses, and mounting up to the brain hinders the functions of Venus." On the other hand, he believed that one romantic innovation introduced by the French actually contributed to the improvement of Aboriginal life: the kiss. Though Professor Karen Anderson has followed up Lescarbot's insight about the impact of the missionaries on marriage among the Native peoples of New France, no one, as far as I know, has advanced our knowledge of the relationship between civilization and osculation.[17]

Virtually all male European visitors to Acadia were struck by the division of labour in Aboriginal communities. Women, it was agreed, "work harder than the men, who play the gentleman, and care only for hunting or for war." Despite this, Lescarbot wrote approvingly, "they love their husbands more than women of our parts." It is interesting that in his discussion of the ease with which Micmac marriages could be dissolved, Father Le Clercq remained detached and uncensorious. "In a word," he remarked laconically, "they hold it as a maxim that each one is free; that one can do whatever he wishes: and that it is not sensible to put constraint upon men." And the priest understood that the maxim applied to both men and women.[18]

Games and the Native peoples' apparent penchant for gambling were described, though not always understood. Then there were science and technology. Father Le Clercq, perhaps the most ethnologically astute of seventeenth-century observers, provided an intriguing account of the ways the Gaspesians read the natural world: their interpretation of the stars and the winds, how they reckoned distance and recognized the changing seasons. The usefulness and limitations of indigenous technology were also commented upon. The efficiency of the birchbark canoe won widespread admiration. "The Savages of Port Royal can go to Kebec in ten or twelve days by means of the rivers which they navigate almost up to their sources," Lescarbot discovered, "and thence carrying their little bark canoes through the woods they reach another stream which flows into the river of Canada and thus greatly expedite their long voyages." While household utensils, manufactured from bark, roots, and stumps, were ingenious, the French realized that the Aboriginals were happy to replace them with metal wares. War and its weaponry drew the somewhat surprised comment that "neither profit nor the desire to extend boundaries, but rather vengeance, caused fairly frequent hostilities between native groups." Torture was graphically described, and condemned, though it was recognized— and judged a sign of savagery—that "to die in this manner is, among the savages, to die as a great captain and as a man of great courage."[19] Much else also caught the attention of these ethnologists: the commonality of property, the importance of gift exchange, the practice of setting aside weapons before entering into discussions with strangers, and the expectation that strangers should do the same. And even though the Natives were "crafty, thievish and treacherous," Lescarbot admitted somewhat superciliously that "they do not lack wit, and might come to something if they were civilized, and knew the various trades."[20]

Though observations and judgments were made with great confidence, indeed, often rather cavalierly, these Europeans were aware that there often existed an unbridgeable communications chasm between the observers and the observed. Like every explorer before them, the French in Acadia attempted to resolve the problem in two ways. The first was to take young Natives back to France for an immersion course in French. ("We had on board a savage," Lescarbot noted in 1608, "who was much astonished to see the buildings, spires and windmills of France, but more the women, whom he had never seen dressed after our manner.") While these interpreters were doubtless helpful in breaking down the "effects of the confusion of Babel," it was hardly a permanent resolution to what Father Biard realized was a fundamental problem. Yet for the French to learn the local languages was time-consuming and the results often frustrating. Learning words was not the same as learning to communicate. "As these Savages have no formulated Religion, government, towns, nor trade," Biard recorded in exasperation, "so the words and proper phrases for all of these things are lacking." The confusion of words with things, of the sign with the referent, was, as Todorov has brilliantly shown, endemic to the European attempt to comprehend America. Acadia was no exception, though I have, unfortunately,

not found any example quite so delicious as the linguistic dilemma encountered by Protestant missionaries in Hawaii. There the Islanders reportedly practised some 20 forms of sexual activity judged illicit—perhaps better, non-missionary. Each had a separate name in the Native language, thus making translation of the Seventh Commandment virtually impossible without condoning the other 19 forms of the joy of sex! The Native peoples of Acadia were apparently much less resourceful—or the celibate Jesuits less well trained as participant-observers.[21]

The natural, ethnographic, and linguistic accounts were not, of course, the work of biological scientists or cultural anthropologists—even taking into account our contemporary skepticism about the objectivity of anthropologists. Rather, they were the observations of seventeenth-century Frenchmen taking inventory of a new land they intended to explore, settle, develop, and Christianize—in brief, to colonize. It is in their works that much of what Braudel called Europe's "own nature" is "most truly revealed." In differing ways it is made emphatically plain by each of these authors that the French objective in Acadia, in the words of Father Biard, was "to make a Garden out of the wilderness." Nor should this be read narrowly as simply meaning the evangelization of the people who lived in Acadia.[22]

In the revealing introduction to his rich and thoughtful *Relation* of 1616, Biard wrote: "For verily all of this region, though capable of the same prosperity as ours, nevertheless through Satan's malevolence, which reigns there, is a horrible wilderness, scarcely less miserable on account of the scarcity of bodily comforts than for that which renders man absolutely miserable, the complete lack of the ornaments and riches of the soul." The missionary continues, offering his scientific conviction that "neither the sun, nor malice of the soil, neither the air nor the water, neither the men nor their caprices, are to be blamed for this. We are all created by and dependent upon the same principles: We breathe under the same sky; the same constellations influence us; and I do not believe that the land, which produces trees as tall and beautiful as ours, will not produce as fine harvests, *if it be cultivated*." Wilderness the expanses of Acadia might be, but a garden it could become, if cultivated. For Father Biard and his contemporaries, "subjugating Satanic monsters" and establishing "the order and discipline of heaven upon earth"[23] combined spiritual and worldly dimensions. Champlain, for whom the Devil and his agents were as real as for Biard, expressed the same objective in a more secular way when he told the local people he met in the region of the Penobscot River that the French "desired to settle in their country and show them how to cultivate it, in order that they might no longer live so miserable an existence as they are doing." The comment is made the more striking when we remember that Champlain knew that some of the inhabitants of Acadia did practise agriculture—though he never suggested that they "cultivated" the land.[24]

It is perhaps not too much to suggest that "cultivation" was a distinctly European concept. "For before everything else," Marc Lescarbot maintained, "one must set before oneself the tillage of the soil." At the first French settlement at Ste. Croix gardens were sown and some wheat "came up very fine and ripened." The poor quality of the soil was one reason for the move across the Bay of Fundy to establish Port Royal, "where the soil was ample to produce the necessaries of life."

But there was more to cultivation than the production of simple foodstuffs. For Lescarbot, at least, the powerful symbolism of planting a European garden in what had been a wilderness was manifest. He wrote on his departure from Port Royal to return to France in July, 1607:

> I have cause to rejoice that I was one of the party, and among the first tillers of this land. And herein I took the more pleasure in that I put before my eyes our ancient Father Noah, a great king, a great priest and a great prophet whose vocation was the plough and the vineyard; and the old Roman captain Serranus, who was found sowing his field when he was sent to lead the Roman army, and Quintus Cinncinnatus, who, all crowned with dust, bareheaded and ungirt, was ploughing four acres of land when the herald of the Senate brought him the letters of dictatorship.... Inasmuch as I took pleasure in this work, God blessed my poor labour, and I had in my garden as good wheat as could be grown in France.

While Lescarbot might be dismissed as suffering from an overdose of Renaissance humanism, it seems more sensible to take him seriously. His florid rhetoric should be seen for what it really was: the ideology of what Alfred J. Crosby has called "ecological imperialism"—the biological expansion of Europe. What Lescarbot, and less literary Europeans, brought to bear on the Acadian landscape was the heavy freight of the European agricultural tradition with its long-established distinction between garden and wilderness. In that tradition God's "garden of delight" contrasted with the "desolate wilderness" of Satan. Though the concept of "garden" varies widely, as Hugh Johnson notes in his *Principles of Gardening*, "control of nature by man" is the single common denominator.[25]

The transformation of the wilderness into a garden is a constant theme in the early writings about Acadia. Father Biard had brought European seeds with him when he arrived in 1611 and at St. Saveur "in the middle of June, we planted some grains [wheat and barley], fruit, seeds, peas, beans and all kinds of garden plants." On Miscou Island (Shippegan Island) Denys discovered that although the soil was sandy, herbs of all sorts as well as "Peaches, Nectarines, Clingstones" and what the French always called "the Vine"—grapes—could be grown. But, as so often is the case, Lescarbot provides the most striking account of what gardening meant. At Port Royal, he "took pleasure in laying out and cultivating my gardens, in enclosing them to keep out the pigs, in making flower beds, staking our alleys, building summer houses, sowing wheat, rye, barley, oats, beans, peas and garden plants, and in watering them." European seeds, domestic animals—chickens and pigeons, too—fences—mine and thine.[26]

Of course, before a garden could be planted, the land had to be cleared. Denys described that work—and its by-product: squared oak timber that could fill the holds of vessels that would otherwise have returned empty to France. If clearing the land did not produce enough space for the garden, then the sea could be tamed, too. In the Minas Basin, where the settlers apparently found cultivating the land too difficult, Dièreville recounted the construction of a remarkable piece of European technology:

> Five or six rows of large logs are driven whole into the ground at the points where the Tide enters the Marsh, & between each row, other logs are laid, one on top of the other, & all the spaces between them are so carefully filled with well pounded clay, that the water can no longer get through. In the centre of this construction a Sluice is contrived in such a manner that the water on the Marshes flows out of its own accord, while that of the Sea is prevented from coming in.

Thus the tidal marshes were dyked for cultivation.[27]

Pushing back the forest, holding back the water, fencing a garden in the wilderness. The rewards would be great—"better worth than the treasures of Atahulpa," Lescarbot claimed. Was the symbolism intentional? Atahualpa, the defeated ruler of Peru, offered his Spanish captors led by Francisco Pizarro a room full of gold and silver in return for his freedom. The Spaniards accepted the ransom and then garrotted the Inca. Hardly a scene from a garden of delights.[28]

The French in Acadia were certainly not the Spanish in Peru. Still, the garden they planned was intended to produce a greater harvest than just sustenance for anticipated settlers. It was to be a garden for the civilization of the indigenous peoples. "In the course of time," Champlain observed on his initial meeting with the people he called Etechemins (Maliseet), "we hope to pacify them, and put an end to the wars which they wage against each other, in order that in the future we might derive service from them, and convert them to the Christian faith." The words were almost exactly those attributed to Columbus at his first sighting of the people of the "Indies": "they would be good servants ... [and] would easily be made Christians." Even the most sympathetic observers of the Native peoples of Acadia were appalled by their apparent failure to make for themselves a better life, a failing that was often attributed to their unwillingness to plan for the future. For Father Biard, Christianity and husbandry obviously went hand in hand. Living the nomadic life of hunters, fishers, and gatherers resulted in permanent material and spiritual backwardness. "For in truth, this people," he claimed, "who, through the progress and experience of centuries, ought to have come to some perfection in the arts, sciences and philosophy, is like a great field of stunted and ill-begotten wild plants ... [they] ought to be already prepared for the completeness of the Holy Gospel ... Yet behold [them] wretched and dispersed, given up to ravens, owls and infernal cuckoos, and to be the cursed prey of spiritual foxes, rears, boars and dragons." In Father Le Clercq's view the "wandering and vagabond life" had to be ended and a place "suitable for the cultivation of the soil" found so that he could "render the savages sedentary, settle them down, and civilize them among us." Though Lescarbot's outlook was more secular, he shared these sentiments completely and expressed them in verse:

This people are neither brutal, barbarous
  nor savage,
If you do not describe men of old that way,
He is subtle, capable, and full of good
  sense,
And not known to lack judgement,
All he asks is a father to teach him
To cultivate the earth, to work the grape
  vines,
And to live by laws, to be thrifty,
And under the sturdy roofs hereafter to
  shelter.

The leitmotif of this rhetoric is obvious: the images of the Christian garden and the satanic wilderness, summed up in the verse from the book of Joel, quoted by Father Biard: "The Land is a Garden of Eden before them, and behind them a desolate wilderness."[29]

Yet it would be quite wrong to assume that these seventeenth-century French visitors to the New World were blind to the potential costs of gardening in the Acadian wilderness. Indeed, there is considerable evidence of nagging suspicions that the very abundance of nature provoked reckless exploitation. Denys witnessed an assault on a bird colony that is reminiscent of the profligacy of Cartier's crew among the birds at Funk Island in 1534. Denys's men "clubbed so great a number, as well of young as of their fathers and mothers ... that we were unable to carry them all away." And Dièreville captured all too accurately the spirit of the uncontrolled hunt when he wrote that:

..., Wild Geese
And Cormorants, aroused in me
The wish to war on them....

He used the same militant language in his admiring account of the "Bloody Deeds" of the seal hunt, and also provided a sketch of another common pursuit: the theft of massive quantities of birds' eggs. "They collect all they can find," he remarked, "fill their canoes & take them away." Scenes like these presaged the fate of the Great Auk, the passenger pigeon, and many other species.[30]

These were the actions of men whose attitude toward the bounty of nature contrasted markedly with that of the indigenous inhabitants of North America. In Europe the slaughter of birds and animals was commonplace, indeed it was often encouraged by law. As Keith Thomas remarks in his study of *Man in the Natural World*—which is largely restricted to Great Britain—"[i]t is easy to forget just how much human effort went into warring against species which competed with man for the earth's resources." Without succumbing to the temptation to romanticize the attitude of North American Native peoples toward their environment—they hunted, they fished, some practised slash-and-burn agriculture—there is no doubt that their sense of the natural world was based on a distinctive set of beliefs, a cosmology that placed them in nature rather than dominant over it. Animistic religion—"everything is animated," Father Le Clercq discovered—a simple technology, a relatively small population, and what Marshall Sahlins has termed "stone age economics" made "war" on nature unnecessary, even unacceptable. "They did not lack animals," Nicolas Denys noted, "which they killed only in proportion as they had need of them." By contrast, the Europeans who arrived in Acadia at the beginning of the seventeenth century belonged to a culture where, in Clarence Glacken's words, "roughly from the end of the fifteenth to the end of the seventeenth century one sees ideas of men as controllers of nature beginning to crystalize." Or, as Marc Lescarbot put it, articulating as he so often did the unstated assumptions of his fellow Frenchmen, "Man was placed in this world to command all that's here below."[31]

The distinction between the "wilderness" and the "garden," between "savagery" and "civilization," between "wandering about" and commanding "all that is here below," is more than a philosophical one, important as that is. It is also, both implicitly and explicitly, a question of ownership and possession. In what has been called "enlightenment anthropology"—though I think that places the development too late—the function of the term "savage" was to assert the existence of a state of nature where neither "heavy-plough agriculture nor monetarized exchange" was practised and from which, therefore, civil government was absent. Moreover, civil

government, agriculture, and commerce were assumed to exist only where land had been appropriated, where "possessive individualism" had taken root. Thus the wilderness was inhabited by nomadic savages, without agriculture or laws, where the land had never been appropriated. Consequently, when Europeans set about transforming the wilderness into a garden, they were engaged in taking possession of the land. "The ideology of agriculture and savagery," in the words of J.G.A. Pocock, "was formed to justify this expropriation."[32]

As European gardeners began slowly to transform the wilderness of Acadia, so, too, as was their intent, they began the remaking of its indigenous inhabitants. And once again, though they rarely expressed doubts about the ultimate value of the enterprise, some Europeans did recognize that a price was being exacted. First there was the puzzling evidence of population decline. In a letter to his superior in Paris in 1611, Father Biard wrote that the Micmac leader Membertou (who himself claimed to be old enough to remember Carrier's 1534 visit) had informed him that in his youth people were "as thickly planted there as the hairs upon his head." The priest continued, making a remarkably revealing comparison:

> It is maintained that they have thus diminished since the French have begun to frequent their country; for, since they do nothing all summer but eat; and the result is that, adopting an entirely different custom and thus breeding new diseases, they pay for their indulgence during the autumn and winter by pleurisy, quinsy and dysentery which kills them off. During this year alone sixty have died at Cape de la Heve, which is the greater part of those who lived there; yet not one of all of M. de Poutrincourt's little

colony has ever been sick, notwithstanding all the privations they have suffered; which has caused the Savages to apprehend that God protects and defends us as his favorite and well-beloved people.[33]

The reality, of course, was more complex than this assertion that God was on the side of the immunized. Though the French were unaware of it, Acadia, like the rest of the Americas, was a "virgin land" for European pathogens. Denys hinted at this when he wrote that "in old times ... they [the Natives] were not subject to diseases, and knew nothing of fevers." Certainly they had not been exposed to the common European maladies—measles, chickenpox, influenza, tuberculosis, and, worst of all, smallpox. (The "pox"—syphilis—Lescarbot believed was God's punishment of European men for their promiscuous sexual behaviour in the Indies.) The immune systems of the indigenous peoples of Acadia were unprepared for the introduction of these new diseases, which were consequently lethal in their impact. Father Le Clercq, at the end of the century, reported that "the gaspesian nation ... has been wholly destroyed ... in three or four visitations" of unidentified "Maladies." Marc Lescarbot probably identified one important carrier of European infections when he stated that "the savages had no knowledge of [rats] before our coming; but in our time they have been beset by them, since from our fort they went over to their lodges."[34]

Disease, radical alterations in diet—the substitution of dried peas and beans and hardtack for moosemeat and other country foods—and perhaps even the replacement of polygamy by monogamy with a consequent reduction in the birth rate all contributed to population decline. Then there was the debilitating scourge of alcohol, another European import for which Native peoples had little, if any, tolerance. Just as they sometimes gorged themselves during "eat all" feasts, so they seemed to drink like undergraduates with the simple goal of getting drunk. Even discounting Father Le Clercq's pious outlook, his description of the impact of brandy on the Gaspesians was probably not exaggerated. The fur traders, he charged, "make them drunk quite on purpose, in order to deprive these poor barbarians of the use of reason." That meant quick profits for the merchants, debauchery, destruction, murder, and, eventually, addiction for the Amerindians. Though less censorious, or less concerned, than the priest, Dièreville remarked that the Micmacs "drank Brandy with relish & less moderation than we do; they have a craving for it."[35]

Estimating population declines among Native peoples is at best controversial, at worst impossible. Nevertheless, there seems no reason to doubt that Acadia, like the rest of the Americas, underwent substantial reduction in numbers of inhabitants as a result of European contact. Jacques Cartier and his successors, who fished and traded along the coasts of Acadia, likely introduced many of the influences that undermined the health of the local people. Therefore Pierre Biard's 1616 estimate of a population of about 3,500 Micmacs is doubtless well below pre-contact numbers, as Membertou claimed. Since it has been estimated that neighbouring Maliseet, Pasamaquoddy, and Abenaki communities experienced reductions ranging from 67 to 98 percent during the epidemics of 1616 and 1633 alone, Virginia Miller's calculation that the pre-contact Micmac population stood somewhere between 26,000 and 35,000 seems reasonable.[36] That was one of the costs of transforming the wilderness into a garden.

If the effects of disease and alcohol were apparent, though misunderstood, then another aspect of the civilizing process was more subtle. That process combined Christian proselytizing, which eroded traditional beliefs, with the fur trade, which undermined many aspects of the Native peoples' way of life. There is among contemporary historians of European-Amerindian relations a tendency to view the trading relationship, which was so central to the early years of contact, as almost benign, a relation between equals. Missionaries, politicians, and land-hungry settlers are credited with upsetting the balance that once existed between "Natives and newcomers" in the fur trade. There can be no doubt that recent scholarship has demonstrated that the Natives were certainly not passive participants in the trade. Far from being naive innocents who gave up valuable furs for a few baubles, they traded shrewdly and demanded good measure.[37]

Nevertheless, it is impossible to read seventeenth-century accounts of the trade and still accept the whole of this revisionist account. These were eyewitness testimonies to the devastating impact of alcohol on the Native traders and their families: murder of fellow Natives, maiming of women and abuse of children, the destruction of canoes and household goods. Beyond this, brandy, often adulterated with water, was used by Europeans "in order to abuse the savage women, who yield themselves readily during drunkenness to all kinds of indecency, although at other times ... they would be more like to give a box on the ears rather than a kiss to whomsoever wished to engage them in evil, if they were in their right minds." The words come from the priest, Father Le Clercq, but the merchant, Nicolas Denys, concurred. That, too, was part of the fur trade.[38]

Moreover, the trade cannot be separated from other aspects of contact that contributed to the weakening of Micmac culture. Fur traders carrying disease and trade goods unintentionally contributed to the decline of both traditional skills and indigenous religious belief. Nicolas Denys's discussion of Micmac burial customs illustrates this point neatly. Like other Native peoples, the Micmacs buried many personal articles in graves so that the deceased would have use of them when they disembarked in the Land of the Dead. The French judged this practice as both superstitious and wasteful—especially when the burial goods included thousands of pounds of valuable furs. They attempted to disabuse the Natives of the efficacy of this practice by demonstrating that the goods did not leave the grave but rather remained in the ground, rotting. To this the Natives replied that it was the "souls" of these goods that accompanied the "souls" of the dead, not the material goods themselves. Despite this failure Denys was able to report that the practice was in decline. The reason is significant and it was only marginally the result of conversion to Christianity. As trade between the French and the Micmacs developed, European goods—metal pots, knives, axes, firearms—gradually replaced traditional utensils and weapons that had once been included in burial pits. The use of European commodities as burial goods proved prohibitively expensive. Denys wrote that "since they cannot obtain from us with such ease as they had in retaining robes of Marten, of Otter, of Beaver, [or] bows and arrows, and since they have realized that guns and other things were not found in their woods or their rivers, *they have become less devout*." Technological change brought religious change. It also led to dependence.[39]

No doubt the exchange of light, transportable copper pots for awkward, stationary wooden pots was a convenient, even revolutionary change in the lives of Micmacs. But convenience was purchased at a price, and the Native peoples knew it. Father Le Clercq was vastly amused when an old man told him that "the Beaver does everything to perfection. He makes us kettles, axes, swords, knives and gives us drink and food without the trouble of cultivating the ground." It was no laughing matter. If, at the outset of European contact, the Native peoples of Acadia had adapted to the trade with Europeans rather successfully, they gradually lost ground, their role of middlemen undermined by overseas traders who came to stay. While Nicolas Denys deplored the destructive impact of itinerant traders and fishermen on the Native peoples, his only solution was to advocate European settlement and the enforcement of French authority. "Above all," he concluded his assessment of the changes that had taken place in Native society during his time, "I hope that God may inspire in those who have part in the government of the State, all the discretion which can lead them to the consummation of an enterprise as glorious for the King as it can be useful and advantageous to those who will take interest therein." In that scheme, when it eventually came to pass, the Micmacs and their neighbours found themselves on the margin.[40]

To these signs that the work of cultivation produced ugly, unanticipated side effects must be added the evidence of near crop failure in the spiritual garden of Acadia. In 1613 a disgusted Father Biard reported meeting a St. John River sagamore (Cacagous) who, despite being "baptized in Bayonne," France, remained a "shrewd and cunning" polygamist. "There is scarcely any change in them after

baptism," he admitted. Their traditional "vices" had not been replaced by Christian "virtues." Even Membertou, often held up as the exemplary convert, had difficulty grasping the subtleties of the new religion. He surely revealed something more than a quick wit in an exchange that amused the Jesuit. Attempting to teach him the Pater Noster, Biard asked Membertou to repeat in his own language, "Give us this day our daily bread." The old sagamore replied: "If I did not ask him for anything but bread, I would be without moose-meat or fish." Near the end of the century, Father Le Clercq's reflections on the results of his Gaspé mission were no more optimistic. Only a small number of the people lived liked Christians; most "fell back into the irregularities of a brutal and wild life." Such, the somewhat depressed Recollet missionary concluded, was the meagre harvest among "the most docile of all the Savages of New France ... the most susceptible to the instruction of Christianity."[41]

It was not just these weeds—disease, alcohol, dependence, and spiritual backsliding—in the European garden in Acadia that occasionally led the gardeners to pause and reflect. Possibly there was a more basic question: Was the wilderness truly the Devil's domain? The Northeastern Algonquian people were admittedly "superstitious," even "barbarian," but certainly not the "wild men" of medieval imaginings, indistinguishable from the beasts. If they enjoyed "neither faith, nor king, nor laws," living out "their unhappy Destiny," there was something distinctly noble about them, too. Despite the steady, evangelical light that burned in Biard's soul, he could not help wondering if the Micmac resistance to the proffered European garden of delights was not without foundation. "If we come to sum up the whole and compare their good and ill with ours," he mused briefly in the middle of his *Relation* of 1616, "I do not know but that they, in truth, have some reason to prefer (as they do) their own kind of happiness to ours, at least if we speak of the temporal happiness, which the rich and worldly seek in this life." Of course, these doubts quickly passed as he turned to consider "the means available to aid these nations to their eternal salvation."

Marc Lescarbot, for whom classicism and Christianity seemed to have reached their apogee in the France of his day, and whose fervour for cultivating the wilderness was unlimited, found much to admire in the peoples of Acadia. They lived "after the ancient fashion, without display": uncompetitive, unimpressed by material goods, temperate, free of corruption and of lawyers! "They have not that ambition, which in these parts gnaws men's minds, and fills them with cares, bringing blinded men to the grave in the very flower of their age and sometimes to the shameful spectacle of a public death." Here surely was "the noble savage," a Frenchman without warts—"a European dream," as J.H. Elliott remarks of the humanists' image of the New World, "which had little to do with American reality."[42]

There was yet another reason for self-doubts about the superiority of European ways over Amerindian ways; the Native peoples struggled to preserve their wilderness, refusing the supposed superiority of the garden. Even those who had become "philosophers and pretty good theologians," one missionary concluded, preferred "on the basis of foolish reasoning, the savage to the French life." And Father Le Clercq found that some of the people of Gaspesia stubbornly preferred their movable wigwams to stationary European houses. And that

was not all. "Thou reproachest us, very inappropriately," their leader told a group of visiting Frenchmen, "that our country is a very little hell in contrast with France, which thou comparest to a terrestrial paradise, inasmuch as it yields thee, so thou sayest, every provision in abundance.... I beg thee to believe, all miserable as we may seem in thine eyes, we consider ourselves nevertheless much happier than thou in this, that we are contented with the little that we have." Thus having demonstrated, 300 years before its discovery by modern anthropology, that having only a few possessions is not the same as being poor, the Algonquian leader then posed a devastating question: "If France, as thou sayest, is a little terrestrial paradise, are thou sensible to leave it?" No reply was recorded.[43]

It is simple enough to imagine one. Even those who could describe as "truly noble" the Aboriginal peoples of Acadia remained convinced that civilization meant cultivation. "In New France," Lescarbot proclaimed, "the golden age must be brought in again, the ancient crowns of ears of corn must be renewed, and the highest glory made that which the ancient Romans called *gloria adorea*, a glory of wheat, in order to invite everyone to till well his field, seeing that the land presents itself liberally to them that have none." The state of nature, a Hobbesian state of nature without laws or kings or religion, would be tamed, "civilized," when men "formed commonwealths to live under certain laws, rule, and police." Here, in Braudel's phrase, Europe's "own nature" was revealed.[44]

Perhaps such thoughts as these filled the heads of the Frenchmen who, according to Micmac tradition, gathered to enjoy one of the curious adventures" of Silmoodawa, an Aboriginal hunter carried off to France "as a curiosity" by Champlain or some other "discoverer." On this occasion the Micmac was to give a command performance of hunting and curing techniques. The "savage" was placed in a ring with "fat ox or deer ... brought in from a beautiful park." (One definition of "paradise," the *OED* reports, is "an Oriental park or pleasure ground, especially one enclosing wild beasts for the chase.") The story, collected in 1870 by the Reverend Silas Tertius Rand, a Baptist missionary and amateur ethnologist, continues: "He shot the animal with a bow, bled him, skinned and dressed him, sliced up the meat and spread it out on flakes to dry; he then cooked a portion and ate it, and in order to exhibit the whole process, and to take a mischievous revenge upon them for making an exhibition of him, he went into a corner of the yard and eased himself before them all."[45]

If, as Lescarbot's contemporaries believed, the wilderness could be made into a garden, then the unscripted denouement of Silmoodawa's performance revealed that a garden could also become a wilderness. Or was he merely acting out the Micmac version of Michel de Montaigne's often quoted remark about barbarians: We all call wilderness anything that is not *our* idea of a garden?

NOTES

1. Fernand Braudel, *The Perspective of the World* (London: William Collins and Son, 1985), 387, 388.
2. Cecil Jane, ed., *The Journal of Christopher Columbus* (New York: Bonanza Books, 1989), 191; Patricia Seed, "Taking Possession and Reading Texts: Establishing the Authority of Overseas Empires," *William and Mary Quarterly* XLIX, no. 2 (April 1992): 199.
3. Marc Lescarbot, *History of New France* (Toronto: The Champlain Society, 1914), III, 246. The theme of my lecture might have benefited had I been able to substantiate the claim, sometimes made, that "Acadie" is a corruption of "Arcadie" — an ideal, rural paradise. Unfortunately the claim, sometimes made on the basis of Verrazzano's 1524 voyage when he described the coast of present-day Virginia as "Arcadie," is unfounded. "Acadie" likely is derived from the Micmac word "Quoddy" or "Cadie," meaning a piece of land. The French version became "la Cadie" or "l'Acadie," even though the French sometimes thought of the area as a potential "Arcadie." See Andrew Hill Clark, *Acadia: The Geography of Early Nova Scotia* (Madison: University of Wisconsin Press, 1968), 71.
4. J.G.A. Pocock, "Deconstructing Europe," *London Review of Books* (December 19, 1991): 6–10.
5. J.H. Elliott, *The Old World and the New, 1492–1650* (Cambridge: Cambridge University Press 1989), cited 10.
6. Alfred W. Crosby, Jr., *The Columbian Exchange: Biological and Cultural Consequences of 1492* (Westport: Greenwood Press, 1972), 67.
7. Tzvetan Todorov, *The Conquest of America* (New York: Harper Colophon, 1985), 146.
8. Kirkpatrick Sale, *The Conquest of Paradise: Christopher Columbus and the Columbian Legacy* (New York: Knopf, 1991), 129.
9. William Cronon, *Changes in the Land Indians: Colonists and Ecology in New England* (New York: Hill and Wang, 1983), 12.
10. Leslie Upton, *Micmacs and Colonists: Indian-White Relations in the Maritimes, 1713–1867* (Vancouver: University of British Columbia Press, 1979), 25.
11. Peter Mason, *Deconstructing America: Representation of the Other* (London: Routledge, 1990), and Antonello Gerbi, *Nature in the New World* (Pittsburgh: Pittsburgh University Press, 1985).

12. *The Works of Samuel de Champlain* (Toronto: The Champlain Society, 1922), 1, 243.

13. Lescarbot, *History of New France*, III, 484–485; Nicolas Denys, *Description and Natural History of the Coasts of North America (Acadia)* (Toronto: The Champlain Society, 1908), 390, 393; Hugh Honour, *The New Golden Land: European Images of America from the Discovery to the Present Time* (New York: Random House, 1975), 36–37.

14. Denys, *Description and Natural History*, 362–369; Le Clercq, *New Relation of Gaspesia* (Toronto: The Champlain Society, 1910), 279.

15. Le Clercq, *New Relation of Gaspesia*, 275; Denys, *Description and Natural History*, 257–340; Champlain, *The Works of Samuel de Champlain*, 247, 368; Denys, *Description and Natural History*, 199.

16. Champlain, *The Works of Samuel de Champlain*, 327; for a discussion of the distribution of Native peoples, see Bruce J. Bourque, "Ethnicity in the Maritime Peninsula, 1600–1759," *Ethnohistory* 36 (1989): 257–284.

17. Sieur de Dièreville, *Relation of the Voyage to Port Royal in Acadia* (Toronto: The Champlain Society, 1933), 130–141; Lescarbot, III, 54, 164, 205; Karen Anderson, *Chain Her by One Foot* (London: Routledge, 1990).

18. Lescarbot, *History of New France*, III, 200–202; LeClercq, *New Relation of Gaspesia*, 243.

19. Le Clercq, *New Relation of Gaspesia*, 135–139; Denys, *Description and Natural History*, 420; Marc Lescarbot, *The Conversion of the Savages*, in Reuben Gold Thwaites, ed., *The Jesuit Relations and Allied Documents* (New York: Pageant Books, 1959), I, 101; *Jesuit Relations*, III, 83; Le Clercq, *New Relation of Gaspesia*, 265, 273.

20. Le Clercq, *New Relation of Gaspesia*, 243; Lescarbot, *History of New France*, III, 333.

21. Lescarbot, *History of New France*, III, 27, 113, 365; *Jesuit Relations*, III, 21; Todorov, *Conquest*, 27–33; Marshall Sahlins, *Islands of History* (Chicago: University of Chicago Press, 1985), 10.

22. *Jesuit Relations*, III, 33–35.

23. Ibid., 33.

24. Champlain, *The Works of Samuel de Champlain*, 295.

25. Clarence J. Glacken, "Changing Ideas of the Habitable World," in William L. Thomas, ed., *Man's Role in Changing the Face of the Earth* (Chicago: University of Chicago Press, 1956), 70–92; Lescarbot, *History of New France*, III, 241, 351, 363–364; Alfred W. Crosby, Jr., *Ecological Imperialism: The Biological Expansion of Europe, 900–1900* (Cambridge: Cambridge University Press, 1986); A. Bartlett Giamatti, *The Earthly Paradise of the Renaissance Epic* (Princeton: Princeton University Press, 1969); Hugh Johnson, *The Principles of Gardening* (London: Michael Beazley Publications, 1979), 8.

26. *Jesuit Relations*, III, 63; Denys, *Description and Natural History*, 303; Lescarbot, *History of New France*, I, xii.

27. Denys, *Description and Natural History*, 149–150; Dièreville, *Relation of the Voyage*, 94–95; see Andrew Hill Clark, *Acadia: The Geography of Early Nova Scotia* (Madison: University of Wisconsin Press, 1986), 24–31.

28. Lescarbot, *History of New France*, II, 317; John Hemming, *The Conquest of the Incas* (London: Penguin Books, 1983), 77–88.

29. Champlain, *The Works of Samuel de Champlain*, 272; Jane, *Columbus*, 24; Le Clercq, *New Relation of Gaspesia*, 115; *Jesuit Relations*, III; Le Clercq, *New Relation of Gaspesia*, 205; Lescarbot, *History of New France*, III, 487.

30. Denys, *Description and Natural History*, 156; Dièreville, *Relation of the Voyage*, 75–77, 102, 122–123.

31. William M. Denevan, "The Pristine Myth: The Landscape of the Americas in 1492," *Annals of the Association of American Geographers* 82, no. 3 (1992): 369–385; Keith Thomas, *Man and the Natural World* (New York: Pantheon Books, 1983), 274; Le Clercq, *New Relation of Gaspesia*,

331; Marshall Sahlins, *Stone Age Economics* (Chicago: Aldine Publishing Co., 1972); Denys, *Description and Natural History*, 403; Clarence J. Glacken, *Traces on the Rhodian Shore: Nature and Culture in Western Thought from Ancient Times until the End of the Eighteenth Century* (Berkeley: University of California Press, 1967), 494; Lescarbot, *History of New France*, III, 137. See also Richard White, "Native Americans and the Environment," in W.E. Swagerty, ed., *Scholars and the Indian Experience* (Bloomington: Indiana University Press, 1984), 179–204.

32. Lescarbot, *History of New France*, III, 94, 127; J.G.A. Pocock, "Tangata Whenua and Enlightenment Anthropology," *The New Zealand Journal of History* 26, no. 1 (April 1992): 35, 36, 41. Pocock bases much of his intricate argument on late seventeenth- and eighteenth-century sources, yet Marc Lescarbot's *History of New France*, first published in 1609, already articulates and assumes, though in a somewhat unsystematic way, a fairly full-blown version of the theory; see also John Locke's chapter "Of Property" in his *Essay Concerning the True Original Extent of Civil Government* (1640). Professor John Marshall has pointed to the importance of domestic animal imports as disease carriers by drawing my attention to Jared Diamond, "The Arrow of Disease," *Discover* (October 1992): 64–73.

33. Lescarbot, *History of New France*, III, 254; *Jesuit Relations*, I, 177.

34. Denys, *Description and Natural History*, 415; Lescarbot, *History of New France*, III, 163; Le Clercq, *New Relation of Gaspesia*, 151; Lescarbot, *History of New France*, III, 227.

35. Le Clercq, *New Relation of Gaspesia*, 254–255; Dièreville, *Relation of the Voyage*, 77.

36. Dean R. Snow and Kim M. Lamphear, "European Contact and Indian Depopulation in the Northeast," *Ethnohistory* 35 (1988): 15–33; Virginia P. Miller, "Aboriginal Micmac Population: A Review of Evidence," *Ethnohistory* 23 (1976): 117–127, and "The Decline of Nova Scotia Micmac Population, AD 1600–1850," *Culture* 3 (1982): 107–120; John D. Daniels, "The Indian Population of North America in 1492," *William and Mary Quarterly* XLIX, no. 2 (April 1991): 298–320.

37. Bruce G. Trigger, *Natives and Newcomers* (Montreal: McGill-Queen's Press, 1985), 183–194.

38. Le Clercq, *New Relation of Gaspesia*, 255; Denys, *Description and Natural History*, 449–450.

39. Denys, *Description and Natural History*, 442.

40. Calvin Martin, "Four Lives of a Micmac Copper Pot," *Ethnohistory* 22 (1975): 111–133; Le Clercq, *New Relation of Gaspesia*, 277; Denys, *Description and Natural History*, 452. See also Wilson D. Wallis and Ruth S. Wallis, *The Micmac Indians of Eastern Canada* (Minneapolis: University of Minnesota Press, 1945), and Bruce J. Bourque and Ruth Holmes Whitehead, "Tarrentines and the Introduction of European Trade Goods in the Gulf of Maine," *Ethnohistory* 32 (1985): 327–341.

41. *Jesuit Relations*, I, 166–167; Le Clercq, *New Relation of Gaspesia*, 193–194.

42. Denys, *Description and Natural History*, 437; *Jesuit Relations*, III, 135; Lescarbot, *History of New France*, III, 189; Elliott, *The Old World*, 27.

43. Le Clercq, *New Relation of Gaspesia*, 104, 125.

44. Lescarbot, *History of New France*, III, 229, 256–257.

45. Rev. Silas Tertius Rand, *Legends of the Micmacs* (London: Longmans, Green and Co., 1894), 279.

Chapter Nine

Averting Disaster

The Hudson's Bay Company and Smallpox in Western Canada during the Late Eighteenth and Early Nineteenth Centuries

Paul Hackett

 *****

THE HBC AND SMALLPOX: BEFORE VACCINATION

Between 1779 and 1783 a massive smallpox epidemic swept through the western half of North America, blanketing a large part of the continent. Beginning in Mexico City during the summer of 1779, the disease there claimed perhaps 40,000 lives.[1] By early 1781 it had reached Santa Fe, and subsequently began moving rapidly northward through the continental horse trade, leaving tens of thousands of new victims.[2] Within a few months smallpox was on the northern Plains, with fatalities amounting to as much as 75 percent or more among some Native American tribes.[3] By the fall of 1781 it had broken out among the Aboriginal peoples living along the Saskatchewan River, in the northern part of what are now the prairie provinces of Canada. It is estimated that more than half of the Natives there died of the disease or of epidemic-related starvation.[4] Too weak to move, the dying lay with the dead,

and the British and Canadian fur traders stationed at posts along the river witnessed mortality on a scale that they could hardly imagine.[5]

To this point in its diffusion from Mexico City, this smallpox epidemic proceeded on its path seemingly unchecked by human intervention; while there were limits to its spread, and some Aboriginal groups were fortunate to escape entirely, these circumstances were part of the natural patterns of such epidemics.[6] Only as it approached Hudson Bay was a systematic effort made to prevent the disease from spreading any farther. It was employees of the Hudson's Bay Company (HBC) who made that effort.

By the early 1780s the HBC was a mature and substantial fur-trading joint-stock company operating over a large part of what is now Canada. Its 1670 charter granted it an exclusive right to the trade of the Aboriginal peoples living within the huge Hudson Bay drainage basin, which stretched from the east coast of Hudson Bay to the Rocky Mountains, and from just north of the Missouri River to the Arctic.[7] Early on, the HBC saw its relationship with the Aboriginal peoples in much more than trading terms: employees were instructed to provide aid and assistance in times of need, and to attempt to prevent conflicts between hostile groups. In part this was self-serving, for any death among the hunters or their families, or any disruption in their normal routine, threatened the profits of the Company's stockholders; however, humanitarian concerns, couched in an air of paternalism, also informed its policies.[8] HBC men came therefore to be involved in issues of epidemic control during the late 1700s. To protect its trade and suppliers of furs, the HBC became as interested in preventing the spread of the disease as London authorities were in arresting smallpox in London.

The HBC followed a basic, successful trading strategy that it had worked out in its very first days of operation. The company settled major posts or factories along the coasts of Hudson and James Bays, and required the Native trappers to bring in their furs or furs they had traded with others [...]. This helped to minimize transport costs for the company and to put the onus for the difficult travel on the trappers. As a consequence of this trading style, the HBC came to differentiate between two basic classes of Indians: One group comprised the "Uplanders" from the interior, a generic term for the Cree, Assiniboine, Chipewyan (Dene), Ojibway, and others, who visited the post with their furs during the brief trading period each summer. The other group included the Lowland Cree, people who resided year-round in the lowlands surrounding the main HBC posts. Among these "Lowlanders" were the Homeguard Cree, who helped to supply the posts with provisions, and performed other critical tasks for the traders. The Homeguard Cree might visit at almost any time of the

year, and had a long history of hunting geese and caribou for the English, as well as transporting goods and mail between the posts.[9] [...]

Early in June 1782 a brigade of Ojibway paddled 16 canoes down the Hayes River to trade at the HBC's chief post of York Factory, located on the coast of Hudson Bay in what is now northern Manitoba [...]. Upon their arrival they told Matthew Cocking, the trader in charge of the post, bleak tales of a deadly eruptive disease that had destroyed large numbers of the people living to the west of nearby Lake Winnipeg. The English trader correctly surmised that the disease was smallpox.[10] An experienced HBC explorer and trader with a long history in the fur trade,[11] Cocking quickly grasped the implications for the local Cree and for his trade with the imminent arrival of infected brigades from the interior who were then preparing to travel to York Factory. From these accounts, it was too late for the HBC men to attempt to preserve the lives of the Uplanders, many of whom were in fact already dead or dying, but it was still possible to help the Lowland Cree. To do so Cocking would have to act quickly, for even as he was learning of the smallpox epidemic, it was nearing his doorsill.

Cocking and the other HBC men had few options available if they were to head off the epidemic. To do nothing would almost certainly doom the Lowland people to the same fate that had befallen those from the interior, since the traders had no means of effectively treating the disease once it became established. And yet, when smallpox had broken out previously at York Factory on two occasions, nothing in the records suggests that the HBC men had done anything but observe the effects of the disease and perhaps attend to the suffering of the victims.[12] The reason for this lack of action is not known; perhaps they were unaware of their options or, more likely, the disease had already made inroads among the Homeguard by the time they were alerted to its presence, and any precautions would therefore have been fruitless. Both outbreaks had occurred during the era prior to the expansion of the HBC into the interior, and it may be that the company's ability to gather information about such matters in a timely fashion was lacking. In 1782, however, Cocking and his men were forewarned about the approaching epidemic and chose to attempt to stem the spread of the disease. By this time two approaches to limiting the spread of smallpox were available: inoculation and quarantine.

Inoculation, or variolation as it was sometimes called, was the intentional infection of an individual with the variola (smallpox) virus in order to produce a mild form of the disease and thereby induce immunity.[13] When done properly, by a skilled physician following an advanced technique such as that developed by Daniel and Robert Sutton, almost all who received the procedure recovered; however, the risks of causing a fatal case of smallpox or of spreading the infection were significant when attempted under uncontrolled conditions by someone who was not trained in the procedure.[14] It was an ancient practice in the East and common in Africa by the beginning of the eighteenth century, but was unknown in the West. Within a few decades of its introduction to Europe and North America, it had become an accepted means of combating smallpox epidemics.[15] Thus, inoculation was commonly practised in Boston by the 1760s, and in 1777, during the American War of Independence, George Washington ordered compulsory inoculations for new recruits in order to avoid the smallpox

that was then decimating his troops.[16] In Great Britain the procedure had been a source of much inquiry in the scientific and medical communities throughout much of the eighteenth century, and it gained acceptance as the century wore on. It is probable that in the early 1780s at least some of the HBC people were aware of the potential benefits of variolation. The medical personnel whom the Company employed at some of the posts, including York Factory's surgeon, Alfred Robinson,[17] would have been familiar with inoculation, as would the more scientifically minded among the other employees.[18] Certainly they would have been desperate to head off the epidemic, and this procedure had been shown to be effective at this task in America and in Britain for several decades. It was, as Crosby suggested, an accepted counter to the disease. Nevertheless, it does not appear that Company employees resorted to variolation at any time prior to the late nineteenth century when faced with the prospect of smallpox among the Indians.

No obvious reason has emerged as to why the HBC men declined to variolate the Indians. In part it may have been that they did not feel comfortable with the procedure, even if they were familiar with the basic concepts. Unskilled application of the technique and inadequate follow-up isolation would certainly have helped to spread smallpox, and at the same time would have placed some of the blame on the Company. In turn, this could have had a disastrous effect on their relationship with the Indians: epidemic disease among the Aboriginal peoples of North America had long been politicized, and it is not unreasonable that the HBC might have faced opposition or even threats from the survivors had they accidentally contributed to the spread of the smallpox.[19] Even without the fuel of a failed variolation program, accusations flew between non-Native parties during this epidemic, opening the door for hostilities.[20]

The HBC practice of not inoculating the Indians seems to have been consistent with British policy in North America. According to Wagner and Allen Stearn, variolation was rarely carried out upon the Canadian Indians.[21] An almost simultaneous episode involving the British army during the American Revolutionary War tends to support this view: in November 1783 a British hospital mate by the name of Gill, stationed on the St. Lawrence River, proposed without authorization to inoculate Britain's Indian allies of the Great Lakes; rather than embracing Gill's suggestion, his superiors moved frantically to track him down and stop him from completing his "imprudent" mission after he had left for the interior with a supply of the smallpox virus.[22] Whatever confidence the British may have had in the safety and efficacy of smallpox inoculation for their own people, this conviction was not extended to the Aboriginal peoples of North America.

The second option available to the HBC was to isolate the Lowland Cree from the disease by imposing a hurried quarantine against communication with the infected Uplanders who would soon descend on the fort. This would be in keeping with emergency measures invoked at least as early as the fourteenth century during the Black Death.[23] Closer to home, the Spanish had begun quarantining ships to the New World almost from the start of contact, and by the 1780s had in place a set of regulations designed to prevent the disease from reaching Mexico City from elsewhere.[24] In much the same way, the French had attempted to block the spread of bubonic plague from Marseilles

to North America during 1721–1724, and (unsuccessfully) of smallpox from the English colonies to Canada in 1731.[25] Similarly, the English had long applied a quarantine period to infected ships visiting Boston in order to stem the flow of epidemic disease.[26] As a public health policy, quarantine had a long history of use throughout the Western world and was easy to invoke, if not always effective.

There appears to have been precedent for the imposition of quarantine by the HBC at York Factory. In 1746–1747 two English ships, the *Dobbs Galley* and the *California*, overwintered at nearby Port Nelson as part of a voyage of exploration. When scurvy broke out on board, the HBC traders went out of their way to make things worse for these unwelcome interlopers who were challenging the Company's authority. According to one of the crewmen, the HBC refused to allow the Lowland Cree to approach with food. In what was more likely an example of profound insensitivity to the suffering of their compatriots than a show of concern for the Natives, "the *Indians* were charged not to come near us, or to furnish us with anything; and this out of a tender Regard for them; because we had a contagious Distemper amongst us, which might communicate itself to them, and to their families."[27] Whatever their motive for doing so, this does suggest that HBC traders were familiar with the practice of quarantine, and when smallpox threatened nearly 40 years later, it was to this rather than inoculation that Mathew Cocking turned.

In keeping with this overall policy, in 1782 Cocking devised an ambitious plan to prevent the Lowland Cree from having any contact with the Upland brigades that arrived at his post. He was already late: two infected Uplanders had reached the plantation on July 8 and were isolated on the grounds.[28] Cocking instructed his men to prevent any other people from coming to the fort, in order to preclude contact with the two victims, and to tell any trading Indians that he would instead send men to meet them at a nearby creek.[29] On August 6, he confined an infected Ojibway trading party to the creek lest "our Homeguards in the Marshes would most probably catch the disorder either by coming to the Fort or by our peoples cloaths who must have gone over in a boat occasionally to them";[30] at the same time, he also declined to send a mail packet to Fort Prince of Wales by some of the local Cree, as he normally would have done, fearing that the disease had already appeared among the Indians there.[31] Subsequently, he ordered the masters of the HBC's Bayside posts to follow his lead. Instructions on how to proceed were spelled out in a letter addressed to Peter Willdridge at Severn, in which the Severn master was told to

> keep a strict look out, that none of the Homeguards [the Lowland Cree] come to the factory but keep them at a proper distance so that none of the Pungee's [Bungee or Ojibway] that come for debt may have any Communication with them. Should you find the disorder has attacked any of them, do all in your power for their preservation. If the Englishmen have been handling any person that may have had the small Pox, you must be careful that they shift, wash and air their Cloaths as well as themselves ere they go near one of the homeguards.[32]

In essence, Cocking argued for a three-pronged approach to the epidemic. The first part was to isolate the Uplanders from the local people, and the situation of York Factory and the other Bayside posts made this a reasonable option. At most of these posts the inland trade arrived via one or perhaps two main rivers, enabling those

in charge to station a few men upriver to intercept the Uplanders before they arrived at the post. Where the trading Indians came overland, the route was also predictable, and men could be stationed to head off those approaching on foot. As the HBC trader controlled the site of the trade, he could influence the behaviour of his visitors, and could keep the different groups separated, at least while they were visiting the post. This precaution hinged on the assumption that the HBC men, who were almost exclusively European born, were no threat to spread the disease directly to the Indians themselves due to existing immunity; it appears to have been a valid assumption for, despite close contact with the Native victims of the disease at several posts, only one employee, a man of mixed Indian and European descent, is recorded as having contracted the disease.[33]

This first aspect of the plan demonstrates a reasonably well-developed understanding of the nature of smallpox and the manner in which it was transmitted. The use of quarantine to prevent contact between individuals who were infected with smallpox and those who were susceptible to the disease suggests that Cocking believed that it was contagious. The same was true of the actions of his fellow HBC men stationed on the Saskatchewan River.[34] Cocking and his associates understood that smallpox was an acute disease, one that produced a lasting immunity, and that they could act accordingly. If a group had completely recovered from the disease, they were not a threat and could visit the post unimpeded. Thus, for instance, when a family from up the Severn River visited Severn master John Hodgson in the spring of 1783, they were allowed to stay and to mix with the Lowland Cree, despite having previously had the disease. In the words of the trader:

[T]hey have seen but one Indian during the Winter, that they are all dead Inland, these are very deeply marked with the small Pox, one of them has lost all his Children by it except one poor Boy, which is both blind and Lame, and they have been obliged to haul him all the Winter.[35]

Hodgson learned that "they are intending to stay here to hunt Geese in the Spring," to which plan he agreed: "as they have all been well of the small Pox since last Fall, there will be no possibility of the homeguards catching it."[36]

Second, Cocking also understood that his men were a threat to spread smallpox, not through contracting the disease, but by virtue of the clothing they wore.[37] The need to "shift, wash and air their Cloaths as well as themselves ere they go near one of the homeguards" was itself sound epidemiologic advice, given that the smallpox virus could survive for extended periods of time on clothing and other personal objects. Similarly, another HBC trader stationed on the Saskatchewan River during this epidemic, William Tomison, had his men wash the furs that they collected with a disinfectant and air them repeatedly in order to prevent the spread of the disease.[38]

Finally, Cocking's plan emphasized the need to provide assistance to those people who had already contracted the disease. In ordering the other post masters to "do all in your power for their preservation," he was simply reminding them of the duty expected of them by their superiors in London. Once smallpox began to manifest itself, the victims were unable to care or to provide food for themselves or their families and, even for those who were unaffected, starvation was a possibility if the hunters were ill or died. When adult males died, this obligation might extend to supporting the widows and orphans at

the post, at least until a suitable substitute family might be found. This aid would have included the provision of food, clothes, medicine, and shelter, when necessary. With the active assistance of the HBC men the death toll was undoubtedly reduced.

For a time this plan worked. Over a span of two months the HBC men kept smallpox from the Lowland Cree living in the vicinity of York Factory, despite the fact that it raged with disastrous results among the Ojibway who lived beyond them in the interior.[39] Indeed, time was on their side: the traders needed only to maintain the separation between the Uplanders and the Lowland Cree for perhaps a month more to cut off the progress of the disease entirely; then the trading season would be over for another year. By the next summer's trading season, with luck, the disease would have run its course in the interior.

The vagaries of smallpox contagion betrayed Cocking. He was defeated by events beyond his control—events, as it turned out, that had nothing directly to do with either the Indians or the fur trade. In a strange twist of fate, the Lowland Cree of Hudson Bay fell victim to a distant war in which they had no stake, becoming, as it were, the most unusual casualties of the hostilities surrounding the American Revolution. In the summer of 1782 three French vessels under the compte de La Pérouse sailed from Haiti to Hudson Bay. Unable to defeat the English fleet in the Caribbean, they were intent on at least dealing a blow to English commerce in North America. On August 8 they attacked and captured Fort Prince of Wales, to the north of York Factory. Later in the month they took York without firing a shot. Because of these attacks the HBC men were dispersed; some were captured, while others, like Cocking, managed to escape to Europe or to the Company's other posts. Fearing additional attacks, the HBC temporarily shut down Forts Prince of Wales, York, and Severn, as well as many of the inland posts, and brought the men to their other establishments on James Bay.[40]

With the posts vacated, the Lowland Cree were left to their own devices. Without Cocking's measures, the disease was in due course passed on to these people, and they suffered terribly. When the traders returned they came back to a much bleaker landscape than they had left, for, according to Donald Gunn, a former HBC employee turned historian, "the bleached bones of those who had become the victims of the plague, were to be seen in great quantities at several points on the shores of the Bay" even as late as 1815.[41] When Humphrey Marten reoccupied York in 1783, he was visited by some of the Nelson River Homeguard who gave "a melancholy account of the havoc death hath made in the North [Nelson] River Indians most of whom are cutt off as are also the Churchill home guard, so that this country for some hundred of miles may too truly be said to be depopulated."[42] At Severn, the master, William Falconer, could count on very few Cree hunters for the spring goose hunt of 1784, although there remained many widows and orphans about the place.[43] Moreover, the epidemic was then free to continue its diffusion within the Lowlands almost to Albany Fort, causing much destruction among the Lowland Cree living a short distance to the northwest.[44] For his part, Mathew Cocking never witnessed the impact of the epidemic among the Homeguard, for he never returned to Hudson Bay after 1782. Although his quarantine-based strategy would in all probability have saved lives had he and his colleagues been able to see it through, in the end it proved unsuccessful.

THE HBC AND SMALLPOX: THE VACCINE ERA

Following the tragedy at Hudson Bay, a new and far more effective tool became available in the fight against smallpox, supplanting both inoculation and quarantine. In 1796 the English physician Edward Jenner first publicly demonstrated that infection with cowpox (or what he termed vaccinia), a related but comparatively mild disease commonly experienced by English milkmaids, had the beneficial side effect of providing immunity to the far more deadly smallpox.[45] The result of this discovery was vaccination, an alternative to variolation that could prevent smallpox without the risk of inadvertently starting a smallpox epidemic if done improperly.[46]

News of Jenner's discovery spread quickly across the Western world, and within a very few years vaccination had been embraced by physicians in many countries. By 1798 immunizations were being performed in London, which soon became the centre for the dissemination of both the practice and the vaccine itself.[47] By 1800, vaccination had reached most European countries, and a year later an estimated 100,000 persons had been immunized against smallpox in England alone. In 1803 King Charles IV of Spain issued a royal order providing for vaccine free of charge in Spain, and arranged for a massive vaccination program in his American colonies.[48] Similarly rapid adoption occurred in other parts of Europe.

*****

For the HBC, vaccination held out immense promise, particularly since they had failed to embrace inoculation. Like all who traded furs in the west during the earlier epidemic, the Company had suffered substantial losses in their trade due to smallpox.[49] It is not clear how soon after Jenner's discovery the HBC began making arrangements to ship vaccine matter to Hudson Bay along with the annual supply of trade goods. However, having quickly aroused interest in England's scientific and medical community, a community with which the HBC had developed close and long-standing ties, the procedure and the potential benefits that vaccination could offer their business interests must have come to the attention of its officers early on. Writing in the 1830s, the HBC governor and committee in London that ran the HBC reminded George Simpson (the governor of the HBC in Canada):

> We have long endeavoured to impress on the Gentlemen in charge of Districts and Posts the importance of introducing vaccination generally among the natives who frequent the different Establishments both as a measure of humanity in endeavouring to avert the calamities arising from that pestilential scourge and with a view to the welfare of the Business which must be seriously injured by the unhealthy state of the country....[50]

Overall, then, the Company favoured a proactive approach to vaccination, ensuring that viable stocks were kept at the posts. At the same time, London encouraged the traders to vaccinate the Indians as a prophylactic measure. The problem appears to have been that the Company's traders favoured a reactive approach, and declined to carry through with London's wishes. With few exceptions over the next half-century, the men failed to employ the vaccine except when pressed by an impending epidemic. Thus, London admonished Simpson in their letter: "we fear that sufficient attention has not been

paid to our observations on the subject and that consequences may be very calamitous."[51] This failure to carry out a proactive vaccination program was to limit the effectiveness of the HBC's public health efforts over the next eight decades.

The first known association of the HBC with vaccination came a decade and a half after Jenner's initial experiments.[52] In 1811, Thomas Douglas, Lord Selkirk, a pragmatic philanthropist and a major shareholder in the HBC, took steps toward founding an agricultural settlement on the banks of the Red River of the north as part of a reorganization of the Company's operations. He advised his representative, Miles Macdonell, to use vaccinations as a tool to gain the Red River Indians' acceptance in preparation for the establishment of his colony:

> A boon of immense consequence may be held out in the communication of the vaccine. On this point it may be necessary to proceed cautiously to avoid misapprehension, but time and patience will convince them, both of the value and the beneficence of the gift. Perhaps by judicious management on the part of the interpreters, they may be made to entertain very high ideas of the power of those who have such a command over nature.[53]

In this, Selkirk was giving shrewd advice, since most of the bands then living in the valley of the Red River had entered the region in the wake of the devastating 1779–1783 epidemic.[54] These people were well aware of the destruction that the disease was capable of causing, and would have been anxious to prevent its recurrence.

Despite the efforts of the HBC's governor and committee in London to press their traders to vaccinate the Indians generally as a preventive, the vaccine they sent out sat unused in the posts' warehouses for the most part. Other than the vaccination of three children at York Factory in October 1813,[55] the next known reference in the HBC records did not come for almost 10 years, and even then the traders proceeded only under the threat of another epidemic. Late in the summer of 1819 the rumour of smallpox and measles among some bands of the Assiniboine and Cree Indians west of Lake Winnipeg began to reach the residents of Selkirk's nascent Red River colony near the junction of the Red and Assiniboine rivers; HBC trader Peter Fidler, who was stationed in the Red River Department, learned of the death of many Assiniboine people by what they supposed was smallpox while returning from a war expedition against the Mandan on the Missouri River.[56] Although, in fact, only measles and whooping cough were abroad at this time, these rumours ushered in the initial vaccine program in western Canada.[57]

Spurred by the rumours, the HBC men in the Red River area began, finally, to vaccinate those who were susceptible. Fidler wrote to the governor and committee in London on August 12, 1820, explaining that "as it was even expected that the small pox might be introduced from the same quarter—in consequence great numbers of the half-Breed Children—Indians + others have been inoculated by Cowpock matter + have succeeded well—and the Indians are now asking more to be operated on in the same manner—as those who have already undergone the operation are all perfectly recovered, without confinement or pain."[58] The trader made no mention of vaccinating the adult colonists, which suggests that they were already immune, perhaps from either previous exposure or earlier vaccination while in Europe. Although

it is possible that Fidler undertook the vaccinations himself, there were several men with medical expertise settled in the colony, men who were quite capable of performing the procedure or of instructing others; they included retired HBC Governor Thomas Thomas, formerly a surgeon, and Alexander Cuddie, the doctor for the colony.[59] Useless against measles, these vaccinations had an important impact by increasing the baseline immunity of the people of Red River when smallpox was eventually reintroduced some 17 years later.[60]

The HBC traders should have expanded this effort into a comprehensive vaccination program in the ensuing years, both beyond the Red River area and to those born after the measles epidemic, but they did not. Instead, as complacency set in, there were but a few haphazard localized attempts before the 1830s to vaccinate some of the men and Indians. For instance, a few people were vaccinated at the HBC's Cumberland House and at Norway House during the summer of 1824, when Chief Factor Dr. John McLoughlin passed through while en route from York Factory to his new appointment in the Columbia District [...].[61] Likewise, in 1826 at Albany on James Bay the retiring chief factor, Thomas Vincent, arranged to have his country-born family vaccinated before they departed for England, and extended the procedure to the children living about the post and the Indians who came to the post to trade.[62] In the end, in the words of another trader, William McKay, "the Experiment made on those vaccinated with the virus sent from Canada has succeeded in the fullest and most satisfactory manner."[63] Such modest successes were, however, the exception to the rule. Without the immediate threat of the disease the traders lacked interest in promoting vaccination among the Indians during this long period when smallpox remained absent, even with supplies on hand or readily available from England or Canada. Indeed, when smallpox returned about 10 years later, it became clear that the HBC men had not taken seriously the instructions of the governor and committee to "introduce vaccination generally among the natives."

The smallpox epidemic of 1837–1838 had its start with the transport system of another fur company. In the late spring and early summer of 1837 the American Fur Company (AFC) steamship, *St. Peter's*, distributed smallpox on its journey from St. Louis to the upper part of the Missouri River, largely because its captain refused to halt or even quarantine the ship despite the presence of an infected individual.[64] In this way the disease appeared among the Yankton and Santee Sioux near Fort Pierre, and the Mandan, Hidatsa, and Arikara around Fort Clark.[65] Unlike the tribes living along the lower part of the Missouri, these groups had not been vaccinated during a massive campaign carried out in 1832 by the American government, which left the entire upper Missouri vulnerable to smallpox in 1837.[66] The epidemic's subsequent diffusion to Forts Union and McKenzie spelled potential disaster for those of the HBC's territory, for among the groups who regularly visited these American posts were Assiniboine, Piegan, and Blackfoot then living to the north of the forty-ninth parallel.

It is illuminating to compare the responses of three of the principal non-Native actors in western North America during this epidemic for what it says about their ability to battle the disease. The AFC, the dominant western trading outfit in American territory, proved unable to stem the tide of the epidemic, even in local situations. For instance, at Forts Clark and

Union, where smallpox was at its worst, the traders had no supply of vaccine, and their request for an emergency supply went unheeded.[67] Therefore, at Fort Clark, Francis Chardon accepted the epidemic with resignation, and merely recorded the details of the events around him. Meanwhile, at Fort Union, Jacob Halsey and Charles Larpenteur attempted to variolate the Indians, with unfortunate results: rather than preventing the spread of the disease, they merely served to accelerate its diffusion.[68]

The response to the epidemic on the part of the American government was more successful than that of the AFC traders, but was still inadequate to counter the threat. With little presence in the west, the federal government failed to vaccinate the Indians living across much of the vast territory in time to stop the epidemic. Although the Indian Vaccination Act of 1832 provided substantial funds to the Office of Indian Affairs between 1832 and 1839, the office lacked the means and information to intervene in a timely manner during this crisis.[69] Instead, reports of the epidemic began to reach Washington only in February 1838, or six months after smallpox had passed through the upper Missouri, and it was only then that officials began to set in motion plans to resume vaccinating the western Indians[70] — and even then, they were forced to call on the help of private citizens, such as the French explorer Joseph Nicollet, touring Minnesota, who co-operated by vaccinating some of the Indians there, while another man travelling to the Pacific Northwest was asked to introduce vaccine among those beyond the Rockies.[71] Most of the damage had already been done.

In contrast, the HBC men had the capacity to deal with the disease in the field from the outset, and they were pressed by necessity to hasten. Smallpox travelled northward in the vanguard of the Blackfoot, Assiniboine, and Piegan as they fled the American posts in terror during the summer and fall of 1837, the disease firmly entrenched among them. By November it was in the Qu'Appelle Valley and on the North Saskatchewan River between Carlton and Edmonton Houses.[72] As in 1819–1820, rumours of a devastating epidemic on the Missouri prompted the traders to act, and this time their swift action played a major role in limiting the spread of an epidemic. By September 20 Chief Trader Dr. William Todd, stationed at Fort Pelly on the Assiniboine River, became acquainted with the "bad disease" then raging on the Missouri. Like Fidler before him, he began vaccinating the local Cree with the stock he had on hand; at the same time, he provided them with vaccine and instructions to carry to their countrymen in order to preserve their lives.[73] To a greater extent than had been true for Fidler, Todd's actions were crucial. Thereafter, he dispatched news of the epidemic and fresh supplies of vaccine to other districts, including the Carlton, Ile à la Crosse, Chipewyan, and Edmonton Districts [...].[74]

Several factors enabled Todd and his fellow HBC traders to respond in a timely fashion, compared to the American government. The first was the presence of company personnel at strategic locations in the west, men who were prepared to act quickly and independently of headquarters if required, unlike the American Indian agents and subagents who only occasionally visited their postings.[75] Second, many of the HBC posts had stocks of vaccine on hand, shipped from London, and in most cases during this epidemic they proved viable. This suggests that the company had been shipping vaccine crusts, or dried lymph, which had a longer shelf life than the more fragile points, which were also in common use at the time.[76]

Finally, the fur company benefited from more efficient communications—both externally, in terms of the rumours that prompted Todd to act, and internally, for the Company's structured communication system helped spread the news of the advancing epidemic.

As news of smallpox passed down the HBC's line of communication during the winter, outstripping the epidemic, fellow traders followed suit, using previously neglected stocks of vaccine. For instance, after being alerted to the effects of the epidemic, John Lee Lewes at Cumberland House vaccinated all the employees at the post and noted in his journal that he would do the same for every Indian "*that will submit to it.*"[77] Thereafter, the terrified Indians flocked to Lewes's post for the procedure, and the trader also sent vaccine and instructions back with many of these people to the Moose Lake outpost and Lac la Ronge.[78] At Fort Frances (or Rainy Lake), Chief Trader John Charles also learned of these occurrences by the winter express. Acknowledging Todd's efforts, and fearing further spread of the disease, he wrote in his journal that "it is to be hoped that all the surrounding Districts will follow his humane example";[79] for his part, he continued, Charles was going to vaccinate the Rainy Lake Indians during the spring, apparently with vaccine that he had on hand.[80] A similar reaction occurred at York Factory, and vaccine, along with instructions from the doctor, was shipped to nearby Severn for the use of post master Robert Wilsons.[81]

With these timely vaccinations, the progress of the epidemic was effectively stopped after only a very limited penetration into Canadian territory. The success of this initial campaign was outlined by James Hargrave, the chief factor at York Factory, in a letter written after the epidemic had reached its final extent:

The apprehensions arising from the presence of Small Pox in the Country have been general; But it is highly gratifying to have to apprise you that its extension by contagious infection has been very limited in comparison. The summer season has passed over without producing a single case among the whole of our voyaging servants; and it does not appear to have as yet made its appearance anywhere within the Company's territories, beyond the limits of the Saskatchewan District.[82]

Clearly, it could have been far worse. In Hargrave's mind this was a "fortunate escape" to be attributed to the efforts of his fellow traders in causing the Indians to be vaccinated and, where possible, in keeping those who were infected from "coming in contact with the people of places still clean of the infection."[83]

Thereafter, during the post-epidemic phase, the HBC took steps to see that the disease would not soon spread within the west again. Having been alerted to the epidemic through their American contacts,[84] London acted vigorously, sending supplies of vaccine for the use of the traders. In their instructions, they noted:

We now forward to each of the Factories Packets of vaccine matter, and we desire that it be distributed throughout the Country, and that the Gentlemen in charge of Posts exert their utmost influence among the different tribes to induce them to submit to inoculation which we have no doubt they will readily do if pains be taken to impress on their minds the great benefit they will derive therefrom.[85]

Over the next two years they sought to vaccinate every susceptible person over much of their domain.

Such an ambitious program was only possible due to the Company's entrenched command structure, its well-developed transport system, and its presence throughout this vast territory—characteristics that set the HBC of the nineteenth century apart from both its trading rival to the south and the American government. Since Mathew Cocking's time the Company had expanded far from Hudson Bay into the interior of the continent, and had posts located throughout much of what is now Canada as well as into American territory. While the American government struggled to get doctors to vaccinate the Indians in 1832 and again in 1837, and failed to serve the upper Missouri, the HBC already had supply lines and personnel in place to perform what amounted to a relatively simple medical procedure. Each year goods, men, and letters, among other things, were sent out from London or from Canada for the benefit of each post. The system was hierarchical: instructions to the factors, traders, and clerks in charge of posts—and vaccine as well—could be sent to a few main supply depots, and from there to second-order posts, and from there to outposts. The HBC also benefited from a pre-existing working relationship with the Indians, which facilitated their co-operation.

As part of this program, the HBC shipped vials of vaccine matter, lancets, and instructions by at least three different routes in 1838. The first supply was sent from Montreal for the use of the posts of the Lake Superior District, including Michipicoten, Batchewana Bay, Pic, Long Lake, Lake Nipigon, and Fort William [...]; the post managers were in relatively frequent contact with the HBC office in Montreal, and it would have been a simple task to arrange shipment early on.[86] Another supply was sent to the Pacific Northwest by ship to supplement that which was already at the posts, and additional vaccine was to be brought from across the Rockies the following summer.[87] Finally, the bulk of the material necessary for the program was to be sent during the summer of 1838 via the annual ships from England to the major posts on Hudson and James Bays.

*****

It is to be expected that, on the whole, the HBC men followed through with their orders to complete this ambitious experiment in public health. There is evidence that, where the procedure did not take on the first attempt, the traders repeated the exercise until the telltale blister of the cowpox infection appeared. On those few occasions when it was apparent that the allotted virus was dead, they obtained new supplies from other posts. When they missed certain individuals by chance, they either performed the operation at the next available opportunity or sent vaccine back with those who had visited their posts. Certainly London was encouraged by the early reports of the program. In March 1839 the governor and committee wrote to George Simpson:

> We are glad to find the vaccination has been practised throughout the country with such beneficial results and have to desire that the Gentlemen in charge of Districts and Posts use their utmost endeavours to extend this most beneficial mode of arresting the progress of Small Pox, which we are concerned to learn has depopulated whole Districts of Country in North America during the past two years.[88]

For their part, the Aboriginal peoples were almost universally anxious to be vaccinated, the devastation of the epidemic of the 1780s and its potential for repeat in the late 1830s being central in their minds, and this made the daunting task of the fur traders possible. While undoubtedly some people were not immunized against smallpox during this program, it is probable that by late 1839 the majority of people, both Native and non-Native, living from Lake Huron to the Pacific and from the American territory to the Arctic, were immune to smallpox—either from previous illness with it, or from vaccination.

Mindful of the possibility of new introductions of smallpox, London continued to ship vaccine to the west after 1839.[89] Nevertheless, and despite an ever-increasing threat of the disease from elsewhere, the HBC men in the field seem to have fallen back into a familiar pattern: rather than carrying out periodic revaccinations among the Indian people, the traders instead chose to remain vigilant, and to act only when the possibility of an outbreak of smallpox was imminent. Every so often the disease would appear on the borders of the company's territory, and then those traders nearest to the outbreak would act. Thus, for instance, when John Swanston of Fort William (present-day Thunder Bay, Ontario) heard rumours that smallpox had broken out below Sault Ste. Marie in September 1842, he immediately began to vaccinate the Indians.[90] Such a policy, although perhaps not intended by the higher officials of the HBC, was nonetheless successful at keeping the disease from becoming widespread in the west again for several decades. Even with the reluctance to act proactively, this made for far more successful public health than Mathew Cocking could offer in the latter part of the eighteenth century. For its part, the policy of quarantine was not abandoned, but in the absence of smallpox epidemics HBC employees turned to it primarily against the other contagious diseases—such as measles, influenza, and whooping cough—that appeared with increasing frequency as the nineteenth century wore on.

CONCLUSION

As a public health institution, the HBC met with mixed success in its attempts to prevent the spread of smallpox during the study period. Initially, the HBC traders chose to employ a strategy of quarantining the Indians in order to prevent the disease from reaching Hudson Bay in 1782. Although it was unsuccessful in the end, its failure was no doubt due to the intervention of the French under La Pérouse (a circumstance that could not be foreseen), for Cocking's plan showed every sign of being successful prior to the dispersal of the HBC traders. Nevertheless, as a public health strategy, quarantine had significant limitations for the HBC's purposes: It required careful and close supervision by the traders in order to ensure that the infected did not come into contact with the susceptible, but during this era the Company stationed only a small number of men at a few posts, most of which were located on either Hudson or James Bay; it was therefore of only limited use during a spatially extensive epidemic. As well, quarantine was highly vulnerable, in that minor mistakes or even chance could leave the defences wide open, as was the case in 1782.

The fact that the HBC men do not seem to have given serious consideration to the use of variolation appears significant, given its relatively common use in Great Britain, continental Europe, and North America. If accepted generally as a preventive for smallpox, it was not a viable option for some who dealt with Aboriginal

peoples; in all probability, the risks of actually spreading the disease and causing additional deaths were judged too great, despite the potential benefit of saving lives, even in the direst state of affairs. This circumstance indicates that, for the HBC and some other organizations concerned with public health, the transition from variolation to vaccination was all the more critical, and was not an evolution in practice. Rather, it was a revolution, as for the first time these agents of public health had at their disposal a means of combating this destructive disease on a grand scale.

For the HBC, the measures taken in 1782 represent a change in their policy toward the control of epidemic disease among the Aboriginal peoples. Previously, the English traders had done little more than observe and attempt to treat the effects of such infections when they arose. Subsequently, they took a far more active role in fighting the spread of epidemics among the Natives. Jenner's cowpox vaccine was a catalyst for this change. Indeed, it was through this procedure that the HBC men were able to gain their greatest triumph in the field of public health, culminating in the vaccination program of 1838–1839. The breadth of this undertaking was immense, encompassing much of the western half of what is now Canada, and yet it was accomplished with relatively little difficulty through the Company's efficient transport and command systems. In this case a fur-trading company was able to carry out a systematic vaccination that not even the United States could parallel.

The use of Jenner's vaccine made the HBC far more successful at controlling smallpox in the west than it otherwise would have been. However, the traders did not continue with the systematic approach that London favoured. Instead, they fell back on the policy of vigilance and prompt response to outbreaks, and for a time this seems to have worked effectively. For several decades following 1839 the disease failed to get more than a toehold in the Canadian west, even as it emerged frequently in epidemic form in American territory. Nevertheless, as the initial widespread immunity gained in 1838–1839 waned, as unexposed children were born, and as a new, unvaccinated Native population emerged, this reactive approach eventually proved vulnerable. Its weakness was exposed in 1869–1870, when smallpox returned to the Canadian Plains and political crisis in Red River prevented the shipment of fresh vaccine supplies throughout the country. Although some traders and missionaries on the eastern part of the Plains were able to limit the spread of the disease, those in the west could do nothing without adequate supplies; consequently, the epidemic claimed several thousands of casualties among the Blackfoot, Cree, and Assiniboine, living in what is now Alberta and western Saskatchewan, who were not so protected.[91] Had London's original policy of comprehensive vaccination been adhered to, rather than the ad hoc vaccinations favoured by the traders, it is possible that this epidemic, too, might have been averted.

NOTES

1. Henry F. Dobyns, *Their Number Become Thinned: American Population Dynamics in Eastern North America* (Knoxville: University of Tennessee Press, 1983), 441; Ann F. Ramenofsky, *Vectors of Death: The Archaeology of European Contact* (Albuquerque: University of New Mexico Press, 1987), 130; Donald B. Cooper, *Epidemic Diseases in Mexico City, 1761–1813*, Latin American Monographs, vol. 3 (Austin: Institute of Latin American Studies, University of Texas, 1965), ix, 56, 68–69, 86.

2. Marc Simmons, "New Mexico's Smallpox Epidemic of 1780–1781," *New Mexico Historical Review* 41 (1966): 319–326, on 321–323.

3. Laura L. Peers, *The Ojibwa of Western Canada*, Manitoba Studies in Native History, vol. 8 (Winnipeg: University of Manitoba Press, 1994), 20; John F. Taylor, "Sociocultural Effects of Epidemics on the Northern Plains: 1734–1850," *Western Canadian Journal of Anthropology*. 7, no. 4 (1977): 55–81; W. Raymond Wood, "Plains Trade in Prehistoric and Protohistoric Intertribal Relations," in *Anthropology on the Great Plains*, edited by W. Raymond Wood and Margot Liberty (Lincoln: University of Nebraska Press, 1980), 98–109, on 100.

4. Joseph B. Tyrrell, ed., *David Thompson's Narrative of His Explorations in Western America 1784–1812*, Publications of the Champlain Society, vol. 12 (Toronto: Champlain Society, 1916), 321; Edwin E. Rich and Alice M. Johnson, eds., *Cumberland and Hudson House Journals 1775–82*, 2nd ser., *1779–82*, 2 vols., Publications of the Hudson's Bay Record Society, vols. 14–15 (London: Hudson's Bay Record Society, 1952), 2, 223–224, 262; Jody F. Decker, "'We Should Never Be Again the Same People': The Diffusion and Cumulative Impact of Acute Infectious Diseases Affecting the Natives on the Northern Plains of the Western Interior of Canada 1774–1839," Ph.D. dissertation, York University, 1989, 86. See also C. Stewart Houston and Stan Houston, "The First Smallpox Epidemic on the Canadian Plains: In the Fur-Traders' Words," *Canadian Journal of Infectious Diseases* 11, no. 2 (2000): 112–115.

5. Tyrrell, *Thompson's Narrative*, 322.

6. See, for example, F.J. Paul Hackett, *"A Very Remarkable Sickness": Epidemic Disease in the Petit Nord, 1670–1846*, Manitoba Studies in Native History, vol. 14 (Winnipeg: University of Manitoba Press, 2002); John F. Taylor, "Sociocultural Effects of Epidemics on the Northern Plains: 1735–1870," M.A. thesis, University of Montana, 1982, 38–39.

7. In practice the charter proved largely unenforceable, and for much of its history the company was beset by competitors from Canada and what is now the United States.

8. Arthur J. Ray, "The Decline of Paternalism in the Hudson's Bay Company Fur Trade, 1870–1945," in *Merchant Credit and Labour Strategies in Historical Perspective*, edited by R. Ommer (Fredericton: Acadiensis Press, 1990), 188–202.

9. Victor P. Lytwyn, *Mushkekowuck Athinuwick: Original People of Swampy Land*, Manitoba Studies in Native History, vol. 12 (Winnipeg: University of Manitoba Press, 2002). Among the Lowland Cree some bands resided farther from the coast and were less frequent visitors to the Hudson Bay posts, perhaps hunting geese during the spring and fall, but not generally appearing during the winter: Victor P. Lytwyn, "The Hudson Bay Lowland Cree in the Fur Trade to 1821: A Study in Historical Geography," Ph.D. dissertation, University of Manitoba, 1993, 63. Some Ojibway bands lived just beyond the lowlands and participated in the posts' deer hunt, but did not spend long periods of time there.

10. Hudson's Bay Company Archives (hereafter HBCA), B.239/a/80, fols. 63, 69, Provincial Archives of Manitoba (PAM), Winnipeg, Manitoba.

11. Cocking, born in 1743 at York, England, was hired by the company as a writer in 1765. During the 1770s he made several important exploratory trips inland from Hudson Bay, reaching as far onto the plains as western Saskatchewan. His resourcefulness was to stand him in good stead as he faced the coming epidemic. In 1775 Cocking took charge of Cumberland House on the Saskatchewan River. Two years later he transferred to Severn as the master of that post; there he stayed until 1781, when he was called on to take charge of York Factory for a brief time, temporarily replacing Humphrey Marten, who departed for England to attend to his health. See Lawrence J. Burpee, ed., "Journal of Matthew Cocking, from York Factory to the Blackfeet Country, 1772–73," *Proceedings and Transactions of the Royal Society of Canada*, 3rd ser., 1908, sect. 2, 89–121; Irene M. Spry, "Mathew Cocking," in *Dictionary of Canadian Biography, 1771–1800*, edited by Francis G. Halpenny (Toronto: University of Toronto Press, 1979), 156–158.

12. There were apparent outbreaks of the disease at York Factory in 1720 and again in 1738, and in both cases the HBC men seem only to have observed the disease, without trying to prevent its spread; see Hackett, *Very Remarkable Sickness*, 55–56, 71. In one instance, during an epidemic of an unknown disease at York in 1757–1758, trader James Isham did gather a few simple statistics concerning the names of the Aboriginal peoples who died, and assembled them into a table with the numbers of survivors: HBCA, B.239/a/44, fols. 41v–42.

13. The method was to transfer small amounts of wound pus from a person with a mild case of the disease to one who had not yet been exposed: Esther Wagner Stearn and Allen E. Stearn, *The Effect of Smallpox on the Destiny of the Amerindian* (Boston: Bruce Humphries, 1945), 53; George Rosen, *A History of Public Health* (Baltimore: Johns Hopkins University Press, 1993), 160.

14. Cyril W. Dixon, *Smallpox* (London: J. & A. Churchill, 1962), 236, 244.

15. Rosen, *History*, 161–163.

16. Dixon, *Smallpox*, 239, 240.

17. Robinson was the surgeon at York Factory during the epidemic and until 1786 (HBCA, Search File "Robinson, Alfred"). He would have served on the post's council under Cocking and advised him on medical matters, but the decision ultimately was the trader's. Houston and Houston ("First Smallpox Epidemic," 113) have suggested that the plan to isolate the Upland Indians was Robinson's, although there appears to be no evidence that this was the case.

18. The relationship between the HBC and the British scientific community was well established by the 1780s. Company employees routinely collected and passed on data concerning the climate and natural history of Hudson Bay. In 1769 the Royal Society and the HBC co-operated to send an astronomer to Hudson Bay to observe the planet Venus passing in front of the sun, while company employees were trained to observe and record this event. One earlier employee,

Christopher Middleton, was a publishing member of the Royal Society; see, for example, Edwin E. Rich, ed., *Observations on Hudson's Bay, 1743, and Notes and Observations on a Book Entitled a Voyage to Hudsons Bay in the Dobbs Galley, 1749* (Toronto: Champlain Society for the Hudson's Bay Record Society, 1949), 168 (n. 1), 327, 328. Another employee, Thomas Hutchins, was a surgeon stationed at Albany Fort at the time of the epidemic, and in 1783 was awarded the Copley Medal by the Society for research in determining the freezing point of mercury: William B. Ewart, "Thomas Hutchins and the HBC: A Surgeon on the Bay," *Beaver* (August/ September 1995): 39–41.

19. For example, the Huron of the Great Lakes region placed a good deal of the blame for the initial round of epidemics during the seventeenth century on the appearance of the Roman Catholic missionaries who went to live in the Native villages, generating hostility toward the French clerics. During the Seven Years' War the fear of smallpox influenced the Great Lakes tribes to forgo sending warriors to the east when called on to aid the French against the British: D. Peter MacLeod, "Microbes and Muskets: Smallpox and the Participation of the Amerindian Allies of New France in the Seven Years' War," *Ethnohistory* 39 (1992): 42–64. On occasion, the consequences could be more direct and drastic—as, for instance, in 1847 with the Whitman massacre in Oregon, which is thought to have been the consequence of accusations by the local Cayuse that the Whitmans, a family of Methodist missionaries, had introduced measles among them.

20. Hackett, *Very Remarkable Sickness*, 94–95. Some 40 years later, rival Northwest Company traders attempted to arouse the ire of the Indians against the HBC by falsely accusing them of being responsible for importing measles and whooping cough (ibid., 141).

21. Conversely, it seems to have been used with some frequency among the Aboriginal peoples of Mexico and South America: Stearn and Stearn, *Effect of Smallpox*, 54.

22. Frederick Haldimand to John Johnson, Quebec, November 3, 1783, 180–181; Johnson to Haldimand, Montreal, November 10, 1783, 182, Correspondence with Officers Commanding at Michilimackinac, MG 21, Add. MSS #21775, microfilm reel #H-1450, Haldimand Papers, National Archives of Canada, Ottawa; R. Mathews to Major Harris, Quebec, November 2, 1783, 125–126, Add. MSS. #21788, microfilm reel #H-1452, ibid. It was their intention to seize and bury the supply of smallpox to prevent its helping to spread the epidemic.

23. Rosen, *History*, 44–45. Isolation had also long been used to prevent the spread of leprosy.

24. Cooper, *Epidemic Diseases*, 99–102; Percy M. Ashburn, *The Ranks of Death* (New York: Coward-McCann, 1947).

25. John J. Heagerty, *Four Centuries of Medical History in Canada and a Sketch of the Medical History of Newfoundland*, 2 vols. (Toronto: MacMillan, 1928), (2), 25–26; Edmund B. O'Callaghan, *Documents Relative to the Colonial History of the State of New York; Procured in Holland, England and France*, 15 vols., edited by John Romeyn Brodhead (Albany: Weed, Parsons, 1853–1861), (9), 1029.

26. Dixon, *Smallpox*, 238.

27. Henry Ellis, *A Voyage to Hudson's Bay by the "Dobbs Galley" and "California" in the Years 1746 and 1747 for Discovering a North West Passage: With an Accurate Survey of the Coast and a Short Natural History of the Country together with a Fair View of the Facts and Arguments from Which the Future Finding of Such a Passage Is Rendered Possible* (London: Printed for H. Whitridge, 1748), 201.

28. The plantation of an HBC fort was an area located between the "the battery and ship," which "was separated from the fort by two rows of high fences, between which were store-houses, the cookery and workshops" (Arthur S. Morton, *A History of the Canadian West to 1870–71, Being a History of Rupert's Land (the Hudson's Bay Company) and of the North-West Territory (Including the Pacific Slope)* [Toronto: University of Toronto Press, 1939], 149).

29. HBCA, B.239/a/80, fol. 79^v.
30. Ibid., fol. 90^v.
31. Ibid., fol. 91.
32. HBCA, B.198/a/28, fol. 3.
33. This was Charles Price Isham, a mixed-blood employee, who contracted the disease on the Saskatchewan River but later recovered: Rich and Johnson, *Cumberland and Hudson House*, 263.
34. Houston and Houston, "First Smallpox Epidemic."
35. HBCA, B.198/V28, fols. 14–15.
36. Ibid., fol. 15.
37. In fact, none of the HBC men who were stationed at the Bayside posts contracted the disease. This is no doubt indicative of the existing immunity of the European-born traders by virtue of earlier exposure to smallpox.
38. Houston and Houston, "First Smallpox Epidemic,"113.
39. Thus, with more than two months having passed since the arrival of the first smallpox victims at York Factory, William Falconer wrote on August 12 that "by this prudent precaution the homeguards here are preserved" (HBCA, B.198/a/28, fol. 3). See also HBCA, B.239/a/80, fols. 93–93^v; Rich and Johnson, *Cumberland and Hudson House*, 298. On August 15 the *King George* arrived at York from England, and the captain noted "a Dreadfull amount of Indians dying in the Small Pox" (HBCA, C.l/386, August 15, 1782); in all probability these were Upland Indians who were too ill to depart the vicinity of the fort for the interior.
40. Richard Glover, "La Pérouse on Hudson Bay," *Beaver* (March 1951): 42–46. For his part, Mathew Cocking was able to slip by the French ships aboard the *King George*, and managed to sail to England on August 23: Spry, "Mathew Cocking"; Edward Umfreville, *The Present State of Hudson's Bay: Containing a Full Description of that Settlement, and the Adjacent Country; and Likewise of the Fur Trade, with Hints for Its Improvement, &c. &c.*, edited by W. Stewart Wallace (reprint, Toronto: Ryerson Press, 1954), 67.
41. Donald B. Gunn and Charles Tuttle, *History of Manitoba from the Earliest Settlement to 1835* (Ottawa: Maclean, Rogers, 1880), 87.
42. HBCA, B.239/a/83, fol. 5.
43. HBCA, B.198/a/29, fol. 44.
44. On the impact of the epidemic in the Albany area, see Lytwyn, *Mushkekowuck Athinuwick*, 164–165, 171; but on its diffusion, cf. Hackett, *Very Remarkable Sickness*, 93–118.
45. Jenner's first experiment with vaccine on a boy, James Phipps, occurred in 1796, but it was not until 1798 that he published his first treatise: Edward Jenner, *An Inquiry into the Causes and Effects of the Variolae Vaccinae, a Disease Discovered in Some of the Western Counties of England, Particularly Gloucestershire, and Known by the Name of the Cow Pox* (London: Sampson Low, 1798).
46. Dixon, *Smallpox*, 261.
47. "Edward Jenner," *Dictionary of National Biography, 110:* 760; Dixon, *Smallpox*, 266, 270, 271.
48. Nicolau Barquet and Pere Domingo, "Smallpox: The Triumph over the Most Terrible of the Ministers of Death," *Annals of Internal Medicine, 1997, 127* (8, part 1), 635–642. The Spanish employed a form of serial vaccination among orphan boys to maintain the supply of vaccine on the long journey across the Atlantic; they were able to control the rate of cowpox infection by using arm-to-arm infections in a sequential manner.
49. Indeed, the epidemic, by virtue of the losses that had been incurred, had played a major role in shaping the emerging fur trade. See Victor P. Lytwyn, *The Fur Trade of the Little North: Indians, Pedlars, and Englishmen East of Lake Winnipeg, 1760–1821* (Winnipeg: Rupert's Land Research Centre, 1986), 44; Morton, *History of the Canadian West*, 334; Richard Glover, "Introduction," in Rich and Johnson, *Cumberland and Hudson House*, lxii.

50. HBCA, D.5/5, fols. 49–49ᵛ.

51. Ibid.

52. I have yet to find evidence of an earlier HBC interest in vaccination, although it is likely that such evidence exists somewhere within the Company's massive archival record. It seems improbable that the HBC would not have at least expressed an interest in a procedure that could preserve their trade prior to 1811, and London had certainly shipped vaccine via Hudson Bay prior to 1813 when it was employed at York Factory. When smallpox returned in 1801 as part of a localized outbreak on the south branch of the Saskatchewan River, the trader, Peter Fidler, arrived only after the disease had passed: HBCA, E.3/2, fol. 71; and see Alice M. Johnson, *Saskatchewan Journals and Correspondence*, vol. 26, *1795–1802* (London: Hudson's Bay Record Society, 1967), 294, 315. About a year later, smallpox broke out at Sault Ste. Marie on Lake Superior after an infected Canadian trader had returned from Montreal. Despite efforts to isolate him, it soon spread to the local Ojibway Indians, with disastrous results; however, "by judicious and early measures" the military surgeon stationed at nearby St. Joseph's Island was able to prevent its further diffusion, perhaps using smallpox vaccine (Henry R. Schoolcraft, *Summary Narrative of an Exploratory Expedition to the Sources of the Mississippi River in 1820: Resumed and Completed, by the Discovery of Its Origin in Itasca Lake, in 1832*, reprinted [Millwood: Kraus Reprints, 1973], 580).

53. "Instructions to Miles Macdonell," 1811, 179–80, Selkirk Papers, PAM.

54. Peers, *Ojibwa of Western Canada*; Chief Albert Edward Thompson, *Chief Peguis and His Descendants* (Winnipeg: Peguis Publishers, 1973).

55. The vaccine had been found in the warehouse of the nearby Severn post: HBCA, B.239/a/124, fol. 73ᵛ.

56. HBCA, B.51/a/2, fol. 13. See also Provencher to Dionne, September 1, 1819, Belleau Papers, MG 7 D13, reel 1, PAM.

57. Years later, the explorer Henry Hind quoted HBC Governor Sir George Simpson in his *Narrative* to the effect that "in the northern parts of Rupert's Land a great mortality took place in 1816, 1817, and 1818, from small-pox and measles. Vaccine inoculation was then introduced by the Hudson' [sic] Bay Company" (Henry Youle Hind, *Narrative of the Canadian Red River Exploring Expedition of 1857 and of the Assiniboine and Saskatchewan Exploring Expedition of 1858* [2 vols. in 1], reprint ed. [Edmonton: Hurtig, 1971], 143); Simpson arrived only after the epidemic, and so his confusion over the diseases and the years is not unexpected.

58. HBCA, A.10/2, fol. 242. See also B.51/a/3, fol. 5ᵛ.

59. Ross Mitchell, *Medicine in Manitoba: The Story of Its Beginnings* (Winnipeg: Stovel-Advocate Press, 1954), 34, 36.

60. The noted American Fur Company (AFC) trader Edwin Denig, who was later stationed at Fort Union on the Missouri, acknowledged that the HBC vaccination efforts began long before the 1837–1838 epidemic: "Another visitation of this malady happened in 1838, but owing to the good management of the Hudson's Bay Company most of the nation [the Cree] were preserved by introducing vaccine matter and persisting in its application for several years previous" (cited in John C. Ewers, *Five Indian Tribes on the Upper Missouri: Sioux, Arickaras, Assiniboines, Crees, Crows* [Norman: University of Oklahoma Press, 1961], 115).

61. Edwin E. Rich, *The Letters of John McLoughlin from Fort Vancouver to the Governor and Committee* (Toronto: Champlain Society for Hudson's Bay Record Society, 1941), xxiv; Paul C. Thistle, *Indian-European Trade Relations in the Lower Saskatchewan River Region to 1840*, Manitoba Studies in Native History, vol. 2 (Winnipeg: University of Manitoba Press, 1986), 62; HBCA, B.49/a/40, fols. 8ᵛ, 9–10.

62. HBCA, B.135/a/128, fols. 30, 50d; B.3/e/8, fol. 3; B.3/a/130, fols. 20, 21ᵛ–23ᵛ; Edwin E. Rich and R. Harvey Fleming, *Colin Robertson's Letters, 1817–1822*, Publications of the Champlain Society

(Toronto: Champlain Society for Hudson's Bay Record Society, 1939), 244–245. In the annual report for the Albany District, Vincent informed London that, after the departure of the doctor from nearby Moose, he had "vaccinated all the rest of half breed children at the place and with the assistance of Mrs. Vincent many of the Indians also who applied to have the operation performed" (HBCA, B.3/e/8, fol. 3). Vincent's decision to have his wife and children accompany him on his retirement to England was a highly unusual one among the HBC men of this era who took Native wives. Given that Vincent's family had not been exposed to smallpox before, taking them to England where the disease was endemic might well have proved a death sentence without the vaccinations.

63. HBCA, B.3/a/130, fol. 20ᵛ. By "Canada," McKay was referring to the provinces of Upper and Lower Canada, which correspond to what are now the southern parts of the provinces of Ontario and Quebec, respectively. Most likely the supply came from Montreal, from which city the HBC regularly shipped some goods at this time.

64. On the diffusion of this epidemic, see Arthur J. Ray, "Smallpox: The Epidemic of 1837–38," *Beaver* 306, no. 2 (1975): 8–13; Clyde D. Dollar, "The High Plains Smallpox Epidemic of 1837–38," *Western Historical Quarterly* 8 (1977): 15–38; Michael K. Trimble, "Chronology of Epidemics among Plains Village Horticulturalists: 1738–1838," *Southwestern Lore* 54 (1988): 4–31.

65. Michael K. Trimble, "Epidemiology on the Northern Plains: A Cultural Perspective," Ph.D. dissertation, University of Missouri-Columbia, 1985, 189; Dollar, "High Plains Smallpox," 20–24; Arthur J. Ray, "Diffusion of Diseases in the Western Interior of Canada, 1830–1850," *Geographical Review* 66, no. 2 (1976): 139–157, on 155–156.

66. Pearson, "Lewis Cass." Pearson argues that the failure to vaccinate the Mandan and Hidatsa, and by extension the tribes living farther up the river, was a conscious decision on the part of Secretary of War Lewis Cass, who considered these groups to be hostile to the United States.

67. D.L. Ferch, "Fighting the Smallpox Epidemic of 1837–38: The Response of the American Fur Company Traders. Part I," *Museum of the Fur Trade Quarterly* 19, no. 4 (1983): 2–7; D.L. Ferch, "Fighting the Smallpox Epidemic of 1837–38: The Response of the American Fur Company Traders. Part II," ibid., 20, no. 1 (1984): 4–8.

68. Ferch (ibid., 1984, 4) condemned variolation as a dangerous procedure, although this was not entirely warranted. Variolation had been used successfully as a control for smallpox for centuries, and by the time of Jenner's discovery it was relatively safe when done properly, although not so safe as vaccination. Done improperly, however, it would spread the epidemic as surely as if the victims had contracted the disease unaided.

69. Pearson, "Lewis Cass." For example, the commissioner of Indian Affairs was uncertain for a considerable time whether the disease had crossed the Rocky Mountains. See U.S. Department of War, *Annual Report of the Commissioner of Indian Affairs* (Washington: Government Printing Office, 1838), 453.

70. Pearson, "Lewis Cass," 24; Department of War, *Annual Report*, 454. It was claimed in the *Annual Report* that a doctor sent by the Office of Indian Affairs was able to vaccinate some 8,000 people, although this hardly seems credible (454).

71. William Clark to C.A. Harris, St. Louis, April 6, 1838, U.S. Office of Indian Affairs, *Letters Received by the Office of Indian Affairs*, reel #751, St. Louis Agency, 1836–1838; Edmund C. Bray and Martha Coleman Bray, *Joseph N. Nicollet on the Plains and Prairies: The Expeditions of 1838–39, with Journals, Letters, and Notes on the Dakota Indians*, Publications of the Minnesota Historical Society (St. Paul: Minnesota Historical Society, 1976), 10, 78, 219.

72. Ray, "Diffusion of Diseases," 156.

73. Arthur J. Ray, *Indians in the Fur Trade: Their Roles as Trappers, Hunters, and Middlemen in the Lands Southwest of Hudson Bay, 1660–1870* (Toronto: University of Toronto Press, 1988), 188–192.

74. Ibid., 190.

75. Typically for those agencies that were more distant from the American frontier, agents made the rounds of their assigned bands only during the summer, and had only limited communication, if any, throughout much of the year. In some cases, such as with the subagent for the Mandan during this epidemic, they visited the Indians only rarely: Pearson, "Lewis Cass," 17.

76. Some indication of how the dried vaccine was prepared is provided by James Hargrave in a letter to Robert Harding (of Churchill), in which Harding was told: "Take half of the enclosed parcel, bruise it upon a piece of glass, dissolve it in a single drop of water, and then use it for vaccination in the same manner as before. Dr. Whiffen says it is perfectly fresh, and sufficient for a dozen cases. If efficacious in one, you will then be enabled to extend it from that one throughout the whole of your Indians" (HBCA, B.239/ b/93, fol. 16d). The last sentence refers to serial, or arm-to-arm, transmission of the cowpox. There were limits to the viability of even dried vaccine, and some problems did occur, including with an earlier supply sent to Harding: ibid., fols. 14, 14d, 16ᵛ. At Carlton House the failures of the cowpox vaccine had an especially demoralizing effect when several people who had previously been vaccinated came down with the disease, although it is not clear how recently they had had the procedure or, indeed, how old was the vaccine supply: HBCA, D.5/14, fol. 360ᵛ. Nevertheless, most stocks appear to have been viable, and where they were ineffective additional supplies were sent from other nearby posts as replacements.

77. HBCA, B.49/a/49, fol. 24.

78. For example, ibid., fols. 24ᵛ, 25ᵛ, 28, 31.

79. HBCA, B.105/a/20, fol. 15ᵛ.

80. Ibid., fol. 16.

81. HBCA, B.198/a/77a, fols. 33, 33ᵛ.

82. HBCA, B.239/b/93, fol. 23ᵛ.

83. Ibid., fols. 23ᵛ–24.

84. HBCA, D.5/5, fols. 49–49ᵛ.

85. Ibid., fol. 49. See also HBCA, A.6/2, fol. 158ᵛ.

86. For instance, on May 30, 1838, Thomas McMurray at Pic post wrote that he had "inoculated the women and children of the Post with vaccine matter received from Montreal" (HBCA, B.162/a/10, fol. 19).

87. In a letter to James Douglas at Fort Vancouver, dated October 31, 1838, the governor and committee expressed their concern over the spread of smallpox to the Indians of the Pacific Northwest and their fear that it would spread to the Indians of the interior. Notwithstanding Douglas's earlier vaccination efforts, they were sending additional vaccine by ship to be distributed throughout the country and, as backup in case the first supply was ineffective, they let him know that "a further supply will be sent across the Mountains next summer [1839]" (HBCA, A.6/25, fol. 10).

88. HBQV, A.6/25, fol. 30.

89. At the 1857 parliamentary inquiry into the Company's charter, HBC Surgeon Dr. John Rae was questioned regarding the Company's role in vaccination, and he replied that "vaccine matter is sent to all the Posts" (HBCA, E.18/4, fol. 94). It would seem, however, that supplies were not always sufficient. A few years after the vaccination program the supply of vaccine was a concern at the Red River Settlement. In March 1841 George Gladman wrote to Chief Factor James Hargrave of York Factory asking him to "mention to Dr. Gillespie that Mr. Bunn has not at present any 'Vaccine lymph' but having requested a supply to be sent him from Canada he will give us a share should it come safely at hand" (Gladman to Hargrave, Red River Settlement, March 27, 1841, in George P. de T. Glazebrook, ed., *The Hargrave Correspondence 1821–1843*, Publications of the Champlain Society, vol. 24 [Toronto: Champlain Society, 1938], 342).

90. HBCA, D.5/7, fol. 229d. Likewise, in 1846 HBC trader John Tod vaccinated the Indians in the vicinity of Fort Kamloops in the Pacific Northwest in the midst of an epidemic: Stearn and Stearn, *Effect of Smallpox*, 92; and the HBC's Dr. John Helmcken vaccinated the Songhees on Vancouver Island during an epidemic in 1862: Robert T. Boyd, *The Coming of the Spirit of Pestilence: Introduced Infectious Diseases and Population Decline among Northwest Coast Indians, 1774–1874* (Vancouver: University of British Columbia Press, 1999).

91. Ray, *Indians in the Fur Trade*, 191–192. On the epidemic, see also James Daschuk, "The Political Economy of Indian Health and Diseases in the Canadian Northwest, 1730–1945," Ph.D. dissertation, University of Manitoba, 2002, especially Chapter 6.

Critical Thinking Questions

1. According to Crosby, how does the process of Demographic Takeover occur? What are its phases, and what requirements need to be met for a demographic takeover to be successful?

2. Which organisms, in your view, are most crucial for demographic takeovers to be successful—plants, animals, micro-organisms, or humans?

3. Again returning to the Recovery Narrative discussed by Carolyn Merchant in Part I of this reader, how does the concept of Demographic Takeover affect the story as it is presented in the recovery narrative? In particular, how does it affect the concept of human agency and action as the key components of the recovery narrative?

4. In what ways do the authors of the primary sources employed by Cook demonstrate European perspectives of the new land that they were colonizing? What were their overall objectives in settling Acadia, and what impact did that settlement have on the environment and Native peoples of the region?

5. Why is the concept of cultivation so important to the European observers of life in Acadia?

6. Cook suggests that the idea of the garden was, for the Europeans, potentially a powerful tool for "the civilization of the indigenous peoples." How did the Europeans see this process unfolding?

7. Why did the vigorous quarantine policy adopted by the Hudson's Bay Company personnel in the face of the smallpox epidemic of 1779–1783 ultimately prove unsuccessful? What accounts for the HBC's preference for quarantine over variolation?

8. In the face of the 1837 smallpox outbreak the HBC adopted a very aggressive policy of vaccination of Native peoples on the Canadian prairies. Why did the HBC try to combat the disease as aggressively as they did, and what was the response among the Native peoples in return?

9. What lessons did the HBC learn from the 1837 epidemic, and how did they affect subsequent relations between the Company and Plains Natives?

FURTHER READING

Alfred W. Crosby, *Ecological Imperialism: The Biological Expansion of Europe, 900–1900* (New York: Cambridge University Press, 1993).

This is a study with a much broader perspective than that implied in Crosby's article in this section. In this book he investigates the means by which Europeans spread not just through North America but throughout the world. He discusses in great detail the so-called "portmanteau species" that rode alongside Europeans as they spread globally and shows, again in persuasive detail, the contributions that those species made to European dominance in large parts of the world. Crosby's work may be read as a foundation for Jared Diamond's, which follows.

William Cronon, *Changes in the Land: Indians, Colonists and the Ecology of New England* (New York: Hill and Wang, 1983, 2003).

When it first appeared this book was something of a watershed. Cronon took Crosby's concept of biological exchange and applied it to New England, synthesizing a story of Native–European interaction in the colonial period that was simultaneously ecological and historical in form. More than 20 years after its appearance, Cronon's argument still seems fresh and relevant as he shows the impact of biological imperialism on a particular region. While this is by no means a micro-history, its narrower focus allows readers to see the strength and elegance of Crosby's theories as they are applied to a particular historical and geographical case.

Jared Diamond, *Guns, Germs, and Steel: The Fates of Human Societies* (New York: Norton, 1997, 1999).

Diamond takes the idea of biological exchanges and carries it further by incorporating it into an even broader analysis that seeks to explain the major directions of human history. While some critics have attacked Diamond's argument as being overly deterministic, there is no doubt that his synthesis of history, geography, linguistics, and genetics represents an important and many would say persuasive version of environmental history on the grandest scale.

Elinor G.K. Melville, *A Plague of Sheep: Environmental Consequences of the Conquest of Mexico* (New York: Cambridge University Press, 1994).

Similar to Cronon's analysis of environmental change in New England, Melville's account of changing agricultural patterns in a Mexican valley in the sixteenth century demonstrates the techniques of environmental history as they are applied to explaining and analyzing environmental change. Melville shows how the arrival and subsequent dominance of Spanish agricultural techniques, based on pastoralist sheep-herding, dramatically affected an environment where Native agricultural techniques, while transforming the landscape, had done so in sustainable ways. The Spanish agriculture was unsustainable in the new environment, and led to significant and irreversible environmental change.

Noble David Cook, *Born to Die: Disease and New World Conquest, 1492–1650*. New Approaches to the Americas Series (New York: Cambridge University Press, 1998).

Cook outlines in this broad but sophisticated effort the ways in which disease facilitated the European conquest of the Americas. Despite the breadth of the topic, Cook is able to show in detail the tremendous impact of Eurasian diseases on the populations of the Americas, describing it, quite rightly, as probably the greatest demographic catastrophe in human history.

RELEVANT WEB SITES

The Columbian Exchange: Plants, Animals, and Disease between the Old and New Worlds.
http://www.nhc.rtp.nc.us:8080/tserve/nattrans/ntecoindian/essays/columbian.htm
This page series is part of the "Native Americans and the Land" series created by the United States National Humanities Center. It is authored by Alfred Crosby and contains a multi-part essay on the Columbian Exchange together with a very extensive list of sources in both on-line and print media.

Exploration, the Fur Trade, and the Hudson's Bay Company
http://www.canadiana.org/hbc/intro_e.html
This is an excellent site that provides a history of the fur trade and HBC in Canada. It is created for a relatively young audience, and sometimes its language is overly basic. However, it nevertheless provides excellent coverage of the HBC's activities in Canada, with a good "Personalities" section that discusses the careers of the company's adventurers, officials, and explorers. There is a very comprehensive page of links to primary sources, and the site offers heavy coverage of pictures, maps, and timelines.

Acadian and Micmac History
http://museum.gov.ns.ca/arch/infos/info.htm
This page is a portal to a series of information sheets created by the Nova Scotia Museum concerning the history of Acadia, Acadians, and the Micmac. They are useful introductions to current work, both historical and archaeological, that is being done on the period and peoples discussed in Cook's "1492 and All That."

Virtual Museum of New France
http://www.civilization.ca/vmnf/
This excellent Web site was created and maintained by the Canadian Museum of Civilization. Its broad coverage includes the lives of Europeans (such as the *habitants*, the *coureurs de bois*, and explorers) who peopled and expanded New France. The emphasis is on day-to-day living and how resources were employed. There is an excellent set of maps, a chronology, and a long list of external links in both English and French.

Pre-industrial Resources and the Changing Culture of Nature

The "New Europes" to be found in North America and other parts of the world were not only zones of settlement for European emigrants. They also served as sources of raw materials for the rapidly expanding European industrial market, which was gradually supplemented by (and also supplanted by) the expanding consumer market as the nineteenth century wore on. Although we may be familiar with the enormous demand for beaver pelts or timber, and the role of organizations such as the Hudson's Bay Company (HBC) in supplying that demand, smaller resource markets are less well understood. This does not, however, mean that they were insignificant, either in European economic terms or in Canadian environmental ones. One striking phenomenon that arises from Lorne Hammond's discussion of the HBC's activities in the Pacific Northwest is the energy expended by the company in *creating* new markets for furs, teeth, oils, feathers, and other exotic natural products found in the region. The source of this supply was guarded closely, indeed ruthlessly, by the HBC: in order to prevent commercial encroachment from the United States, for example, it was company policy to trap and hunt to extinction fur-bearing animals across wide areas, to create fur barrens, borders into which it would be pointless for competing trappers or hunters to penetrate. That the Company should engage in what Hammond calls "direct ecological warfare" may not surprise us, but what is remarkable is the breadth and degree of planning that characterized the campaigns. It is clear that, in adopting such brutal tactics, by the early nineteenth century at least the idea of inexhaustible plenty that had been so common among the early Euroamerican settlers was largely absent in the minds of HBC personnel.

At the same time, as Hammond shows, the HBC began to work to protect and conserve stocks of its most lucrative resource, the beaver, whose numbers were in decline in the second quarter of the nineteenth century. These conservation efforts may be seen as the counterpoint to the eradication campaigns undertaken by the company; in both activities we can see the growth of a managerial ideology, based on the premise that wildlife resources could be husbanded by the application of policy. On the one hand, if they were valuable, then animal resources could be protected and encouraged to expand in numbers and range. If, however, they were profligate breeders, they could be exploited maximally, or if they were predatory or otherwise nuisances, then they could be eradicated entirely.

In addition to welding North American ecosystems into European-driven markets for "products," European settlers also undertook to integrate the new environment into their cultural framework. Colin Coates outlines the way in which this process occurred in Lower Canada in the first half of the nineteenth century, though it occurred in much of the rest of Canada as well. He demonstrates the powerful psychological factors that were at work as elite European settlers tried (unsuccessfully, as it turns out) to establish a particular form of existence, one that mirrored the lifestyle of the landed gentry in England at the time. It is revealing that one participant in the endeavour referred to her actions as "neatifying" rather than "beautifying," thereby reflecting the idea that the Canadian environment was already beautiful, perhaps even outstripping England's in this regard, but that it was also riotous and uncontrolled as well.

Very little sentimentality is detectable in the events described by William A. Dobak in the last reading in this section. The annihilation of the Plains buffalo is, together with the astonishing eradication of the passenger pigeon, the totemic example of environmental destruction in the Americas. But all too often the near-extinction is told as a tale that occurred purely in the United States when, in reality, truly enormous kills were common north of the forty-ninth parallel as well. According to Dobak, the destruction of the buffalo arose from the imposition of market forces by the Hudson's Bay and North West companies. The market involved the Native peoples and Métis as active agents in the pemmican and buffalo robe trade. As Dobak shows, even more ruthless economic calculations were at work than meets the eye initially: Buffalo robes aroused little interest among HBC traders, and there was only low demand for them in the company's primary European market. Yet the robe trade was encouraged, partly to keep the Plains Natives active in the associated pemmican trade, which the company *did* require, and partly to keep them from drifting into trading partnerships with the American companies to the south. But Dobak demonstrates convincingly, too, that Native hunters were active participants in the trade, and that spiritual factors may have led them to believe that the origins of the buffalo were supernatural, and therefore the buffalo themselves were inexhaustible.

The overarching theme that ties these readings together is the idea of the commodification of nature. Euroamericans were seemingly interested primarily, perhaps overwhelmingly, in the profit to be generated by the new environments that they encountered as they spread westward across Canada. Gone is the earlier ideology of dominance, within which nature must simply be controlled. By the second quarter of the nineteenth century it had been replaced by an ideology of simple exploitation for commercial purposes. That exploitative mentality, more than anything else, led to the great anthropogenic extinctions of the nineteenth century.

Chapter Ten
Marketing Wildlife:
The Hudson's Bay Company and
the Pacific Northwest, 1821–1849

Lorne Hammond

Wildlife is an aspect of the fur trade that has received little attention in the literature, yet animals were the foundation of the trade.[1] Their presence lured trapper and trader, and the demographic fluctuations of wildlife, along with characteristics of different species, marketing, and the vagaries of fashion, were unpredictable variables of the business. [...]

In the Pacific Northwest, as in many other parts of the world, commerce was the first external agent in the exploration and assignment of values and utility to wildlife. Commerce predated scientific inquiry, European settlement, the legislative process, and the cultural influence of the writer and painter.

One influential force in the field of commerce was the Hudson's Bay Company (HBC), a group with a managerial structure that distinguished it from other fur trade companies. It was a company whose management was hierarchical, an arrangement complicated by distance since its headquarters were in London, England. Major HBC shareholders were represented by a governor, deputy governor, and committee. These individuals communicated their wishes about company operations in North America to George Simpson, the North American governor. Simpson had responsibility to provide the committee with detailed annual letters on conditions in each department and post within the districts under his control. The people in each department who were responsible for major posts were called chief factors. Chief traders and clerks ran the smaller posts. Coordinating this system involved complex supply lines and frequent communication via services such as the express canoe, which took only passengers and mail. Evidence suggests that this bureaucratic structure conditioned responses to managing and standardizing wildlife as products.[2]

In 1821 HBC merged with its competitor, the North West Company of Montreal, and gained significant new territory. This new area was designated the Columbia Department and it produced 8 percent of the 18.5 million hides and pelts exported from North America from 1821 to 1849.[3] [...] The Columbia Department included the northern interior posts of New Caledonia and much of present-day British Columbia, Oregon, and Washington. The latter two were jointly occupied with the United States under a temporary agreement regarding the disputed boundaries and ownership of the Oregon Territory, that land north of the Adams-Onis line of 1819 and west of the continental divide.

NATURE'S INVENTORY

Although it was not the first fur trading company in the Pacific Northwest, the Hudson's Bay Company was the largest and the most successful. The Company's rapid expansion into the region during the years 1821–1849 offers a good picture of the process of assessing wildlife as a product. When Governor George Simpson took his first tour of the Columbia Department in 1824–1825, diversification in trapping and hunting the region's wildlife was just beginning. He reported on the commercial potential of wildlife, such as the mountain goat.[4] As an experiment Simpson joined his voyageurs in eating the first two he saw; he described them as "tough." From then on there was a regular flow of potential products from the hinterland posts to London, from grizzly bears to the small hoary marmot. In 1826 he issued instructions to have sample skin and horns of a mountain goat saved for a naturalist, but the first specimens went to London for a determination of their commercial potential. In 1825 Chief Factor John McLoughlin forwarded samples of swan skins and sturgeon's bladders from his post on the Columbia River at Fort Vancouver "to know what they are worth." Seventeen years later he was still sending samples, including "oulachan oil," spermaceti (a wax found in the head cavities and blubber of the sperm whale), "sea horse teeth" (walrus tusks), and sea lion hides as possible products "for trial in the English market."[5]

Once these samples reached London, the Company sought informed opinions as to their relative quality, value, market, and potential competitors. Some items were sold at auction so that buyer response could be evaluated. The auction served to process market information, a means of introducing potential new products as "odd lots" on a test basis.[6] For example, wolf and wolverine appeared first in the HBC catalogs as "Sundries," but as demand increased they were sold as distinct categories of fur.

Beaver pelts for the felt hat industry are the most well-known item HBC supplied to the European market, but the company also imported a large variety of other wildlife products. Fur hats, muffs, cuffs, collars, and coats were only part of the trade. There were military contracts for bear skins; cutlers bought deer horns or stag horns for pen knife handles; and dentists used "sea horse teeth" (walrus tusks) to make dentures.[7] Castoreum, from the glands of beaver, sold as a scent lure for beaver traps and as a component in medicinal preparations and perfumes.[8] Isinglass, a pure gelatin extracted from the sturgeon's float bladder, helped clarify wine and beer. Feathers, such as the down of the trumpeter swan, sold for powder puffs.

Establishing a demand among the public for a wildlife product was difficult. Ross Cox, who traded in the Pacific Northwest before the Hudson's Bay Company, recalled how fickle the market for bearskins had been. At one point, he remembered, the North West Company found itself with a glut of bearskins and no buyers. So the company had a "hammercloth" made up, with a coat of arms in silver, and gave it to a British prince. It hung below the driver's seat on the front of his coach, covering the toolbox and advertising the rank of the coach's occupant. As the company had hoped, the bearskin was a fashion hit when it was shown at the king's next levee. Within three weeks the North West Company's warehouse was empty.[9] This and other lessons demonstrated the benefits of subtle promotion.

Although HBC's exploration of nature was primarily a commercial enterprise, the company's endeavours also promoted increased scientific knowledge. Hudson's Bay Company maintained a small museum of natural history in London and assisted institutions, artists, and scientists in their research. John Richardson, a prominent natural historian, was one such scientist.[10] His observations on wildlife illustrate the ways in which scientific study of an animal could conflict with its identity as a commercial product.

Richardson became involved in a disagreement about the red fox that reveals the company's competing interest for a recognizable and easily promoted product despite contrary opinions from scientists. The colour of individuals in a litter of the red fox is determined by a single pair of genes. Individuals may be born a colour other than red, similar to the occurrence of different colours of hair among the children in a family. Genetic variations in colour are further complicated by age, region, and climate.[11] The cross phase is a greyish-brown coat with dark black markings down the spine and across the shoulders. The silver is actually black with a white-tipped tail and silver frosting effect produced by the outer hairs of the coat. Silver is the most highly valued phase, partly because it is almost unknown in the European red fox.[12]

The silver fox was a mysterious animal, and market prices, thriving on mystery, reflected that. A prime silver fox sold for about $40, half the price of a good sea otter.

For HBC to insure a continuing supply of silver fox furs, it had to bow to the caprice of genetic variations within litters and accept less valuable colour phases as well. Refusing pelts of the other colour phases would discourage fox trapping generally and therefore reduce the number of silver foxes trapped. Thus the genetic characteristics of the red fox created constraints both on commercial harvesting and on marketing.

Although the fur trade industry preferred to distinguish between the colour phases as if they were separate forms of fox, the scientific community had a different opinion. In 1829 Richardson commented on red and cross foxes: "I am inclined to adhere to the opinion of the Indians in considering the Cross Fox of the fur traders to be a mere variety of the Red Fox, as I found on inquiry that the gradations of colour between characteristic specimens of the Cross and Red Fox are so small, that the hunters are often in doubt with respect to the proper denomination of a skin, and I was frequently told 'This is not a cross fox yet, but it is becoming so.'"[13]

There is a middle ground between the cross and red fox in their aesthetic variations, as twentieth-century biologist Alexander William Francis Banfield noted. He commented that differing perceptions of colour phases are the result of "selective pelting, and variation in the identification of the cross fox."[14] Certain pelts could therefore be graded subjectively either as cross or red fox at the trading post or during preparation for auction. Richardson disagreed with the distinction made by fur traders, although he was admittedly uncertain about how the colour variations occurred. But it was in the industry's best interest to ignore the opinion of scientific observers as well as Native peoples, and benefit from the flexibility of choosing the label for distinguishable but similar products.

The Hudson's Bay Company disagreed both with Richardson and Native peoples over the nature of another animal—the bear. The 1825 returns for the Columbia Department distinguished among three forms of bear skins: black, brown, and grizzly. The common black bear has several colour phases other than black, including cinnamon, honey, white, and blue. The bears of the interior of the Pacific Northwest were evenly divided between cinnamon-brown and black phases. Richardson noted that the perceptions of fur traders about the species differed from that of the Native peoples: "The Cinnamon Bear of the Fur Traders is considered by the Indians to be an accidental variety of this species (black bear), and they are borne out in this opinion by the quality of the fur, which is equally fine with that of the Black Bear."[15] For the bear, like the fox, colour phase characteristics constrained the supply of wildlife, but aided in establishing product identity.

COMPETITIVE STRATEGIES

Other species of wildlife arrived on the market more as a result of competitive strategy than on the merits of the product. Faced with competition from American trapping parties, especially in the Snake River basin of the disputed Oregon Territory, HBC responded by employing its own parties, in part as protection of what it viewed as its territory. These parties of Métis and Iroquois trappers worked constantly as the company engaged in a rigorous extirpation of beaver and river otters, creating "a fur desert" or a "cordon sanitaire" to destroy any inducement to American trappers and traders.[16] American traders were regarded as precursors of their

government's colonization policy, drawn to the area by the economics of the wildlife resource. As Governor Simpson explained, "The greatest and best protection we can have from opposition is keeping the country closely hunted as the first step that the American Government will take towards Colonization is through their Indian Traders and if the country becomes exhausted in Fur bearing animals they can have no inducement to proceed thither."[17] The result was not a competition for Native trade but direct ecological warfare.

The Snake River party bypassed trade with Native communities and moved directly into harvesting the resource, as did other trapping parties working on the Umpqua River and in California's Buena Ventura Valley. Both the American and Hudson's Bay Company trapping parties quickly realized that an early arrival at the trapping grounds was critical to the spring hunt. Until 1824 Snake River beaver had been trapped only from June through August, yielding a harvest of inferior summer pelts.[18] In 1825 spring hunts began, which not only collected the more valuable winter coats but also moved the hunt into the reproduction cycle of the beaver. Beavers mate in January and February and give birth between late April and the end of June.[19]

Harvesting intensity increased when first a fall hunt and then year-long trapping expeditions were instituted. In 1825 the Snake River party reorganized for this competition; Simpson commented that "the country is a rich preserve of Beaver and which for political reasons we should destroy as early as possible."[20] However, despite the Snake River party's specialized purpose (beaver comprised 67 percent of its returns) and strategic value, it collected only 10 percent of the Columbia's 443,010 beaver pelts from 1825 to 1849. The party's

43,113 beaver pelts were far fewer than the 132,000 taken within New Caledonia's sustainable harvesting system, one based on beaver ponds owned by individual Native families. But as part of the battle for control of the Oregon Territory, the symbolic and political value of those furs was more important than their economic value.

The fur traders also fought an ecological war with a natural predator.[21] The Hudson's Bay Company viewed competition from predators such as wolves as destructive to the company's interest. The attitude both of Simpson and the HBC governing committee in London toward predators is clearly documented and reflects a consensus among most senior members of the fur trade. In 1822 the committee instructed Simpson that wolves on the plains should be hunted in the summer and their hides prepared for use as leather: "If the wolves are not destroyed they will either kill or drive away the Buffalo; it is therefore desirable to destroy them, if the skins will pay for the expenses it will also be the means of employing the Indians."[22]

In the Columbia annual fur returns, the first entry for wolves appears for the outfit of 1827 when five wolves were traded. By 1830 the number of wolf pelts had climbed to 69, and in 1831 as the Company began exercising control over its expanded network of posts, the number was 468. In 1833, responding to Simpson's report of wolves preying on the cattle of the Red River Colony, the governor and committee went further, issuing these instructions concerning controlling wolves: "[S]ending re your request 'two ounces of Strychnine' which is considered the most powerful agent for destroying wild animals (it is used in the East Indies for killing Tigers and Leopards) ... three or four grains are a sufficient dose for a full grown beast; the best way to apply it is to make an incision in a piece of flesh in which the Strychnine should be inserted, and to place the bait in situations that the animals frequent."[23]

In 1839 McLoughlin requested poison for the company farms in the Pacific Northwest and for general sale to settlers. The committee replied by sending "a small quantity of Strychnine made up in dozes [*sic*] for the destruction of Wolves; it should be inserted in pieces of raw meat placed in such situations that the shepherd's dogs may not have access to them, and the native people should be encouraged by high prices for the skins to destroy wolves at all seasons."[24]

Although the Columbia Department wolf kill reached a high of 1,653 in 1847, it fell to only 76 in 1853. Whether they were poisoned to protect domestic cattle, or hunted down by parties of Native peoples, settlers, or individual trappers, it is clear that from the company's perspective wolves were a threat to its operations. Their instructions about how to deal with wolves also reveal a much larger process: the exchange of information concerning large predators among companies operating in different parts of the world. The wolf became a product in the fur trade only as a means of insuring and underwriting its extermination.

Another animal, the highly prized land or river otter, had a role as currency, paid to the Russians in exchange for a lease. Under the terms of the 1840 Hamburg Agreement between the Hudson's Bay Company and the Russian American Company (RAC), 2,000 Columbia river otter were taken to Sitka, Alaska, to pay the RAC for the use of the Alaskan panhandle. The lease contained a provision for a further 3,000 otter to be traded at the Hudson's Bay Company's option. Total otter exports from North America declined from 1840 to 1847 by roughly the amount of the lease. Despite the cost of bringing extra skins over the Rocky Mountains to Sitka, and the delay caused by taking payment in bills of exchange drawn on St. Petersburg, the Hudson's Bay Company believed it had a bargain. It calculated the actual cost of leasing the Alaskan panhandle for one year at the equivalent of a middle level manager's annual salary.[25] The RAC also benefited because, unlike the Hudson's Bay Company, it had official access to the lucrative Chinese market. HBC continued to use otter instead of cash to make the lease payment until 1856.

WILDLIFE CYCLES AND MARKETS

In the preceding cases the Hudson's Bay Company exercised active control over which animals it chose to pursue and present as products. Having decided to supply a form of wildlife, the company often found itself mediating between the forces of nature and the vagaries of the market. Two factors created the greatest difficulties: the seemingly separate worlds of wildlife demography and fashion. The company's ability to buffer these forces varied, depending on which species were involved. With lynx and muskrat the Hudson's Bay Company was successful in minimizing the impact of wildlife demography on the marketplace.

Although the dynamics of wildlife demography are still not fully understood, trappers and sportsmen have long recognized the existence of wildlife cycles. There are various theories about why these cycles exist, ranging from the cycles of sunspot activity to a combination of ultraviolet radiation and malnutrition to periodic epidemics. Current debate focuses

on intricate mathematical probability equations. Lloyd Keith has commented on the evolution of this discussion:

> Theories and hypotheses confound the natural-history literature. Most are untested and as a result unanimity is lacking on many aspects of animal-population cycles. For thirty-odd years these periods of abundance and scarcity have been passed off as resulting from chance alone; lacking sufficient precision or amplitude to be acceptable; being so multifactorial as to defy appraisal; and eventually regarded as phenomena easily reduced to ridicule. Not all attitudes are disparaging or skeptical. There are zealots in the field of biology who see cycles in virtually all tabulations of natural and social interactions. From the production of pig iron to tent caterpillars, and from ozone quantity to Nile floods, cycles have been regarded as the skeletons or the souls of numerical data.[26]

Nevertheless, Keith has concluded that there is a "ten-year cycle," a regular but imprecise fluctuation for several forms of grouse, the snowshoe rabbit, and its predator, the lynx. There is also evidence for such a cycle in the fox, the muskrat, and its predator, the mink.[27] These demographic fluctuations of the animals on which the fur trade depended threatened the stability of the market, but they also provided the Hudson's Bay Company with opportunities for speculation.

The lynx and the muskrat provide two examples of how the company responded to wildlife cycles. These species were wildcards in the marketplace because of their periodic sudden and massive population explosions. High spring water levels or low winter water levels exaggerated the muskrat's reproductive cycle, the former causing an increase in population and the latter a decrease. The lynx's cycle was dependent on the increase of its food source, the snowshoe rabbit.[28] Although the lynx was never a particularly popular fur, its rapid increases tended to flood the market and undercut prices for cats in general.[29]

The muskrat (referred to within the trade as "the rat"), a seemingly insignificant marsh dweller sold in the marketplace as the musquash, had a large impact on the fur trade. In the wild, like the beaver, it seeks shallow water where it feeds on roots and grasses. Unlike the beaver, it can have several litters in a year, resulting in dramatic population fluctuations.[30] These population fluctuations meant a periodic flooding of the market, which occurred in 1828 and 1834 when over one million muskrat skins arrived at auction in London. The muskrat's demographic cycle profoundly influenced the market because it could be used as a cheaper substitute for more expensive furs, especially in hat making. Muskrat sold for between 3 and 13 pence each, depending on supply. By contrast, beaver sold for 17 to 32 shillings per pound.

Although the company realized the increases were part of a recurring pattern, they could not predict the onset of the cycles. Their experience taught them how long peaks would last and what caused them to end. To support beaver prices the company repeatedly used private export contracts to divert muskrat surpluses from the London market. The one million skins taken in the cycle year of 1829 threatened the trade's stability, but the company knew this increase would have limited duration and be followed by an equally drastic collapse in population. They reacted by limiting the importation and paying close attention to demographics. Governor Simpson wrote to the committee that "it is

evident that the importation of next year will be small compared with that of the present, owing to a mortality which has seized the tribe [muskrats] in several parts of the Country, we beg leave to suggest that a part only be exposed to sale this year, and the remainder held until the following, when 'tis probable they will command better prices."[31]

A large increase in lynx and muskrat shipped in 1839 drew this reaction: "We do not wish that more than half the quantities shipped ... this year, but that the surplus be laid aside for shipment the following season by which time those animals will in all probability become scarce as we rarely find they continue numerous three years in succession."[32]

The bottom of these cycles was equally important to the company's position in the market. After such a "crash" the wholesalers in the industry, anticipating a shortage, rushed to take control of the previous excess. Simpson wrote in 1831 that he suspected that the heavy buying by Astor's fur company was such a speculation: "It is in the anticipation of such scarcity Messrs. Astor & Company have purchased so largely, not for immediate consumption, as we have reason to believe they were at the time this purchase was made large holders of the American Fur Company's Musquash, but on speculation with a view to benefit by the demand which the probable scarcity will occasion. And as this falling off in quantity must increase the prices for all description we have not thought it advisable to destroy the small inferior skins."[33] The Hudson's Bay Company had little or no control over the entry of American muskrat onto the London market or to markets on the European continent. One example of how swings in muskrat population created competition based on speculative opportunities was the 1834 adjustment HBC made to its shipping schedule.[34] In response to the American Fur Company's entry into the English market that year, HBC changed its transportation schedule to beat the arrival of American furs.[35]

Muskrat returns were also the subject of speculation by smaller dealers and houses. When an 1824 contract for a large sale to an American house fell through "on a flimsy pretext," the committee consoled itself with the knowledge that consumers would be forced to buy on the English market.[36] Then the following year, as a condition of the sale of 150,000 skins to Henry Carey & Co. of New York, the skins were not to be resold on the English market. When this contract fell through, the company reminded Carey that he was liable for any loss incurred in reselling the skins.[37] Again, in the 1840s, the large numbers of muskrat harvested prompted the company to restrict the succeeding year's imports. Glut continued to be a problem for the trade; in 1846 one American dealer alone brought 600,000 skins into the English market.[38] So, although historians such as John Galbraith stress the element of competition between British and American fur traders on the frontier, these divisions were not as clear in the marketplace, where furs moved back and forth through private contracts and speculative practices.[39]

The muskrat cycle also created problems for the departments and posts in North America. The large numbers of easily trapped muskrat drained posts of their inventories of trade goods during the first year of a high cycle and created an ongoing inventory problem thereafter:

> The immense trade made for two or three successive years in the article of Rats was the cause of the increased demands from all parts of the country for goods, which led to the overstock we now have of many articles. This overstock will however be carried off by the outfits of next and the following years....[40]

Overall, the system was self-regulating and able to absorb muskrat surpluses, whose roughly three-year cycle was known to the traders. The large numbers of muskrat disrupted the marketplace, but the company had a series of responses that could be implemented either in London or at the shipping depot. Although speculation on the cycles could be lucrative, HBC adapted to muskrat demography as it attempted to create an orderly system of business and left most of the speculation to others.

The beaver was one species for which HBC tried to develop a unique long-term strategy of conservation and resource management. This plan was part of an attempt to gain a competitive edge in the marketplace. Unlike the muskrat, the demographic problem presented by the beaver was one of declining population, particularly east of the Rockies, due to overtrapping and epidemics such as tularemia. Of all the species harvested, beaver was the main focus of the trade. Exploration, competition, and management of beaver stocks was foremost in shaping the managerial policy of the Hudson's Bay Company.[41] The company's expansion into the Columbia Department in the 1820s was part of a drive to conserve and replenish the depleted stocks east of the Rockies by relying on new sources for the market in the Columbia.

AN EFFORT AT CONSERVATION

The company engaged in one of North America's earliest conservation experiments using those districts in which it had undisputed possession. The policy's strength lay in the advantage that the company's size and market longevity gave it. The policy's weakness lay in the assumption that market demand would continue unchanged. The conservation policy was also difficult to implement because of the incentive system that comprised part of the company's managerial organization. The wintering partner system of the North West Company, linking salaries and pensions to the success of the concern, was adopted by HBC in 1821 and carried with it a liability: it encouraged a personal profit motive and a frontier machismo.

Through conservation strategies, the company tried to combine the advantage of scale of operation with monopoly control over the "nursing" or "recruiting" of beaver stocks. This tactic enabled HBC to supply sufficient beaver to "meet extended consumption and secure to the Company the entire control of the Trade, as, [in] the countries exposed to opposition the expenses are so heavy that those who now pursue it will not then be able to meet us in the home market."[42] If the plan had succeeded it would have meant the end of the Columbia Department's frontier competition, "and at once put an end to opposition in those Countries where we have no exclusive privilege, as the high prices of Beaver alone enables the small Traders to continue."[43]

The committee supplied Simpson with a mathematical argument for the Northern Council that showed the benefit of this policy over a seven-year period in a hypothetical district. If the district produced 1,200 pelts annually without depleting stocks, this would yield 8,400 pelts for export. If the conservation policy were implemented, restricting the harvest to 400 pelts for each of the first two years, 600 the following, 800 the next,

2,800 in the fifth year, 14,000 in the sixth, and 20,000 in the seventh year, the total harvest would be 39,000 pelts with 38,800 animals remaining in the district.[44] The committee acknowledged that this model was optimistic, but even with a 50 percent error the benefit was clear: They would be able to bring to market twice the volume of fur. Even at half the going price this plan would squeeze their competitors, who faced the same outfitting costs as HBC for trade goods.[45]

Arthur Ray has attributed the failure of this conservation strategy to opposition by American, Métis, and Native trappers. Ray has argued that on the prairies "without a monopoly it was not possible to manage the fur trade on an ecologically sound basis since the primary suppliers of fur pelts, the Indians, did not really support the Hudson's Bay Company's conservation programme."[46] But the conservation strategy faltered due to a structural flaw in the company; ingrained habits and a related system of prestige could not be easily changed in practice. In the past the company and other traders judged individual traders according to the size of returns they generated, and their pensions were tied to past profits. Under the conservation policy, Simpson stated, "We must judge of their talents by the quality alone not the quantity."[47] Convincing the HBC commissioned officers and traders in council to agree to decrease the returns was one thing. Putting the changes into practice was another: "Many of them give it their best attention, which a few, who either cannot or will not understand either their own interests or the interests of the country and native people, give it but very little attention. They all, however, while assembled here, talk of the subject as if fully convinced of its importance, and make fair promises of giving it their best support, but I fear that many of them lose sight of it before they reach their wintering grounds."[48]

In 1841 Simpson, faced with an accelerating decline of beaver stocks throughout the districts, placed the blame on the commissioned gentlemen: "All our endeavours I am sorry to say, have been fruitless, owing very much in my opinion to the disinclination of many Gentlemen in charge of Districts & posts, to occasion a reduction in the returns, even as a measure of preservation to the country, from an over anxiety as to the appearance of turning their charges to profitable account, and in some cases perhaps, from a mistaken notion that by curtailing the returns they were injuring their own immediate interests."[49] Few traders were willing to sacrifice their short-term income in order to benefit their successors or, although they would not say it in council, the Company's long-term position. Although Ray's argument is correct, it ignores what Simpson clearly viewed as managerial resistance, structurally rooted in the profit system of the company.

The company understood that it was not possible to increase the consumption of beaver in Europe unless there was a material drop in price. Until it could achieve this, the company restricted importation in an effort to maintain the high prices needed to subsidize conservation. In 1830, a decade before the beaver market collapsed, the committee told Simpson:

> It is more profitable to keep the importation moderate until the animals become so numerous as to enable you to double the importation, which then might be sold so cheap as to force a larger consumption either by means of exportation or by making Hats so cheap as to induce a larger class of the

people of this Country to use Beaver hats. We consider the effect on the market this year, holds out the strongest inducement to preserve in the plan of nursing the Country, as it shews that such an increase of quantity as has been made this year only diminishes Profits, and that it would have been better if the animals had been allowed to live and multiply.[50]

The policy was a calculated gamble, given the declining number of American trapping parties in the Columbia in the late 1830s and their problems obtaining financing to bring supplies from St. Louis. For a time it appeared to be working, but the company had underestimated the extent to which both beaver populations had declined and consumer preference had changed as silk hats grew in popularity. The slow reaction of the Hudson's Bay Company to the erosion of its traditional markets resulted from the Company's reluctance to abandon a policy that would have worked in a static market.

In the Columbia Department, most of which was not under the conservation policy, production grew rapidly. Simpson's production target was 20,000 beaver pelts a year, a number reached in 1831 and surpassed by the peak of 28,949 pelts in 1833. Beaver production hovered between 18,000 and 21,000 until 1844 when it dropped to 10,812. Production rallied for two or three years and then dipped to 5,991 in 1850. Columbia district beaver production falls into two main periods: 1826–1843 (the years of Simpson/McLoughlin management) and the declining years after 1844. Production dropped before the political settlement of the Oregon Boundary, the turning point in political interpretations of the Company's fortunes. The cause of this drop had more to do with European fashion than with political events in North America.

DEATH OF A HAT

Beaver, the historic staple of the fur trade, suffered a serious collapse in the marketplace. During the 1840s the silk hat, symbolic of changing culture in the age of machinery and steam, caught consumers' attention and sense of fashion. The first blow to the market occurred in the Columbia auction on August 31, 1842. Hudson's Bay Company Secretary William Smith, forwarding catalogues of the sale to James Keith in Lachine, Quebec, wrote that demand for beaver had fallen off considerably because of the silk hat's popularity.[51] In their spring letter to George Simpson the committee wrote: "From an extraordinary freak of fashion, the article [beaver], moreover, has of late fallen much into disuse in hat making, silk hats being principally worn at present; the consequence is that its value has greatly decreased in the market, as will be seen by the accompanying sales catalogues. This depression however is but temporary, as no doubt exists that beaver hats will soon again come into more general use, when of course an amendment may be expected in the price.... The martens on the other hand, as you will observe by the late sales, have commanded very high prices."[52]

The committee was wrong, and beaver prices began to plummet at auction after auction. While the Columbia furs were being prepared for auction in the warehouse, the male population of London was turning out for the summer promenade dressed like Prince Albert in the new silk fashion. Prices even continued to drop after an effort by HBC to carefully present "best Beaver" at the August 1843 auction.[53] The committee wrote to McLoughlin that "[t]he continually decreasing price, when considered in connexion with a constantly decreasing supply, holds out no cheering prospect for the future, unless the tide of fashion change, and the consumption of Beaver in the manufacture of hats become more general than it has been for some time past. We hope that the low price may have some effect in bringing about an alteration in the public taste, but no hope of this must lead us to neglect any means, by which our great expences may be safely curtailed ... doubly so when prices are declining and returns annually diminishing."[54]

Prices dropped further at the January auction. In March 1845 the committee told Simpson that prices probably would continue to fall. Rather than publicly accept reduced bids, the bulk of the furs were returned to the warehouse while the company searched without success for a discreet private buyer with whom they could clear the total inventory. By 1845 the silk hat was firmly established both in England and on the Continent. The committee described the situation in a letter to Simpson: "The best description of which [silk hats] may be purchased at retail shops in London about fifty percent cheaper than the first quality beaver hat."[55] Wildlife demography came into play the next year. The 1846 market was inundated with muskrat from the United States, a beaver substitute. Having lost the higher-priced market and unable to compete with muskrat prices for the lower market, the Company finally conceded that a revival of the product was very unlikely.

Meanwhile, the warehouse inventory of unsold beaver pelts continued to grow. Auction catalogues show that the Columbia auction, which usually included a selection of furs from the two previous years, grew to a backlog that included unsold beaver from five separate years. In 1847 the company could delay no longer and cleared the inventory at whatever price was necessary to remove it.[56] Beaver, which sold for 30 to 35 shillings per pound in 1821, sold for 3 to 4 shillings. HBC began to consider alternate uses for the pelt: "We are not without hope that great cheapness may have the effect of forcing the article into consumption in some form or other, as the ingenuity of purchasers will naturally be stimulated to the means of applying it to new purposes."[57]

Beaver was no longer a viable product, and the company began to experiment with methods that could make it one again. The experiments undertaken were ingenious. Using a new process, they shaved and dyed pelts to resemble fur seals, an operation also used later during the fur seal vogue of 1890–1910. The skins exhibited in 1847 at the annual Leipzig Fair, in the German state of Saxony, evoked little response from the fur industry; many fashionable furs, including marten, went unsold. By 1847 the price of a beaver hat had dropped to that of a silk hat, but consumer response was still negligible.[58]

The company continued to experiment, preparing 1,000 fully dressed and dyed beaver pelts to be sold as "fur," a novel use for the pelt that had never before been attempted. The marketplace was Canada, but only 500 pelts were sent because of production problems. Further eroding the market, several London fur dealers who were aware of the company's plans sent earlier shipments. In another attempt to rally the market, unsold stock was sent to dealers in the United States where the beaver hat was still in use; a test shipment also went to China, with disappointing results. After these attempts, the 1849 importation from North America was severely restricted to 20,000 skins, formerly the average output of the Columbia Department alone. The beaver, slowly gaining acceptance as a "fur," ceased to be important to the European hat industry.[59]

The Hudson's Bay Company's plans for market dominance had failed due to a combination of wildlife demography and European fashions. However, the Company gained experience in experimenting with and promoting alternate uses for its products, and in knowing when to cut its losses. At the same time, HBC was turning to the promotion of a "new" product, drawn from their inventory of wildlife. While the steady decrease in the price of beaver made Company officials such as Archibald Barclay gloomy, there was consolation in the corresponding increase in the value of the marten. The marten's value was strong, but the best evidence of its importance is found in the tremendous volume produced. During the experimentation with beaver as a fur, Warehouse Keeper Edward Taylor expressed concern that large quantities of beaver, contrasted with the smaller numbers of marten, might cause the market for both to collapse. Another member of the Company, Edward Roberts, told Simpson: "The fur is beautiful and when dyed looks as well as sea otter.... I showed a specimen to Nicholay the Queen's furrier, who has a high opinion of the fur and thinks it likely to come into extensive use for trimmings, and also for muffs, and does not think it will come into competition with Marten, so as in any way to affect the value of that article which our friend Taylor is very much afraid of."[60] Ever aware that sales depended on the perception of distinctly separate products, the Company sought to maintain this separation even down to the level of a fur becoming a cuff on a garment.

Before 1838 the marten received only passing notice in the discussions of auction trends. Then increased demand pushed prices steadily upward. The declining population cycles of the Columbia

Department's marten and lynx in 1839–1840 stimulated prices further. Simpson considered the declines only temporary: "By the knowledge which has been acquired by experience, of the habits of these latter animals, however, there is every reason to believe that this diminution in their numbers is merely temporary, arising either from migration to other quarters, or from disease, but that as soon as these causes shall be removed, they will become as plentiful as formerly, and assist in retrieving the present unpromising aspect of affairs in this district."[61]

The decreased supply reduced the furriers' inventories, and prices continued to climb. By 1843 demand was again strong, and prices began to set new levels as they attracted speculative buyers. Increased prices in 1844 more than matched the losses incurred by declining beaver stocks. The Columbia posts encouraged trappers to switch from beaver to marten and other small furs. The committee proposed new incentives to offset the losses that beaver was taking in the marketplace: "Beaver has again fallen in price, but, as a stimulus to exertion in hunting martens, lynxes, musquash and all other furs, increased prices may be offered, as we can afford to be liberal in that way, inasmuch as all those furs are at present much in demand and have advanced, musquash as much as 40 percent on the price of last year, as you will perceive by the catalogues."[62]

The Columbia Department board of management received instructions to concentrate on marten in the fall of 1846 as the Company benefited from high prices. The increased supply was not large enough to flood the market the way muskrat, rabbit, and lynx frequently did. Marten returns from the department increased rapidly after 1846. Almost 45 percent of all the marten harvested in the Columbia

Department from 1825 to 1849 were taken in the following three years. This increased harvest of Columbia marten points toward a rapid response to the market by Native trappers. The Company used its experience with the population cycles of other species to play the market as closely as possible. HBC cautioned in the fall of 1847 that imports from the northern department should not be so excessive as to weaken prices.[63]

New Caledonia's fur-based economy responded quickly to new markets due to the flexibility of its species mix. Because of low demand and the bottom of the marten's demographic cycle, in 1840 only 1,251 pelts came to market. The 1845 production was 7,383, reaching a high of 9,586 in 1846 before declining to a low of 2,652 in 1849. The taking of so many marten, while clearly a response to market forces, also represents the coincidence of high prices with a demographic upswing. Unlike the muskrat, which had a low market value and a high demographic peak, the marten did not exist in sufficient numbers to cause a price collapse. The large numbers traded represent the response of Native trappers to its increased value.

This change in the trade must have had structural and cultural implications for Native peoples living in New Caledonia. Were depleted family-tenure beaver ponds abandoned in the search for a more valuable and, in the marten's case, more mobile commodity?[64] The region's species mix, rich both in marten and lynx, allowed New Caledonia to shift easily from one staple fur to another. The New Caledonia cyclical peak in marten matched the changing market's preference for the fur. Unlike the muskrat, the size of the marten population was easily absorbed by the market. However, the marten soon began one of its endemic declines. James Douglas, in charge of the Columbia Department in 1847, commented:

> A heavy decline in Beaver and Martens. The former apart from the measles, which also severely afflicted the natives of the District wherein the Steam vessel carries on trade, was partly the effect of the reduction in prices; the decrease in the latter is either caused by want of exertion in the hunters, or which is more probable it arises from a scarcity of the animal producing that valuable fur. From the great abundance of martens for some years past, in all parts of the Indian Country, and the general decline which we notice with regret—this year, at all the Marten Posts in this District it is feared that we are on the eve of one of those fluctuations to which the Marten trade independently of hunting is almost periodically subject, and if so there will be a further decline in the returns of that fur next year and for some years after, until from some unknown cause they again multiply and reappear in their native forests in the utmost abundance.[65]

The decline in marten was not serious enough to create problems for the Company. Douglas noted that marten probably would rebound at the end of three years in the same manner as the muskrat.

The market and the post had shifted from the traditional staple fur to the marten. In time other small furs, such as the mink, would become substitutes for the marten.[66] After intervening in the management of business at all levels in order to survive the crisis, the committee returned by 1850 to its traditional role in the daily routine of directing the system of trade. Districts that overtrapped or had too many low-grade or damaged marten in their returns were reprimanded. The staff at posts responded to the demands of the buyers of the new staple: they stopped the practice of cutting paws from marten pelts that caused a shilling depreciation in the value of the skin; pale martens were no longer classed as damaged but as high quality because of their end purpose as trimming and collars. Native trappers followed the price incentives and brought in small furs instead of the beaver. The committee remarked about the Columbia Department that "from the abundance of small furs and the increased industry of the natives, we are inclined to look with greater hope to the future."[67] Daily operations of the Company continued much as they had when the product was beaver. An early naval visitor to Fort Victoria in 1848 commented in a letter to the London *Times* that the beaver had "hardly any value now."[68] A fundamental change had taken place in the upper levels of the Company. The balanced symbiotic relationships of the Company with its trappers and European buyers had shifted in favour of the marketplace.

CONCLUSION

The Hudson's Bay Company succeeded in establishing a centralized system to the fur trade, smoothing demographic bumps, improving quality and shipping methods, and establishing recognizable products. It survived the unexpected collapse of its historic and traditional market, the felt hat industry, because of the biological and ecological diversity of the wildlife harvest gathered by its network of posts. Information about markets and wildlife became the predominant feature of company operations as the process of deciding to restrict imports or to redirect trappers became more efficient. It learned that if one product could be abandoned after so many years, then the adoption of an entirely new group of products was conceivable.[69]

There is evidence in trade leaflets of a gradual shift in organizational mentality. The oldest leaflet (1799) shows animals in descending order of value. Mid-century leaflets alphabetize the lists to a certain extent. Beaver still heads the list, before bear or badger, despite its declining economic importance, but mythical terms such as "sea horse teeth" give way to the more familiar "walrus tusks." By 1870 all animals appear alphabetically, badger before beaver, reflecting a systemized inventory.[70]

The implications for the wildlife of this case study are difficult to assess. The emphasis on smaller furs meant a wider incursion into the river–forest ecosystem. There is evidence of complex interactions between species and within specific populations that even now are only vaguely understood. None of the Pacific Northwest's more than 60 forms of wildlife recorded in HBC records are extinct. Most have declined because of habitat loss, a loss not directly attributable to the fur trade.

To the fur trade, wildlife was a harvest. Nature provided a collection of potential products, each containing characteristics that influenced how it could be offered as a product. A primary feature of the fur trade management was mediating between demographic fluctuations and the world of fashion and markets. It was not wildlife management in the twentieth-century conception, but in this nineteenth-century interest in conservation and wildlife cycles there is a curious resonance with subsequent efforts. In our focus on the fur trade, we know much about traders and Natives, but very little about what drove the global market for wildlife.

NOTES

1. Most fur trade literature makes passing reference to wildlife and then moves immediately to European–Native trade relations. The literature that does address animals is concerned with wildlife as an aspect of Native culture, such as Calvin Martin's *Keepers of the Game: Indian-Animal Relationships and the Fur Trade* (Berkeley: University of California Press, 1978), the rebuttal by Shepard Krech III, *Indians, Animals, and the Fur Trade: A Critique of Keepers of the Game* (Athens: University of Georgia Press, 1981), or Adrian Tanner's *Bringing Home Animals: Religious Ideology and Mode of Production of the Mistassini Cree Hunters* (New York: St. Martin's Press, 1979).

    There are many discussions of wildlife in the literature of environmental history, although the fur trade is strangely missing from Alfred W. Crosby's pivotal *Ecological Imperialism: The Biological Expansion of Europe, 900–1900* (Cambridge: Cambridge University Press, 1986). See Christine and Robert Prescott-Allen, *The First Resource: Wild Species in the North American Economy* (New Haven: Yale University Press, 1986); Thomas R. Dunlap, *Saving America's Wildlife* (Princeton: Princeton University Press, 1988); Peter Matthiessen, *Wildlife in America* (New York: Viking, 1987); Lisa Mighetto, *Wild Animals and American Environmental Ethics* (Tucson: University of Arizona Press, 1991); Farley Mowat, *Sea of Slaughter* (Toronto: McClelland & Stewart, 1984); and Morgan Sherwood, *Big Game in Alaska: A History of Wildlife and People* (New Haven: Yale University Press, 1981).

2. How this differed from other more flexible and locally controlled companies, such as the individual trapper of the Rocky Mountain rendezvous or the merchants of the St. Louis fur trade, is important. Fur trade historiography suffers from political and archival compartmentalization. Canadian scholars, including this author, rely primarily on the papers of the Hudson's Bay Company. For the researcher, the recent microfilm publication of the *Papers of the St. Louis Fur Trade* (Bethesda: University Publications of America, 1992) offers an opportunity for a

comparative study. The papers include an introduction by William R. Swagerty and an essay by Janet Lecompte.

3. This is based on a statistical examination of fur importation handbills. The error between these estimated shipments and a ledger kept by James Douglas for each trading year showed an average 4 percent difference in the annual shipments from 1825–1849. Unless otherwise noted all figures cited come from a series of databases constructed by the author (these are mainframe codebook-based datafiles run under SAS, SPSS, and Paradox statistical analysis packages). One contained the annual fur returns of wildlife traded each year at the posts of the Columbia department. It was based on James Douglas, *Fur Trade Returns for Columbia and New Caledonia Districts, 1825–1857*, Provincial Archives of British Columbia, Victoria, British Columbia (hereafter PABC), A/B/20V3. For a full analysis of the James Douglas ledger, see Lorne Hammond, "Studies in Documents: Historians, Archival Technology, and Business Ledgers," *Archivaria* 28 (Summer 1989): 120–125. Another database contains the figures published in fur importation handbills listing North American fur imports to London, *Fur Trade Importation Book 1799–1912*, Hudson's Bay Company Archives, Series 1, Winnipeg, Manitoba (hereafter HBCA), A.53/1. The third is a listing of sales of Columbia wildlife drawn from the warehouse keeper's annotated auction catalogue, Auction Catalogues of Fur Produce, HBCA, A.54.

4. George Simpson to the Governor, Deputy Governor and Committee, March 10, 1825, HBCA, D.4/88, Para. 20; Frederick Merk, *Fur Trade and Empire: George Simpson's Journal* (Cambridge: Belknap Press, 1968), 32–33.

5. John McLoughlin to the Governor, Deputy Governor, and Committee, October 6, 1825, Para. 49, in *The Letters of John McLoughlin: From Fort Vancouver to the Governor and Committee, First Series, 1825–38* (London: Hudson's Bay Record Society (HBRS), 1941), 16; John McLoughlin to the Governor, Deputy Governor, and Committee, October 31, 1842, Para. 20, in *The Letters of John McLoughlin: From Fort Vancouver to the Governor and Committee, Second Series, 1819–1844* (London: HBRS, 1943), 81.

6. James R. Beringer, *The Control Revolution: Technology and Economic Origins of the Information Society* (Cambridge: Harvard University Press, 1986), 144–153.

7. Governor, Deputy Governor, and Committee to Simpson, March 11, 1825, HBCA, A.6/21, Para. 23. Earlier in 1825 the denture market was overstocked because of Greenland fisheries activity.

8. Robin F. Wells, "Castoreum and Steel Traps in Eastern North America," *American Anthropologist* 74 (June 1972): 479–483; "Castoreum," *The Museum of the Fur Trade Quarterly* 8 (Spring 1972): 1–5.

9. Ross Cox, *The Columbia River, or Scenes and Adventures during a Residence of Six Years on the Western Side of the Rocky Mountains among Various Tribes Hitherto Unknown; Together with a Journey across the American Continent* (Norman: University of Oklahoma Press, 1957), 243.

10. John Richardson documented plants, lichens, birds, mammals, and fish during John Franklin's Arctic expeditions between 1819 and 1827. His lasting achievement is the multi-volume *Fauna Boreali-Americana; or the Zoology of the Northern Parts of British America: Containing Descriptions of the Objects of Natural History Collected on the Late Northern Land Expeditions, under Command of Captain Sir John Franklin, R.N.* (New York: Arno Press, 1974); *Arctic Ordeal: The Journal of John Richardson, Surgeon-Naturalist with Franklin* (Montreal: McGill-Queen's University Press, 1984); C. Stuart Houston, "John Richardson—First Naturalist in the Northwest," *Beaver* (November 1984): 10–15.

11. In the Columbia Department from 1820–1849 the average proportions were: red phase 58 percent (8,402 pelts), cross phase 31 percent (4,430 pelts), and silver phase 11 percent (1,608 pelts). See also the discussion between naturalists John Bradbury and Thomas Nuttall in James P. Ronda, *Astoria and Empire* (Lincoln: University of Nebraska Press, 1990), 133, 316–320; and M. Novak,

J.A. Baker, M.F. Obbard, and R. Malloch, eds., *Wild Furbearer Management and Conservation in Northern America* (Toronto: Ministry of Natural Resources, 1987), 751.

12. John Richardson cites A. de Capell Brooke as noting that only three or four silver foxes were taken annually on the Lofoten Islands of Norway and that they are not found elsewhere. Richardson, *Fauna Boreali-Americana*, 94. In "Platinum Mutations in Norwegian Silver Foxes," *Journal of Heredity* 30 (June 1939): 226–234, Otto L. Mohr and Per Tuff say that much of Norwegian stock is imported from Canada.

13. Richardson, *Fauna Boreali-Americana*, 93.

14. Alexander William Francis Banfield, *Mammals of Canada* (Toronto: National Museum of Natural Sciences, University of Toronto Press, 1974), 299.

15. Richardson, *Fauna Boreali-Americana*, 15.

16. E.E. Rich, ed., *Peter Skene Ogden's Snake Country Journals, 1824–1825 and 1825–1826* (London: HBRS, 1950); K.G. Davies, ed., *Peter Skene Ogden's Snake Country Journal, 1826–27* (London: HBRS, 1961); Glyndwr Williams, ed., *Peter Skene Ogden's Snake Country Journals, 1827–28 and 1828–29* (London: HBRS, 1971).

17. George Simpson to John McLoughlin, July 9, 1827, HBCA, D.4/90, Para. 6.

18. Peter Skene Ogden to the governor, chief factors, and chief traders, October 10, 1826, in Merk, *Fur Trade and Empire*, 285.

19. Banfield, *Mammals of Canada*, 161.

20. Simpson to the governor, deputy governor, and committee, March 10, 1825, HBCA, D.4/88, Para. 26.

21. For a discussion of attitudes toward wolves in America after 1880, see Dunlap, *Saving America's Wildlife*.

22. Governor, Deputy Governor, and Committee to Simpson, February 27, 1822, HBCA, A.6/20, Para. 48.

23. Governor, Deputy Governor, and Committee to Simpson, June 7, 1833, HBCA, A.6/23, Para. 40.

24. Pelly, Colvile, and Simpson to McLoughlin, December 31, 1839, in *Letters of John McLoughlin, Second Series, 1839–1844*, 164n.

25. From 1827–1837 the average price for a river otter was 18 shillings 11 pence (18/11), less 1/5 for insurance and other charges from the Columbia Landing warehouse. This left 17/6 as the average worth. The R.A.C. agreed to pay 23/- each, for a 5/6 profit on each to the HBC. The additional 3,000 otters from the Northern Department had an average value of 26/5, less the 1/5 charge, leaving a net average value of 25/-. The R.A.C. agreed to pay 32/- each, giving HRC a 7/- profit on each skin. Although the Panhandle cost £1,750 (2,000 × 17/6) a year, the hidden profit on the Columbia otters was £550 (2,000 × 5/6 and £1,050 on the Northern Department otters (3,000 × 7/-), making the cost only £150. "Memorandum," HBCA, F.29/2, fo. 182. The conversion rate for 1840–1850 was $4 to the pound in urban centres and $5 to the pound where there were currency shortages, such as in the West.

26. Lloyd Keith places the origins of the field at a conference held in July 1931 on the Matamek River in Labrador under the patronage of Copley Amory of Boston. Lloyd B. Keith, *Wildlife's Ten-Year Cycle* (Madison: University of Wisconsin Press, 1963), vii.

27. Keith, *Wildlife's Ten-Year Cycle*. For related work on the fluctuation of muskrats, see Paul L. Errington, *Muskrat Populations* (Ames: Iowa State University Press, 1963), 522–538.

28. For the scientific literature, see Charles S. Elton, "Periodic Fluctuations in the Numbers of Animals: Their Causes and Effects," *Journal of the Society for Experimental Biology* 1 (October 1924): 119–163; "Plague and the Regulation of Numbers in Wild Animals," *Journal of Hygiene* 24 (October 1925): 138–163; "The Ten-Year Cycle in Numbers of the Lynx in Canada," *Journal of Animal Ecology* 2 (November 1942): 215–244; *The Ecology of Invasions by Animals and Plants*

(London: Methuen, 1958); Charles S. and Mary Nicholson, "Fluctuations in Numbers of the Muskrat (*Ondatra Zibethica*) in Canada," *Journal of Animal Ecology* 2 (May 1942): 96–126.

29. Ian McTaggart Cowan, "The Fur Trade and the Fur Cycle: 1825–1857," *British Columbia Historical Quarterly* 2 (January 1938): 19–30, points out that figures for the lynx are slightly confused because they do not distinguish between the more southern bobcat, a generalist predator with alternate food sources, and the true lynx, a specialist predator whose numbers fluctuate in direct response to its prey, the snowshoe rabbit. However, the composite totals for the two show the lynx had demographic peaks in the returns of 1829–1830, 1837–1840, and 1848–1850.

30. There is a relationship between the latitude of a muskrat population and the number of litters as well as the number of young per litter. Muskrat in Louisiana may have three to six litters a year with an average of 2.4 young in each litter, while muskrat in northern Canada may have only two litters, but each averaging 7.1 young. Banfield, *Mammals of Canada*, 198–199. See also David J. Wishart, *The Fur Trade of the American West, 1807–1840: A Geographical Synthesis* (Lincoln: University of Nebraska Press, 1979), 36.

31. Simpson to the Governor, Deputy Governor, and Committee, June 30, 1829, HBCA, D.4/96.

32. Governor, Deputy Governor, and Committee to Simpson, March 1840, HBCA, A.6/ 25. Para. 23.

33. Simpson to the Governor, Deputy Governor, and Committee, July 18, 1831, HBCA. D.4/98, Para. 5. Astor had approached the company about contracting for muskrat in 1827, but the company wanted a guarantee that he would take 70,000 to 100,000 a year at a fixed price for five to seven years. This was a dangerous contract given the fluctuations. See William Smith to George Simpson, May 30, 1827, HBCA, A.6/21.

34. The 1835 total import of 1,111,646 muskrats came to a market that had been paying 9.5 pence per skin, so the total value of the stock was $44,000. But the prices given for the 1837 shipment of 838,549 skins was 3 pence each, for a book value of $10,000. This demonstrates the failure of the market to absorb these quantities. HBCA, A.54.

35. Governor, Deputy Governor, and Committee to Simpson, March 5, 1834, HBCA, A.6/ 23, Para. 28.

36. Governor, Deputy Governor, and Committee to Simpson, June 2, 1824, HBCA, A.6/20, Para. 54.

37. Due to the depressed market the committee decided to bend on the issue and accepted $300 compensation for the loss when the muskrats were resold. William Smith to Henry Carey & Co., June 16, 1825, February 1, 1826, and June 12, 1826, HBCA, A.6/21.

38. Archibald Barclay to George Simpson, February 3, 1846, HBCA, A.6/27.

39. John S. Galbraith, *The Hudson's Bay Company as an Imperial Factor* (New York: Octagon Books, 1977); Mary E. Wheeler, "Empires in Conflict and Cooperation: The 'Bostonians' and the Russian-American Company," *Pacific Historical Review* 40 (November 1971): 419–441; Frank E. Ross, "The Retreat of the Hudson's Bay Company in the Pacific North West," *Canadian Historical Review* 18 (September 1937): 262–280; Herman J. Deutsch, "Economic Imperialism in the Early Pacific Northwest," *Pacific Historical Review* 9 (December 1940): 377–388.

40. Simpson to the Governor, Deputy Governor, and Committee, August 10, 1832, HBCA, D.4/99, Para. 3.

41. The idea of managing beaver stocks is old. An early North American reference to the idea is a letter of Jesuit Father Paul le Jeune dated August 28, 1636. Reuben Gold Thwaits, ed., *The Jesuit Relations and Allied Documents* (New York: Pageant Book Company, 1959) 9: 165–167.

42. Simpson to the Governor, Deputy Governor, and Committee, July 10, 1828, HBCA, D.4/92, Para. 9.

43. Governor, Deputy Governor, and Committee to Simpson, January 16, 1828, HBCA, A.6/21, Para. 10.

44. Governor, Deputy Governor, and Committee to Simpson, February 23, 1826, HBCA, A.6/21, Para. 36.

45. Governor, Deputy Governor, and Committee to Simpson, October 25, 1832, HBCA, A.6/22, Para. 14.

46. Arthur J. Ray, "Some Conservation Schemes of the Hudson's Bay Company, 1821–50: An Examination of the Problems of Resource Management in the Fur Trade," *Journal of Historical Geography* 1 (January 1975): 58. Any monopoly HBC had developed was only in isolated areas because the company had difficulty getting co-operation even from its own staff. The number of furs brought in influenced not only an individual's prestige but calculation of his pension.

47. Simpson to the Governor, Deputy Governor, and Committee, August 20, 1826, HBCA, D.4/89, Para. 30.

48. Simpson to the Governor, Deputy Governor, and Committee, August 10, 1832, HBCA, D.4/99, Para. 27.

49. Simpson to the Governor, Deputy Governor, and Committee, June 20, 1841, HBCA, D.4/99, Para. 31.

50. Governor, Deputy Governor, and Committee to Simpson, March 3, 1830, HBCA, A.6/22, Para. 2.

51. William Smith to James Keith, September 3, 1842, HBCA, A.6/26.

52. Governor, Deputy Governor, and Committee to Simpson, April 1, 1843, HBCA, A.6/26, Para. 23.

53. Archibald Barclay to James Keith, September 4, 1843, HBCA, A.6/26.

54. Governor, Deputy Governor, and Committee to John McLoughlin, September 27, 1843, HBCA, A.6/26, Para. 9.

55. Governor, Deputy Governor, and Committee to Simpson, March 11, 1845, HBCA, A.6/26, Para. 3.

56. Auction, September 1, 1847, HBCA, A.54/182.

57. Governor, Deputy Governor, and Committee to chief factors Peter Skene Ogden, James Douglas, and John Work, September 8, 1848, HBCA, A.6/27, Para. 2.

58. Governor, Deputy Governor, and Committee to Simpson, June 5, 1847, HBCA, A.6/27, Para. 6; Governor, Deputy Governor, and Committee to Simpson, April 7, 1847, HBCA, A.6/27, Para. 3.

59. Governor, Deputy Governor, and Committee to Simpson, April 5, 1848, HBCA, A.6/27, Paras. 4, 5, 7; Governor, Deputy Governor, and Committee to Simpson, April 4, 1849, HBCA, A.6/28, Para. 4.

60. Edward Roberts to Simpson, February 3, 1846, HBCA, D.5/16, fos. 168–169, cited in Caroline Skynner, "History of the Beaver and Beaver Hat," unpublished paper, HBCA, PP. 1984, 27.

61. Simpson to the Governor, Deputy Governor, and Committee, November 25, 1841, HBCA, D.4/110, Para. 7.

62. Governor, Deputy Governor, and Committee to Simpson, March 11, 1845, HBCA, A.6/26, Para. 3.

63. Governor, Deputy Governor, and Committee to Simpson, September 18, 1847, HBCA, A.6/27.

64. Ian McTaggart Cowan, in "The Fur Trade and the Fur Cycle: 1825–1857," *British Columbia Historical Quarterly* 1 (January 1938): 27, states that the marten and the lynx both undergo mass movements after the decline of a rabbit cycle.

65. Board of Management to the Governor, Deputy Governor, and Committee, November 6, 1847, *Fort Victoria Letters, 1846–1851*, edited by Hartwell Bowsfield (Winnipeg: HBRS, 1979), 23; James Douglas to Archibald Barclay, July 22, 1851, *Fort Victoria Letters*, 200–201.

66. "Mink and Musquash will no doubt rise ... as those who cannot afford to pay a high price for Martens will content themselves with inferior furs of the same class," Archibald Barclay to George Simpson, December 28, 1849, HBCA, A.6/28.

67. Governor, Deputy Governor, and Committee to chief factors Peter Skene Ogden, James Douglas, and John Work, Fort Vancouver, September 8, 1848, HBCA, A.6/27, Para. 3.

68. "North West Coast—Visit of H.M.S. *Constance*," *Times* (London) (May 1849): 7.

69. For a discussion of non-fur trade economic activities, see Richard S. Mackie, "Colonial Land, Indian Labour and Company Capital: The Economy of Vancouver Island, 1849–1858," M.A. thesis, University of Victoria, 1984.

70. The concept is drawn loosely from Robert Darnton's examination of class through a description of a parade in pre-industrial France. See Robert Darnton, "A Bourgeois Puts His World in Order: The City as a Text," in *The Great Cat Massacre and Other Episodes in French Cultural History* (New York: Vintage Books, 1985), 106–143. See also Keith Thomas, *Man and the Natural World: Changing Attitudes in England 1500–1800* (London: Allen Lane, 1983). All examples used here are drawn from Fur Trade Importation Book, 1799–1912, HBCA, A.53/71.

Chapter Eleven
Like "The Thames towards Putney"
The Appropriation of Landscape in Lower Canada

Colin M. Coates

As they fanned across the globe in the eighteenth and nineteenth centuries, Europeans invoked aesthetic principles in order to establish dominion over foreign territories. By imposing European perspectives, they made the lands accessible to future expansion.[1] This creation of "new" worlds overlooked the legitimacy of Aboriginal sovereignty and often foretold disastrous results for the inhabitants.

In a few areas, British migrants confronted existing European settler societies, for instance, Afrikaners in southern Africa and Canadiens in Quebec. There, the invaders' reaction was not fundamentally different from their attitude toward areas of Aboriginal occupation. Analyzing explorers' accounts of southern Africa, Mary Louise Pratt writes: "The European improving eye produces subsistence habitats as 'empty' landscapes, meaningful only in terms of a capitalist future and of their potential for producing a marketable surplus."[2] Travel and exploration literature addressed new landscapes in Old World terms, if only to make sense of them to the home audience for which they were written.

Of more direct consequence than these publications were cases where individuals acquired property abroad and attempted to introduce new land uses and perspectives. Shortly after the Conquest of 1759–1760, English merchants and government officials began to acquire seigneuries in the St. Lawrence Valley. The literature on Canadian seigneurialism has focused on the structural, economic links between seigneur and *censitaire*, dealing little with the cultural aspects of proprietorship.[3] Yet given the cultural and economic significance of land owning in eighteenth-century Britain, the ideological import that landscape held for upper-class Britons cannot be underestimated. This article provides a case study of what occurred when English cultural proclivities were brought to the North American colony.

In 1819 John and Elizabeth Hale purchased the 130,000-acre seigneury of Sainte-Anne de la Pérade, located between Quebec and Trois-Rivières on the north shore of the St. Lawrence. The Hale family's appropriation of the seigneury involved more than a simple monetary transaction. They perceived an opportunity to fashion a new landscape in the St. Lawrence Valley.

An examination of their actions reveals some aspects of the English imperial project in the colony of Lower Canada. This article explores the intellectual processes involved in appropriating, or establishing dominion over, new lands. The Hale family imposed their own cultural understandings as they attempted to make a home for themselves in this "new" land. They applied lessons of history, morality, and aesthetics in their quest for dominion.

This article draws on the literature dealing with the way perceptions of the landscape reflect ideals of social structure.[4] The Hale family's correspondence describing their activities and aspirations, their commissioned maps of the seigneury, and Elizabeth Hale's sketches all reveal similar ideological or cultural contents.

PICTURESQUE SAINTE-ANNE

By the late eighteenth century, about 100 years after the first European settlement at the confluence of the Sainte-Anne and the St. Lawrence rivers, the landscape could be described explicitly according to European conventions. This result of generations of habitant farming and forest removal was of particular interest to elite English visitors to and residents of the colony. The habitants had cut trees around their buildings (often too many for English tastes) and had established extensive fields. Distancing indigenous wildlife and vegetation from their settlements, French farmers had refilled the landscape with their buildings, fences, crops, and animals. Without fully resembling an English agrarian landscape, the habitants' farmland was nonetheless recognizable and often "picturesque."

As with other scenes along the St. Lawrence River, the area near Sainte-Anne impressed English visitors. In 1785 Robert Hunter, Jr., compared the region near Sainte-Anne with England: "Here and there you see some charming plantations, by the side of the river, which you might imagine to have been planned by a [Capability] Brown, but being natural exceed him as much as nature surpasses art.... This stage is quite romantic, and beautiful beyond description. What a show would such a noble river make running through England, with the noblemen's seats to adorn it instead of the habitants' huts."[5] Hunter suggested that the landscapes of the area surpassed the designs of the leading garden architect of the day in England. Canada, in particular this part, naturally offered more than invention could in the mother country, but English social structure could improve the prospects further.

Other visitors also saw Sainte-Anne as a picturesque spot. Precisely because he was imposing a new aesthetic, John Lambert stressed the picturesque qualities of the area, because, he stated, "[T]he inhabitants ... are no great admirers of the beauties of Nature." Lambert described the post road in Sainte-Anne, "wind[ing] along the summits of the lofty banks which overlook the river, or along the borders of delightful valleys."[6] From the other side of the St. Lawrence, the entire region was a pleasing tableau for Elizabeth Simcoe: "The mouths of the rivers Batiscan and St. Anne are seen on the opposite shore, with distant blue hills. This is the finest point on the river and a good military position."[7] For his part, Joseph Bouchette stressed the picturesque setting of the seigneurial manor-house of Sainte-Anne: "The manor-house, agreeably situated near the point formed by the Ste. Anne and the St. Lawrence, is surrounded by excellent gardens and many fine groups of beautiful trees.... At the mouth of the Ste. Anne lie the isles ... being well clothed in wood they afford several very pleasing prospects."[8]

Throughout Lower Canada, the scenery encouraged artists to take pencil or brush to paper to capture it for posterity (and for the British empire). In doing so, artists applied eighteenth-century aesthetic conventions. For instance, George Bulteel Fisher and George Heriot published their visions of "a Canadian Arcadia," depicting local scenery in reference to European ideal landscapes. Numerous military artists, educated at the Woolwich Academy, brought their trained perspectives to bear on the landscape of Lower Canada. While celebrating North American scenery, their art placed it in an English frame.[9]

Their artistry made particular reference to the cult of the picturesque. Initially harking back to the classical landscapes of artists such as Claude Lorrain and Salvator Rosa, the picturesque was a concept that evolved over the decades, through debates between British authors such as William Gilpin, Richard Payne Knight, Uvedale Price, and Humphry Repton. Many different landscapes could be celebrated according to its conventions; some artists preferred forest scenery, others parks.[10] Nonetheless, the concept involved a number of specific connotations. Picturesque landscapes are ones that make the viewer think of paintings; in other words, where nature imitates art. The framing of the perspective by vegetation or hills, contrasts between shadows and light, a variety of natural phenomena, the presence of ruins or decaying trees that lend themselves to reflections on mortality: all these are possible elements of the picturesque. The observer may regard the scene, either natural or painted, in a leisurely fashion, the eye slowly taking in the entire view. It is a diverse but unified landscape that invites the viewer to occupy it or travel through it.[11]

Depending on the context in which it appeared, the picturesque bore different, if related, ideological implications. First, in the British context, this genre accentuated harmony between human beings and nature, and between people of all classes. It is no coincidence that the picturesque reached its highest point of popularity in England during a time of rapid agrarian change. The period of enclosures and of technological innovations led to changes in social structure and to increased rural unrest. The picturesque movement, with its emphasis on compositional unity, represented an escape from the social problems of the period.[12]

Indeed, even if some theorists of the picturesque preferred wild and unenclosed scenery to agrarian landscapes,[13] the approach was not always opposed to land ownership and to agricultural improvement. As art historian Anne Bermingham argues, "[a]lthough the picturesque celebrated the old order—by depicting a pastoral, preen-closed landscape—some of its features—the class snobbery, the distancing of the spectator from the picturesque object, and the aestheticization of rural poverty—suggest that at a deeper level the picturesque endorsed the results of agricultural industrialization."[14] In practice, the very landowners who directed agrarian changes in the eighteenth century also celebrated the picturesque landscape. It was one thing to have a picturesque park surrounding the manor-house—it was another to have productive, improved farmland lying just out of sight. Even Gilpin, who preferred "wild" scenery, accepted the importance of the landlord's dominating presence: "A noble park ... is the natural appendage of an ancient mansion."[15]

Thus, the introduction of picturesque sensibilities on noble estates in England had a special resonance. Moving away from the rigid lines of earlier periods, landscape theorists emphasized highly structured but natural-appearing landscapes. Humphry Repton, Capability Brown's heir as leading garden architect, incorporated picturesque qualities in the noble landscape. This was an important endeavour, he saw, as landscape gardening was "among those [arts] which distinguish the pleasures of civilised society from the pursuits of savage and barbarous nations."[16] Repton tried to put picturesque attitudes into practice, though he did not wish, for practical reasons, to translate canvases into landscape. He emphasized the perspective, which could be seen from the manor-house; the landscape should celebrate noble property. From the manor-house, one should see only objects that belong to the owner: "[T]he views from a house, and particularly those from the drawing-room, ought rather to consist of objects which evidently belong to the place. To express this idea, I have used the word *appropriation*, by which I mean such a portion of wood and lawn as may be supposed to belong to the proprietor of the mansion, occupied by himself, not so much for the purposes of gain as of pleasure and convenience."[17] Fences should be hidden, in order to frame the landscape with trees and bushes. Although Repton disapproved of such actions, some nobles would go as far as to relocate entire peasant villages because they interfered with the lord's perspective.[18] In this way, even if almost everything was artificial, it should give an appearance of being natural. By extension, noble property, in spite of the socio-political struggles involved in expanding and maintaining it,[19] was a natural phenomenon.

Taking the same picturesque sensibility overseas modified somewhat its implications. In foreign contexts, the imposition of picturesque categories held imperialistic connotations. As I.S. MacLaren has argued, "The picturesque sustained the sojourning Briton's sense of identity when he travelled beyond European civilization. It was his aesthetic baggage, as it were, carried in the belief that it could organize the new world and sustain his hopes of controlling, governing it and its denizens." However, perceiving picturesque landscapes could be difficult, MacLaren shows, where the tundra or the plains made it impossible to establish discrete, framed perspectives.[20] Nonetheless, explorers did their best to depict scenery according to convention. In heavily wooded areas, the picturesque offered a break from the impenetrable forests, and was associated with prosperous, improved landscapes.[21] Abroad, picturesque sentiment could thus celebrate agrarian development. By making landscapes accessible, artists, like politicians and generals, created new spaces for imperial expansion.

THE HALES IN LOWER CANADA

John Hale and his wife Elizabeth Amherst Hale were well placed to bring an English aristocratic aesthetic to the colony. Both from Yorkshire families, the Hales and Amhersts had built their careers in large part on the appropriation of Quebec. General James Wolfe sent John Hale's father to take the news of the 1759 military victory to the king. John, the eldest child in the Yorkshire gentry family, himself came to North America in the 1790s as aide-de-camp to the Duke of Kent.

Elizabeth's uncle, General Jeffrey Amherst, was supreme British military commander in North America and received the French army's capitulation at Montreal. On his return to England, he renamed his family estate in Kent "Montreal"[22] and planted on it "American woods," reminders of his role in conquering this part of the empire. In recompense for his contributions, the king promised him the Canadian estates of the Jesuits, one of the largest landholders in the colony. Distrusted by the English, the Jesuit Order was about to experience a period of difficulties throughout Europe, culminating in its suppression in 1773.[23] In Quebec, Governor James Murray reached a compromise with the Catholic authorities: the local Jesuits were able to maintain possession of their properties until the death of the last of their order. The last Jesuit, Father Jean-Joseph Casot, outlived Lord Amherst by a few years, and the claims to the estates fell to Amherst's nephew. The latter's sister, Elizabeth Amherst, married John Hale in 1799. Mostly for career reasons, but in part to oversee the transfer of the estate to the second Lord Amherst, the Hales moved to Quebec, whence they provided intelligence on the state of affairs.

Elizabeth and John Hale, who maintained a lengthy correspondence with Lord Amherst, were quickly disillusioned by the political wranglings involved in the transfer of the estates. Declaiming the untrustworthiness of Governor Robert Shore Milnes and his "Yankee" attorney-general Jonathan Sewell, they complained about the obstacles Amherst faced. Finally, Amherst accepted an annuity from the king in exchange for his pretensions to the Jesuit estates. The Hales, in contrast, were more fortunate in receiving land grants. The Crown awarded them property in Upper Canada. Elizabeth Hale explained to her brother, with no irony whatsoever, "[s]urely if any one has right to lands in a Country it must be the Children of those who were at the Conquest of it."[24]

*****

The Hales' disillusionment with Canadian politics was partially compensated by their enthusiasm for the landscape. In one of her first letters, Elizabeth exclaimed to her brother: "This country is very beautiful," although she persisted in thinking that the family estate in Kent was "the prettiest spot I know."[25] Quebec City was particularly attractive from a picturesque point of view. Its topography provided a multitude of perspectives from which to admire the surrounding scenery.

Like many other well-educated women of her period, Elizabeth Hale was a trained artist.[26] Claiming no talent for portraits,[27] she preferred to sketch scenery, her aesthetic sense strongly influenced by the canon of the picturesque. In 1802 she advised her brother on improvements to the landscape at Montreal estate in Kent, specifically telling him to avoid geometric lines in favour of "natural" variation: "I used to think a Beech tree would have a very good effect where the single Poplar stood, opposite the Bow room window, & would not shut out any view. The line of hills in that part is rather too strait [sic] & wants something to cut it."[28] She sent sketches of the new country back to her brother and her aunt, continuing the aesthetic exchange between the two landscapes. She collected botanical specimens for transplanting in the mother country, and she and her husband requested seeds in order to grow an English garden in their back yard in Quebec. With satisfaction Elizabeth invited her brother to visit in 1799: "I wish you c[oul]d see our pretty house & garden, it is just like an English one."[29]

Like many other members of the gentry in England during the period, the Hales were particularly keen on country living. They rented country homes for Elizabeth and the children during the summer; John joined them in the evening after work or on weekends. Escaping the heat of the city and of colonial politics, they retreated into an idyllic, stable world with obedient servants. At this time, however, they did not trust the French Canadians (or for that matter, the "Yankees"). They were fearful that the entire French-Canadian population would support a French invasion if one ever came to pass.[30] As war with the United States approached in 1812, Elizabeth Hale returned to England along with the children. During her time in England, she undoubtedly came into contact with the views of landscape designer Humphry Repton. In 1812 Repton completed his commission to provide suggestions for the renewal of Montreal estate in Kent.[31]

On the Hale family's return to Lower Canada after the war, they seemed more secure in their position in the colony. With the successes of the War of 1812, the generally strong showing of French-Canadian support, and the end of the Napoleonic threat, the English elite in Quebec breathed a sigh of relief. The Hales now began to consider the possibility of establishing themselves in the colony. They were less concerned than previously about the possibility of Quebec falling back into the hands of the French: "Canada is now no longer likely to become again a French Province; and though it may in a few years be one of the United States of America, it is not probable that the same proscriptions would in that event attend the English settlers, as might have been apprehended from the French Government."[32] By 1818 they had begun considering the purchase of a family estate in the New World. They continually stressed to their relatives in England that such an investment would not preclude their return to the mother country, given the right circumstances.

As they looked for an estate, their interests coincided with the demographic and financial fortunes of the Lanaudière family, whose seigneurial title to Sainte-Anne dated from the 1670s. Only one of Charles-Louis de Lanaudière's children had survived into adulthood. Following Charles-Louis's death in 1811 and his widow's in 1817, Marie-Anne de Lanaudière, at 42 years of age, found herself the sole heir to Sainte-Anne seigneury. After overseeing a number of concessions to habitants, she decided to dispose of the property. Elizabeth Hale reported that the seigneury "is sold in consequence of being now the property of a Lady not likely to marry & who cannot bear the trouble of looking after it."[33]

By the summer of 1819, then, the Hales had a "grand Speculation in view."[34] Various elements of the sale recommended themselves to the Hales: the timber resources, the accessibility of the seigneury from Quebec, the promise of a return of 4.5–5 percent on their investment, and the escape from the capital's summer heat. Impressed by the beauty of the site and their belief in the effectiveness of agricultural "improvement," the Hales offered £12,000 for the seigneury. More than anything, the purchase of the seigneury represented the investment of their future in the colony, a property with possibilities for their nine surviving children.

*****

Once the purchase was finally concluded in 1820, the Hales were quite content. At least in part because Sainte-Anne happened to fill English aesthetic criteria, it had earlier attracted Elizabeth's eye. As she considered the prospect of purchasing the seigneury, she recalled, "I was there many years ago & admired the place, when there was no chance of its being sold."[35] Typical in many ways of the British gentry's attitude toward the moral qualities of rural living,[36] the Hales believed that investment in the countryside was the most legitimate use of their capital. They revelled in the prospect of owning so much more land than could ever have been possible in England. "Only consider," Elizabeth Hale exclaimed to her aunt, "what an immense tract of land [is] 60 square miles!"[37] For Elizabeth, their ownership of the land made all their endeavours worthwhile: "I was always partial to the Country & when the place is your own & all the alterations & improvements made for the benefit of your children, your interest in it is naturally increased & your attention almost engrossed by them."[38] They were thrilled with the exotic seigneurial tributes they could exact: every twelfth fish, part of the maple syrup production, the right of passage over the Sainte-Anne River, not to speak of the more lucrative yearly *cens et rentes*, the occasional *lods et ventes*, and the mill toll.[39] Timber sales also promised great profits, which John Hale was quick to exploit. In 1822–1823 he contracted with local habitants for some 2,300 pine and spruce logs, to be cut in the unconceded parts of the seigneury.[40]

In addition to its economic benefits, the seigneury also allowed the Hales to ignore present troubles. One of its appeals was the escape it represented from the tribulations of contemporary England, such as the Peterloo massacre. About a year after they purchased Sainte-Anne, Elizabeth Hale wrote to her brother: "[H]ere we enjoy peace & quiet at present & all the comforts of a Country life far from all political alarms."[41] Referring to the popular unrest related to Queen Caroline's trial for adultery, she commented that "some of them [the English ministers] might envy us enjoying the quiet of the Country amongst well-disposed habitants."[42] Sainte-Anne offered a rural refuge from the vagaries

of imperial and, for that matter, Colonial politics.

But the seigneury did not only allow a retreat into a rural idyll. It also represented an investment of the family's future in the continent, an attempt to remain in step with a particular historical narrative, which Elizabeth Hale revealed in 1819: "[F]or my own part I cannot but consider America as the rising Country & thus our Children or Grandchildren will be very glad to have property here—in history we find that as one Country begins to decline another rises & generally to the Westward & I cannot but think that our children may see something of the kind during their lives."[43] With its promises of present peace and future prosperity, Sainte-Anne would become home to the Hales.

THE APPROPRIATION

It was one thing to purchase the land; it was another to take possession of it. For the Hales, the appropriation of their land proceeded along some lines typical of all seigneurs. Other elements of their approach were more limited to their class and cultural background.

Shortly after the purchase, as a first act of possession, Elizabeth Hale compared the seigneury to a fashionable London suburb: "The village is remarkably pretty & the whole scenery reminds me very much of the Thames towards Putney," she claimed to her brother in England, "only that the St Anne's is a finer river & nothing like Mud."[44] To enjoy fully the seigneury's many qualities, an inventory of present and absent elements had to be compiled.

Establishing proprietorship over land first implied taking stock of nature.[45] John Hale observed nature in a scientific spirit: he studied the sexual reproduction of crickets for a paper he later presented in

Quebec.[46] Elizabeth Hale made the link between ownership and inventories of nature even more clearly in writing to her brother: "If we get the estate I have mentioned before to you, I shall make a point of endeavouring to collect seeds & roots."[47]

In the Hales' perception of landscape, and those of the English gentry in general, trees wielded the highest metaphorical value.[48] The forests in the deepest parts of the seigneury held the promise of economic benefits for the seigneur. Closer to the seigneurial manor-house, Elizabeth Hale deplored the lack of proper trees. In the garden, there were no fruit trees, not even a fine gooseberry bush.[49] To remedy this shortcoming, the Hales planted an orchard of plum and apple trees.[50] But they were not content to stop there. Oaks, because of their identification with British patriotism and, by extension, with the British gentry,[51] represented the most significant absence: "I am sorry to say," Elizabeth wrote, "there is not an Oak to be seen here or in the neighbourhood, but I intend trying some for the honor of my native country."[52] The Hales believed that the natural beauty of Sainte-Anne could greatly be improved by adding the proper English elements.

Thus, garden plants were imported to improve the domaine. Elizabeth Hale's sister-in-law in England sent flower seeds, and son Jeffrey on naval duty promised to keep looking for exotic specimens.[53] In 1822 the Hales planted hedges, possibly to disguise some of the many fences that were anathema to the proper noble landscape.[54] For Elizabeth, the activities represented the improvement of the landscape: "I am very busy planting & neatifying which is (altho' not to be found in [Samuel] Johnson['s dictionary]), more descriptive of my occupation than beautifying would

be."[55] Having moved into the manor-house, the Hales expected to enhance the perspectives they could command, planning "at our leisure to build on higher ground."[56] Though they never replaced the manor-house, their vista was appropriately noteworthy. An 1822 visitor recalled the setting: "[A] roomy, lightsome house, built mansion-like, one hundred yards from a trout-stream, the St. Anne de la Pérade, on the upper edge of a large park-like meadow, which runs down to the St. Lawrence. From the house we saw the river, a woody islet or two, close in shore, and had between them a momentary glimpse of the passing steamers."[57] The Hales' desire to catalogue, reshape, and dominate the domaine reflected their hopes for the seigneury as a whole.

A second, rather typical, element of their appropriation was to commission maps of the seigneury. This was necessary to fix the exact boundaries of the seigneury, to locate *censitaires'* lands, and to permit future modifications in tenure. More than an objective representation of geography, maps also have political purposes.[58] Large-scale maps, such as seigneurial surveys, contain their ideological implications in their project: to justify and take stock of the hierarchical distribution of land.

An 1825 map [...] presents a logic of appropriation. Except for the boundaries of the concessions, the rivers, and some indications of the density of buildings in the village, relatively few details appear on the map. In the northern part of the seigneury in particular, a great blank appears where the seigneurs could write their own history. In fact, the surveyor had begun to do so, providing the title in that space. John Hale's occupations coincide with and explicate his status as seigneur. The few details noted near the northwestern boundary of the seigneury promise future expansions: small rivers, lakes, pine forests, and prairies. The Hales were attentive to the surveyors' findings. Elizabeth Hale wrote of the results to her brother in 1820: "The Surveyors have been at work & about 5 Leagues from our house came to a beautiful little Lake covering about 6 acres of Land with large Trees & a very rapid river full of Trout—they found a very compleat [sic] Beaver dam which Mr Hale intends seeing but the road not being yet made I have no chance of getting there—He describes the Land as very rich & very fine Timber without under wood."[59] A second survey from 1830 places the names of each plot-holder within the boundaries of his or her land. The exception to the rule is telling: The Hales' land is referred to only as the "Domaine." Although they occupied no other lots in Sainte-Anne, the Hales' centrality is thus implicit in the construction of the map at the same time that their name is omitted.

With surveyors trampling through the back country and family members busy making physical changes to the landscape, the Hales went about establishing proprietorship. Elizabeth Hale also expressed the family's appropriation of Sainte-Anne in her sketches. Only days after acquiring the seigneury, she sent her brother a drawing of a scene a short distance from the manor-house: "the stream you see is a branch of the river St Ann's which falls into the St. Lawrence about a quarter of a Mile from the house & being wooded on each side is very pretty."[60] A watercolour painter of some talent, Elizabeth Hale's art sometimes had a more explicit political aim. In 1824, anxious to establish closer relations with Governor Dalhousie, she sent him sketches of Quebec City: "I request your Lordship's acceptance of the enclosed sketches, & trust they may serve to remind you of a

place where the return of your Lordship & Lady Dalhousie will be most anxiously expected."[61]

Elizabeth Hale applied her talents to the scenery of Sainte-Anne, casting the landscape as a picturesque, summer-time idyll. A small sketchbook, dated between 1823 and her death in 1826,[62] reveals the particular scenes she found worthy of illustration. [...] In theory, Elizabeth Hale could have sketched many different landscapes in the seigneury, showing habitant farms, bucolic labouring scenes, or wilderness views. In practice, she did not stray very far from the manor-house, and consciously or not she chose to depict views in which the family's proprietorship formed a symbolic presence. Of the 24 sketches pertaining to Sainte-Anne seigneury, 6 portray different views of the manor-house. [...] Other sketches implicitly celebrated other seigneurial privileges: the family's sawmill or the mere act of fishing on the Sainte-Anne. The Hales' proprietorship is subtly present in all these drawings. In this sense, her art reflects Repton's views of the picturesque, which should celebrate proprietorship.

As in her correspondence, trees occupied a prominent place in Elizabeth Hale's drawings. The predominance of deciduous trees in her views of an area where pines were not uncommon illustrated her desire to portray the landscape in English terms. Like other artists of the picturesque, she did not attempt to provide a strict, empirical reproduction of the landscape; rather, a natural-looking but nonetheless ideal landscape was more important.[63] Aesthetic perspectives determined Elizabeth Hale's ability to see Putney along the Sainte-Anne as much as did the area's natural characteristics.

Equally important for this analysis is what Elizabeth Hale chose not to depict.

Without being unoccupied, her landscapes were certainly not overpopulated with human beings (or animals). There are few people in her landscapes, especially French Canadians. Some people fish, others canoe and pole across the Sainte-Anne River. Generally, however, the local habitants do not form an important presence. Although the fences suggest agricultural activity, and fishing could be important in an economic sense, there is little acknowledgment of habitants' labour. No one actually works the land. Indeed, it is possible, given their clothing, that many of the figures represent members of the Hale family itself. Although fishing may have been work for the habitants, it was recreation for the Hales. Everything, or almost everything, was leisure in this ideal landscape.

Outnumbered by the sketches of the manor-house is Hale's one drawing of the village of Sainte-Anne. [...] Like the peasants who had actually shaped the landscape, the village occupied a shadowy part of Hale's picturesque world. In her letters to England, Elizabeth Hale seldom mentioned the village or the activities of habitants, except when it came to decrying their deficiencies and expressing a desire to "improve" them. She pitied her daughter's isolation at Sainte-Anne: "We have not a single being above the capacity of a common farmer in the neighbourhood."[64] Rather than expressing interest in Canadian habitants, Elizabeth Hale saw the future promise of the seigneury as relying more on livestock and immigrant Yorkshire farmers.

Part of their belief that the domaine farm could be so quickly made profitable was due to their sense that it had been neglected by the Lanaudières: "We are very busy opening & deepening drains, which had not been touched for these 12 years certainly — & cleaning the land which had been suffered to get full of weeds, so that a very large field has been left without any produce this year."[65] With a picturesque perspective in mind, the Hales dreamed of the changes they could effect. The Hales' appropriation of the environment focused on, but was not restricted to, the area around the manor-house. They wanted to "improve" the land. For them, "improvement" implied the importation of English farmers, livestock, and techniques. For instance, the commons had to be enclosed: "We might make great use of it if we had the right to enclose it when cleared, a great deal of Timber has already been cleared there but being in common it is no one's business to take care of it or improve it."[66] Believing that expansion would ensure the profitability of the investment, John Hale considered purchasing the neighbouring seigneury of Dorvilliers (Sainte-Anne East).[67] The intellectual principle justifying the Hale's appropriation of the seigneury was the belief that they were improving it.

Of course, the Hales' superior attitude toward the previous seigneur was compounded by their class disdain for the habitants. They complained that habitants did not know how to use manure, and they tried to show them how to run a farm. In an attempt to introduce English techniques, John Hale taught the Canadians "the advantages of Drill Husbandry for Turnips, and of Summer Fallows. I have likewise astonished them with a Hay Stack, which was never seen here before."[68]

In this limited sense, the habitants were important to the Hales' world view, but only as pupils to be taught or at best as actors in a play the seigneurs initiated. Entertaining a visitor in the early 1820s, John Hale walked with him through the village: "It was a promenade," the guest recalled, "of unconstrained greetings and pleasant looks. Red worsted caps and uncouth hats were doffed at every turn."[69] Being seigneur of Sainte-Anne fulfilled John Hale's desire to see acknowledged his social prestige, a feeling undoubtedly enhanced by his many disappointed attempts to receive patronage.

But the Hales could not rely solely on their efforts to impose new agricultural techniques on the Canadian peasants. Forming a population of some 2,178 in 1822, local habitants complained of land shortages.[70] The Hales maintained the high rents of the previous seigneurs, and posited that the future growth of the seigneury depended on the immigration of Yorkshire farmers. John Hale sent an advertisement to the *Farmers' Journal* in Yorkshire, requesting 20 or 30 families to settle in his seigneury.[71] He particularly tried to attract the Chandler family.[72] Ultimately, the Hales wanted to create a rival for the village of Sainte-Anne. They projected the clearing of a second domaine 6 or 8 miles from the St. Lawrence in order to establish another village: "We might perhaps be able to make a Protestant village of it from the many settlers who come here & then have a Protestant Church."[73] For the Hales, then, this was the promise of Sainte-Anne. They could reshape the landscape, physical and human, and create a "new" England, or rather a new Yorkshire, in the St. Lawrence Valley.

Ultimately, the development of the seigneury would coincide with that of their family. As their family grew and prospered, so would Sainte-Anne. "I sometimes amuse myself," Elizabeth Hale confided to her brother in 1823, "with looking forward to the improvements [son] Edward will make some years hence when he returns from India—we have not yet received much from it, but it goes on

well & we still hope will be advantageous to the young ones."[74] A permanent desire for improvement should, she thought, transform the local landscape.

RESULTS

However, the Hales' picturesque landscape never fully took root in Sainte-Anne. Of the 72 concessions John Hale granted between 1821 and 1831, only 4 went to people with English names.[75] There would be no Protestant village in the seigneury of Sainte-Anne. The manuscript census for 1825 lists few Protestants and no Chandler family.[76] Elizabeth Hale died of cancer in 1826. In 1827 John Hale, who relished his position as seigneur, applied to the governor to have the northern part of Sainte-Anne transmuted to free and common socage.[77] By 1833 the Anglican minister at Trois-Rivières warned that the few Protestants in the area "are much in danger of being seduced from the Protestant faith or of having their children brought up in the Romish Communion."[78] The Hales' dreams did not come to fruition.

As long as John Hale lived, there seemed to be no question of selling the seigneury.[79] Although his son George planned at various points to move into the manor-house, no Hales appeared in the government censuses of the area after 1831. Within a year of John's death, son Edward began speaking of placing it on the market. By the 1830s the habitants had greatly reduced their wheat crops, which in turn limited the seigneurial mill toll. John's brother Edward commented in 1840, "I really don't know how those manage who depend on their seign[euria]l property for a lively hood [sic]."[80] A calculation of seigneurial profits between 1841 and 1844 showed average net proceeds of about £286 per year, for a return of 2.4 percent on the initial investment.[81] In 1844, uncle

Edward Hale disparaged the materialistic turn of seigneurs at the same time that he agreed Sainte-Anne should be placed on the market: "The taste for this kind of property is I fear on the decline ie. all look to the £.S.D. return without (as formerly) attaching the least value to the imaginary dignity of Seigneur, but something must be done, this property cannot be sold as the U[pper] C[anada] Land in lots."[82] They were not able to find a buyer.

Despite their willingness to sell Sainte-Anne, the Hale children had nonetheless imbibed many of the cultural proclivities of their parents: They botanized and sketched and fanned across the British empire. Son Edward Hale built a new estate near Sherbrooke, and in 1846 he commissioned Cornelius Krieghoff to commit it (and the ancestral manor-houses at Sainte-Anne and in England) to canvas.[83] Edward Hale's name for his estate reflected in many ways his mother's expectations of hers: he called it "Sleepy Hollow."

On a small scale, the Hales attempted to apply in the St. Lawrence Valley the lessons of English imperialism that were reshaping the world in the early nineteenth century. This article suggests that this project was not merely political or economic; it also involved an aesthetic appropriation. Though the Hales' ideal landscape did not emerge, the aesthetic impulse behind it flourished elsewhere in the colony. Members of both English and French elites reflected picturesque sentiments as they commissioned and purchased paintings and manor-houses.[84]

Elizabeth Hale's sensibilities and work were of particular significance in the appropriation of Sainte-Anne seigneury. Bounded by her picturesque sense, she could not see Sainte-Anne, or by extension North America, as a fundamentally different world. Although some scholars argue that the Canadian landscape freed certain English gentlewomen from the constraints of their class and sex, Elizabeth Hale's case emphasizes rather the imperialist implications of her perception.[85]

The Hales' attitude toward their property ultimately bore a close resemblance to other European imperialist projects of the period.[86] Celebrating the social and economic satisfactions of their proprietorship, the Hales' vision of the land left little place for its local inhabitants. They ignored the specificity of the new context in order to recreate their homeland. As Northrop Frye observes, "in general the picturesque eye was an idealizing one, assimilating past experience in Europe to a future when the new world would look more like the old one."[87] For the Hales, Sainte-Anne offered both an escape into a rural idyll and a New World future. This suggests tension within imperialist projects like the Hales', which must promise both a secure, local prospect and an idealized, foreign-determined vision.

NOTES

1. Bernard Smith, *European Vision and the South Pacific: A Study in the History of Art and Ideas* (London, 1960); Doug Owram, *Promise of Eden: The Canadian Expansionist Movement and the Idea of the West 1856–1900* (Toronto, 1980); Paul Carter, *The Road to Botany Bay: An Exploration of Landscape and History* (New York, 1988).

2. Mary Louise Pratt, *Imperial Eyes: Travel Writing and Transculturation* (New York, 1992), 61.

3. Allan Greer, *Peasant, Lord and Merchant: Rural Society in Three Quebec Parishes, 1740–1840* (Toronto, 1985); Sylvie Dépatie, Mario Lalancette, and Christian Dessureault, *Contributions à*

l'étude du régime seigneurial canadien (LaSalle, 1987); Franchise Noel, *The Christie Seigneuries: Estate Management and Settlement in the Upper Richelieu Valley, 1760–1854* (Montreal, 1992).

4. Denis E. Cosgrove, *Social Formation and Symbolic Landscape* (London, 1984); Anne Bermingham, *Landscape and Ideology: The English Rustic Tradition, 1740–1860* (Berkeley, 1986).

5. Louis B. Wright and Marion Tinling, eds., *Quebec to Carolina in 1785–1786; Being the Travel Diary and Observations of Robert Hunter, Jr., a Young Merchant of London* (San Marino, 1943), 25.

6. John Lambert, *Travels through Canada, and the United States of North America, in the Years 1806, 1807 & 1808,* 2nd ed. (London, 1814), 425, 460.

7. J. Ross Robertson, ed., *The Diary of Mrs. John Graves Simcoe* (Toronto, 1911), 348.

8. Joseph Bouchette, *A Topographical Dictionary of the Province of Lower Canada* (London, 1832), "Ste. Anne, seigniory," no pagination. This is the English translation of a passage from Bouchette's 1815 book, *Description topographique de la Province du Bas-Canada* (Montreal, 1978), 326–327.

9. Gerald Finley, *George Heriot: Postmaster-Painter of the Canadas* (Toronto, 1983), 196–197; Didier Prioul, "Les paysagistes britanniques au Québec: de la vue documentaire à la vision poétique," in Mario Beland, dir., *La Peinture au Québec, 1820–1850: Nouveaux regards, nouvelles perspectives* (Quebec, 1991), 50–59, and Prioul, "Georges Bulteel Fisher," ibid., 155–158.

10. John Dixon Hunt and Peter Willis, eds., *The Genius of the Place: The English Landscape Garden, 1620–1820* (New York, 1975), reprint excerpts illustrating some aspects of the debate.

11. Carter, *Road to Botany Bay*, Chapter 8; I.S. MacLaren, "The Aesthetic Mapping of Nature in the Second Franklin Expedition," *Journal of Canadian Studies* 20 (Spring 1985): 40–46; I.S. MacLaren, "The Limits of the Picturesque in British North America," *Journal of Garden History* 1 (January–March 1985): 97–111; I.S. MacLaren, "The Pastoral and the Wilderness in Early Canada," *Landscape Research* 14 (Spring 1989): 15–19; Patricia Jasen, "Romanticism, Modernity, and the Evolution of Tourism on the Niagara Frontier, 1790–1850," *Canadian Historical Review* 72 (September 1991): 288–290.

12. John Barrell, *The Dark Side of the Landscape: The Rural Poor in English Painting, 1730–1840* (Cambridge, 1980); Hugh Prince, "Art and Agrarian Change, 1710–1815," in Denis Cosgrove and Stephen Daniels, eds., *The Iconography of Landscape: Essays on the Symbolic Representation, Design and Use of Past Environments* (Cambridge, 1988), 98–118. On the escapism that the picturesque allowed upper-class women, see Marian Fowler, *Below the Peacock Fan: First Ladies of the Raj* (Markham, 1987), 45–47.

13. John Barrell, *The Idea of Landscape and the Sense of Place, 1730–1840: An Approach to the Poetry of John Clare* (Cambridge, 1972), 79.

14. Bermingham, *Landscape and Ideology*, 75.

15. Cited in Hunt and Willis, eds., *The Genius of the Place*, 338.

16. Humphry Repton, *Theory and Practice of Landscape* [1803] in John Nolen, ed., *The Art of Landscape Gardening* (Boston and New York, 1907), 71.

17. Ibid., 150–151.

18. Cosgrove, *Social Formation and Symbolic Landscape*, 214.

19. E.P. Thompson, *Whigs and Hunters: The Origin of the Black Act* (Harmondsworth, 1977).

20. MacLaren, "Limits of the Picturesque," 100; MacLaren, "Aesthetic Mapping of Nature."

21. D.M.R. Bentley, "Afterword" to Catharine Parr Traill, *The Backwoods of Canada* (Toronto, 1989), 297–298.

22. C.P. Stacey, "Jeffery Amherst, 1st Baron Amherst," *Dictionary of Canadian Biography* (DCB), vol. 4 (Toronto, 1979), 20–26.

23. Roy C. Dalton, *The Jesuits' Estate Question 1760–1888: A Study of the Background for the Agitation of 1889* (Toronto, 1968).

24. Thomas Fisher Rare Book Library, University of Toronto, Hale Papers, Manuscript Collection 90 (UT-Hale), E.F. Hale to Lord Amherst, November 14, 1811.

25. UT-Hale, E.F. Hale to Lord Amherst, July 22, 1799, and October 1, 1800.

26. On amateur noblewomen artists during this period, see Germaine Greer, *The Obstacle Race: The Fortunes of Women Painters and Their Work* (New York, 1979), Chapter 14; Maria Tippett, *By a Lady: Celebrating Three Centuries of Art by Canadian Women* (Toronto, 1992), 5–9.

27. UT-Hale, E.F. Hale to Lord Amherst, August 4, 1804.

28. Ibid., August 12, 1802.

29. Ibid., July 22, 1799.

30. Ibid., John Hale to Lord Amherst, February 2, 1803.

31. NA, Documentary Art Division, Report concerning Montreal in Kent a Seat of the Right Honorable Lord Amherst, etc., etc., etc. by H. Repton, 1812.

32. NA-Hale, reel A1085, John Hale to Lord Amherst, May 16, 1817.

33. UT-Hale, E.F. Hale to Lord Amherst, June 24, 1819.

34. Ibid., June 3, 1819.

35. Ibid., June 25, 1819.

36. Keith Thomas, *Man and the Natural World: Changing Attitudes in England, 1500–1800* (London, 1983), 13.

37. UT-Hale, E.F. Hale to Lady Amherst, June 25, 1819.

38. NA-Hale, reel A1085, E.F. Hale to Lord Amherst, October 19, 1820.

39. UT-Hale, E.F. Hale to Lord Amherst, December 27, 1819 and July 6, [1822].

40. Archives nationales du Québec à Trois-Rivières, Greffe Joseph–Casimir Dury, CN401-33.

41. NA-Hale, reel A1085, E.F. Hale to Lord Amherst, no date [late October 1820?].

42. Ibid., no date [July 20, 1820?].

43. UT-Hale, E.F. Hale to Lady Amherst, July 17, 1819.

44. Ibid., E.F. Hale to Lord Amherst, August 21, 1819.

45. Suzanne Zeller, *Inventing Canada: Early Victorian Science and the Idea of a Transcontinental Nation* (Toronto, 1987), 3–9.

46. John Hale, "Observations on Crickets in Canada," in *Transactions of the Literary and Historical Society of Quebec* 1 (Quebec, 1829): 254–255.

47. UT-Hale, E.F. Hale to Lord Amherst, July 28, 1819.

48. Stephen Daniels, "The Political Iconography of Woodland in Later Georgian England," in Cosgrove and Daniels, eds., *The Iconography of Landscape*, 43.

49. UT-Hale, E.F. Hale to Lord Amherst, December 27, 1819.

50. NA-Hale, reel A1085, E.F. Hale to Lord Amherst, no date [late October 1820?].

51. Daniels, "The Political Iconography of Woodland," 48.

52. NA-Hale, reel A1085, E.F. Hale to Lord Amherst, no date [July 20, 1820?].

53. UT-Hale, E.F. Hale to Lord Amherst, June 2, 1822; NA-Hale, vol. 4, Jeffrey Hale to mother, June 17, 1820, ff. 250–251.

54. UT-Hale, E.F. Hale to Lord Amherst, October 29, 1822; NA, Picture Division, Hale Sketchbook, 1–2, neg. no. C-1307, which shows that the hedges were planted to obscure a fence.

55. UT-Hale, E.F. Hale to Lord Amherst, June 2, 1822.

56. Ibid., E.F. Hale to Lady Amherst, September 17, 1819.

57. John J. Bigsby, *The Shoe and Canoe or Pictures of Travel in the Canadas* (New York, 1969 [1850]), I, 48.

58. G.N.G. Clarke, "Taking Possession: The Cartouche as Cultural Text in Eighteenth-Century American Maps," *Word & Image* 4 (April–June 1988): 455–474; J.B. Harley, "Deconstructing the Map," in Trevor J. Barnes and James S. Duncan, eds., *Writing Worlds: Discourse, Text and Metaphor in the Representation of Landscape* (London, 1992), 231–247.

59. NA-Hale, reel A1085, E.F. Hale to Lord Amherst, December 15, 1820.

60. UT-Hale, E.F. Hale to Lord Amherst, September 20, [1819].

61. NA, Dalhousie Muniments, MG 24, A12, reel A537, E.F. Hale to Lord Dalhousie, April 5, 1824.
62. W. Martha E. Cooke, *W.H. Coverdale Collection of Canadiana: Paintings, Watercolours and Drawings: Manoir Richelieu Collection* (Ottawa, 1983), 98; Gilbert Gignac provides an insightful discussion of the artistry of the sketchbook: "Elizabeth Frances Hale, 1774–1826," in Béland, dir., *La Peinture au Québec*, 277–282.
63. Prioul, "Les paysagistes britanniques au Québec," 51.
64. NA-Hale, reel A1085, E.F. Hale to Lord Amherst, December 15, 1820.
65. Ibid., October 19, 1820.
66. Ibid., February 24, 1820.
67. Ibid., J. Hale to Lord Amherst, February 24, 1820.
68. Ibid., July 20, 1820.
69. Bigsby, *The Shoe and Canoe*, I, 49.
70. Colin Coates, "The Boundaries of Rural Society: Batiscan and Sainte-Anne de la Pérade to 1825," Ph.D. dissertation, York University, 1992, Chapter 4.
71. NA-Hale, vol. 4, J. Hale to Revd. Richard Hale, June 6, 1824, f. 340.
72. Ibid., Jeffery Hale to John Hale, January 7, 1821, f. 259.
73. Ibid., reel A1085, E.F. Hale to Lord Amherst, January 11, 1820.
74. Ibid., May 20, 1823.
75. Archives nationales du Québec à Trois-Rivières, Répertoires des greffes de Joseph-Casimir Dury, CN401-33, et Augustin Trudel, CN401-91.
76. NA, Census of 1825 for Sainte-Anne, reel C718. Rather, the Chandlers purchased their own seigneury of Nicolet in 1821, with similar aspirations to establish an English community. Richard Chabot, "Kenelm Conor Chandler," *DCB*, vol. 7 (Toronto, 1988), 165–167.
77. NA, Lower Canada Land Papers, RG 1, L3L, vol. 100 (reel C2531), John Hale to Earl of Dalhousie, March 10, 1827, ff. 49453–4.
78. Quebec Diocesan Archives, Anglican Church of Canada (Lennoxville), Three Rivers, B24, Parish Report, Rev. Sam Wood, January 26, 1833.
79. A memorial plaque dedicated years later in the Anglican cathedral in Quebec City to four Hale patriarchs listed John Hale's status as seigneur before his government positions. In a similar vein, his obituary commented that John "lived some years at the Domain [of Sainte-Anne], on the best terms with the inhabitants, amongst whom he introduced several improvements." *Quebec Gazette* (December 26, 1838).
80. McCord Museum, Hale Family Papers, M20483, Edward Hale [uncle] to Edward Hale, November 26, 1840.
81. Ibid., George Hale to Edward Hale, April 16, 1845.
82. Ibid., Edward Hale [uncle] to Edward Hale, November 5, 1844.
83. J. Russell Harper, *Krieghoff* (Toronto, 1979), 18.
84. France Gagnon-Pratte, *L'architecture et la nature à Québec au dix-neuvième siècle: les villas* (Quebec, 1980), Chapter 3; Janet Wright, *Architecture of the Picturesque in Canada* (Ottawa, 1984).
85. Marian Fowler, *The Embroidered Tent: Five Gentlewomen in Early Canada* (Toronto, 1982). See also the articles in Lorraine McMullen, ed., *Re(Dis)covering Our Foremothers: Nineteenth-Century Canadian Women Writers* (Ottawa, 1990). For analyses closer to this discussion of Hale, see F.K. Stanzel, "Innocent Eyes? Canadian Landscape as Seen by Frances Brooke, Susanna Moodie and Others," *International Journal of Canadian Studies* 4 (Fall 1991): 97–109; Bentley, "Afterword."
86. Pratt, *Imperial Eyes*.
87. Northrop Frye, "The Canadian Scene: Explorers and Observers," in R.H. Hubbard, *Canadian Landscape Painting, 1670–1930* (Madison, 1973), 3.

Chapter Twelve
Killing the Canadian Buffalo, 1821–1881

William A. Dobak

The figure of the mounted, buffalo-hunting Plains Indian has engaged the imagination of Euroamericans for more than 200 years. Anthony Henday, on a trip west to drum up trade for the Hudson's Bay Company in 1754, found the Native peoples of eastern Alberta living among "[b]uffalo in great droves," and the HBC's Matthew Cocking used the same words a generation later, when he saw them some 200 miles to the east, along the North Saskatchewan River. Although horses had arrived in the Northern Plains barely 50 years earlier, by 1794, Native peoples were "so advantageously situated that they could live very happily independent of our assistance," wrote the North West Company trader Duncan McGillivray.

> They are surrounded with innumerable herds of various kinds of animals, whose flesh affords them excellent nourishment and whose skins defend them from the inclemency of the weather, and they have invented so many methods for the destruction of animals, that they stand in no need of ammunition to provide a sufficiency for these purposes.

Yet before 80 years had passed, the buffalo were gone from the Canadian Plains, and Native peoples were living on reserves. The trade in tanned buffalo robes was largely responsible—the Canadian historian Robert Beal has called it "the fundamental factor"—but the decades-long provision trade, in which Native and *métis* hunters furnished pemmican for the Hudson's Bay Company's boatmen, also played a part, as production statistics suggest. What happened, to the people and to the buffalo, has been described by Canadian writers, but is not generally familiar to American readers.[1]

The story of the extermination of the buffalo in the northern United States, on the other hand, is well known. A few years after the Sioux War of 1876, commercial hide-hunters went to work. Since they were interested in the hides as leather, rather than as robes, they hunted year-round, even in summer when the buffalo's coat was thin and robe-hunters would have let the herds alone. In November 1881, the Northern Pacific Railroad reached the Yellowstone River at Miles City, and direct shipment of hides became possible. "In 1882 the number of hides and robes bought and shipped was about two hundred thousand," a dealer in

Minneapolis recalled, "and in 1883 forty thousand. In 1884 I shipped ... the only car load of robes that went East that year, and it was the last shipment ever made." William Hornaday, of the Smithsonian Institution, assessed the damage in 1887: "Many think that the whole great body [of buffalo] went north into British territory, and that there is still a goodly remnant of it left in some remote region ... which will yet return to the United States. Nothing could be more illusory than this belief."[2]

Whether the destruction of the buffalo in the United States resulted from deliberate government policy or from market forces alone, commercial hide-hunters did the final damage. More fundamental, though, is the question of whether Native peoples themselves "had established a society in ecological equilibrium, one whose population did not exceed the carrying capacity of its habitat and so maintained a healthy, functioning ecology that could be sustained over the long term." Concentrating on the Southern Plains, Dan Flores has concluded that several factors militated against equilibrium, foremost among them being sheer human population pressure on the Plains, and from outside the region, the expanding capitalist market system and its demand for tanned buffalo robes, which were exchanged for European trade goods.

Native beliefs in a supernatural sanction for buffalo hunting probably played a part as well.[3]

The Canadian Plains are a good proving ground for these findings. No year-round trade in raw buffalo hides, which depended on breech-loading rifles to kill the animals and railroads to ship the hides, developed there. The animals were gone years before the tracks of the Canadian Pacific arrived. In the absence of catastrophic grass fires or epidemic disease, one must conclude

that the buffalo's disappearance was due to human agency. What sort of pressure, exerted by how many people, sufficed to devastate the Canadian herds and to drive the survivors south of the forty-ninth parallel, indeed, south of the Missouri River? How did the international market economy, in the form of the robe trade, affect the number of buffalo killed? What may have been the effect of the Indians' world view on their hunting practices? What, finally, does this suggest about the ecological balance in the North American grassland between human predators and their prey?[4]

An attempt to explain how and why Native groups were able to drive the buffalo out of Canada must begin with an estimate of the number of buffalo, derived from the approximate extent and carrying capacity of the range. Vegetation maps of the Northern Plains show grassland extending from the aspen parkland along the North Saskatchewan River, southward across the drainage basin of the Yellowstone, with an approximately equal area of grass on either side of the international boundary. The predominant species are grama, needlegrass, and wheatgrass.[5]

The method Flores used to figure the nineteenth-century bison population of the Southern Plains was to take the number of domestic grazing animals listed in the 1910 U.S. Census and adjust it by means of a "grazing quotient," reflecting the varying range-use efficiency of different species. In 1910, the total number of cattle, horses, and mules in the six large counties of Montana, where the buffalo had ranged in the last third of the nineteenth century, was 527,468. Adjusting for the grazing quotient of horses and mules (1.25, a cow's being 1.0) gave a figure of 563,758, which, given the more efficient grazing of buffalo, yielded a possible population of 675,000.

Doubling that (a crude expedient, in the absence of comparable data for Canada) gave a total of 1,350,000 for the northern herd during the late 1860s.[6]

The trouble with using this method for Montana arises from the size of Montana's counties in 1910. There were only four in the state's entire northern tier, and they included a variety of terrain and vegetation, some parts of it less hospitable to buffalo than others. Fortunately, a bulletin of the Montana Agricultural Experiment Station defines 22 different types of rangeland plant associations, locates them on a map, gives the total acreage for each type, and lists the carrying capacity of each. Figuring the acreage and carrying capacity of the five vegetative types that predominate in the part of Montana where buffalo were still found during the last third of the nineteenth century gave a figure of about 950,000 buffalo, or 1,900,000 on both sides of the line in the early 1870s. This latter, higher, figure may reflect the nineteenth-century carrying capacity of the range more accurately than the figure derived from the census totals. Figures based on the 1910 census may be on the low side, for 1910 fell in the middle of a seven-year dry spell, when the carrying capacity of the range was reduced. The drought, though, does not appear to have been as severe in Montana as it was elsewhere on the Plains.[7]

The buffalo had ranged far more extensively, however, only a few decades before their destruction. Large expeditions from the Red River Settlement did not need to travel far to load their carts with meat during the 1820s, and in the fall of 1829 the Hudson's Bay Company's employees at Brandon House, on the Assiniboine River, lived on buffalo killed close by. The buffalo range extended east, early in the century, into the wheatgrass-bluestem-needlegrass

prairie in the valleys of the James and Sheyenne rivers, which indicates a much larger buffalo population in the 1820s. The agricultural census of 1910 for North and South Dakota (where counties were much smaller than in Montana) and northeastern Wyoming listed 1,157,022 horses and mules and 2,151,005 cattle, which, with the figures for Montana and Canada just cited, suggests a carrying capacity of about 5,600,000 buffalo for the entire Northern Plains, from the North Saskatchewan River to the Pine Ridge Escarpment. Comparison of this figure, for the whole of the northern range as it existed in the 1820s, with either of the estimates of the buffalo population of the Saskatchewan–Yellowstone region in the late 1860s, indicates that outright predation had diminished the number of buffalo by about two-thirds during those 50 years.[8]

Such a loss, based on an estimate of 1,900,000 buffalo for the late 1860s, seems likely. Before detailing the killing of the buffalo, though, other possible explanations for their disappearance should be considered. These are: restriction of their range, competition from other species, drought, and disease.

Diminished rangeland did not deplete the herds. The grassland extended farther north in the nineteenth century than it does now, because of grass fires. In 1858, the Canadian geologist Henry Hind noted that in southwestern Manitoba "small 'hummocks' of aspen, and clumps of partially burnt willows, were the only remaining representatives of an extensive aspen forest which formerly covered the country." Fire, whether set by humans (purposefully or accidentally) or by lightning, could alter the range of the buffalo for a season, and inflict hardship on people who depended on the animals for their livelihood. "The prevailing conflagration of the plains has driven the buffaloe from this part of the country towards the River Missouri," Francis Heron, the Hudson's Bay Company trader at Brandon House, recorded in October 1828, "and all the Indians who have no other source of subsistence, are consequently of necessity bending their course thither in pursuit of them." Observations of grass fires go back to the earliest written records; trading post journals do not record an unusual number, which might have accounted for the buffalo's disappearance from Canada during the 1870s.[9]

Although fire maintained, and even increased, the area of grassland, the British explorer John Palliser, who crossed the Plains several times in the late 1850s, noticed that the buffalo nevertheless seemed to be overgrazing the range: "The grass in this arid soil, always so scanty, was now actually swept away by the buffalo, who, assisted by the locusts, had left the country as bare as if it had been overrun by fire." Overgrazing may have been caused by crowding due to pressure from human hunters, but grasshoppers were indeed competing for the grass; Hind mentioned them almost as often as he did the buffalo. Where the insect swarms had passed, he wrote, "the grass was cut uniformly to one inch from the ground." U.S. Indian agents' reports during the 1870s show that grasshoppers were active on both sides of the international line. Along Milk River in 1874, the grasshoppers "ate every green thing close to the ground." But competition from insects, like grass fires, was a recurring phenomenon in the Plains, and could hardly account for the buffalo's disappearance from the country between the Saskatchewan and the Missouri.[10]

When the Plains tribes adopted the horse as a beast of burden, they replaced an animal (the dog) that, like themselves, thrived on the flesh of the buffalo, with an animal that competed with the buffalo for grass. But because the Assiniboines, Blackfeet, and Crees owned fewer horses than did tribes farther south, competition for forage between horses and buffalo was less intense in the Northern Plains. None of the northern tribes had vast horse herds of the kind accumulated in the Southern Plains, and the Crees and Assiniboines, in the northeastern reaches of the buffalo range, had the fewest horses of all. (Whenever possible, they used the pre-horse hunting technique of driving the buffalo into "pounds," or corrals, where they killed all the animals they could.) Hudson's Bay Company horses, and those belonging to *métis*, spent most of their time in the parkland north of the Plains, and normally did not compete with the buffalo for forage.[11]

Reckoning the effect of climatic conditions on the carrying capacity of the range is difficult because of fragmentary and conflicting evidence. For instance, in 1875, the U.S. Indian agent at Cheyenne River, a few miles up the Missouri from Pierre, complained of "continuous drought," while the agent at Fort Berthold, some 200 miles to the north, called it "the wettest season on record in this country." Traders at Hudson's Bay Company posts on the margin of the Plains, who kept journals of daily occurrences, confined themselves mostly to remarks like "fine clear weather," and commented on the annual first sighting of geese, the breaking up of ice in the river, and, at the end of the year, the freeze. Generally speaking, though, the equestrian culture of the Plains Indians, and the buffalo herds upon which it depended, nourished during the second half of the Neo-Boreal climate period (the Little Ice Age), when cooler temperatures and more moisture produced abundant grass.

Dendrochronological evidence suggests that the second quarter of the nineteenth century, in particular, saw above-average precipitation, followed by a drying out in the 1850s, especially in the Northern Plains, and generally drier conditions during the rest of the century. This would have restricted forage for the buffalo at a time when the herds were under ever-growing pressure from human hunters.[12]

It is possible that brucellosis (a bacterial disease that affects the reproductive system and often causes spontaneous abortion), introduced by the herds of domestic cattle that were beginning to move into the region by the mid-1870s, may have infected the buffalo. An observer in western Saskatchewan noted in 1877 "that very few calves of this season were to be seen." This might well indicate the presence of brucellosis. In a letter, a missionary mentioned diseased cattle during the summer of 1879 (by which time the buffalo were gone), and a memoirist writing 32 years after the event recalled a report of "sickness among the buffalo" in 1878. References are sparse, though; in the absence of clear evidence that disease reduced the number of buffalo, or that grasshoppers, fire, or catastrophic drought diminished the extent of their range, human predation remains the only likely cause for their disappearance.[13]

Because Native peoples killed buffalo not only to meet their own needs for food, clothing, and shelter, but to provide a surplus of tanned robes for trade, some idea of their numbers is necessary in order to estimate the rate of attrition they imposed on the herds. The Plains tribes were the despair of White officials who had to enumerate them in written reports for higher authority, whether for the Hudson's Bay Company or the Office of Indian Affairs. "They have so strong a predilection for the plains," the HBC trader at Fort Assiniboine wrote in 1825, "that it is impossible to give even an idea of their numbers." In later decades, U.S. Indian agents on the upper Missouri would echo the plaint. Nevertheless, experienced observers made some estimates for the three tribes who pressured the buffalo from the north: the Blackfeet, the Assiniboines, and the Crees. Twentieth-century scholars can choose figures that seem consistent over time.[14]

Closest to the Rocky Mountains lived the Blackfeet, a confederation of three groups, the Piegans, the Kainah (Bloods), and the Siksika (Blackfeet). In 1853, the American James Doty, who had the job of finding their camps and notifying them of an impending treaty conference, estimated a population of 6,650 Blackfeet and 2,520 Gros Ventres. Governor Isaac Stevens, in charge of the conference, wrote that Doty "had the opportunity of making an actual count of more than half these Indians, and his estimate cannot be far from the mark." The Blackfeet population rose to more than 9,200 before a smallpox epidemic in 1870 reduced it to 7,500.[15]

The American Fur Company trader Edwin Denig, who had married an Assiniboine woman, estimated in 1854 that 2,340 Assiniboines traded regularly at Fort Union, and that between 1,125 and 1,350 more of them lived in British territory, between Red River and the Saskatchewan. The tribe numbered about 3,200 in 1858, according to Hind. By 1874, their number had reached nearly 4,700.[16]

The Crees had first encountered European traders on the shores of Hudson Bay in the late seventeenth century. One hundred years later, some Crees had moved beyond Red River, westward toward the Saskatchewan, in search of new fur-hunting territory. Those who moved on to the Plains acquired horses and began to hunt buffalo. An era of specialization had begun, in response to the penetration of the Canadian West by two rival fur-trading enterprises, the Hudson's Bay Company and the North West Company. The result was described in 1809 by the trader William McGillivray: "The principal aid given by these Indians to the Fur Trader is to kill Buffalo and Deer, and prepare the flesh and tallow for the Company's servants who without this provision, which could not be obtained in any other part of the Country, would be compelled to abandon the most lucrative part of the trade." (Pemmican, in other words, was necessary for European traders to travel to the country where they could obtain beaver pelts and "fine furs" like marten, mink, and otter, which were their real object.) In 1858, Palliser estimated a Cree population of 11,520.[17]

These figures suggest a total population of about 24,000 among the three largest Indian groups hunting the buffalo from the north in 1860. Possession of horses, and year-round residence in the Plains, gave the tribes a new degree of freedom from Europeans and their trade goods. Buffalo in herds could be hunted with bows and arrows; guns, powder, and shot, all obtainable only from traders, were more effective against solitary deer and moose in the woods. Nearly 30 years later, after the merger of the Hudson's Bay and North West Companies, the HBC's George Simpson complained that the "Plains Tribes ... continue as insolent and independent if not more so than ever; they conceive that we are dependent on them for the means of subsistence." The Plains equestrian way of life, only a few generations old, would last as long as the buffalo that sustained it; the intrusion of market forces of European provenance, as was the horse, would end it.[18]

The Plains tribes supplied meat and pemmican to the trading posts in their country. The Hudson's Bay Company was not much interested in bulky buffalo robes, which fetched an average price that varied between $3.00 and $4.50. The HBC's primary market was in Europe, where the winters were warmer than in Canada, and there was little demand for buffalo robes; yet more and more robes were traded in the years after 1821, largely to keep the provisions trade—pemmican and dried meat—of the Plains tribes, and to dissuade them from taking their furs to the American Fur Company's trading posts. The Plains Indians were "more regular in their visits to the American establishments on the Missouri than to us," the HBC's Governor George Simpson wrote in 1831; "this arises from buffalo being more numerous in that quarter than in the vicinity of our Posts, and from their finding a market for their skins and robes on the spot instead of being at the trouble of dragging them overland a distance of several hundred miles."[19]

Since the Hudson's Bay Company needed pemmican to feed its boatmen on their trips north to Lake Athabasca and the Great Slave Lake, the council at its annual meetings in 1832 and 1833 decided to accept more buffalo robes "in order to withdraw the Plains tribes from the American Establishments on the Missouri." In 1836, the council adjured its traders in the Swan River District, between Red River and the Saskatchewan, to increase the robe trade, "which for the want of such encouragement falls into the hands of the American Traders on the Missouri." As late as 1870, traders in all three districts were "directed to use their best endeavours to increase the Returns of Buffalo Robes." Toward the end of the 1870s, though, dealers in Montreal were complaining of the inferior quality of the robes they were receiving. Finally, robes became "a drug on the market." The collapse of the Montreal robe market coincided with the buffalo's disappearance from Canada.[20]

The economics of tanning and using buffalo skins can be worked out with figures from a population estimate in the Blackfeet agent's annual report for 1858. Alfred Vaughn, the agent, estimated the entire population, including the allied Gros Ventres, as 9,400, including 3,100 women. It was unusual for a population estimate to include the number of women—most gave the number of "lodges," of "warriors," and an approximate total—but the number of women is important because women skinned and butchered the buffalo, dried the meat, and tanned the skins. If each woman tanned an average of 20 skins, the total would have been 62,000. Allowing one skin per person per year for tipi covers, two robes per person, and two skins for garments, domestic use would amount to 47,000, leaving 15,000 robes and skins for trade. In the real world, of course, the estimated 4,240 children would have required fewer skins for garments, but the vicissitudes of outdoor skinning and tanning would certainly have claimed some skins, as well. At any rate, these figures indicate that about 25 percent of the robes and skins may have been available for trade. In 1858, an estimated 20,000 robes were shipped from Fort Benton, the site of the Blackfeet agency; returns that year from Edmonton House, the centre of the Hudson's Bay Company's Blackfeet trade, included 1,473 robes and tanned buffalo skins.[21]

The Blackfeet agent reckoned that adult women made up nearly 32 percent of the population in 1858. An 1861 estimate of the Southern Cheyennes, who traded buffalo robes at Bent's Fort on the Arkansas River,

numbered 480 women in a population of 1,380, or nearly 35 percent. If adult women accounted for one-third of the population of Plains tribes, there would have been about 8,000 of them among the 24,000 buffalo-hunting Natives north of the Missouri River. If each woman was able to tan 20 hides a year—not a remarkably high figure—it would have represented an annual kill of 160,000 animals, a little more than Edwin Denig's estimate of 150,000. An average daily consumption of 3 pounds of meat per person would have required 87,600 cows weighing 300 pounds each to feed the tribes for a year, which would have been easily provided, even given the conditions of outdoor butchering. Immediate use, for lodgeskins, clothing, and sleeping robes, would have taken 125,000 skins, leaving 35,000 to trade. The robe trade, therefore, may have taken the skins of nearly 30 percent more buffalo than the tribes' minimum requirements.[22]

Besides the three tribes, there was another population group whose activities must be taken into account. The Métis (francophones of mixed descent) and Native English (anglophones of mixed descent) of the Red River Settlement were the most often-counted people hunting the Canadian buffalo. Their numbers increased rapidly in southern Manitoba after the Hudson's Bay Company absorbed its competitor, the North West Company, in 1821. In that year, the population of the Red River Settlement stood at 419, including 154 women. By 1823, the reorganization of the fur trade had added a number of retired or discharged traders and their families, increasing the population to "about 600." It quadrupled in the next eight years, to 2,427 in 1831, and reached 4,073 in 1840. The settlement had 6,523 residents in 1856, and the Canadian census of 1871 listed 5,757 Métis and 4,083 Native

English, besides 1,565 Whites and 558 Indians, an 1871 total of 11,963 residents; about equal in number to the Crees, or as many as the Assiniboines and Blackfeet combined.

There were also communities of *métis* along the North Saskatchewan River. In 1862, Lac Ste. Annes, a day's ride from Edmonton House, was "a considerable Settlement" of more than 1,000 inhabitants.[23] A census 10 years later, listing residents in the Hudson's Bay Company's Saskatchewan District by ethnic group, gave a total of 2,154 Métis (793 adults, 1,361 children) and 256 Native English (101 adults, 155 children). This thriving population of new arrivals exerted additional pressure from the north on the buffalo herds.[24]

Alexander Ross, a retired fur trader, is the only published authority for the early years of the Red River hunts, which began in 1820. The organization of the summer hunt of 1840, which Ross described in detail, resembled that of the Plains Indians' summer and fall hunts, and included sanctions against any hunter who attempted "to run buffalo before the general order," prescribing destruction of property for the first two infractions and flogging for the third. "At the present day," Hind wrote in 1858, "these punishments are changed to a fine of 20 shillings for the first offense." (Twenty shillings—£1 sterling—was the equivalent of two-thirds of a month's pay for an unskilled labourer and a whole month's pay for a tradesman's apprentice in the fifth year of his apprenticeship.) According to Ross, the first hunt, that of 1820, included 540 Red River carts, two-wheeled vehicles of wood and rawhide construction that could haul a 900-pound load when drawn by an ox; 20 years later, the 1840 hunt included 1,210 carts.[25]

Ross wrote that the first day's run of the 1840 hunt yielded 1,375 tongues, and "that not less than 2,500 animals had been killed," an average of slightly fewer than four for each of the 650 hunters Ross claimed took part in the hunt that year. These figures are astonishingly meagre. A kill of 2,500 cows, carefully butchered, might have produced 375 tons of meat, allowing 300 pounds per cow. Yet the hunters preserved only 375 90-pound bags of pemmican and 240 60-pound bales of dried meat, amounting to 24 tons. (Even if the bales of dried meat weighed 100 pounds, as another source suggests, the result would have been only 29 tons.) That would have been enough to load 53 ox-carts (64, if the dried meat were in 100-pound bales). Even allowing for weight loss in drying, and for what the hunters needed to eat between kills, the amount of meat saved represents only a small fraction of what was killed. It may be that the hunters were profligate in the early stages of the hunt; it is likely that rain ruined meat that was being dried. Ross wrote that "a thunder-storm, in one hour, will render the meat useless," and travellers' journals mention afternoon thunderstorms frequently. A pack of "no fewer than 542" dogs accompanied the hunt, and wolves scavenged whatever carcasses had not been butchered by nightfall. In any event, if 2,500 buffalo yielded 24 or 29 tons of pemmican and dried meat, a total of 545 tons would have required the slaughter of anywhere from 46,000 to 56,000 head.[26]

However many animals the Red River hunters killed, buffalo had become so scarce by 1840 that the hunt of that year had to travel west of Pembina for nearly three weeks before it found any. Overhunting was diminishing the eastern extent of the buffalo range. When he ascended the Missouri River in 1832, the

German traveller Prince Maximilian saw his first buffalo between the mouths of the Niobrara and White rivers, but was told "that the herds of buffaloes had left the Missouri, and had been followed by the Sioux Indians, so that we must expect to see only a few of them on the river." Thirty years later, the logs of steamers bound up the Missouri recorded sighting the first buffalo near the Moreau River, above Fort Pierre. By 1868, travellers upriver saw no buffalo until they were above Fort Union, past the mouth of Poplar River.[27]

How many buffalo were killed, over the years, to be traded as robes and provisions? The Hudson's Bay Company required an average of 40 tons of pemmican a year to feed its boatmen during the 1840s; 65 tons a year in the 1850s; and 106.5 tons a year in the 1860s. To try to figure the number of buffalo that were turned into pemmican would involve determining an average weight of fresh meat per animal, allowing for weight loss while the meat was drying, and factoring in the half-dried meat, half-fat composition of pemmican. The number of robes, dressed skins, and tongues delivered to trading posts presents a clearer, though still rough, picture of the number of buffalo killed beyond the subsistence needs of the people of the Northern Plains.[28]

Table 12.1 shows that during its most active participation in the buffalo robe trade, from 1841 to 1870, the Hudson's Bay Company handled the product of nearly 17,000 buffalo in an average year. In boats that moved upstream by sail and oar, and downstream with the current, robes, skins, and tongues were carried down to Hudson Bay, along with furs, which were the company's real object.

The American Fur Company, on the other hand, supplied its Missouri River posts by steamboat, which carried robes on the voyage downstream. The steamer *Yellow Stone* reached Fort Union in 1832, and took on board a cargo of 7,200 robes. That same year, the AFC built Fort McKenzie at the mouth of the Marias River, in the Blackfeet country. Its trade increased from 9,000 robes in 1834 to more than 13,000 in 1844, its last year in operation.[30]

In the absence of a record of American trade on the Missouri River that is anywhere near as complete as the Hudson's Bay Company's records, we can still hazard an educated guess at the total volume of the buffalo trade. In 1855, Isaac Stevens estimated that trade with

Table 12.1: Hudson's Bay Company Buffalo Robes, Dressed Skins, and Tongues, 1821–1879[29]

	Robes and Dressed Skins	Tongues		Robes and Dressed Skins	Tongues
1821–1825	7,113		1856–1860	67,126	16,827
1826–1830	15,676	1,923	1861–1865	64,427	16,441
1831–1835	20,519	3,385	1866–1870	59,984	2,566
1836–1840	41,416	13,620	1871–1875	56,896	2,944
1841–1845	73,278	25,657	1876–1879	46,778	1,576
1846–1850	69,603	18,581	1821–1879	595,746	126,108
1851–1855	72,930	22,588			

Table 12.2: Returns of Robes and Skins from HBC's Saskatchewan District and AFC's Upper Missouri Outfit[32]

Year	U.M.O.	H.B.C.	%	Year	U.M.O.	H.B.C.	%
1830	34,158	4,229	12.4	1853	29,595	8,007	27.0
1831	32,319	3,951	12.2	1859	41,850	3,273	7.8
1835	32,900	4,345	13.2	1860	41,975	2,041	4.9
1837	24,700	3,408	13.8	1862	71,900	6,336	8.8
1846–1847*	39,750	9,265	23.3	1864	33,600	4,377	13.0

* average

the HBC amounted to "about one-sixth" of the trade "at the American posts along the Missouri." A comparison of the returns of robes and skins for the few years when statistics are available, though, shows that the HBC's trade averaged about one-eighth of the American trade.[31]

The number of robes declined in 1837 due to the smallpox epidemic of that year. The Upper Missouri returns of 1859–1864 are based on sales in New York City credited to the Upper Missouri Outfit, plus an additional 30 percent reflecting robes held in St. Louis to supply dealers west of the Appalachians. The disparity in figures is only partly due to the HBC's reluctance to trade high-bulk, low-value robes. All the posts in the Saskatchewan District stood at the northern margin of the buffalo range, while the Upper Missouri Outfit collected robes from the Crows along the Yellowstone and from some bands of Sioux, as well as from the Assiniboine, Blackfeet, and Crees.

*****

The average number of robes and skins traded each year in the districts reporting to York Factory—the Saskatchewan, Swan River, and Red River districts—was

11,730 for the period 1831–1870. (The Saskatchewan District's average was 6,119.) The American trade, then, would have averaged 93,840, or a total of more than 115,000 buffalo a year for the entire region, from the North Saskatchewan south to the Platte River, beyond the subsistence needs of the Native peoples.

This predation, for robes and pemmican, was all the more damaging because of the hunters' preference for the skins and meat of cows. Cows' skins were lighter and easier to tan than were bulls', and their flesh more palatable. The best robes were taken during the winter, between the rutting season of the previous summer and the following spring, when births occurred, so many cows must have died while carrying calves. Since buffalo cows rarely bear more than one calf a year, the hunters' purposeful selection steadily reduced the herds' ability to maintain their numbers.[33]

The effect of the slaughter on the size of the herds was not readily apparent to persons who lived in the heart of the buffalo range. "Buffalo are very numerous," Denig wrote in 1854, "and we do not, after 20 years' experience, find that they decrease .... It may be," he added, "that the range of these animals is becoming more limited."

Only on the edge of the buffalo country was the change noticeable. The Crees and Assiniboines, on the northeastern rim of the region, complained of *métis* incursions. To a well-travelled observer such as the missionary Pierre-Jean DeSmet, the future seemed clear. In 1846, he predicted that "the plains of the Yellowstone and Missouri and as far as the forks of the Saskatchewan, occupied today by the Blackfeet, will be within the next dozen years the last retreat of the buffalo."[34]

As the buffalo range contracted, inter-tribal competition stiffened. "The plains where the buffalo graze are becoming more and more of a desert, and at every season's hunt the different Indian tribes find themselves closer together," DeSmet observed in 1846; "whenever they meet, it is war to the death." Buffer zones were narrowing as the tribes converged, and the area in which the buffalo might exist undisturbed shrank accordingly.[35]

During the 1860s and 1870s, political and economic pressures originating far outside the buffalo range pushed and pulled new groups of people into the region, and exacerbated pressure on the herds. As many as 3,000 Eastern Sioux moved north of the forty-ninth parallel after the Minnesota uprising of 1862. Some eventually returned to the United States, but 10 years later, those who had remained numbered about 1,780. At the same time, other Sioux were moving west on the American side of the line. Just as the Teton Sioux had moved toward the valleys of the Platte and Republican rivers a generation earlier, the Yanktonai and other Sioux groups were moving into the Upper Missouri and Milk River country. A special inspector from the Office of Indian Affairs reported in 1871 that the newly arrived Sioux were forcing the Blackfeet to move north. The Sioux, the inspector

wrote, "were only the forerunners of larger numbers, who were on their way, attracted by the large herds of buffalo in that region." By 1875, the U.S. Indian agent at Fort Peck, Montana, reported more than 4,100 Sioux living on that reservation.[36]

While this was going on, *métis* from Red River were taking up year-round residence in the Plains. The Hudson's Bay Company was reluctant to buy buffalo robes, but in 1844, the American Fur Company opened a post at Pembina, a site the HBC had abandoned 20 years earlier after a boundary survey showed that it stood just south of the line. The market boomed. Trains of Red River carts hauled as many as 25,000 robes a year overland to St. Paul, on the Mississippi River. Robe prices continued high through the 1860s, and about 2,500 *métis* left Red River and set up *hivernement* (wintering) settlements in the heart of the buffalo range.[37]

After 1870, the transfer of jurisdiction from the Hudson's Bay Company to Canada, and subsequent loss of their land holdings in the Red River Settlement, sent still more *métis* to the new settlements in the West. Natural disasters, too, could bring increased pressure on the buffalo herds. In both 1875 and 1876, grasshoppers destroyed the crops of 800 or so *métis* settlers who were attempting to farm in the Upper Saskatchewan District, and forced them to return to buffalo hunting or starve. Finally, in 1877, perhaps as many as 5,000 Western Sioux fled north, harried by the United States Army. Taken together, all these groups represented a gradual addition, during a 15-year period, of more than 60 percent to the roughly 24,000 Assiniboines, Blackfeet, and Crees who were already hunting the Canadian buffalo.[38]

The Hudson's Bay Company's chief factor at Edmonton House was glum. "Buffalo are fast decreasing and the population fast increasing," he had written in 1867, "the supply of food to be got in the District is inadequate to the demand, hence the Saskatchewan can never be depended upon for large supplies of provisions, as in former years. At present there is little good going on in the Plains." While the herds thinned out north of the Missouri River, there were still buffalo farther south, and U.S. Indian agents in Montana continued to report "sufficiently abundant" herds, with only one hint of future scarcity. In 1875, the agent for the Blackfeet wrote: "A large number of buffalo ranged in the hunting-grounds of these people during the past winter, but the amount was insignificant compared with that of former years. A few years and the hunt will afford but a scant, precarious means of living."[39]

The continuous high level of robe production during the 1870s was largely the work of *métis* hunters who lived along both branches of the Saskatchewan and their tributaries. Recent historians have taken issue with the traditional interpretation of buffalo hunting as an instance of *métis* "primitivism." Gerhard Ens, in particular, has called the *métis*'s part in the robe trade "an adaptive, innovative response to new economic opportunities." When robe prices fell during the mid-1870s, though, the reaction was merely to produce more robes, which drove prices down even further. Increased production was as short-sighted a policy as was the Cree and Assiniboine proposal to exclude outsiders from their part of the buffalo range a generation earlier.[40]

The buffalo hunts of the late 1870s were disasters. "Buffalo are nowhere near us," the Hudson's Bay Company trader at Fort Pitt complained in February 1877. A month later he commented on the starving people: "I do not know what the poor

Beggars will live on when there are ... no Buffalo to run out to. We are in a manner bound to maintain them, so far." Although the Canadian government had assumed responsibility for Native peoples, it was incapable of relieving their hunger. "No Buffalo," the trader at Edmonton House recorded in the post journal during the summer of 1879, "the Half Breeds living on dogs," and added a few weeks afterward, "all on the plains [are] starving." That year, for the first time, the HBC began to feed its western boatmen pork from pigs raised in Manitoba, besides pemmican from buffalo killed on the Plains. By 1881, with the disappearance of the buffalo, pork had replaced pemmican altogether.[41]

How were Native peoples able to drive the buffalo from the Canadian Plains? The most obvious answer is sheer population pressure, augmented by commercial demand, which claimed the skins of 30 percent more buffalo than Natives required for their minimum needs. (Immigration of Sioux and *métis* into the Canadian Plains during the years between 1862 and 1876 more than offset losses from the smallpox epidemic of 1870.) Cows were especially subject to attrition, for their skins were lighter and more easily worked than those of bulls, and their meat was more tender. The ability of equestrian hunters to select cows would have restricted reproduction severely.[42]

Another part of the answer, though, may lie in the spiritual world view of Native peoples. The nineteenth-century American author Richard Dodge generalized that Plains Indians "firmly believed that the buffalo were produced in a country under the ground." The British traveller and army officer William Butler offered a Canadian version in 1873: "Southwest from the Eagle Hills ... lies a lake whose waters never rest; day and night a ceaseless murmur breaks the silence of the spot. 'See,' says the Red man, 'it is from under that lake that our buffalo comes. You say they are all gone; but look, they come again and again to us. We cannot kill them all—they are there under that lake.'"[43]

Ascribing an underground origin to the buffalo does not seem to have been common in the Northern Plains, though. John Maclean, a Hudson's Bay Company trader who contributed to the *Journal of American Folklore*, reported that the Blackfeet Culture Hero created buffalo, gave the Blackfeet bows and arrows, and sanctioned the killing of buffalo. George Grinnell learned that the Culture Hero created the first buffalo from lumps of clay and taught the Blackfeet how to build pounds. The buffalo was not the only animal with religious significance. Although fur traders' accounts indicate that the Blackfeet acquired horses about the middle of the eighteenth century, the animals became so important to the tribe's way of life that at least three supernatural explanations of their origin evolved.[44]

The anthropologist Robert Lowie, interviewing Assiniboines in Alberta and Montana, found that their sanction for killing buffalo came when their Culture Hero "went to a hilltop, and cried out, 'Let everyone get ready for the buffalo hunt!' All got their horses ready. Some buffalo were coming across the hills." (In the story, buffalo already existed, and so familiar were Lowie's informants with horses that they included them, as well.) The Culture Hero taught the Assiniboines how to kill and use the buffalo, and told them afterward: "From now on your people will subsist on such food. The buffalo will live as long as your people. There will be no end of them until the end of time."[45]

Religious beliefs aside, another explanation for the tribes' failure to curtail the slaughter may lie in the gregarious nature of the buffalo and the limited outlook of people whose entire lives were spent in the buffalo range. Surely one reason why the buffalo seemed inexhaustible was the size of the herds when found. The extent of the range may have shrunk, but when met with, the herds must have looked as large as ever. In July 1859, along the South Saskatchewan, the Scottish sportsman James Carnegie, Earl of Southesk, encountered "[i]mmense herds ... stringing all across the country ... [T]he plains were alive as far as the eye could reach." Ten weeks later, near Edmonton House, he reflected: "The plains are all strewn with skulls and other vestiges of the buffalo, which came up this river last year in great numbers.... Large as were the herds I saw in July, they were nothing to what I have heard and read of, and there is reason to believe that I then beheld all the buffaloes belonging to the two Saskatchewan valleys and the intervening country, pressed from various quarters into one great host."[46]

DeSmet had made a similar observation 10 years earlier: "The buffalo is disappearing and diminishing each successive year on the prairies of the upper Missouri. This does not, however, hinder them from being seen grazing in very numerous herds in particular localities; but the area of land that these animals frequent is becoming more and more circumscribed. Besides, they do not remain in the same place, but change pasturage acording to seasons." Denig, as quoted earlier, wrote, "It may be that the range of these animals is becoming more limited," but it was an afterthought to the statement, "Buffalo are very numerous, and we do not, after 20 years' experience, find that they decrease." For Denig, "20 years' experience" meant continuous residence at Fort Union, in the heart of the buffalo range; DeSmet travelled widely, and had much more recent experience of the outside world. The Indians, for the most part, had none. Those leaders who made government-sponsored visits to the eastern United States and returned to tell of tall buildings and innumerable people were often mocked as fools and liars, and lost their status and influence.[47]

"It is most difficult to make the Indian comprehend the true nature of the foreigner with whom he is brought in contact," wrote William Butler [in] 1871:

> English, French, Canadians, and Americans are so many tribes inhabiting various parts of the world, whose land is bad, and who are not possessed of buffalo—for this last *desideratum* they (the strangers) send goods, missions, &c, to the Indians of the Plains. "Ah," they say, "if it was not for our buffalo where would you be? You would starve, your bones would whiten the prairies." It is useless to tell them that such is not the case, they answer, "Where then does all the pemmican go that you take away in your boats and in your carts?"

Other European and Canadian observers reported similar instances of skepticism regarding White men's tales of cities, telegraphs, and railroads.[48]

If the judgments of these outside observers are accurate, then the insular outlook of Native peoples should surely be taken into account in trying to explain why they continued to hunt buffalo as they did. Hunters, both Native and European, continued to see large herds because the density looked the same, even though fewer buffalo occupied a smaller range. The reaction of the Crees and Assiniboines to scarcity of buffalo was to try to exclude competition; if such a response seems hardly adequate, it must be remembered that they saw the buffalo as an infinitely renewable resource of supernatural origin.

Evidence indicates that the Native peoples of Canada, living on the northern edge of the grassland, killed buffalo at an unsustainable rate. An increase of nearly two-thirds in the region's human population during a 60-year period was a contributing factor to the overkill, as was (eventually, after a period of Native "independence") the ascendance of market forces originating outside the buffalo range. While commerce demanded more buffalo products, horses enabled Native hunters to select cows, impairing the herds' ability to maintain their numbers. In order to supply the provision and robe trade, Native peoples each year killed perhaps one-third more buffalo than would have sufficed for their own needs. In any case, the techniques of outdoor butchering and tanning, and fickle weather, required the slaughter of many more animals than can be calculated by a mathematical formula.

To ask if the Plains Indians in general, and the Native Americans of Canada in particular, could have "established a society in ecological equilibrium" is to ask the unanswerable, for these peoples never existed in isolation. The Hudson's Bay Company's Henry Kelsey reached the eastern margin of the Plains as early as 1690, one-half century before horses arrived there, and he was followed in each succeeding decade by other travellers from the bay. The Verendryes approached the Northern Plains by way of the Great Lakes toward the end of the 1730s, the same decade in which horses entered the region from the south and west. The first devastating smallpox epidemic came less than 50 years later, and was followed by others in 1837 and 1870. The peoples of the Plains were usually off-balance throughout the 150-year span of their equestrian

culture. If not approached directly by
Europeans, they were continually in
contact with new tribal groups moving into
the Plains, intent on trade or on domination
of the buffalo range. There was hardly a
generation between the 1730s, when horses
arrived in the Northern Plains, and the

1870s and 1880s, when the buffalo finally
died, during which the Native peoples of
the Northern Plains could have attained
anything like equilibrium. The forces with
which they had to contend made "a healthy,
functioning ecology," sustainable "over the
long term," impossible to achieve.[49]

NOTES

1. Lawrence J. Burpee, ed., "York Factory to the Blackfeet Country: The Journal of Anthony Hendry [sic], 1754–55," *Transactions of the Royal Society of Canada* 1 (1907): 336; "An Adventurer from Hudson Bay: Journal of Matthew Cocking, from York Factory to the Blackfeet Country, 1772–73," in *Transactions*, edited by Lawrence J. Burpee, 3rd ser., vol. 2, sec. 2 (1908), 109; John S. Milloy, *The Plains Cree: Trade, Diplomacy and War, 1790 to 1870* (Winnipeg, 1988), 29. The standard historical work, *The North American Buffalo: A Critical Study of the Species in Its Wild State* (Toronto, 1951 and 1970), was written by Canadian Frank G. Roe before the Hudson's Bay Company opened its archives to scholars; the pages cited in this article are the same in both editions, and the author used both editions. A good brief overview of the last days of the buffalo in Canada is John E. Foster, "The Métis and the End of the Plains Buffalo in Alberta," in *Buffalo*, edited by John E. Foster, Dick Harrison, and I.S. MacLaren (Edmonton, 1992), 61–77. Robert F. Beal, "The Buffalo Robe Trade," in "The Métis Hivernement Settlement at Buffalo Lake, 1872–1877," (unpublished report in author's possession) Robert F. Beal, John E. Foster, and Louise Zuk, 82. (I use "*métis*" to refer to persons of mixed blood, including members of both linguistic groups, and Métis to refer to francophone mixed bloods of Canada.) Beal also comments wryly on the way Canadian historians moralize about the slaughter of buffalo in the United States and about American "whiskey traders" in the Canadian Plains.
2. William T. Hornaday, "The Extermination of the American Bison," *Annual Report of the Smithsonian Institution for 1887* (Washington, 1889), Pt. 2, 511, 513.
3. Andrew C. Isenberg, "Toward a Policy of Destruction: Buffaloes, Law, and the Market, 1803–83," *Great Plains Quarterly* 12 (Fall 1992): 227–241. Richard White states the case for market forces in "Animals and Enterprise," in *The Oxford History of the American West*, edited by Clyde A. Milner II, Carol O'Connor, and Martha A. Sandweiss (New York, 1994), 247–248; Dan Flores, "Bison Ecology and Bison Diplomacy: The Southern Plains from 1800 to 1850," *Journal of American History* 78 (September 1991): 465–485, quotation on 476.
4. Journals for the following Hudson's Bay Company posts (listed from east to west) indicate the regular occurrence of grass fires, and contain no rumours of epidemic disease among the buffalo: Brandon House, 1828–1830; Fort Ellice, 1871–1872; Fort Pelly, 1832–1834, 1837–1838, 1853–1854, 1874, and 1876–1878; Carlton House, 1821–1826 and 1834–1839; Fort Pitt, 1830–1832; and Edmonton House, 1854–1873 and 1877–1879.
5. *National Atlas of the United States of America* (Washington, 1970), 90–91; *National Atlas of Canada* (Ottawa, 1973), 45–46.
6. Flores, in "Bison Ecology and Bison Diplomacy," chose the 1910 census because overgrazing was no longer as intense as it had been during the 1880s, and farmers' small holdings had not yet occupied large tracts of the range. On 470–471 he says that rainfall "was at median, between-droughts levels." *Thirteenth Census of the United States: 1910* (Washington, 1913), 6: 958–960 (hereafter, *U.S. Census* [1910]). According to the *Annual Report of the Commissioner of Indian Affairs* (hereafter cited as AR-CIA) for 1871, House Ex. Doc. no. 1, Pt. 5, 42nd Cong., 2nd Sess., 833, there was still "abundance of buffalo" west of the 105th meridian.

7. *Vegetative Rangeland Types in Montana,* Bulletin 671 (Bozeman, 1973). Waldo R. Wedel mentions the drought of 1906–1912 in his *Central Plains Prehistory: Holocene Environments and Culture Change in the Republican River Basin* (Lincoln, 1986), 45. David M. Meko presents evidence about rainfall in Montana in his article, "Drought History in the Western Great Plains from Tree Rings" which appeared in the 1982 *Proceedings of the American Water Resources Association, International Symposium on Hydrometeorology* (xerox copy of this piece in author's possession), and in "Dendroclimatic Evidence from the Great Plains of the United States," in *Climate Since AD 1500,* edited by R.S. Bradley and P.D. Jones (London, 1992), 312–330.

8. Brandon House Journal, October 25 and 16, 1829, B.22/a/23, Hudson's Bay Company Archives (hereafter HBCA). *U.S. Census* (1910), 7: 284–288, 538–544, 952–953. The totals represent five Wyoming counties north of the North Platte and east of the Bighorn Mountains, and all counties in North and South Dakota except those in the eastern tier.

9. Henry Y. Hind, *Narrative of the Canadian Red River Exploring Expedition of 1857 and of the Assiniboine and Saskatchewan Exploring Expedition of 1858,* vol. 1 (London, 1860), 308; Brandon House Journal, October 26, 1828, B.22/a/22, HBCA.

10. Irene M. Spry, ed., *The Papers of the Palliser Expedition* (Toronto, 1968), 146; Hind, *Narrative,* 1: 298; AR-CIA 1874, House Ex. Doc. no. 1, Pt. 5, 43rd Cong., 2nd Sess., 575.

11. Data on Assiniboine, Blackfeet, and Cree horse ownership are in John C. Ewers, *The Horse in Blackfoot Indian Culture,* Bureau of American Ethnology Bulletin 159 (Washington, 1955), 20–28; see also Beal, "Buffalo Robe Trade," 96. "Pounding," or corralling, was possible only when the buffalo approached the edge of the woodland, where trees and brush furnished the necessary construction material.

12. AR-CIA 1875, House Ex. Doc. no. 1, Pt. 5, 44th Cong., 1st Sess., 738, 744. On p. 43 of his Central Plains Prehistory, Wedel states that "[a]bout 1883, strengthening westerlies resulted in drier and warmer conditions on the plains, ending the Little Ice Age and producing today's climate." Merlin P. Lawson, *The Climate of the Great American Desert: Reconstruction of the Climate of the Western Interior United States, 1800–1850* (Lincoln, 1974), 28–31.

13. Roe, *North American Buffalo,* 477; Stacy V. Tessano, "Bovine Tuberculosis and Brucellosis in Animals, Including Man," in Foster, Harrison, and MacLaren, eds., *Buffalo,* 217; John E. Foster, "End of the Plains Buffalo," in Foster, Harrison, and MacLaren, eds., *Buffalo,* 72; Maurice F.V. Doll, Robert S. Kidd, and John P. Day, *The Buffalo Lake Métis Site: A Late Nineteenth-Century Settlement in the Parkland of Central Alberta* (Edmonton, 1988), 76–77. Beal, in "The Buffalo Robe Trade," 102, says that hunters' preference for cows "would very quickly destroy almost the entire calf drop."

14. Fort Assiniboine Report, 1824–1825, B.8/e/l, HBCA. See also the Blackfeet agent's complaint in AR-C1A 1866, House Ex. Doc. no. 1, 39th Cong., 2nd Sess., vol. 2 (serial 1284), 203: "They live for the most part in the British possessions, and only come here to receive their annuity goods or to commit some depredation. Many have never been here at all."

15. Isaac I. Stevens, "Report of Explorations," Senate Ex. Doc. no. 78, 33nd Cong., 2nd Sess., vol. 1, 151–152; AR-CIA 1870, House Ex. Doc. no. 1, Pt. 1, 41st Cong., 3rd Sess., 190; AR-CIA 1871, House Ex. Doc. no. 1, Pt. 5, 42nd Cong., 2nd Sess., 843. In Doll, Kidd, and Day, *Buffalo Lake,* 76, John P. Day suggests that the post-smallpox resurgence of population in the Canadian Plains during the 1870s put additional pressure on the buffalo.

16. Edwin T. Denig, "Indian Tribes of the Upper Missouri," *46th Annual Report of the Bureau of American Ethnology* (1928–1929), 397, 431; Hind, *Narrative,* 2: 152; AR-CIA 1874, 572.

17. Milloy, *Plains Cree,* 26. "Servants" means HBC employees, both clerks and artisans. For population estimate, see Arthur J. Ray, *Indians in the Fur Trade* (Toronto, 1974), 185.

18. Milloy, *Plains Cree,* 29; Ray, *Indians in the Fur Trade,* 207.

19. In Harold A. Innis, *The Fur Trade in Canada* (Toronto, 1956), 288.

20. Minutes of Council, York Factory, 1832, 1833, and 1836, B.239/k/2, HBCA; Minutes of Council, Norway House, 1870, B.154/k/l, HBCA; J.A. Graham to L. Clark, September 11, 1877, and November 20, 1879, Carlton House, Correspondence Inward, B.27/c/2, HBCA.

21. AR-CIA 1858, Sen. Ex. Doc. no. 1, 35th Cong., 2nd Sess., 432; Vaughn's estimate seems to be the first one to categorize the population by age and sex. Ewers, *Horse in Blackfoot Indian Culture*, 131–132, gives figures on buffalo-skin use. Denig, "Indian Tribes," 541, gives 18–20 robes as a woman's average annual production. Shipments of buffalo products for Fort Benton are listed in "The [James H.] Bradley Manuscript," *Contributions to the Historical Society of Montana* 8 (1917): 156, and York Factory District Fur Returns, B.239/h/2, HBCA, for the Saskatchewan District. Edmonton House hide returns are taken from Edmonton House Journal, December 22, 1858, B.60/a/30, HBCA. Vaughn's estimate is only slightly greater than James Doty's in 1853. During the six years 1853–1858, robe shipments from Fort Benton averaged about 22,000, while returns from the Saskatchewan District averaged 8,460 robes and dressed skins and 3,419 tongues. Given the consistency of production, these population estimates are probably accurate.

22. For Cheyenne census figures, see Donald J. Berthrong, *The Southern Cheyennes* (Norman, 1963), 155; Denig, "Indian Tribes," 410, 578; Ewers, *Horse in Blackfoot Indian Culture*, 168.

23. William L. Morton, *Manitoba* (Toronto, 1967), 61, 145; Hind, *Narrative*, 1: 176; Alexander Ross, *The Red River Settlement: Its Rise, Progress, and Present State* (Minneapolis, 1957), 409; Roe, *North American Buffalo*, 414; John E. Foster discusses the various strains of *métis*— francophone and anglophone, Ojibwa and Assiniboine-Cree, and whether oriented toward the Great Lakes or Hudson Bay—in "Some Questions and Perspectives on the Problem of Métis Roots," in *The New Peoples: Being and Becoming Métis in North America*, edited by Jacqueline Peterson and Jennifer S.H. Brown (Winnipeg, 1985), 73–91.

24. Edmonton House Report, 1862, B.60/e/9&10, HBCA; "Census of Whites, Halfbreeds and Indians at the Several Posts in the Saskatchewan District, 1st January 1872," Item 155, Richard Hardisty Papers, Glenbow-Alberta Institute.

25. Ross, *Red River Settlement*, 246, 249; Hind, *Narrative*, 2: 111; John Palliser, *Further Papers Relative to the Exploration of British North America* (New York, 1969), 56–57; Innis, *Fur Trade in Canada*, 311–312. David G. Mandelbaum, *The Plains Cree* (Regina, 1979), 56, and Regina Flannery, *The Gros Ventres of Montana* (Washington, 1953), 44–45, 56, 69, show elements of communal hunting among Plains tribes similar to the rules of the Red River hunt.

26. Ross, *Red River Settlement*, 246, 258, 264; Roe, *North American Buffalo*, 58; Hind, *Narrative*, 1: 312; Ray, *Indians in the Fur Trade*, 264. Smaller groups may have been more efficient; see the account of George Belcourt, a Roman Catholic priest who accompanied a party of hunters that filled 213 carts with the product of only 1,776 buffalo, in *Historical and Statistical Information Respecting the History, Condition and Prospects of the Indian Tribes of the United States*, edited by Henry R. Schoolcraft (Philadelphia, 1851–1857), 4: 101–110. Average weight of meat from a cow derived from Edmonton House Post Journals, January 31, 1855, March 17, 1860, February 6 and December 29, 1869, and January 27 and February 16, 1870, B.60/a/29a, 31 and 37, HBCA.

27. Reuben G. Thwaites, ed., *Early Western Travels* (Cleveland, 1904–1907), 22: 298, quotation on 300; "C.J. Atkins' Logs of the Missouri River Steamboat Trips, 1863–1868," *Collections of the State Historical Society of North Dakota* 2 (1908): 261–417, buffalo sightings on 276, 295, and 353.

28. Pemmican statistics are in the annual Minutes of Council: York Factory, 1832–1850, B.239/k/2; York Factory, 1851–1870, B.239/k/3, HBCA.

29. York Factory District Fur Returns, 1821–1845, B.239/h/4; 1846-75, B.239/h/5; and 1875–1892, B.239/h/6, HBCA.

30. Fort Tecumseh Journal, June 24, 1832, in *Papers of the St. Louis Fur Trade, Part 1: The Chouteau Collection*, edited by William R. Swagerty (Bethesda, 1991, microfilm), roll 19; John E. Sunder, *The Fur Trade on the Upper Missouri, 1840–1865* (Norman, 1965), 20, 62; John C. Ewers, *The Blackfeet: Raiders on the Northwestern Plains* (Norman, 1958), 64, 66.

31. Isaac Stevens to James Doty, September 28, 1855, Letters Received from Blackfeet Agency, Office of Indian Affairs, RG75, National Archives, Washington, D.C.

32. The data for the Upper Missouri Outfit comes from several sources: Figures for 1830 and 1831 can be found in Swagerty, ed., Book S, *Papers of the St. Louis Fur Trade, Part 2: Fur Company Ledgers and Account Books*, roll 4; for figures for 1835 (which represent forts Assiniboine, McKenzie, and Union), and 1837 (forts Clark, McKenzie, and Union), see David J. Wishart, *The Fur Trade of the American West, 1807–1840: A Geographical Synthesis* (Lincoln, 1979), 58–59; figures for 1846–1847 are taken from an estimate based on the Fort Pierre Letterbook in Swagerty, ed., *Chouteau Collection*, roll 29; figures for 1853 are from Swagerty, ed., *Chouteau Collection*, roll 33; estimates for 1859–1864 are based on figures in the Pierre Chouteau & Company Sales Book in Swagerty, ed., *Fur Company Ledgers and Account Books*, roll 21. The data for the HBC's Saskatchewan district can be found in York Factory District Fur Returns, 1821–1842, B.239/h/l and 1842–1868, B.2-39/h/2, HBCA. The skins referred to are tanned buffalo skins with the hair removed, like those used for tipi covers.

33. Roe, *North American Buffalo*, 860–861; Joel A. Allen, *The American Bisons, Living and Extinct* (Cambridge, 1876), 184, 187; Beal, "Buffalo Robe Trade," 84, 103.

34. Denig, "Indian Tribes," 410, 460; Hind, *Narrative*, 1: 360; Hiram M. Chittenden and Alfred T. Richardson, eds., *Life, Letters and Travels of Father Pierre-Jean DeSmet, S.J.* (New York, 1905), 3: 948.

35. Chittenden and Richardson, eds., *DeSmet*, 3: 948.

36. Peter D. Elias, *The Dakota of the Canadian Northwest: Lessons for Survival* (Winnipeg, 1988), 37; AR-CIA 1871, 849; AR-CIA 1875, 680.

37. Gerhard J. Ens, "Kinship, Ethnicity, Class and the Red River Métis: The Parishes of St. Francois Xavier and St. Andrews," Ph.D. dissertation, University of Alberta, 1989, 96–97, 209, 216, 224–225. The population estimate is based on the number of emigrant families and mean family size.

38. P.R. Mailhot and D.N. Sprague, "Persistent Settlers: The Dispersal and Resettlement of the Red River Métis, 1870–1885," *Canadian Ethnic Studies* 17, no. 2 (1985): 1–30. Doll, Kidd, and Day, *Buffalo Lake*, 51–59. A recent estimate of the number of refugee Sioux in the late 1870s is in Robert M. Utley, *The Lance and the Shield: The Life and Times of Sitting Bull* (New York, 1993), 200.

39. William J. Christie to governor, January 2, 1867, Edmonton House Correpondence, B.60/b/2, HBCA; AR-CIA 1874, 572; AR-C1A 1875, 802.

40. John P. Day cites W.L. Morton and G.F.G. Stanley as exponents of the "primitive" interpretation in Doll, Kidd, and Day, *Buffalo Lake*, 13; Ens, "Kinship, Ethnicity, Class," 206; Foster, "Métis and the End of the Plains Buffalo," 73.

41. Doll, Kidd, and Day, *Buffalo Lake*, 70; Edmonton House Post Journals, June 29 and July 26, 1879, B.60/a/41, HBCA; Minutes of Council, Norway House, July 9, 1879, and July 7, 1881, B.154/k/l, HBCA.

42. Roe, *North American Buffalo*, 348, 860.

43. Richard I. Dodge, *Our Wild Indians* (Hartford, 1882), 293; William F. Butler, *The Wild Northland* (New York, 1922), 59–60.

44. John Maclean, "Blackfoot Mythology," *Journal of American Folklore* 6 (1892): 166; George B. Grinnell, *Blackfoot Lodge Tales: The Story of a Prairie People* (New York, 1892), 142–143; Ewers, *Horse in Blackfoot Indian Culture*, 291–298.

45. Robert H. Lowie, "The Assiniboine," *Anthropological Papers of the American Museum of Natural History* 4 (1910): 1–270, 129.

46. James Carnegie, Earl of Southesk, *Saskatchewan and the Rocky Mountains* (Rutland, 1969), 92–93, 245–255. I was first struck by the importance of "limited vision" a few years ago, when a friend

told me that travellers through the Pacific Northwest who stay on hard-surfaced roads have no idea of the extent of clear-cutting because loggers always leave a wooded strip along the highways; only by flying over the country can one see how much has been cut. See also Beal, "Buffalo Robe Trade," 102–103, and Bill McKibben, "An Explosion of Green," *The Atlantic Monthly* 275 (April 1995): 61–83, especially 70–74.

47. Chittenden and Richardson, eds., *DeSmet*, 3: 1188–1189; Denig, "Indian Tribes," 410. John C. Ewers, "When the Light Shone in Washington," in his *Indian Life on the Upper Missouri* (Norman, 1968), 75–90, tells the tragic story of an Assiniboine leader's trip east and its consequences.

48. William F. Butler, *The Great Lane Land: A Narrative of Travel and Adventure in the North-West of America* (London, 1872), 361–362. See also Isaac Cowie, *The Company of Adventurers* (Toronto, 1913), 301; Denig, "Indian Tribes," 403, 595; and the remarks of a trader and missionary quoted in Milloy, *Plains Cree*, 106.

49. Flores, "Bison Ecology and Bison Diplomacy," 476.

Critical Thinking Questions

1. How did the Hudson's Bay Company go about creating markets for its goods in Britain and the rest of Europe, according to Hammond?
2. In what ways was the HBC vulnerable to cycles, both in the market and in the environment? How did personnel seek to minimize the impact of those cycles? Were they successful, in your view?
3. Were the Company's efforts at conservation of their resources successful in your view? Why or why not? In what ways did those conservation efforts differ from the kinds of conservation measures that are practised today?
4. In what ways did the English families settling in Lower Canada attempt to recreate the English landscape along the St. Lawrence River? How were these efforts to modify the landscape different from, say, the activities of the earlier European settlers in Acadia?
5. Why did the attempt to recreate "the Thames toward Putney" in Lower Canada fail, according to Coates?
6. What were the primary causes of the destruction of the Canadian buffalo, according to Dobak? Are these reasons persuasive in your view? Why, in particular, does he place so much emphasis on human actions for the demise of the buffalo?
7. How was the Canadian buffalo turned into a commodity to be traded between Native peoples and the Hudson's Bay Company? How did this accelerate the destruction of the buffalo as a species?
8. Dobak argues that cultural factors as much as economic, political, or social factors may have been at work in the destruction of the Canadian buffalo. Is this argument convincing to you?
9. Dobak's arguments, if valid, seem to undermine the idea of Native peoples as inherently conservationist in their outlook. But defenders of that perspective often reply that Native behaviour was modified by their contact with Europeans, and this modification caused them to behave in unsustainable ways. To what extent is this a viable response, in your view?

FURTHER READING

George Colpitts, *Game in the Garden: A Human History of Wildlife in Western Canada to 1940* (Vancouver: UBC Press, 2002).

This excellent study examines the human–wildlife relationship in western Canada from pre-contact times to the middle of the twentieth century. Colpitts carefully charts the changing nature of that relationship, as well as the myths that were created (especially that of wildlife "superabundance") in order to romanticize and popularize the Canadian West in the minds of potential settlers. He also discusses the origins of the western Canadian conservationist impulse and shows how conservationism was not necessarily geared toward protecting natural resources but the communities that depended on their exploitation.

Theodore Binnema, *Common and Contested Ground: A Human and Environmental History of the Northwestern Plains* (Toronto: University of Toronto Press, 2001).

Binnema's study is an environmental history of the northwestern great plains, from roughly AD 200 to 1806 (the year of Lewis and Clark's expedition), but it is more as well. He shows the fluid nature of prairie life and the way in which environmental change was an ongoing and often dramatic determinant of cultural and social development. Of particular interest are Binnema's discussion of the ecology of the Plains, his placement of the Native peoples within that ecology, and his analysis of the imbalances created by the arrival of Europeans.

Clinton L. Evans, *The War on Weeds in the Prairie West: An Environmental History* (Calgary: University of Calgary Press, 2002), Chapter 4: "Dominion of the West, 1867–1905."

This chapter from Evans's study examines the remarkable success of invasive weeds on the Canadian prairie, and the attempts of government and farmers to combat them. Evans places the combat within the context of human control of nature, and shows that the management impulse extended well beyond wildlife.

C. Stuart Houston, "John Richardson—First Naturalist in the Northwest," *Beaver* (November 1984): 10–15.

Houston traces the career and activities of John Richardson, a naturalist whose work was stimulated by the materials collected by the Hudson's Bay Company in the Pacific Northwest. Houston traces the linkages between the HBC's activities, their popularization of the Pacific Northwest, and the increased scientific interest in the region. In addition, he shows the tensions that arose between scientists and the company personnel concerning the identification and marketing of wildlife.

Neil S. Forkey, "Damning the Dam: Ecology and Community in Ops Township, Upper Canada," *Canadian Historical Review* 79, no. 1 (1998): 68–99.

Forkey traces the attempt by the government of Upper Canada in the second quarter of the nineteenth century to expand the Ops Township community. Even though the local environment was unhealthy (it was a breeding ground for malarial mosquitoes), the government not only offered financial incentives to settle in the area but encouraged a transplanted mill owner to construct dams to provide power to his mills. The ponds created served as a breeding ground for mosquitoes, and Forkey traces the community responses to the dam (which included rioting after the 1837 Rebellion). He also places the narrative in the framework of "contested property" and the tensions that are created when the public and private usage of natural resources are set against one another.

RELEVANT WEB SITES

Buffalo Tales: The Near-Extermination of the American Bison
http://www.nhc.rtp.nc.us/tserve/nattrans/ntecoindian/essays/buffalo.htm
This page series is part of the "Native Americans and the Land" series created by the United States National Humanities Center. It is authored by Shepard Krech, whose work appears in Part II of this reader. The "Buffalo Tales" site is an excellent resource in understanding the eradication of the prairie bison, and it includes a detailed discussion of the Natives' relationship to the bison, the reasons underlying the bison's near-annihilation at the hands of European hunters, and the effects of the bison's loss on the prairie ecosystem. There is an excellent set of resources, both on-line and print, listed as well.

History of the Prairie Bison
http://www.albertabuffalotrail.org/bison.html
This Web site provides a good but brief history of bison. It explores the impact of the fur trade and the arrival of railroads on the bison. It has a map and links to historical attractions that have to do with Albertan history and buffalo. It also lists ranches in Alberta that can be visited to learn more about bison ranching today.

The Visual Diary of Elizabeth Simcoe
http://collections.ic.gc.ca/ElizabethSimcoe/default.html
This superb historical site was created as part of the Canada's Digital Collections initiative. It tells the story of the remarkable Elizabeth Simcoe and her journeys through Lower and Upper Canada at the end of the eighteenth century. Her observations on the environment of the British colonies are acute and are reinforced by her many watercolour paintings of the landscapes through which she travelled. This is an excellent companion to Coates's essay on landscape in Lower Canada.

Hudson's Bay at Fort Victoria
http://collections.gc.ca/fortvictoria/
Another site created under the auspices of Canada's Digital Collections. It covers the establishment and history of the Hudson's Bay Company factory at Fort Victoria in the mid-nineteenth century. Daily life, people, trading activities, and trade goods are all explored in this excellent site. There are many links to maps and other visual holdings in the British Columbia Provincial Archives.

Industrialization

The effects of Britain's industrialization were felt in her colonies long before they industrialized themselves. In the case of the Dominion of Canada, the effects were centred on the establishment and expansion of an industrialized transportation system founded on the steam engine. Rail networks rapidly expanded throughout the Maritimes and Ontario and Quebec, and steam-powered vessels ferried goods along canals that had originally been cut for the navigation of sailing ships. The construction of railways across the Prairies, as in the United States, allowed for the "opening of the West," and there was a good deal of nation-building rhetoric that surrounded the construction of the Canadian Pacific Railway in the aftermath of Confederation. In reality, though, the transportation networks were initially constructed not to ferry people or build nations but to move goods, primary or semi-finished resources to be exact, to the Atlantic ports and thence to the insatiable industries of Britain, where they were converted into manufactured items for resale on the continent or the colonies. In many ways the Canadian colonies before Confederation were already conditioned to function in the system—industrial modes of transportation and manufacture simply accelerated the rate by which resources could be extracted and the volume of manufactured goods that could be produced in return. As time passed a nascent industrial base was created in Canada, partly as a response to American industrialization to the south, partly as a means by which domestic supplies of industrial products could be at least partially guaranteed, and partly to fill a semi-processing niche. In this latter case the semi-processing industries, such as smelting or lumber milling, converted purely raw materials into semi-processed goods that could be more efficiently shipped to Great Britain.

The environmental consequence of processing resources—in this case lumber in mills—is the focus of R. Peter Gillis's essay. He investigates the discourse that occurred between business interests on the one hand and an emerging conservationist and anti-pollution mentality on the other. In the latter part of the nineteenth century, waste from lumber mills was both the most obvious and the most serious pollutant in Canada. What is striking is the similarity in the terms of the debate concerning that crisis at the end of the nineteenth century with the debates that surround environmental crises today. This is especially true in the context of the debate over another form of emissions reduction—carbon dioxide and other greenhouse gases. It is in a situation such as this that environmental history as an investigative tool can serve a very useful function indeed: while it is true that no self-respecting historian, environmental or otherwise, would ever claim that the past repeats itself, it is still the case that we may examine past phenomena to detect patterns of behaviour and determine whether similar patterns are at work, in general terms, in the present.

With that in mind, can the reader learn anything from the nineteenth-century "question of sawdust" that Gillis discusses here? Mill owners, for example, claimed that more environmentally friendly practices would reduce their competitiveness and eat into their profits. When faced with legislation curbing their polluting practices, the same mill owners threatened to close their operations, throw thousands out of work (thus leaving government with chaos on its hands), and move their activities elsewhere. As you read this, can you detect any resonance between that historical debate and our current debate concerning Canada's Kyoto obligations, which many in the business sector say can be achieved only through major employment reductions and a fall in the standard of living of Canadians?

Another interesting dimension raised by Gillis is the concept of stakeholders' rights: the problem of sawdust, for example, became serious not simply because of any environmental impact, but when its presence affected other human activities such as navigation and fishing. This idea of environmental degradation becoming problematic only when it affected human quality of life is raised also by Ken Cruikshank and Nancy B. Bouchier in their investigation of industrialization and urbanization on Hamilton's waterfront. The focus of their analysis, however, is socio-economic rather than purely economic. They ask the question: How is it that the poorest areas of a city become the most polluted? Is it simply a matter of political disempowerment? As Cruikshank and Bouchier show, the answer is far more complex and nuanced than can be found in a direct class-based analysis. Culture, memory, a sense of community, and even a willingness to surrender to the pull of "home" over the knowledge of environmental danger all play important roles.

Chapter Thirteen
Rivers of Sawdust
The Battle over Industrial Pollution in Canada,
1865–1903

R. Peter Gillis

 Industrial pollution of rivers and lakes is a very explosive political issue in late twentieth-century North America. Fouled waterways have raised the ire of naturalists, environmentalists, recreationalists, and a host of concerned citizens as health hazards are created, recreational sites ruined, traditional fisheries destroyed, and the natural beauty of many areas reduced. These problems, however, usually are viewed as a fairly recent phenomenon, the product of an aging industrial society where manufacturing enterprises are concentrated in large, somewhat obsolete, facilities. Yet deterioration of the natural environment was an important issue for a substantial and influential group of individuals in both Canada and the United States during the late nineteenth century. Highly critical of the rapacious way in which particular business interests used natural resources, this group looked to government action and regulation to control such abuses. American historians have created a considerable literature that analyzes these events in that country, but in Canada little work has been done in the field.[1] This has resulted in the

oversight of a vital and interesting part of early conservationist impulse in Canada and neglect of an important perspective on controversies within the Canadian community over resource and industrial development strategies.

This article will attempt to shed some light on these events by focusing on one celebrated and controversial cause involving early conservationists in Canada—the dumping of sawmill refuse into rivers and lakes. Drawing on British precedents, early legislation in British North America had long contained provisions forbidding the contamination of waterways with mill waste.[2] Up until the 1850s, because the number and size of lumber-producing plants remained relatively small and the regulatory system primitive, these laws remained unenforced. The Reciprocity Treaty of 1854 dramatically changed this situation. As the United States market opened up to British North American lumbermen, there was a rapid expansion of the sawn lumber industry, especially in the Province of Canada.[3] The inevitable result was the construction of more and larger mills, the creation of a greater amount of mill waste because of increasing production, and consequently the fast deterioration of rivers and lakes where sawmills were located.

The refuse problem was particularly severe in water-powered mills, which were designed, unlike their steam counterparts that used the waste for fuel, so that edgings, butts, and sawdust dropped through the floorboards into the water. Once there, this refuse was carried along by the current until it was washed into bays and other shallow areas where it sank, forming shoals of rotting material that obstructed navigation, destroyed fish spawning grounds, made the water objectionable to drink, and contributed to the formation of various gases, which emitted obnoxious smells from the water and were prone to explode due to spontaneous combustion. In a few short years after large-scale lumber manufacturing began in earnest in British North America, mill refuse was perceived by a vocal and influential part of the population of these provinces as a definite nuisance to health, navigation, and recreation. It would remain so until 1902 when the largest sawdust polluter in the country, the J.R. Booth Company of Ottawa, was forced to install equipment to stop such materials from entering the Ottawa River, thus paving the way for enforcement of the anti-dumping regulations.

The controversy, which brewed for almost 40 years, drew into political conflict business and conservationist interests that were to marshall all the major arguments currently employed in similar, present-day disputes. The lumbermen talked of the declining quality of eastern Canadian timber, the loss of their competitive edge if forced to invest in remodelling their mills, and the economic cost to the country if they were forced to close out their businesses. The preservationists, concerned with the natural environment, charged that the mill operators were ravaging the environment to the detriment of other citizens and that they were wealthy enough to afford systems to eliminate the dumping of refuse. This, they stated, was the true test of the forest industries' public spirit and business morality.

The lines between these two groups were drawn firmly by the late 1860s and the ensuing public debate demanded from politicians an increasingly sophisticated response to the issue. Starting from a position in 1866 that largely accommodated the lumbermen, successive governments were forced by public pressure into an interventionist stance by the mid-1890s. The Department of Marine and Fisheries, the federal agency responsible for enforcing the anti-pollution regulations after 1867, responded to public sentiment against refuse dumping by actively inspecting mill sites and amassing detailed scientific data to rebut the mill operators' arguments about the negligible effects of the pollution.[4] Thus challenged, the lumbermen hired their own consultants to prove their case and experimented with new ways of using the waste. Ironically, mill owners such as W.C. Edwards, J.R. Booth, and E.H. Bronson from the Ottawa Valley were leaders of the early conservation movement in Canada. Their views were much more utilitarian, however, than those of individuals who opposed sawdust dumping; they emphasized the efficient and planned use of forest resources in support of large-scale industrial development. In this way, the "sawdust question," as it was called, became one essential catalyst that differentiated various proponents of the early Canadian conservation movement. It also helped to draw together another diverse group of interests—including sportsmen, public health advocates, individuals concerned with recreation, naturalists, scientists, and proponents of water navigation improvements—who offered a conservationist viewpoint that went beyond narrow business needs and efficiency in exploitation. They did not argue that the forest industries should be put out of business, but rather that they should be carried on in a manner that respected the rights of other users of rivers and streams, preserved sanitary conditions, and protected the fish population and natural beauty of such areas. Essentially they represented a less utilitarian, more public-oriented approach to resource use that derived from an urban environment where there was some appreciation that efforts had to be made if living conditions were to be maintained and improved. Not formally organized, this collection of interests did not form a dominant strain within the Canadian population in the late nineteenth century, but was influential enough to force concerted action by the federal government against the obvious abuse of sawdust pollution. At the same time, their demand for action against the lumbermen played some part in encouraging a more scientific approach toward evaluating man's effect on his environment.

Canadian historians rarely have looked at the question of sawdust pollution, especially in its broader context as a crucial part of the debate over early conservationist notions,[5] mainly because there has been no detailed analysis of the impact of the sawdust question on the area that was the cockpit of the debate—the Ottawa Valley. It is to this analysis that this paper turns.

I
_____

The agitation against sawdust pollution actually started among sport fishermen in the Province of Canada who were alarmed at the rapid decline in fish populations. They were particularly worried about the salmon, which was suffering from deteriorating water quality as a result of more extensive settlement.[6] Lobbying by this group persuaded the legislature of the Province of Canada to pass two "Fishing Acts" in 1857 and 1858, providing the first means by which fishing regulations could be enforced. These remained moribund, however, until 1865 when the legislature passed another Fisheries Act. This legislation included a substantially strengthened section dealing with water pollution, which brought together and expanded ideas expressed in older statutes. It read:

> Lime, chemical substances or drugs, poisonous matter (liquid or solid), dead or decaying fish or any other deleterious substance, shall not be drawn into or allowed to pass into, be left or remain in any water frequented by any of the kinds of fish mentioned in this Act, and sawdust or mill rubbish shall not be drifted or thrown into any stream frequented by salmon, trout, pickerel or bass under penalty not exceeding one hundred dollars.[7]

The period of the early 1860s witnessed an awakening of Canadians to the decline in the quality of their natural surroundings. Immediately after passage of the new Act, Dr. E. Van Cortland, officer of health for the City of Ottawa, launched a petition. In the spring of 1866 he wrote to the Department of Crown Lands, the agency responsible for fisheries until Confederation, requesting that lumber companies operating on the Ottawa River be restrained from dumping mill waste in that stream. His reasons for the ban were threefold: that refuse was a threat to public health; that it was destroying the spawning grounds for fish; and that it posed a potential threat to navigation.

Van Cortland's plea broadened the basis of support for government action on sawdust dumping beyond sport fishermen and, indeed, it set forth the exact concerns that would be espoused by anti-pollution advocates through to 1902. The letter, coming as it did from a respected local official, struck a responsive chord with the provincial Coalition ministry. On August 25, 1866, the Department of Crown Lands issued a circular that informed all mill owners in the Province of Canada that they "must adapt their premises to the disposal of waste materials in such a manner as shall obviate further injury to rivers and streams."[8] The reaction from the mill operators along the lower Ottawa River, at whom the new regulations were primarily aimed, was swift. On September 8, the leading lumbermen of Ottawa-Hull, including Gilmour and Company, Levi Young, the Bransons and Weston Company, Perley and Pattee, and Ezra Butler Eddy, sent a long and pointed memorial to the governor general, Lord Monck. The men involved were among the wealthiest entrepreneurs in the province and their trade in lumber products, which was oriented to both Great Britain and the United States, was the largest non-agricultural industry in British North America. Politically, the lumbermen formed a powerful influence in both the Conservative and Reform parties. Now, they stated flatly that, as their mills were driven by water power, it was quite impossible to prevent sawdust from falling into the Ottawa River. The mills were built out over the water with wide cracks left between the floorboards precisely so that the waste would pour into the stream. To install equipment to collect this sawdust, they argued, would double

the cost of producing their lumber. The only other solution, they foresaw, was to abandon their present mills and build new steam factories, but the attendant loss of business involved would be large and the operators agreed that they would rather quit the trade.[9]

Of course, the lumbermen's response was in part an elaborate bluff. Most would not abandon their lucrative trade in pine timber. Beyond a doubt their water-powered mills were cheaper to operate than many steam mills and as a result gave these entrepreneurs some advantage in a highly competitive trade. But primarily the operators were establishing a case in hopes that the Department of Crown Lands would accept the compromise solution they had to offer.

The lumbermen seized on one aspect of the problem, the threat to navigation, with which they had considerable sympathy, given the fact that the vast majority of their product was shipped by water. They suggested that grinding machines could be installed, which would reduce mill waste to sawdust form. Boats and barges could then sail on the Ottawa River without fear of foundering on heavy and dangerous waste. In addition, the operators requested the government to appoint an arbitrator to investigate the matter and to suspend action against them until this gentleman reported.[10]

John Merril Currier, Conservative member of the provincial legislature for Ottawa and prominent lumberman, presented a copy of the memorial to the Department of Crown Lands and also met with his colleague, the Hon. Alexander Campbell, the Conservative commissioner. A conscientious and philosophical man, Campbell was a guiding spirit in securing early conservation measures.[11] He was John A. Macdonald's former law partner in Kingston, and the prime minister's good friend and political confidant. He had pushed the new Fisheries Act through the Assembly. Now, under intense pressure, Campbell suspended indefinitely all lawsuits pending against the Ottawa River lumbermen for dumping waste in the river and appointed Horace Merrill, superintendent of the Ottawa River Works for the provincial Department of Public Works, as arbitrator in the matter.[12] Merrill, who had numerous private business connections with local lumbermen, produced a report on December 12, 1866, that largely adopted the mill owners' recommendations. Grinders would be installed that would reduce all waste to sawdust form; then the waste would be dumped in the river. But even Merrill could not ignore the detrimental effect mill waste was having on fish life. He candidly admitted that his recommendation was only a utilitarian compromise and not a positive solution for a grave problem.

It did, however, prove politically acceptable to the Hon. Mr. Campbell. In early 1867, he ordered that all lumbermen in the Province of Canada could continue to dump mill waste in rivers as long as they installed grinders. If they did not have grinders, they would be prosecuted. Throughout 1867, mill owners on the lower Ottawa were busy installing these new machines under the watchful eye of Detective E. O'Neill of the Fisheries Service. By early 1869, O'Neill was able to report that grinders were in general use in all mills in the area, except at J.R. Booth's operations where most of the slabs were used in making lath, and at Hamilton Brothers at Hawkesbury, a firm that had secured an exemption from the regulations because it dumped its waste into a series of rapids.[13]

This grinding scheme did not, however, secure much breathing space for the lumbermen. Early in 1870 a private members' bill entitled "An Act for the Better Protection of Navigable Streams and Rivers" was introduced into the House of Commons. It stipulated that no sawdust or mill offal was to be thrown into any navigable stream. Its sponsor was Richard Cartwright, Liberal MP for Lennox, Ontario.[14] Although the new bill did provide for certain exemptions if they were deemed in the public interest, it attacked the lumbermen at a most vulnerable point. Good navigation was generally deemed an essential part of a sound industrial policy. If indeed the lumbermen were obstructing navigation on various streams, they were retarding, at the same time, the industrialization of the new nation.

The bill failed to pass in 1870, but it was reintroduced in both 1871 and 1872. During this time the lumbering community along the lower Ottawa galvanized its friends in Parliament to battle against the legislation. Arguments were made that lumbering, the largest non-agricultural industry in the country and a major backbone of its economy, must be given some special privileges, including the right to dump its waste in all rivers and steams. As in the previous controversy, dire threats were made about closing the mills and throwing thousands out of work. This time, as well, lumbermen from the Maritimes, who would also be affected by the new legislation, were persuaded to lend their influence against the bill.[15]

All was to no avail, however, as the Act moved through the various legislative stages to become law in 1873.[16] Most MPs were inclined to agree with Cartwright's position that mill waste was an enormous and avoidable nuisance. Few were willing to deny that sawdust had had an effect on all navigable streams flowing into the

Great Lakes and they were particularly incensed at the condition of the Ottawa River, which was described as a "national disgrace."[17] Some felt the threat to the Ottawa River was doubly dangerous because it put in jeopardy the Georgian Bay Ship Canal—a proposed short route from Chicago, across the French River system and down the Ottawa Valley to Montreal—a route that many Canadian businessmen still felt was a key to the economic development of central Canada. Finally, many MPs were convinced that the lumbermen had grossly inflated the cost of installing the machinery to dispose of the waste properly. Indeed, they argued that sawdust could be converted into a cheap source of fuel.[18]

The lumbermen attempted to counter these arguments with evidence about the state of the Hudson River, which they considered to be still navigable despite years of lumbering along its banks.[19] In the end, however, when the bill passed, they accepted the result and placed their trust in the Conservative federal administration of John A. Macdonald, which they hoped would be lax in enforcing a law sponsored by a Liberal private member. However, the Macdonald government fell in 1873 as a result of the Pacific Scandal.[20] The new Liberal-Reform government of Alexander Mackenzie included Richard Cartwright as minister of Finance. Cartwright did not move immediately, but he did have his revenge. In the spring of 1875 the Department of Justice, along with the Department of Marine and Fisheries, which was responsible for administering the Act, began proceedings against J.R. Booth, one of the most powerful and prominent Ottawa lumbermen.

Booth had been recalcitrant about installing the original grinders in 1867 and had constantly broken the regulations since

then. Now he was the first to be convicted under the new act. His fine was not large, $20.00, but the threat of continued court action, escalating fines, and then finally an eventual injunction against sawdust dumping soon brought both Booth and his fellow lumbermen to heel. Having done nothing to live up to the regulations, they appealed to the governor-in-council for exemptions under the Act. This suited the Department of Marine and Fisheries. It had no wish to destroy the Canadian lumber industry but rather hoped to improve upon the performance of the companies in abiding by its fisheries regulations. During 1876, because the companies had proven obstinate, the department exacted its pound of flesh. In order to be exempted from the provisions of the Act, a company had to prove that the dumping of sawdust in a particular stream "was not injurious to the public interest." It declared that, along the lower Ottawa, it was readily apparent that sawdust pollution was harming the river and its tributaries. Therefore, the agency asked the lumbermen to prove that the cost involved in converting the water-powered mills so that no sawdust entered the river was prohibitive; that if the mill owners were forced to undertake these renovations they would have to close down, throwing men out of work; and, finally, whether the mill waste served other interests that had been neglected by overconcentration on the destruction of fish life and navigation.

During the early months of 1877, in order to form an opinion on these issues, the Department appointed John Mather, one of the best mill engineers in Canada, to investigate conditions in each of the sawmills along the lower Ottawa. Mather presented his report in June 1877. It was a far more critical document than that prepared by Merrill a decade before.

He stated categorically that in every mill, disposal of waste could be handled differently than at the present time. Waste burners could be installed with conveyors or tramways to carry the sawdust to them. This technology was already in use at various steam-mills and could readily be applied to the water-powered facilities, though such installation would be costly. He concluded that for this reason, and the fact that the burners would push fire insurance rates up 2 percent, the lumbermen were resisting these changes. As evidence, Mather estimated that 12,300,000 cubic feet of sawdust were being dumped in the Ottawa River alone each year and that this amount was increasing annually, thus posing a serious threat to fish life and navigation. The problem had become serious enough, in Mather's view, to merit a major effort to enforce existing laws that curbed the dumping of waste, despite both the costs involved for the lumbermen and the attendant risk that they might close down their operations.

On December 20, 1877, the mill owners met with the Hon. Sir Albert James Smith, minister of Marine and Fisheries, and with Mather to determine if the two sides could agree on a suitable plan for limiting the dumping and to consider generally the Mather Report. W.G. Perley, prominent lumberman and Tory MP for Ottawa, acted as spokesman for the industry. The lumbermen took the same position they had held in 1866. They were of the view that slabs and other heavy waste should be kept out of the river, but argued that the sawdust was not obstructing navigation, that fish life was of negligible importance in comparison with the lumber industry, and that, therefore, the dumping of such waste should be permitted. The minister replied that he was not going to try to close the sawmills along the Ottawa River; he

was, however, going to delegate Mather to supervise at each mill renovations designed to put a stop to all heavy waste entering the River and to improve the general system of disposing of all sawdust and chips. Furthermore, the owners were to institute these modifications immediately and bear their cost. The lumbermen concurred in this arrangement.[21]

Thus, during the winter and spring of 1877, each mill operator along the lower Ottawa was busy installing more chippers and grinders under various plans approved by Mather. At the same time, the Department of Marine and Fisheries received several petitions from inhabitants along the river pleading that the mill slabs not be chipped since they collected them for firewood. Such opposition did not, however, delay work on the renovations and by January 15, 1879, Mather reported that nearly every mill along the lower Ottawa was equipped in the approved fashion. The companies immediately were granted exemptions from the Act, and these exceptions were officially embodied in an order-in-council on June 23, 1880.[22]

II

At this point, the lumbermen thought that a final arrangement on the sawdust problem had been reached. In reality, the Department of Marine and Fisheries, despite considerable effort, had made an unsatisfactory compromise in dealing with the problem of mill offal along the Ottawa. Essentially, in the period 1866 to 1878, officials had managed only to enforce regulations that kept slabs and heavy waste out of the water, and even this was proving problematical since complaints were still being received about the dumping of heavy waste into the Ottawa. In fact, the lumbermen had managed to continue their operations in very much the way they had before the anti-dumping legislation. Thus, the condition of the Ottawa River had continued to deteriorate. In both 1866 and 1873 the mill owners had found it impossible to resist the political pressure and public outcry aroused in favour of the anti-pollution measures. After passage of the Acts, however, they had managed, through negotiations with government officials, to limit the problem to the issue of navigation and then to persuade authorities that chipped waste was causing no problem. As well, the lumbermen had been sufficiently organized to present a united front through which they threatened to close all their mills and throw thousands of men out of work, a bluff the federal government was not willing to call. Finally, the operators' lawyers and agents sought support from the local population. Because no one wanted to see the mills closed down, petitioners flooded the Department of Marine and Fisheries with requests for continuance of sawdust dumping.[23] These manoeuvres demonstrated the lumbermen's effective use of their political and economic power. While appearing to negotiate earnestly with federal officials to solve a grave problem, they were actually approaching the issue in such a manner as would allow them to sidestep it. Indeed, while posing as public-spirited men, the mill owners were managing events for their own profit.

But if the lumbermen thought that their trouble over sawdust dumping had ended with the 1880 order-in-council, they were sorely mistaken. True, federal authorities had lost interest in the question after the Conservatives were returned to power in 1878. Sir John A. Macdonald's government completed the arrangements started by the Liberals, but then let the matter drop. However, the problem itself was too serious

and obvious in nature to remain forgotten. Complaints continued to be received from boat operators and other users of the river about the dumping of not just sawdust but also heavy waste.[24] The issue was pushed to the fore when the Province of Ontario passed its own fisheries act in 1885. Rumours were rife that provincial authorities would use injunctions to stop the dumping of mill offal into various streams, including the Ottawa. Once again, the lumbermen used their economic and political power to obstruct this special legislation even as they blithely assured the Ontario government that "it [the industry] was doing everything in its power to remedy the matter."[25] The Ottawa Valley mill owners felt compelled, however, to approach the federal government again for a general order-in-council insuring their right to dump sawdust into the Gatineau and Ottawa rivers. The order was granted on April 17, 1885, but its issuance once again drew attention to the abuse. Late the next year the deputy minister of Fisheries, John Tilton, condemned the lumbermen's total disregard of existing regulations. Prompted by powerful renewed criticism in the Ottawa Valley of the whole mill waste problem, Tilton was particularly critical of the continued dumping of heavy waste into the Ottawa River.[26]

The issue was taken up by a long-time resident of Ottawa and Tory senator, Francis B. Clemow. Clemow was a prominent member of the Ottawa business community, manager of the Ottawa Gas Works, and frequent partner with local lumbermen in a variety of ventures.[27] He was, however, willing to risk the wrath of these entrepreneurs by championing measures to prevent the dumping of mill offal. His concern was centred on the risks that continued dumping of sawdust posed to the natural beauty of the Ottawa Valley, its fish life, its business development through navigation, and the health of its citizens. This threat was emphasized for Clemow by the growing frequency of explosions in the river resulting from concentrations of methane gas produced by the rotting refuse. Thus the senator presented the perfect combination of early preservationist and industrial promoter—he was a supporter of the Georgian Bay Ship Canal—which could galvanize some political action on the matter.

Clemow's interest in the sawdust question was compounded by a pressing international problem of sawdust fouling the St. John River, its tributaries, and the waters of Maine. As a result, a Select Committee of the Senate was formed to study the total extent and effect of dumping refuse into Canadian streams.[28] The committee requested the government to make available Henry A. Gray, assistant chief engineer, Department of Public Works, to conduct a detailed survey of the river between the Chaudière Falls and Grenville during the summer of 1887. He took soundings and borings, and interviewed those who had lived along the river for many years. He found the sawdust to be a definite hazard to navigation, blocking channels with accumulations in some areas up to 80 feet deep. Furthermore, Gray found the substance difficult to dredge because of its loose composition and speculated that it was difficult to see how fish life could survive in such water. Finally, the engineer described the methane explosions, which on occasion shook the river.[29]

Gray's official report was not released by the Department of Public Works until the spring of 1889. On the basis of his findings the Senate Committee recommended, on May 15, 1888, that "the government take steps to prevent such deposits in the future."[30] The lumbermen were quick to react. The Gray Report snatched from them the main basis of their argument—that the sawdust was not obstructing navigation. They organized their own committee under E.H. Bronson, president of the Bronsons and Weston Lumber Company and Reform member of the Ontario Legislature for Ottawa. Bronson began immediately to raise money for the mill owners' own survey of the river to disprove Gray's findings. He summed up the issue succinctly to a smaller operator, W. McClymont:

> This sawdust question is not as to whether we [the mill owners] shall be permitted to put a little sawdust into the river as you state you are doing, or whether we shall put in all we make, but is a question as to whether we shall be allowed to put any *at all* [emphasis his] in the river.
>
> If the decision is that we must keep it all out, the only alternative will be for us all to make any necessary alterations in our mills to keep all the sawdust out of the river, and put in refuse burners to consume it. This would cost you and all the rest of us ten times as much every year as this examination will cost.[31]

The lumbermen were fully aware of the gravity of their situation and they wanted a prominent engineer to lend weight to their survey. After much discussion, they settled on the well-known civil engineer, inventor, and father of Standard Time, Sir Sandford Fleming, to organize the operation. He was not the lumbermen's first choice for the job. They had hunted in Montreal, Toronto, and the northeastern United States for a well-known and respected engineer who would prove sympathetic to the business concerns involved. Fleming appears to have become available at the eleventh hour and was considered to have the proper credentials.

Fleming undertook the survey in the summer of 1888—and reported, predictably, that, with the single exception of the mouth of the Rideau Canal, accumulations of sawdust were producing no major obstructions to navigation along the lower Ottawa River. To bolster this report, Branson sought from various American lumbermen and politicians information on the effects of sawdust on other rivers. Most replies indicated that sawdust was not the great nuisance that some pretended it to be. With this evidence, plus the Fleming Report, Bronson was confident that the lumbermen could persuade the federal government to leave the regulations as they were. As he informed Hector Langevin, Minister of Public Works:

> [It is] our firm conviction ... that the high water each year clears the river channel of any sawdust deposits that the slow current the preceding fall may have permitted to lodge therein.[32]

Bronson did not have to convince the prime minister, Sir John A. Macdonald. In the summer of 1889 Macdonald was an aging and tired man who was rather bewildered by the whole debate over the sawdust question. He wanted to make sure that there was "no suspension of sawmills" as a consequence of the problem and took the rather novel approach of arguing that lumbering should continue until all the trees had been cut along a river; at that point the river could be restocked with fish at government expense.[33]

In taking this stance, Macdonald was adopting the traditional view of lumbering as a transitory industry that simply removed the forests in preparation for settlement. It was an opinion that many Ottawa Valley lumbermen no longer shared. Faced with large investments in mills and declining wood stocks, they were beginning to push for more conservation-based measures regarding forest use in order to protect the pineries.[34] Now, however, in order to protect their interests, they were quite willing to abandon these recently acquired views. They told the government that logging had gone on in the valley for 95 years without harm and that, since two-thirds of the merchantable timber was now gone, the lumbermen should be left to remove the last third unmolested. In such vital matters, they had little use for Clemow's preservationist ideas. Conservation for them was utilitarian and business-oriented, as witnessed by the operators' declaration:

> As regards the beauty of the landscape ... we admit that floating sawdust does not improve the general appearance of the river, but it must be remembered that this is a utilitarian age and that the interests of any important industry, the success of which affects the well-being of so many people, are invariably held to be paramount to the gratification of mere aesthetic taste, satisfactory and desirable as that may be under proper conditions.[35]

But others had very different views on the water pollution issue, and public pressure was placed on the government to take some action. Clemow continued to spearhead a drive to exact from the Conservative government a promise to do something about the sawdust problem. He proposed legislation that received renewed support after September 15, 1890, when it was revealed that a study by W.A. McGill, assistant analyst in the Department of Inland Revenue, revealed that the water of the Ottawa River, while not yet dangerous, was in a state that would support morbific bacteria.[36]

The battle shaped up around amendments to the Fisheries Act and whether or not the Tory administration (now without Macdonald, who had died on June 6, 1891) could be persuaded to enact tougher pollution measures. The minister of Marine and Fisheries was Charles Hibbert Tupper, a Nova Scotian who had little interest in any measures except those relating to commercial coastal fisheries. As he observed, the Ottawa River could continue to receive sawdust because "there are no lobsters in it."[37] In taking this stand, Tupper was reneging on what Clemow considered to be a government promise made in March 1890 to pass special clauses to protect the navigable rivers of Canada. The senator was enraged. He smelled a deal designed to protect the powerful Ottawa lumbermen at the expense of his city's health, beauty, and commercial destiny. The stench of political trading was given especial credence because the chronic offender in sawdust dumping, J.R. Booth, had just rebuilt his mills with no provision being made for the disposal of waste, except in the Ottawa River! Clemow charged publicly that the Department of Marine and Fisheries was ignoring vast amounts of information collected by government officials that documented the existence of the nuisance. Further, he launched an attack on Booth, declaring that he had come to Ottawa poor and raped the countryside to make a fortune. He concluded that the country owed the Booths of the world nothing and demanded that after May 1, 1895, all exemptions for sawdust dumping be terminated. Clemow was supported in his battle by a petition from 715 Ottawa area residents, which contended that the mill refuse in the lower Ottawa was a hazard to navigation, prevented use of river properties, and posed a menace to public health.[38] In the furor that resulted, the Senate acted on Clemow's motion to amend Tupper's Fisheries Act and the Commons moved to support the amendment.

The new provision declared that after May 1, 1895, all exemptions for sawdust dumping were to cease unless good and sufficient reasons could be given by the mill owners for continuing the practice. There was still room for some discretion in the matter, but it was much narrower, especially given the fact that Colonel F.H.D. Vieth, Fisheries overseer, was instructed to visit each mill to establish definitely where the offal was deposited, if it harmed spawning beds, where the sawdust came to rest, and if it was harmful to navigation. The lumbermen were notified of this arrangement on August 3, 1894. A month later William Smith, deputy minister of Marine and Fisheries, again wrote to operators deploring their non-co-operation with Vieth.[39] The mill owners adopted their classic stance of outright refusal to comply in preparation for another round of negotiations. Smith, however, warned them that their delay was seriously prejudicing any case that the lumbermen might be able to establish for an extension of their exemptions.

This threat brought a flood of letters of apology and justification from the lumbermen. This correspondence showed that a few operators were taking the matter more seriously. Several, including the E.B. Eddy Company and the W.C. Edwards Company, indicated that they were utilizing sawdust both for fuel and manufacturing purposes. Indeed, Eddy's was buying sawdust from other operators.[40] Such changes were not, however, particularly connected to a fit of conscience. Rather, first-class wood stocks were declining in the Ottawa Valley and there was greater need to utilize more parts of the tree, as well as to diversify into new product lines such as pulp and paper. Such measures were part of the eastern forest industry's attempt to make its basis more durable, to promote forest conservation measures, to protect standing timber, and generally to make its operations more efficient. In this sense the threat of stronger pollution regulations spurred on the diversification process. But such measures were still the exception and most mills, including the large Booth and Bronsons and Weston enterprises, had made little progress and simply stated that the delay had been occasioned by meetings among the operators to establish a common response to the department's demands.

The Ottawa Valley industry replied to Smith on September 22, 1894. It indicated that as far as navigation was concerned, the operators stood by the conclusions of the Fleming Report. Further, it stated that the fishery business on the Ottawa was, at best, a very small one; if it needed to be saved, according to the lumbermen, the govenment should be responsible for putting in fish ways and for restocking. To dwell on absolute dumping bans was to threaten what was, by contrast, a large and substantial industry. Reflecting E.H. Bronson's views, the response quoted scientists and analysts who claimed that sawdust did not harm rivers and asked pointedly why federal officials were doing nothing to stop the dumping of sewage in navigable rivers—a nuisance that really did harm the streams and posed a definite health hazard. Finally, the valley operators reminded the government of the jobs that would be lost if the mills closed and argued that the industry had passed its peak on the Ottawa, with the consequence that smaller numbers of trees would be cut by lighter gauge saws, which would automatically reduce the amount of sawdust entering the Ottawa River. As well, several lumbermen, including W.C. Edwards, informed the government that the cost of installing large burners was $13,000–15,000 per burner and that fire insurance rates would be forced up between 4 percent and 5 percent, both extra financial burdens on the industry.[41]

The lumbermen, while expanding the grounds of argument over their right to pollute, did not trust to chance in preserving their exemption. Booth and Hiram Robinson, a partner in the Hawkesbury Lumber Company and a known Tory, met Tupper's successor as minister of Marine and Fisheries, John Costigan of New Brunswick. Being from a lumbering area, Costigan was more conversant with the sawdust problem. H.K. Egan, Robinson's partner, promised to supply the minister with any data that he required to justify passage of special remedial legislation.[42] A new dimension was also introduced into these private discussions. Clemow had acquired a particularly vocal ally in R.J. Wicksteed, an Ottawa barrister who was an avid advocate not only of protecting the Ottawa River from pollution but also of ending the lumbermen's special privileges in regard to refuse dumping. Booth and Robinson now indicated to Costigan that Wicksteed was intent upon closing some of the mills through use of the fisheries regulations. This alarmed Costigan, who informed the new Conservative prime minister, Sir John Thompson, of the situation. Thompson, who was the same individual consulted by Sir John A. Macdonald on the subject, obtained directly from Smith a full appraisal of the situation.[43] The deputy minister indicated that he was pleased with the progress made by firms such as E.B. Eddy and the W.C. Edwards Co. to ease the pollution situation, but also revealed his agency's lack of resolve to push the new legislation when he stated that "he could not really justify the closing down of the other mills over the sawdust issue." He feared the impact of the closures on the Ottawa Valley and national economy and introduced the new issue of smoke pollution. The deputy informed the prime minister that "not sufficient research has been done into the effects of smoke pollution from massive numbers of sawdust burners or the impact such burners would have on insurance rates." Sensing the political realities of the matter and the Department of Marine and Fisheries' lack of resolve, the mill owners played up the unpopularity of the waste dumping measures in the Ottawa Valley and in the Maritimes where the new regulations would also be in effect. Despite stout efforts by Clemow, particular pressure was brought by the Ottawa Board of Trade to suspend all efforts to prosecute offending mill owners under the new law and, indeed, for the Tories to seek remedial legislation to enshrine the dumping exemptions in law.[44] The two factions disputing the issue were driven almost to overt violence at board meetings and the government simply delayed action on remedial legislation.

As the May 1, 1895, deadline approached, however, the Department of Marine and Fisheries finally revealed where the government really stood on the matter. It had continued its study of the matter with E.E. Prince, the department's leading fish specialist, reporting definitively that sawdust did affect fish spawning and did chemically foul the water. As well, other officials, including Vieth, produced evidence to refute Fleming's findings on the effect of sawdust on navigation and testimony to prove the deterioration of fish species in the Ottawa River.[45] Nevertheless, in mid-March the department, on Costigan's behalf, issued directives indicating that the government would seek remedial legislation to reinstate the dumping exemptions and that until this was passed, inspectors would not enforce the existing law. Further, the directives stated that if proceedings were taken by members of the public, the department would refund any fines levied against the mill owners. The measures were justified as necessary to prevent hardship in the forest industries. News of this position on the part of the government prompted lumbermen in both the Maritimes and the Ottawa Valley to flood Marine and Fisheries with requests for exemptions that went beyond those already in place. The government, however, restricted its directives to existing exemptions. Basically, the minister indicated that no evidence had come to his attention to convince him that dumping exemptions should not be extended through remedial legislation!

The Tories were defeated in the general election of 1896 without having changed the Fisheries Act. The new Liberal government of Wilfrid Laurier appeared to be no less sympathetic to the lumbermen; W.C. Edwards was a close friend of the prime minister and other operators, such as E.H. Bronson, were powerful and influential Reformers. Nevertheless, the fact remained that the Ottawa River and its tributaries were now heavily clogged with waste. As well, public pressure for government action continued to be strong. Indeed, methane explosions were now commonplace, endangering boaters to the extent that in late 1897 a prominent Montebello area farmer was drowned when dumped in the river as a result of such a shock.[46] This deteriorating situation prompted the new deputy minister of Marine Fisheries, François Gourdeau, to take action. Promoted from within the department, he was the type of public servant who was tired of seeing public research and effective regulations hamstrung by lack of political will to face down powerful business interests. Gourdeau viewed the situation along the lower Ottawa as a national disgrace and was determined to stop the dumping of sawdust. In order to meet some of the political pressure, he postponed the effective date of the new fisheries regulations until June 30, 1898, but he also officially informed the lumbermen that no exemption would be allowed beyond the extension date.[47] Basically, the operators were being given a year to alter their plants or they would be shut down.

Gourdeau might have been thwarted in achieving his objectives had he not received powerful support from his own minister, Sir Louis Davies, a P.E.I. native not vulnerable to the lumber lobby, and from the minister of Public Works, Joseph Israel Tarte, an ultramontanist and advocate of colonization. Tarte, in particular, was influential in Cabinet; with his reputation as a political bagman and fixer, his support at first appears somewhat strange. But the minister, like Clemow before him, had an interest in preserving the natural environment for recreational and aesthetic reasons as well as navigational purposes. In a letter to Davies on August 14, 1901, Tarte congratulated the minister on his department's strong stand against sawdust dumping and categorically declared that "there is no justification in lumber merchants and millionaires, to ruin [sic] such a magnificent highway."[48]

Only under this direct and powerful threat did the influential lumbermen on the lower Ottawa hasten to move quickly to solve the sawdust problem. Mill owners, in general, began to emulate the Eddy and Edwards firms in getting rid of the waste.

Some found outlets for the offal as fuel for other industries, replacing expensive American coal. Others installed pulp and paper plants and other wood products operations that utilized sawdust and chips. Still others experimented with the use of the mill waste to produce carbons and acetylene. Finally, if there was no other outlet, burners were put in the mills to dispose of the sawdust. Some of these solutions had been recommended since 1866 but stoutly resisted by business interests that wished to maximize profits while not interfering with the competitive cost advantages of their water-powered mills. Predictably, it was J.R. Booth who dragged his feet in obeying the new regulations. As the largest operator in Canada, Booth's problem was much more difficult to solve. The old entrepreneur, in a characteristically individualistic stance, refused to make piecemeal renovations to his mills but rather made plans for a new pulp plant that would utilize the sawdust as fuel.[49] This was a longer-range solution than the Department of Marine and Fisheries envisioned and tension mounted after mid-1901.

Up until 1900, Booth had shown good faith in attempting to live up to the regulations by finding buyers for his sawdust and participating in various experiments for by-products. In that year, however, a massive fire at the Chaudière wiped out many of the establishments that had bought Booth's sawdust and left him with a disposal problem. Though he had laid down plans for his pulp mill, it was not due for completion until at least 1903. Booth viewed this as the best solution because it would not only take care of the mill refuse but also employ more men. He acknowledged, however, that he was breaking the law every day in dumping the offal, and the Department of Marine and Fisheries could not ignore the situation since public complaints were now mounting. W.C. Edwards became a mediator between the firm and the government. He convinced the department to refrain from action until experiments in using sawdust to produce carbon had been completed on his assurance that Booth would then put in place "efficient measures ... for the disposal of the sawdust."[50]

By mid-1901, the experiments had come to nothing and Booth had done little else to remedy the waste problem. The government, therefore, had little choice but to begin prosecution against the lumberman. Booth pleaded for leniency, indicting that, since his sawing season was already underway, he could not be expected to put in burners and furnaces until the coming winter. He further indicated that he would pay fines until such time as an injunction was issued; then he would close his mills, throwing 1,500 to 2,000 men out of work. Gourdeau and the Department of Marine and Fisheries would not be deterred, however, and instructed court proceedings to be undertaken.[51] Booth commenced to pay fines, and on September 13, 1901, Gourdeau instructed the Department of Justice to seek an injunction against sawdust dumping on behalf of the government.

The situation reached the point of confrontation; in the City of Ottawa, it was an embarrassing time for both the Liberal federal government and John Rudolphus Booth. Edwards, the local Reform fixer, intervened personally with Laurier in an attempt to reconcile the dispute. The prime minister ordered the minister, Davies, to postpone court action against Booth for one week. Then, on September 24, 1901, Laurier further indicated to Gourdeau that

I desire you to give instructions to ... counsel for the Department of Marine and Fisheries against Mr. Booth for putting sawdust in the Ottawa River, to have that prosecution withdrawn as Mr. Booth has undertaken, as soon as the sawing season is over, to put up a burner and to entirely stop the illegal practice complained of, at the beginning of the next season.[52]

The prime minister had intervened to give Booth his extra time, but at the price of a promise to absolutely stop the dumping of sawdust. The battle over sawdust pollution finally had been won. There were still some doubts that Booth would actually keep his word. Indeed, Tarte wrote to the new minister of Marine and Fisheries, James Sutherland, on September 8, 1902, stating that Canada's leading lumberman "appears to have made a promise to Sir Wilfrid that he does not plan to keep."[53] But, by the spring of 1903, inspectors reported that Booth had taken care of 95 percent of the waste problem at his mills. With J.R. Booth obeying the regulations, there was no reason for anyone else in Canada not to do so.

Thus, 35 years after the first public petition to stop the dumping of mill waste into the Ottawa River, government officials had finally been given the necessary political support to stop the abuse. It must be noted, however, that the victory was won against an Ottawa Valley sawn lumber industry that was already in decline and that was retooling into pulp and paper and wood by-products, procedures that demanded more efficient utilization of mill wastes. Nevertheless, the lower Ottawa had become a symbol for those who wished to end sawdust pollution. If laws could be flaunted right under the windows of the Parliament Buildings, then there was little hope of enforcing pollution regulations elsewhere in the nation. The battle was joined there and became the bellwether for testing the fisheries regulations. At the same time, the controversy became an important episode in defining the early conservationist impulse in Canada. The Ottawa lumber barons and associates elsewhere in eastern Canada had attempted to appropriate that movement to their own business needs by defining its precepts as solely utilitarian and oriented to efficient, planned use of the forest resource with an emphasis on preserving commercial species for full utilization. The anti-pollution proponents, however, served to challenge and widen that definition. They introduced into the pollution debate the factors of maintaining fish life, preserving natural beauty, encouraging recreation, ensuring public health, and promoting water navigation. In this way they were the early harbingers of the preservationist wing of the conservation movement, which was to have its greatest success a few years later in the expansion of wildlife sanctuaries and national parks, as well as the multiple use concept of resources. Perhaps the most important contribution of the whole sawdust question was the use made by anti-pollution advocates of government officials and facilities to research, investigate, and regulate the dumping of waste. Once again, this was an early example of a more activist government that would regulate business activities for a politically defined public interest. In short, from the microcosm of the sawdust question emerge themes about the relationship of Canadians with their environment—natural, social, and economic—in the late nineteenth century, which provide important new perspectives on the development of the country in the period of "national transformation."[54]

NOTES

1. See, for example, H. Clepper, ed., *Origins of American Conservation* (New York: Ronald Press, 1966); S. Hays, *Conservation and the Gospel of Efficiency: The Progressive Conservation Movement, 1890–1920* (New York: Atheneum, 1969); D.H. Strong, *The Conservationists* (Menlo Park: Addison-Wesley, 1971); and R. Nash, *Wilderness and the American Mind* (New Haven: Yale University Press, 1967). The standard Canadian studies are R.S. Lambert and P. Pross, *Renewing Nature's Wealth: A Centennial History of the Public Management of Lands, Forests and Wildlife in Ontario, 1763–1967* (Toronto: Ontario Department of Lands and Forests, 1967); H.V. Nelles, *The Politics of Development: Forests, Mines and Hydroelectric Power in Ontario, 1849–1941* (Toronto: Macmillan, 1973); and J. Foster, *Working for Wildlife: The Beginning of Preservation in Canada* (Toronto/Buffalo: University of Toronto Press, 1978); as well as R.P. Gillis, "The Ottawa Lumber Barons and the Conservation Movement, 1880–1914," *Journal of Canadian Studies* 18 (February 1974): 14–30; and R.P. Gillis, B. Hodgins, and J. Benidickson, "The Ontario and Quebec Experiments in Forest Reserves, 1883–1930," *Journal of Forest History* 26, no. 1 (January 1982): 20–33.

2. Statutes of Lower Canada, 6 Vict., Cap. XVII, "An Act for Better Preventing the Obstruction of Rivers and Rivulets in Canada East," 1843; and Statutes of Upper Canada, 7 Vict., Cap. XXXVI, "An Act to Prevent Obstructions in Rivers and Rivulets in Upper Canada," December 9, 1843.

3. The best discussion of the development of the North American lumber trade is found in A.R.M. Lower, *The North American Assault on the Canadian Forest* (Toronto: Ryerson, 1938).

4. The detail of these activities and, therefore, much of the documentation for this article is drawn from Public Archives of Canada, Department of Fisheries, Record Group 23, file 1669, vols. 1–3.

5. See, for instance, P. Mitchell, "Déja Vu: Refuse in Rivers and Harbours," *Nature Canada* 7 (October–December 1978): 33–34; and G. Allardyce, "The Vexed Question of Sawdust: River Pollution in Nineteenth-Century New Brunswick," *Dalhousie Review* 52 (1972): 177–190.

6. Lambert and Pross, *Renewing Nature's Wealth*, 151–154.

7. Province of Canada, Statutes, 29 Vict., Cap. 11, "An Act to Amend Cap 62 of the Consolidated Statutes of Canada and to Provide for the Better Regulation of Fishing and Protection of Fisheries."

8. PAC, RG23, file 1669, digest of papers on the sawdust question, 1866–1918.

9. Ibid.

10. Ibid., record of transmittal and interview, J.M. Currier and Hon. A. Campbell, September 12, 1866.

11. Lambert and Pross, *Renewing Nature's Wealth*, 151–154 and 178.

12. PAC, RG23, file 1669, report of Horace Merrill, December 12, 1866.

13. Ibid., plan submitted by Merrill, March 4, 1867; report on ability of Victoria Foundry and machine shops to provide slab cutters, April 18, 1867; report by Detective E. O'Neill that mills taking steps to install machinery, November 19, 1867; and final reports by O'Neill, August 19–20, 1868. It is worth noting that in line with the political morality of the time, Merrill was a leading shareholder in the Victoria Foundry, the company that made the chipping equipment.

14. Canada, House of Commons, *Debates*, 1871, 171.

15. PAC, RG23, file 1669, various requests for exemption, 1871–1876.

16. Canada, Statutes, 36 Vict., Cap. 65, "An Act for the Better Protection of Navigable Streams and Rivers."

17. Canada, *Debates*, 1871–1872.

18. Ibid.

19. Ibid., see Currier's statement, 1871.

20. Details given in D.G. Creighton, *John A. Macdonald: The Old Chieftain* (Toronto: Macmillan, 1965), 158–160.

21. PAC, RG23, file 1669, various petitions from mill owners and conviction of J.R. Booth under Act, June 22, 1875; report of John Mather, June 1877; summation of meeting with Smith, December 20, 1877.

22. Ibid., representation of Ottawa mill owners for formal exemption, February 3, 1880, and official order-in-council, June 23, 1880.

23. Ibid., circular letter to all mill owners as result of complaint by Capt. Bowie of Steamship *Peerless*, November 2, 1886; and petition from residents of counties bordering the Ottawa River, January 29, 1878, February 19, 1878, and March 7, 1878.

24. Ibid. Complaint by Capt. Bowie of *Peerless*, November 2, 1886; and *Antoine Rattee v. Booth, Perley and Bronson*, Ontario Appeal Reports, vol. 14, L.R. 15, p. 188. Rattee was a boat owner below the Chaudière Falls in Ottawa. He contended in court and eventually won a claim that mill waste in the river prevented him from properly operating his boats and detracted from the value of his property. The mill owners considered him a shyster who lived off the damages collected from them. The Ontario Appeal Court thought otherwise, however, and decided in Rattee's favour. He petitioned frequently against the waste problem.

25. PAC, Bronsons and Weston Company Records, MG28III37, vol. 106, Bronsons and Weston to Thomas Murray, M.P.P., March 19, 1885; and RG23, file 1669, Ottawa committee of lumbermen to minister of Marine and Fisheries, September 22, 1894, recalling Ontario situation.

26. PAC, MC28III37, J. Tilton, deputy minister of Fisheries to Bronsons and Weston Comp., November 3, 1886; and RG23, file 1669, Bowie Complaint.

27. J.K. Johnson, ed., *The Canadian Directory of Parliament* (Ottawa: Public Archives 1968), 124.

28. Canada, Senate, *Sessional Paper*, no. 43C, submitted December 16, 1890; and RG23, file 1669, Notation, May 15, 1888.

29. Canada, Department of Public Works, report of chief engineer's office, Ottawa, 1888.

30. Canada, Senate, *Debates*, 1891, Senator Clemow on background to the Sawdust Bill, which he is sponsoring.

31. PAC, MG28III37, E.H. Bronson to Messrs. W. McClymont and Co., May 26, 1888.

32. PAC, RG23, file 1669, summation of Fleming Report, January 30, 1889; MG28III37, E.H. Bronson to Judge Brown, Glen Falls, New York, June 12, 1888; Bronson to T. Pringle, Montreal, November 29, 1888; Bronson to city clerk, Muskegon and Bay City, Michigan, February 4, 1890; and Bronson to Hector Langevin, minister of Public Works, May 22, 1888.

33. PAC, Sir John A. Macdonald Papers, MG26A, Macdonald to A.W. McLelan, lieutenant-governor of Nova Scotia, July 4, 1889; Macdonald to John Thompson, minister of Justice, July 1889.

34. For a full discussion of these issues, see Gillis, "The Ottawa Lumber Barons."

35. PAC, RG23, file 1669, Ottawa committee of lumbermen to minister of Marine and Fisheries, September 22, 1894.

36. Ibid. There is no doubt that the state of the river in this respect resulted from the dumping of human waste in it, but there was a feeling at the time that sawdust was also a contributing factor.

37. Canada, House of Commons, *Debates*, Sir Charles Hibbert Tupper on Amendments to Fisheries Act, pp. 4557–4566.

38. Canada, Senate, *Debates*, 1894, pp. 720–725; and PAC, RG23, file 1669, Petition A. Rattee, W.R. Bell M.D., F.X. Vaiade M.D., P. St. Jean M.D., and 715 others.

39. Canada, Statutes, 58–59 Vict., Cap. 27, "An Act Further to Amend the Fisheries Act"; and PAC, RG23, file 1669, Marine and Fisheries circular to mill owners, August 3, 1894.

40. PAC, RG23, file 1669, Ottawa committee of lumbermen to minister of Marine and Fisheries, September 22, 1894; report from W.C. Edwards on his mills, September 19, 1894; Buell, Hurdman and Co. to deputy minister of Marine and Fisheries, September 21, 1894; Bronsons and Weston Co. to deputy minister of Marine and Fisheries, September 22, 1894; E.B. Eddy Co. to deputy minister of Marine and Fisheries, September 22, 1894; and MG28III37, vol. 115, E.H. Bronson to A.M. Low, secretary, International Sulphide, November 18, 1890, December 17, 1890, January 12, 1891, and to Geo. M. Fletcher, president, International Sulphide, May 28, 1891.

41. PAC, RG23, file 1669, Ottawa committee to minister, September 22, 1894, and W.C. Edwards to Charles Hibbert Tupper, September 24, 1894.

42. Ibid., Hiram Robinson to Charles Hibbert Tupper, August 10, 1894.

43. Ibid., H.K. Egan (partner in the Hawkesbury Lumber Co.) to John Costigan, April 5, 1895; and deputy minister of Marine and Fisheries to Sir John Thompson, September 27, 1894.

44. Ibid., resolutions passed by citizens of Hawkesbury, November 19, 1894; Ottawa Board of Trade to minister of Marine and Fisheries, February 26, 1895; and Departmental memorandum by John Hardie, acting deputy minister of Marine and Fisheries, March 11, 1895.

45. Ibid., E.E. Prince, report regarding sawdust question on the Ottawa River, [July] 1894; W. Wakeham, report on Ottawa River sawdust nuisance, July 11, 1894; note from minister of Marine and Fisheries to Samuel Wilmot, July 25, 1894; survey by F. Vieth of Quebec, New Brunswick, and Nova Scotia mills, October 1894; report by F. Vieth on navigation of Ottawa River, [Autumn] 1894; departmental memorandum, John Hardie, acting deputy minister of Marine and Fisheries, March 11, 1895; and MG28III37, vol. 123, E.H. Bronson to Sen. Sanford, July 4, 1895.

46. *Ottawa Citizen*, November 12, 1897, report on death of John Kemp.

47. PAC, RG23, file 1669, circular from F. Gourdeau, deputy minister of Fisheries, September 27, 1897.

48. Ibid., see Hon. I. Tarte, minister of Public Works, to Hon. Sir L. Davies, minister of Marine and Fisheries, August 14, 1901.

49. Ibid., J.R. Booth to F. Gourdeau, deputy minister of Marine and Fisheries, August 13, 1901.

50. Ibid.

51. Ibid., Gourdeau to Booth, August 14, 1901; report of Fisheries overseer, R.C. MacQuaig, conversation with J.R. Booth, August 17, 1901; Gourdeau to Booth, August 23, 1901; Gourdeau to E.L. Newcombe, deputy minister of Justice, August 24, 1901; Newcombe to Gourdeau, August 26, 1901; and Gourdeau to Newcombe, September 13, 1901.

52. Ibid., Sir Wilfrid Laurier to Sir Louis Davies, September 17, 1901; PAC, Sir Wilfrid Laurier Papers, MG26G, Laurier to F. Gourdeau, September 24, 1901.

53. PAC, RG23, file 1669, Hon. I. Tarte to James Sutherland, minister of Marine and Fisheries, September 8, 1902.

54. The term "nation transformed" has been fastened on the late nineteenth and early twentieth centuries by two leading scholars in the field, R.C. Brown and R. Cook, in *Canada 1896–1921: A Nation Transformed* (Toronto: McClelland & Stewart, 1974).

Chapter Fourteen

Blighted Areas and Obnoxious Industries

Constructing Environmental Inequality on an Industrial Waterfront, Hamilton, Ontario, 1890–1960

Ken Cruikshank and Nancy B. Bouchier

In 1947, with apparent concern about the environmental hazards associated with a wartime industrial boom, town planner E.G. Faludi created a master plan for the port city of Hamilton, Ontario, a place affectionately known as "Steeltown." Faludi's plan was to undergird Hamilton's first comprehensive zoning regulations. It aimed to isolate residential districts from industry by designating existing neighbourhood areas as "declining," "blighted," and "slum," while identifying appropriate locations for the placement of "light," "heavy," and "obnoxious" industries.[1] Unfortunately, it proved insensitive to dilemmas faced by many working-class Hamiltonians who sought affordable housing nearby their industrial workplaces. According to the 1947 plan, a number of Hamilton's working-class neighbourhoods either did not exist by definition, or were blighted areas in need of transformation. Both views had serious social and environmental consequences for local residents.

The failures of Faludi's 1947 plan were not unique by any means. From the first years of Hamilton's rise as a major industrial port city, both private and public decision making had created environmental inequalities for the city's residents. Throughout the twentieth century, town planners like Faludi had repeatedly promised to make the city's growth more orderly and attractive, yet their initiatives did not solve the environmental and social problems caused by the city's urban growth and industrial development. Indeed, the efforts of planners often ratified and even exacerbated the burdens born by the city's working class residents. The environmental costs and benefits of urban and industrial growth were not equally shared or agreed upon by all who lived there.

Twentieth-century urban planners like Faludi sought to harness public authority in order to allocate separate spaces for different kinds of human relationships with nature. In thinking about urban nature, as Matthew Candy argues, historians need to be attentive to nature both as a "biophysical fabric" and as a cultural construct. Nature and natural processes played a role in shaping the development of cities, but they inevitably interacted with human conceptions of nature.[2] Hamilton's planners saw the environmental and social problems of the city in spatial terms. In their view, problems inevitably arose when private and public decisions jumbled together industrial, residential, and other land uses, and when urban dwellers did not have access to rural or even wilder forms of nature. They and their political allies presented their vision as a corrective to the chaotic, and therefore unhealthy, development of the city: they were allocating urban space in a more rational manner for the benefit of all residents.[3]

UNPLANNED HAMILTON: TO WORLD WAR I

The first step in planning, according to the famed American city planner and landscape architect John Nolen, was to undertake "a careful study of the underlying physical, business and social conditions of the city." Practical visionaries, urban planners prided themselves on working with, and building upon, the existing natural and human environments of the cities they aimed to transform.[4] By the time Hamilton's elites embraced town planning, their city had acquired a particular urban form based upon the area's natural terrain and a number of key private and public decisions. This urban form created different relationships between particular social groups and their environment, and different perceptions of what constituted a healthy or unhealthy

environment. We cannot make sense of the planners or their visions, or assess their influence upon environmental inequality, without first understanding the city's early unplanned history.

Settlements around Burlington Bay (renamed Hamilton Harbour in 1919) developed in ways that both accommodated and defied the physical landscape. Located at the western end of Lake Ontario, the bay area held a potentially strategic position within Ontario's transportation network. To the southeast of the lake lay the Niagara peninsula and access to the northeastern United States via Buffalo, New York. To the northeast lay Toronto and access to eastern Canada, while to the west shippers could reach southwestern Ontario and the American Midwest via Detroit. Although well situated, not all of the land around the bay could be developed easily. Deep ravines cut the northern shoreline off from inland settlements and transportation routes, and steep shale bluffs made it difficult to land goods on the shore. A high and relatively narrow ridge known as Burlington Heights, which separated the bay from an extensive marshland known as Cootes Paradise, undermined the usefulness of the western shoreline. The marshland had to be dredged and a passage carved through the heights to create the Desjardins Canal, which offered limited and still imperfect access to farming communities farther west. Of all the shorelines, therefore, only the flat, low-lying land on the southern shore, which was inundated with creeks, ravines, and inlets, offered accessible land for settlement and port development. Even here, however, settlement initially proved unpromising, because a thin sandbar on the eastern side of the bay prevented ships from entering the enclosed harbour from

Lake Ontario. By cutting a canal through the beach strip in the late 1820s, Hamilton's political and business leaders overcame the natural limits of the area, and transformed the southern shore into a bustling port.[5]

The residents of the port town never enjoyed a golden age of environmental equality. The terrain of the south shore influenced where people chose to live and how they used the land. The homes and enterprises of Hamilton lay squeezed between the bay and the Niagara escarpment (known locally as the Mountain), with its 300-foot limestone face, a couple of miles inland to the south. Within this area, as historians Michael Doucet and John Weaver have shown, wealthier residents claimed the high ground near the escarpment, and on a ridge that meandered from the escarpment toward the Desjardins Canal on the west end of the harbour. On that ridge, the city's mercantile and political leaders had their homes built on well-drained land that afforded them vistas of the city and harbour seemingly appropriate to their social ambitions. These homes would be high and dry, with drainage flowing to the lower lands on either side of the ridge, or toward the harbour. The town centre, with its administrative buildings and merchant houses, would develop nearby. Hamilton's working classes, on the other hand, settled on the low, flat, and poorly drained lands east of the ridge, or north and northeast close to the shoreline of the bay.[6]

The arrival of the Great Western Railway in the 1850s, connecting Hamilton to Toronto and the border cities of Niagara Falls and Windsor, reinforced environmental inequality. Because most elite homes were built well away from the harbour, the railway could be located as close as possible to the docks and warehouses of the port without disrupting the quality of life of Hamilton's wealthier residents. The Great Western Railway located its main yards, repair shop, and rolling mill along the western harbour front. The presence of this railway complex, in turn, attracted many of the region's metalworking firms to this area. Other industries, including oil refineries, soap factories, tanneries, and meat packers, located farther to the cast, at the end of several marshy inlets where they had easy access to rail and water connections and to supplies of water.[7]

The decision by Hamilton's city council in the mid-1850s about where to get their city's water shaped the environmental development of the city's waterfront tremendously. Following a string of fires and a cholera outbreak in the 1850s, the city sought a water supply more safe and reliable than area wells and creeks. Although several engineers recommended bringing drinking water in from the bay, a lead engineer consulted with two experts from New York's Croton waterworks and convinced city aldermen to build a waterworks system that took in water from Lake Ontario from a place 3 to 4 miles from the city's centre. This enabled city residents and factories alike to use the harbour as a sink for their wastes, without much affecting the city's drinking water.[8] Hamilton's water-borne waste particularly affected the ecology of the shallow inlets along the harbour's southern shore. Hamiltonian John William Kerr, the fisheries' inspector for western Lake Ontario, reported in the 1860s that fish caught in the Sherman Inlet tasted of coal oil emitted from two refineries at the water's edge alongside the railway tracks.[9] He also noted that dead ducks, muskrats, and fish covered in oil floated on the water's surface. Kerr estimated that about 360 gallons of sulphuric acid ran into the inlet every week.[10] Although the Fisheries Act of 1868 authorized him to stop industries that hurt the fishery by dumping waste into the water, his efforts to prosecute offenders found little support from the local business community, political leaders, or authorities in the provincial ministry.[11] Even so, he won a few convictions against companies, including an oil refinery, sheep skin tannery, and a meat packer.[12] As importantly in Kerr's eyes, he persuaded some local industrial manufacturers to improve their industrial waste-storage systems and at least dilute the waste with water before dumping it into the inlets.[13] These appear to have been minor victories, however, slowing the pace of—rather than preventing—the polluting of the inlet waters. By the mid-1880s, Kerr complained of an equally serious source of pollution—the "filth" flowing into the inlets and the shores of the bay from the city's sewer pipes.[14]

Environmental historian Joel Tarr argues that late nineteenth-century medical health concerns over household sewage often directed popular attention away from the wastes created by industry.[15] Inspector Kerr's observations and his reports to the Board of Health suggest that this also happened in Hamilton, where sewage floated on the water for all to see.[16] Although preliminary work on Hamilton's sewer system had begun in the 1850s, by 1876 only one house in ten had access to sewer frontage. Twenty years later, sewers served more than half the city's houses, and outdoor privies had become largely a relic of the past by the First World War.[17] Fecal and other matter doubtless had made its way to the bay as runoff and overflow from outdoor privies, but the gravity-based water-carriage sewer system ensured that residential and industrial waste flowed efficiently into the bay's waters.[18] This system first served Hamilton's wealthier

residential areas. By the late nineteenth century, the local officer of medical health advocated bringing sewers to other areas of the city to help solve the health problems plaguing working-class neighbourhoods.[19] The extension of the sewer system aimed to serve people of all social classes and benefit those workers living in poorly drained sections of the city. Yet those who lived near the shoreline and shallow inlets of the waterfront encountered new problems caused by city sewer building. Examining one sewer outlet in the north end of the city in 1886, a delegation from the Board of Health "found an accumulation of the most disgusting and filthy matter, which was being covered with earth."[20] It recommended that the city purchase the adjoining property and increase the pay of the men working at the outlet! Such conditions concerned local political leaders for reasons of public health, and because they feared landowners would sue the city for damages to their waterfront properties.

Although the city had a clean source of water pumped in from Lake Ontario, ice harvested for the refrigeration of Hamilton's food and for customers as far away as Buffalo came from the increasingly dirty harbour. The Board of Health "solved" the problem of polluted ice through regulation: it monitored ice quality, requiring ice to be harvested only in certain areas far from sewer outlets and other sources of pollution.[21] City officials found the damage to private properties caused by the city's residential waste a more difficult problem to solve. Worried about "nuisance" lawsuits brought against the city by owners of damaged property, Hamilton's city council bought up some of the most seriously polluted waterfront properties. The city then began to divert sewage away from privately owned inlets near more populated areas into new publicly owned ones that industry already had damaged badly. In the 1880s, for example, city engineers diverted the city's sewage from the small inlet at the end of Ferguson Street into the adjacent Coal Oil Inlet (at Wellington Street), a place already heavily polluted by a tannery and soap factory. They also diverted sewage into the Sherman Inlet, the very area where Kerr had prosecuted oil refiners and meat packers. When the city lost two nuisance lawsuits in the 1890s, it moved its sewage pipes to inlets even farther eastward.[22]

The way that city officials dealt with the heavily polluted land along the sewer outlet at the end of Ferguson Street would become a model for how Hamilton dealt with environmentally damaged areas of its waterfront. By the turn of the century, a local newspaper editor and alderman, John Morrison Eastwood, championed the reclamation of the city's north-end shoreline and inlets. He spearheaded a campaign to clean up the area by building a revetment filled with dredged material and other waste products.[23] To this end the city directed its scavengers to deliver "clean" garbage (free of rotting vegetable matter that would attract rodents) to the site. Using coal ash and all manner of refuse as fill, the city created a park for north-end residents on the reclaimed land. By the First World War, the city had filled much of the Coal Oil Inlet running east of Wellington Street, and it began similar work on parts of the Sherman Inlet. In contrast to Eastwood's park, however, much of this newly reclaimed land eventually was used for industrial and residential development.[24]

By placing sewer outlets farther to the east and reclaiming land from the inlets, civic leaders could encourage industrial development in the areas east of the Sherman Inlet and north of the main railway line. In 1902 and 1903, not coincidentally after voters rejected a municipal bonus to attract an industry, the council annexed more than 650 acres of land and created a special district with a low tax rate (based upon a rural rate) to promote the industrial development of the area.[25] Their plan worked; many industrialists took advantage of the tax break and placed their factories along the waterfront's northeastern shore. This happened just as increased competition and new industrial processes were creating demands for better coordinated flows of material, and as larger industries sought locations that had direct access to both railway and port facilities. In 1895, a company that would later amalgamate with others to become the Steel Company of Canada (Stelco), placed its blast furnace on Huckleberry Point between Sherman and Harvey's Inlets.[26] International Harvester, an agricultural tool manufacturer, joined the large steel manufacturer on the annexed industrial land.

City aldermen believed that these new industries would be less likely to complain of pollution damaging their property than private individuals. Indeed, new industries might even help the reclamation of the inlets. This happened in 1910, when the steel company began dumping its slag to fill the Sherman Inlet for the other manufacturers found there.[27] The shallow inlets, however, both before and as they were being filled, represented serious environmental hazards for the people residing nearby. Fire insurance maps for the city of Hamilton show that the city's early industries—such as tanneries and oil refineries—first sprang up alongside the inlets in the old port section of town. Residential neighbourhoods soon followed them.[28]

*****

In 1911, the areas most seriously affected by the polluted waters and shoreline were along Wellington and Sherman streets, the former of which had proportionately more labourers and fewer professionals, proprietors, or even clerks than most other Hamilton neighbourhoods. Around the seriously damaged and heavily polluted Sherman Inlet, occupational profiles reflect more diversity, the kind more typically found elsewhere in the city. Yet Sherman-area white-collar workers were underrepresented. Unlike other parts of the city, especially the southeastern neighbourhoods built upon the well-drained land nearer to the escarpment, the Sherman neighbourhood had few professionals or proprietors, other than local grocery store or restaurant owners. Whatever their social background, however, everyone who lived in the area faced exposure to significant environmental hazards.

[…] In the middle of the first great wave of immigration of southern and eastern Europeans, these neighbourhoods had the same British and Western European ethnic profile as the rest of the city. Yet things unfolded slightly differently east of the Sherman Inlet. There residential populations remained small. Since statistics for those neighbourhoods are not as reliable as older areas of the city, they must be examined with some care.

The story of Brightside, a small neighbourhood that had fewer than 50 households in 1911, suggests the nature of the city's development through the northeastern areas of the harbour.[29] In 1910, real estate developer W.D. Flatt proposed to develop Brightside at the base of Lottridge and Stipes Inlets, just east of Sherman Inlet, close to the Steel Company of Canada's site located just north of the railway tracks.[30] While he assured potential buyers that the

land was not as marshy as the surrounding inlets, and that their homes would enjoy the bracing winds of the harbour, Flatt was clear about the kind of neighbourhood that he aimed to create. The names selected for its streets—names like Sheffield, Birmingham, and Manchester—suggest the industrial character that the place was to take. Advertisements for the Brightside development emphasized the amounts of time and money homeowners would save by living close to their factory workplaces.[31] By 1911, it attracted industrial workers, and, in spite of the British street names, more people of Southern and Eastern European descent could be found there than in any other part of Hamilton. Those who moved there may have enjoyed an environment temporarily better than what workers had in the more developed Wellington and Sherman neighbourhoods, but they moved in at the very time when one property owner in the area was suing the city for damage to her property caused by the city's sewage.[32]

Those wealthier Hamilton residents who lived away from the harbour front and toward the escarpment distanced themselves from the environmental degradation accompanying the extension of the sewer system, and the industrial progress of a city that proudly advertised itself as the "Birmingham of Canada."[33] They mainly viewed the harbour as an environmental amenity—something pretty to look at from a distance—not as an environmental hazard. Between 1875 and 1900, many members of Hamilton's elite had grand summer homes built on the narrow beach strip separating the harbour from Lake Ontario.[34] By 1900, half the households listed in the city elite's blue book social directory spent their summers there.[35] Many more of them belonged to the socially restricted Royal

Hamilton Yacht Club, which placed its magnificent clubhouse alongside the beach strip canal.[36]

The city of Hamilton actively supported the development of the beach strip as an exclusive resort area for its wealthiest citizens by protecting it from development and by preventing the construction of boathouses and other facilities that would "interfere materially with the enjoyment of the Beach promenading."[37] After 1900, the construction of an electric street railway between the beach strip and the city began to alter the character of the area, as Hamiltonians who were less well heeled found an occasional weekend visit to its amusement park and swimming beach within their financial grasp. As late as the 1920s, however, middle-class social reformers voiced concerns that working-class children and their families— especially people from the industrial north end of town—rarely visited the beach strip, apparently because they were unwilling or unable to spend the money needed to get there by streetcar.[38]

In 1924, Hamilton's reform-minded newspaper, the *Herald*, voiced concerns because people from these same working-class families continued to swim in the bay near their north-end homes. Just the year before, city engineers and the local medical officer of health deemed the water in the area to be too polluted for safe swimming.[39] Yet despite such concerns about water quality, political leaders who sought to represent working-class constituencies pushed for more accessible recreational areas for workers and their families along the harbour. Eastwood Park remained a significant achievement for them, although its revetment wall meant that the area was not meant as beachfront land. Those who dove off this wall, like the competitors in the first British Empire Games held by

Hamilton in 1930, did so right next to a sewer outlet. Many children from the north end spent their days swimming amid the busy wharves and alongside sewer outlets with no sense of the dangers involved.[40] A grandson of one of Hamilton's most prominent boat builders and livery owners recalled fondly swimming there in the days of his youth: "I know people from other parts of the city were warning kids within an inch of your life, don't go near the water. Don't go near the water. In fact somebody said to me once, the north end kids hardly ever drown. It was the visitors from other parts of the area who did.... We just grew up with it. You grow up in your environment."[41] A small beach located where the Wellington inlet once had run inland offered a place for small children to play, although the conditions of the water in this area probably were not healthy, and it soon was condemned by the medical health officer in the early 1920s.

Better swimming could be had at Wabasso (later LaSalle) Park, a beach area on the bay's north shore that Hamilton had annexed from a neighbouring municipality for a nominal fee. City planners hoped that this area always would be less polluted than the industrially developed southern shore of the harbour, but, like the beach strip, Hamilton's working-class families had a hard time getting there. Renting a rowboat to make it across the bay was out of the financial reach of many people, as one north end resident recalls: "The north enders [didn't have boats,] these were all rented out to what we'd call 'rich people.' Like the people who lived up the hill on Bay Street. They were 'rich people.'"[42] The high cost of ferryboat rides also ensured that such a trip was a special occasion for working-class families from the north end: "[I]f we could have a big day on Sunday morning, the family would go down ...

and they'd take you across to LaSalle Park and boy could we have a picnic. And the ferry would pick you up and bring you back. Five cents a piece. Man, that was big money in those days ... it was a special treat if we went."[43] At the southern shore at the other end of the bay, however, those who worked in the waterfront's easternmost factories could spend their leisure time at Stewart Park, east of the industrial districts. At least until the First World War city sewage outlets had not touched this area's inlets.[44] But Hamilton's working-class families found it hard to get there too.

While swimmers found the harbour's waters to be increasingly dirty, working-class communities still enjoyed access to large undeveloped parts of the shoreline for other recreational pursuits. Those willing to take a hike along the water's edge could enjoy muskrat hunting or angling for the many types of fish that still could be found in and around the most easterly inlets that remained unfilled, and in the western marshlands of the bay known as Cootes Paradise. Working-class residents also ice fished with spears on the bay during the winter months, in spite of city efforts to ban the activity.[45] One north-ender born in 1907 remembered the fishing hut where his uncle spent the winter months during the building trade's off-season: "There was all kinds of fish huts. And I've gone in there with him lots of times, fishing for fish. And you'd just look down there and it'd be clear as a bell. You'd swear that it wasn't any more than 2 feet deep. Instead of that it was about 12 feet deep. You'd see the fish swimming down there. And you'd put a decoy down and attract the fish. Catch them."[46] Sometime during the First World War, a boathouse community (known as the "shack town") sprang up along the narrow shores along the western end of the bay and into the Cootes Paradise

marsh. Although doubtless prompted by housing shortages in the city, these makeshift homes gave their dwellers ready access to fishing, hunting, and swimming, along with escape from the crowded living conditions of some of Hamilton's older industrial neighbourhoods.[47]

By the First World War, then, a number of key developments shaped the city that Hamilton's earliest generation of planners were to build upon. Urban planners exaggerated the chaotic nature of the unplanned development; a certain logic or rationality did govern the development of the city. Changes in industrial structure and a number or key public policy decisions had encouraged the development of a large industrial district east of the old port and north of the original railway lines. There new industrial and working-class residential districts formed. In response to industrial and residential pollution, factories had reclaimed several northeastern inlets. Industrial location and pollution were just beginning to take their toll on working-class recreational access to Hamilton's harbour, making the provision of beaches and swimming facilities for north-end families a social planning problem for municipal officials. Without any formal town planning, the social and environmental design of the industrial city already had begun to take shape.

LEGITIMATING ENVIRONMENTAL INEQUALITY

In 1917 and 1919, two planners reflected on the future of Hamilton, seeking to rationalize its development while beautifying it. By that time the city's population had risen to 100,000, and, with the annexation of eastern areas, Hamilton's size had doubled. A new wave of immigration, while not diminishing the predominantly Anglo-Celtic origins of most of its people, brought in the first significant numbers of immigrants from other ethnic backgrounds. Public health officials and social reformers warned of the dangers of juvenile delinquency and crowded and filthy housing conditions in the new industrial city. They used data from the 1913 *Report of a Preliminary and General Social Survey of Hamilton* to support their case.[48] This sense of rapid change encouraged local politicians to sponsor town planning experts to turn their professional gaze upon Hamilton and think about solutions to the problems plaguing the city.

In 1917, Canada's pre-eminent town planner, Noulan Cauchon, a man influenced heavily by the City Beautiful movement, produced a grandiose urban design for Hamilton. It featured garden suburbs, a high-speed electric commuter railway, and a boulevard from the bay to the escarpment, ending at a Greek theater built where a gravel pit had disfigured the escarpment.[49] His elaborate parks system would provide "the lungs of the city," and would include a "wilder and freer" parkland around the western end of the harbour and the Cootes Paradise marsh. This would be an area that, in the words of historian Nicholas Terpstra, "allowed access to the unsullied realm of nature for citizens bound up in the urban realm of culture."[50]

Unsullied nature, however, was to be carefully constructed and framed by the arches, colonnades, and balustrades of a proposed new northwestern highway entrance to the city. What did such a plan mean for the rest of the city? In a subsequent report, Cauchon proposed enhancing the industrial district that had developed in the city's northeast end.[51] He contended that the city should reclaim all of the

eastern inlets, filling them to make room for even more industrial development. Much of Hamilton's waterfront, beginning in the old north-end port and extending almost to the beach strip, thus would be transformed. In the process it would become inaccessible to local people. Cauchon proposed that a narrow scenic lagoon separate the industrial district from recreational areas on the beach strip. He intended that only three areas—the very westernmost end of the harbour, the north shore, and the beach strip—should be kept for residential and recreational purposes. Even then, he argued that nature had to be carefully constructed to produce the "appropriate" aesthetic and recreational response. The rest of the waterfront would be left to development by the city's growing industries.

In 1919, the chief engineer for Toronto's Harbour Commission, E.L. Cousins, produced a second planning report on the bay, this time for the Hamilton Harbour Commission, the special federal agency created in 1912 to govern the harbour.[52] Surprisingly, given the commission's mandate to develop Hamilton's port capacity, Cousins emphasized the importance of balancing the harbour's recreational, residential, and economic uses. He maintained that a large industrial city's citizenry needed plenty of areas set aside for their healthy outdoor recreation. He declared that the beach and the north shore of the harbour were areas that, "by their physiography lend themselves admirably to aesthetic treatment"; they were areas to be saved for parks and outdoor amusement.[53] Like his predecessor Cauchon, Cousins supported other waterfront development for port and industrial purposes, such as reclaiming the southeastern inlets and creating more shoreline through infilling. Cousins, too, sought a "clear cut line between the industrial area and beach development." But instead of creating a scenic lagoon between the two, the more practical engineer favoured separating the areas with a narrow ship canal and turning basin. Even more enthusiastic about reclamation than Cauchon, Cousins wanted to make room for public recreation and more housing on the beach strip. He planned to fill 172 acres of its bay side and reclaim enough land from all sides of the harbour—even at the foot of the bluffs of the north shore—for a scenic parkway for automobiles along the waterfront.

The town planning reports of both Cauchon and Cousins reflected a faith that, through the careful and precise designation of the harbour's various functions, industry, homes, and recreation could continue to coexist there. Neither planner saw any difficulty, for example, in having only a narrow lagoon and ship canal separating heavy industry on the southeastern shore from a recreational resort area on the beach strip. Neither thought that the massive industrial site proposed for the southern shore would endanger people's recreational enjoyment along the other shores. They were not alone in their belief. Hamilton's city council, which had acquired LaSalle Park on the north shore for public swimming, felt the same way. So too did groups like a new local angling club, which suggested stocking the area with fish for its sport. How could pollution possibly spread from the south shore across the water? The goal of preserving an "unsullied realm of nature" meant reconstructing natural areas to make them aesthetically pleasing and suitable spots for passive or organized public recreation. In the western end of the harbour, Cauchon and Cousins worked to create places of refuge from the city, with carefully tended gardens that would

encourage contemplative recreation. These designs limited the creation of places for active recreation—amusement parks, boating facilities, and carefully supervised swimming areas—to the waters of Lake Ontario at the beach strip.

Although the two town planners worried about some social issues, especially the need for housing for the city's ever-growing population, their overall visions focused little on the future of existing working-class communities. Their plans for the industrial development of the northeast sanctioned the existing processes of reclamation and advocated even more extensive infilling of the harbour. They cut both traditional and newer working-class neighbourhoods off from their access to the water, leaving them in the shadow of industries situated on the acres of reclaimed land. In spite of their lip service to varied kinds of recreational areas, the large waterfront parklands that already existed—like Eastwood and Stewart parks—disappeared from their planning. So, too, did a smaller beach in the north end. The inlets farthest east, which had not yet been damaged by development, would also eventually disappear. The residential communities and parkland Cauchon and Cousins envisioned on the bay's north shore likely would not attract—nor were they designed for—the working-class families of Hamilton's waterfront neighbourhoods. The neglect of working-class interests was lamented in a letter to the editor of the *Hamilton Spectator*. "To the plutocracy of the third manufacturing city of the Dominion," the author wrote, "it possibly does not appear that their poorer neighbours suffer a hardship having to gravitate to Wabasso park, by paying a boat fare or traverse the road about nine miles, for a day's outing. The former have their autos, the latter ... more often than not a sadly depleted pocket book to meet the expenses of such an outing."[54]

Both the audacity and expense of Cauchon's and Cousins's designs for the harbour ensured that the city never would adopt either plan fully. Nevertheless, their views of industry, nature, and recreation proved influential for generations to come. They helped transform much of the city's north end into an industrial district and sanctioned the reclamation and infilling of the harbour. In effect, the vision of town planners legitimated designs of many industries, including the Steel Company of Canada's. By building porous retaining walls and dumping slag into the harbour to create land for its steel plant, the infilling program permitted it to expand southward gradually toward the railway lines, and northward into the bay. Throughout the 1920s and 1930s, members of the city's parks board championed at least part of the planners' vision: They sought to create what they envisioned as a carefully regulated, aesthetically pleasing, and morally clean park along the western shoreline of the harbour.[55] They engineered land deals ensuring that the Cootes Paradise marsh adjoining the western harbour and much of the northwestern shoreline would become public land, beyond the reach of industrial or real estate developers.[56] They hoped to create a much more extensive development in the western end of the harbour. It would include a picnic area, a restaurant, a bandstand, a model yacht pond, botanical gardens, a zoo, and an automobile park.[57] Depression-era finance, however, precluded the creation of all but a sedate rock garden constructed by relief workers out of an abandoned gravel pit. It would form the basis of Hamilton's world-famous Royal Botanical Gardens, which had both cultivated gardens and a nature preserve. There Hamilton residents could birdwatch, but could not hunt or fish.[58] Although the prohibition on hunting in the area applied equally to all, it most negatively affected working-class families that relied upon area game for their dinner tables. They could not afford—either because of time or money—to travel long distances from the city to hunt.[59] This restricted their access to a principal resource offered by the harbour environment. During the 1920s, 1930s, and 1940s, the city ended up demolishing the boathouse homes in the working-class shack town neighbourhood that had developed along the western edge of the bay to make way for the extensive park complex that never quite materialized. The city compensated some boathouse dwellers—those who had a legitimate claim to reside on the land and were not squatters—for the loss of their homes. But residents received no compensation for their loss of ready access to hunting, fishing, and swimming, which had made living there so appealing.[60]

The town plans of Cauchon and Cousins also sacrificed north-end waterfront recreational areas to industry, something that north-end families fought against to preserve the limited access to the waterfront that they still possessed. People in working-class neighbourhoods lobbied their aldermen to resist the planners' visions. The aldermen, in turn, persuaded a reluctant city council to create a new swimming beach at Bay Street in the city's old north-end waterfront. Although the Board of Health had condemned another beach in the area because of its dirty water in the early 1920s, the city council eventually conceded defeat when local residents persisted in swimming there. Without officially recognizing the area as a beach, the council gave way to public demand in the 1930s, agreeing to provide lifeguard and concession services there. Farther to the eastern end of the harbour, the parks board sold Stewart Park to the city council in exchange for park space elsewhere. Stewart Park remained open space for only a few more years before being transformed into industrial land.[61]

By the Second World War, the pollution from the city's growing waterfront industrial district defied the careful functional mapping of the harbour envisioned by planners, a vision that put everything—shipping, industry, and recreation—into orderly, self-contained places. The environmental effects of wartime industrial growth could not be contained. [...] Between 1912 and 1957 the expansion of Hamilton's industrial giant Stelco alone increased the number of tons of industrial waste entering the harbour enormously.

This increase in industrial wastes happened while an aggressive program of infilling decreased the bay's area by roughly one-quarter. The tonnage of phenols, cyanide, ammonia, sulphuric acid, and oils emitted from the plant increased dramatically. Stelco, however, cannot be singled out as the sole culprit. During the interwar years, newspaper reports highlighted the bad situation by pointing to acid and oil wastes, sometimes diluted with water, which also flowed from the Hamilton By-Product Coke Ovens and two smaller metal fabricating companies on the waterfront.[62] City engineers also recorded that dyes, fats, and grease—presumably from textile mills and meat-packing companies in the area—flowed from private sewers directly into the harbour.[63] Small oil slicks made this pollution quite visible. Particularly during the spring thaw, as an editorial in local paper put it sarcastically, "the ice of the bay melts to the amorous kiss of warm spring suns and the congealed oils and waste release their aromatic delights."[64] Yet public controversy

over the polluting of the harbour typically ended with the conclusion that, until the city addressed the dumping of untreated residential sewage into the harbour, it could take no concerted action against industrial polluters.[65]

For sure, the accumulation of residential and industrial wastes undermined recreational uses of the harbour. A provincial investigation in 1943 estimated 70 million gallons of industrial waste and 25 million gallons of residential sewage entered the bay every day, much of it untreated.[66] A few years later the local health officer reported that the coliform count in parts of the harbour had increased 700 percent since 1923.[67] Perhaps, as the town planners might have realized, beaches on the harbour's south shore were unsustainable; local health authorities had closed all of them by the end of the Second World War. Even bathers on the eastern beach strip complained that they could not swim in the bay without becoming coated in oil. This led them to take their plunges into the colder waters of Lake Ontario.[68] Combined with traffic congestion from the steady stream of cars and trucks crossing the beach strip en route between Buffalo and Toronto, pollution led wealthier Hamiltonians to give up on summering there. As for the north shore, local health authorities outlawed swimming at LaSalle Park in 1946. The city's medical officer of health warned that this meant pollution was "getting beyond the current which travels down the middle of the bay to the canal," which, at one time, he argued, "acted as a barrier, keeping the polluted water confined to the Hamilton side of the current."[69] An editorial writer in the *Hamilton Spectator* noted with some resignation that from a distance the bay's waters looked attractive and inviting, yet "nearby it is seen to be dirty and flecked with foulness, the pollution of its waters being the price we pay for modern urban life."[70] While noting that doing more to control the city's waste products would necessarily increase municipal and manufacturing costs substantially, the writer argued that the effort was important for civic pride: "We pay for things in other ways besides money.... A little but essential sacrifice will prevent Hamilton's appearance from growing as shabby as a European refugee's."[71]

To some extent, all Hamiltonians suffered from the public's loss of access to the waterfront and the polluting of the bay's waters. Nevertheless, those who lived closest to the harbour, or who now lived next to the industries that cut off access to the harbour, were the most clearly affected. In their colourful rendition of the days of their youth entitled *Tales from the North End*, raconteurs Lawrence and Philip Murphy write in some depth about how the rise of heavy industry, infilling, and the fencing off of the waterfront affected both the lives of area residents and the cohesiveness of their working-class community.[72] For many north-enders, the harbour had ceased to be much of an environmental amenity. Increasingly, it became an environmental hazard. The neighbourhoods created on or near inlets that had been filled in by the end of the Second World War with slag, coal ash, and other potentially toxic materials became even more distinctly working class in comparison to the rest of the city than had been the case at the turn of the century.

[...] From the Sherman Inlet area eastward, these neighbourhoods, especially the industrial suburb of Brightside and another housing project in the vicinity of Gage's Inlet, had more semi-skilled factory workers and more Italian and Eastern European residents than most other parts of the city. Although these neighbourhoods grew up "on the other side of the tracks,"

squeezed between filled inlets and not far from a waste-disposal site, they provided tightly knit communities for working-class families in the city's industrial northeastern end. For those Hamiltonians who found it necessary to live close to their workplaces, their homes were not in an orderly and inspiring setting, but in some of the city's most environmentally hazardous areas.

Powerful and tax-averse interests undermined the grandiose features of the visions offered by town planners. The full development of their plans, however, would have offered even less to Hamilton's working-class neighbourhoods than their partial implementation. The planners had envisioned even more infilling of the kind that generated environmental hazards and attracted heavy industry. Planners had given little consideration to local access to waterfront resources: their view, partially implemented, still ensured that some working-class residents lost access to their hunting and fishing resources, while others even lost their homes. Even imperfectly realized, the abstract vision of the planners had real social consequences: they legitimated and deepened environmental inequality in the city.

HAMILTON'S "MASTER PLAN"

During the Second World War, a second wave of enthusiasm for town planning swept through Hamilton. Like other places throughout Ontario, Hamilton's political leaders hired town planning consultants to draw up a master plan to ensure their city's future growth in a new era of city zoning.[73] Hamilton's city council, which by this time had a committee on bay pollution, already had engaged in some limited, piecemeal zoning: For example, it prevented some industries from locating in specific neighbourhoods. The new comprehensive zoning regulations authorized by the provincial government, however, aimed to support an overall vision of the city, something that the local press claimed was "far from being a mere blueprint of Utopia. It deals realistically with those problems which hinder the development of Hamilton and offers practical solutions."[74] While intended to be more precise and practical than the grand urban visions of earlier planners, master plans and zoning regulations continued an important tradition. They carefully designated spaces for industry, homes, and recreational areas. By doing so, planners aimed to bring order and rationality to what they viewed as chaotic development. The master plan thus aimed to provide "a complete picture" of the city, with the zoning bylaw to be "its tool" for the managing and directing of Hamilton's growth and development.[75]

Beginning in 1944, E.G. Faludi, planning director of the firm Town Planning Consultants, worked with the City's Planning Committee to prepare a master plan for Hamilton.[76] He began by organizing interviews with various representatives from the city's business, labour, professional, and social welfare organizations. He also gathered statistical information to assess the condition of the city's neighbourhoods. Not surprisingly, this process reinforced existing divisions within the city. Faludi identified 16 residential districts composed of some 32 neighbourhoods. He used 10 criteria to determine their condition—including features such as the physical condition of buildings, the number of derelict properties, the acreage covered by buildings, the number of buildings used for non-residential purposes, the quality of sanitation services, the availability of

recreational facilities, the amount of heavy traffic moving through the area, and the population density. Faludi concluded that 26 percent of Hamilton's neighbourhoods were sound, 49 percent were declining, and 26 percent were blighted.[77]

Of the north-end areas, Faludi considered all but two blighted. He did not deem the remaining two—Brightside and the Gage Inlet area—to be "recognizable neighborhood communities" at all. But he was wrong. In the case of Brightside, hundreds of Eastern European and Italian factory workers took great pride in their homes and neighbourhood. In 1977, some three decades after their homes had been demolished to make way for more industrial development, some 900 people returned to their old neighbourhood for a reunion. There they saw old friends and had the chance to reminisce about their once-vibrant community, fondly remembering their fruit and vegetable gardens, local grocery, and local tavern, the Brightside House.[78] Ironically, property owners from the north side of Brightside's Birmingham Street had inadvertently helped industry encroach into their neighbourhood when they petitioned the city's board of control to fix the Lottridge Inlet nuisance by getting Stelco to fill it in.[79] In Faludi's view, neighbourhoods like Brightside, so close to industrial factories, ought never to have existed in the first place.

How would the master plan deal with the blighted communities in the new age of urban zoning? Faludi suggested an "industrial" designation for the land between the shoreline and the railway tracks from the area of the old Coal Oil Inlet (where the city first began its reclamation work) eastward.[80] In all, 352 acres of blighted residential areas and more than 1,000 acres of nearby vacant land would become zoned for industrial use. In the areas of Wellington and Oliver streets, light industry would replace residential homes. The new zoning would permit heavy industry east of Sherman Street.[81] The 1,000 acres of vacant land identified included inlets and harbour shoreline to be reclaimed for industrial purposes. Therefore the master plan sanctioned reclamation and the infilling of the harbour by industry and the Harbour Commission, and the complete transformation of Hamilton's northeastern end into an industrial district. Faludi also considered the question of "obnoxious industries," and offered a traditional spatial solution to the problems they posed. These industries, he argued, should exist beyond the city limits, still farther to the east. Hamilton's final zoning regulations reflected the town planner's proposals and designated the entire area east of the Wellington area as "light" or "heavy industrial."

In the town planner's view, only the old north-end port area could be kept as a residential area. Faludi proposed a substantial redevelopment of this area, largely through cleaning old buildings and constructing new homes. He wanted to leave the Bay Street beach as a recreational area, but public health officials closed it for swimming just as the planner was developing his report. Faludi also wanted Eastwood Park to remain as a waterfront recreational area. In suggesting this, however, he ignored the fact that the docks and naval barracks built during the war already had cut the park off completely from the water. In its final zoning regulations, Hamilton's city council designated the entire blighted north end as a maximum-density residential district. This opened the way for the construction of apartment buildings and multiple unit housing, as well as single homes on small lots.

The idea that the waterfront might also serve as a recreational site for Hamilton's citizens seems to have been lost in the 1947 master plan and its accompanying zoning regulations. Although Faludi argued that the city needed a waterfront park with a bathing beach, swimming pool, and restaurant, he planned these for the shores of Lake Ontario—not the bay—in an area at the southern end of the beach strip. Hamilton's Harbour Commissioners concurred with the master plan's suggestion that the harbour and its waterfront no longer be considered suitable for residential or recreational uses. A 1957 Harbour Commission plan, proposed on the eve of the opening of the St. Lawrence Seaway, reveals the extent to which recreational uses of the harbour had taken a back seat to industry and shipping. It proposed the continued infilling of the southeastern shore for industrial land and wharfage. In contrast to most earlier planning documents, the Harbour Commission also proposed the extensive development of the western part of the bay. It aimed to fill in much of the southwestern shore and construct a high-level bridge straight across the harbour to the north shore. This would serve as a trucking route for Toronto-bound vehicles.[82] The plan also called for filling in the bay side of the beach strip at the other end of the harbour to provide even more wharfages. Finally, it called for 50 acres of reclaimed waterfront property just north of the canal to house a thermonuclear power plant. While many aspects of this 1957 plan never took form, it suggested the extreme nature of planners' visions for Hamilton's waterfront in the years just following the war.[83]

In the years immediately following the Second World War, therefore, planners and their allies saw little room for natural areas in the city. Previous efforts to allocate separate spaces in the city for different kinds of human relationships with nature had been undermined; using one part of the harbour as an industrial and residential septic tank had serious implications elsewhere. Nevertheless, post-war planners retained the belief that the different functions of the city and the waterfront could be compartmentalized, creating separate spaces for industrial, residential, and recreational uses of nature. The problem, in their view, was that previous planning efforts had been incomplete, resulting in multiple and conflicting uses in different parts of the city. They saw effective zoning as the means to achieve a carefully ordered city.

What did planning and zoning mean for the working-class residents living in the north end of Hamilton? On the one hand, the plans did not seem to change very much since the tensions between industrial development and residential neighbourhoods predated their existence. On the other hand, the official zoning of many neighbourhoods as industrial had important repercussions for the lives of people living there. As Yale Rabin and others argue, zoning decisions legitimated industrial encroachments on neighbourhoods. They reduced the value of homes that remained in the area, and they made it unlikely that banks would provide loans to area residents for home improvements or maintenance.[84] Whether they had been designated as blighted before the zoning was inconsequential, neighbourhoods were bound to become so after being designated as industrial. Such a designation reduced the values of homes in Hamilton's Brightside, making it easy for the Steel Company of Canada to snap them up from anxious homeowners who wanted to cut their property losses. This enabled Stelco to expand its massive complex

both southward into the community and northwards. Its aggressive program of infilling provided new land for a 1,000-foot dock, 83 new coke ovens, a new blast furnace, and four 250-ton open-hearth furnaces. This infilling and zoning also helped transform a once relatively small harbourfront foundry into Hamilton's second integrated iron and steel maker.[85]

Zoning at least forced some working-class families to move away from environmentally degraded areas, although those who had to move did not necessarily see the move as positive. Many people felt strongly about where they lived, as one steelworker who, in 1978, was told that the area where he had grown up and recently purchased his own home was "appalling" and slated to be demolished. "But I've lived around tracks all my life and so has my wife," he complained. "It doesn't bother me ... My house looks good inside and out. It's well kept and so are my neighbours.'"[86] For still other working-class families, industrial zoning further reduced the quality of their neighbourhoods. While zoning was intended to eliminate the mixing of land uses, the effect of the policy was to sanction further industrial development in some neighbourhoods, and to reduce the likelihood that those areas would acquire the same recreational facilities or other forms of community improvement offered in formally recognized residential areas.

While a vision of the harbour as an industrial port meant that all Hamilton residents lost an environmental amenity, working-class residents of the areas closest to the harbour felt the effects of this loss most sharply. Their children were more likely to swim in the seriously degraded water of the harbour, or to play in yards and vacant lots that once were polluted inlets, filled in with potentially toxic or dangerous materials. Further, their families were the least likely of Hamilton's population to be able to afford to escape the city to cleaner spaces in nature, or to pay for indoor recreational facilities. The master plan and the designation of urban space through zoning—advocated and supported by urban planners—legitimated and even deepened environmental inequalities in the city.

Formal urban planning did not create environmental inequality in Hamilton; as the city developed in the nineteenth century, those holding social, political, and economic power had the resources to acquire what they viewed as the most healthy places in the city for their homes, offices, and recreational spaces. By the time planners surveyed the city, key public and private decisions had ensured that the waterfront would develop as an industrial zone, particularly in the northeast of the city. Planners supported and legitimated policies consistent with the urban form the city already had acquired. They called for the creation of an industrial zone in the northeast, the reclaiming of the inlets and the filling in of the harbour with residential and industrial waste, the location of obnoxious industries to the east beyond city limits, and the identification of the north shore and beach strip as potential recreational sites. The eastern limits of the city changed over time, as did the recreational potential of the northern and western waterfronts. In sanctioning existing and emerging divisions within the city, planners did nothing to counter existing and emerging environmental inequalities.

At least into the 1960s, urban planners believed they could divide the city into tidy spatial compartments, so that Hamilton residents could establish a different relationship with nature in the city's different parts. This truly was abstract thinking, which worked better on paper maps than on real social and natural landscapes. Industrial areas were never just separate places for industry, they also were neighbourhoods of working-class families. Industrial and residential areas produced wastes, and containing those wastes to any one part of the city proved difficult. Planners gave little if any thought to the impact of extensive infilling of the harbour and its inlets on the overall waterfront environment. All Hamilton residents now suffer from the limits of the planners' visions, which legitimated the development of their city as a particular kind of industrial port, but the working-class citizens who live, work, and play near the water's edge suffer the most.

NOTES

The authors wish to thank the anonymous reviewers for their helpful suggestions, and we acknowledge the support of the Social Sciences and Humanities Research Council of Canada and McMaster University's Arts Research Board for their financial support of this research. This paper represents part of a larger study of the interaction of environmental change, regulation, and popular use of Hamilton Harbour.

1. F.G. Faludi, *A Master Plan for the Development of the City of Hamilton* Hamilton (City Planning Committee, 1947); "Now That Was a Plan!" *Hamilton Spectator* (January 5, 1980).

2. Matthew Glandy, *Concrete and Clay: Reworking Nature in New York City* (Cambridge: MIT Press, 2002), 7.

3. For an excellent study of the changing ideas of urban planners, see M. Christine Boyer, *Dreaming the Rational City: The Myth of American City Planning* (Cambridge: MIT Press, 1983).

4. Boyer paraphrases Nolen in *Dreaming the Rational City*, 74. For an overview of Canadian literature on urban planning, see "Town Planning," in Harold Kalman, *A History of Canadian Architecture*, vol. 2 (Toronto: Oxford University Press, 1994), 643–676; Alan Artibise and Gilbert Stelter, eds. *The Usable Urban Past: Planning and Politics in the Modern Canadian City* (Toronto: Macmillan, 1979); Gerald Hodge, *Planning Canadian Communities: An Introduction to the Principles, Practice, and Participants* (Scarborough: Nelson Canada, 1991); Thomas I. Gunton, "The Evolution of Urban and Regional Planning in Canada: 1900–1960," Ph.D. dissertation, University of British Columbia, 1981; Walter Van Nus, "The Fate of the City Beautiful Thought in Canada, 1893–1930," in *The Canadian City: Essays in Urban History*, edited by Gilbert A. Stelter and Alan F.J. Artibise (Toronto: Macmillan, 1979), 162–185.

5. S.B. McCann, "Physical Landscape of the Hamilton Region," in *Steel City: Hamilton and Region*, edited by M.J. Dear et al. (Toronto: University of Toronto Press, 1987), 30–33; L.J. Chapman and D.F. Putnam, *The Physiography of Southern Ontario* (Toronto: University of Toronto Press, 1973). On Hamilton's growth and development, see John Weaver, *Hamilton: An Illustrated History* (Toronto: Lorimer, 1982); Nicholas Terpstra, "Local Politics and Local Planning: A Case Study of Hamilton, Ontario, 1915–1930," *Urban History Review/Revue d'Histoire Urbaine* 19 (October 1985): 114–128; Michael Doucet and John Weaver, *Housing the North American City* (Montreal and Kingston: McGill-Queen's University Press, 1991). For an overview of environmental changes in the harbour related to development, see Mark Sproule-Jones, *Governments at Work* (Toronto: University of Toronto Press, 1993), 135–142.

6. Doucet and Weaver, *Housing the North American City*, 446–466; Weaver, *Hamilton*, 31–32, 60.

7. Weaver, *Hamilton*, 59–68, 96–99; R. Louis Gentilcore, "The Beginnings: Hamilton in the Nineteenth Century," in Dear, *Steel City*, 108–118.

8. William and Evelyn M. James, *"A Sufficient Quantity of Pure and Wholesome Water": The Story of Hamilton's Old Pumphouse* (London: Phelps, 1978).

9. Hamilton Public Library (hereafter HPL), Special Collections, John William Kerr Diaries, v. 5, p. 5, June 14, 1871, April 1866, vol. 2, 4, 8. The Kerr Diaries provide much evidence of early industry and industrial development on and along inlets, which tends to be ignored in other analyses. See, for example, incidents in the larger watershed area, v. 7, pp. 2–26, September 29, October 1, 1876.

10. Kerr Diaries, v. 4, p. 8, May 14, 1870; v. 6, p. 3, April 17, May 21, 1873; v. 7, p. 3, October 26, 1876.

11. Ibid., v. 6, p. 3, May 30, 1873. Kerr often found himself between a rock and a hard place in his prosecution of fishing violations as well. In June 1870, for example, he confided that he "never felt smaller in his life" after Judge Logie refused to hear his evidence against a local fisherman (v. 4, pp. 9, 20, June 1870).

12. Ibid., v. 6, pp. 4–7, July 1, 1873; v. 7, p. 2, September 29, 1876.

13. Ibid., v. 5, p. 5, June 14, 1871.

14. Ibid., v. 13, p. 35, March 24, April 20, 1886; v. 15, p. 17, March 6, 1888.

15. Joel Tarr, "Industrial Wastes, Water Pollution and Public: Health, 1876–1962," in Joel A. Tarr, *Search for the Ultimate Sink* (Akron: University of Akron Press, 1996), 356–357.

16. Hamilton Board of Health *Minutes*, vol. 1, September 7, 1886, 47. See, for example, vol. 1: September 20, 1886, 4; April 5, 1887, 53; October 5, 1887, 54.

17. Doucet and Weaver, *Housing the North American City*, 442–443; Elizabeth Bloomfield, Gerald Bloomfield, and Peter McCaskell, *Urban Growth and Local Services* (Guelph: University of Guelph, 1983), 104–105.

18. For one Hamiltonian's account of the negative effect of sewage on the bay's health, aesthetic, and recreational qualities, see the letter to the editor, *Hamilton Spectator* (November 23, 1887).

19. See, for example, Hamilton Board of Health *Minutes*, vol. 1, December 1, 1884, 15; December 6, 1892, 164.

20. Hamilton Board of Health *Minutes*, vol. 1, September 20, 1886.

21. "About Ice," *Hamilton Herald* (January 21, 1895); "Hamilton Ice Dealers Have Fine Harvest," *Hamilton Spectator* (January 31, 1920).

22. "Coal Oil Inlet," Hamilton Harbour Scrapbook, vol. 1, part 1, p. 201: Ontario, Divisional Court of High Court of Justice, *Susan Stipes versus City of Hamilton*, January 29, 1912, reprinted in Hamilton City Council *Minutes* (hereafter HCCM) (1919), May 2, 1919, 106–110.

23. On ashes being dumped into Land's Inlet, see Kerr Diaries v. 12, p. 25, October 22, 1884. Eastwood of the *Hamilton Times* never won office as Liberal provincial candidate in Hamilton East. On his local political and social work, see S. Patricia Filer, "John Morrison Eastwood," in *Dictionary of Hamilton Biography*, Vol. III, edited by T. Melville Bailey (Hamilton: W.I. Griffin, 1992). See also, Annual Reports of the Board of Health, 1905–1906, 20; 1906–1907, 30.

24. "Coal Oil Inlet Must Be Cleaned," *Hamilton Spectator* (August 22, 1906); "Four Nuisances Found at Inlet," *Hamilton Spectator* (May 31, 1907); "Coal Oil Inlet," *Hamilton Spectator* (March 3, 1908); "The North End of Hamilton Is Coming into Its Own at Last," *Hamilton Times* (November 17, 1910); "Great Improvements Have Been Made in the North-End of the City this Year," *Hamilton Spectator* (November 19, 1910); "Passing out of Coal Oil Inlet," *Hamilton Spectator* (July 28, 1911); "Hamilton Harbour as It Was and Is," *Hamilton Herald* (June 2, 1922); "To Stop Garbage Dumping," *Hamilton Herald* (May 11, 1933).

25. Weaver, *Hamilton*, 88–89.

26. William Kilbourn, *The Elements Combined: A History of the Steel Company of Canada* (Toronto: Clarke Irwin, 1960); June Corman et al., *Recasting Steel Labour: The Stelco Story* (Halifax: Fernwood Publishing, 1993).

27. "Passing out of Coal Oil Inlet," *Hamilton Spectator* (July 28, 1911).

28. *Hamilton Ontario Fire Insurance Map* (1878), National Map Collection of Canada (hereafter NMC1 Microfiche #88304-5, 16617, 1" = 50'; *Hamilton Ontario Fire Insurance Map* (January 1898; revised 1911), 1" = 50' Lloyd Reeds Map Collection, Mills Library, McMaster University, Hamilton Ontario; *Hamilton Ontario* (1927, revised 1933) NMC microfiche #10742; Hamilton, Ont. (revision slips, special revision embracing manufacturing risks only Feb. 1941), NMC microfiche #10590; *Insurance Plan of the City of Hamilton, Ont.* (Underwriters' Survey Bureau Ltd., May 1927 revised to October 1947), 1" = 1,000' NMC Microfiche #9880; *Hamilton, Ontario* (revision slips Oct. 1947) NMC microfiche #103599. See also, "Maps for the Environment," in Diane L. Oswald, *Fire Insurance Maps: Their History and Applications* (College Station: Lacewing Press, 1997), 83–88.

29. "Brightside," *Hamilton Spectator* (July 12, 1975); Lloyd Reeds Map Collection, Mills Memorial Library, McMaster University, Hamilton, Ontario: Property Surveys, W.D. Flatt, "Brightside, 1910."

30. Lloyd Reeds Map Collection, Mills Memorial Library, McMaster University, Hamilton Ontario. (NMC 0015354). D. Nicholson, *Map of the City of Hamilton* (Hamilton: Canadian Records Company, 1912), 1:3,600 ft.; J.D. Nicholson, *Greater Hamilton. Comprising the Township of Barton. All Present Subdivisions and Proposed Layouts by the City Corporation and Parts of the Townships of Saltfleet, Binbrook, Glanford, Ancaster, West Flamboro, East Flamboro* (Hamilton: The Ramsay Thomas Company, 1913), 1:2,000 ft.

31. HPL Special Collections, *Hamilton Times* Advertisement Scrapbook, 18, "Brightside" (March 4, 1911).

32. Ontario, Divisional Court of High Court of Justice, *Stipes versus City of Hamilton*, January 29, 1912, reprinted in HCCM (1919) (May 2, 1919): 106–110.

33. See, for example, *Hamilton, the Birmingham of Canada* (Hamilton: Times Printing Co., 1892).

34. Ken Cruikshank and Nancy B. Bouchier, "'The Heritage of the People Closed against Them': Class, Environment, and the Shaping of Burlington Beach, 1870s-1980s," *Urban History Review* 30 (2001): 40–55.

35. *The Toronto, Hamilton and London Society Blue Book: A Social Directory, Edition for 1900* (Toronto: Wm. Tyrrell & Co, 1900).

36. Harry L. Penny, *One Hundred Years and Still Sailing: A History of Hamilton Yachts, Yachtsmen, and Yachting, 1888 to 1988. Centennial Yearbook* (Hamilton: Royal Hamilton Yacht Club, 1988).

37. "The Local Legislators," *Hamilton Spectator* (August 4, September 1, 1885), "The Beach," *Hamilton Spectator* (June 17, 1880); "Burlington Beach," *Hamilton Spectator* (May 7, 1881).

38. "Bay Is Unsafe to Bathe in," *Hamilton Spectator* (July 9, 1924). For an example of a city council scheme to get Hamilton's working-class children to the beach strip, see: *Hamilton Herald* (July 9, 15, 16, 24, 29, 1929). More generally, see Nancy B. Bouchier and Ken Cruikshank, "Dirty Spaces: Environment, the State and Recreational Swimming in Hamilton Harbour, 1870–1946," *Sport History Review* 29 (1998): 59–76.

39. "Where Children Swim," *Hamilton Herald* (July 9, 1924); William Gore, G. Naismith, William Storrie, *Report on Sewage Disposal* (Hamilton, Ontario, 1923); and E.R. Gray, *Report on Sanitary Intercepting Sewers, Pumping Stations and Drainage Required in the City of Hamilton for Co-ordination with an Adequate System of Sewage Disposal* (Hamilton, 1923).

40. For a rare photograph of young children swimming from the busy docks of the north end, see "Where Children Swim," *Hamilton Herald* (July 9, 1924).

41. Interview, August 1, 1999, Royal Hamilton Yacht Club, kindly supplied by interviewer Andrew Stevenson.

42. Interview, Worker's Arts and Heritage Centre (hereafter WAHC), Hamilton, Ontario. Rob Kristofferson's interview with a north-ender (b. 1927) who was a former NASCAR, Sawyer-Massey, and Westinghouse employee.

43. Ibid.

44. "The Stewart Park," *Hamilton Herald* (August 21, 1911).

45. There are a number of photographs of ice fishing on the bay in John Boyd Collection, National Archives of Canada, Ottawa, Ontario. See, for example, PA-84012-4.

46. Interview, WAHC, Rob Kristofferson's interview with a north-ender (b. 1907) who was a former Westinghouse employee.

47. Nancy B. Bouchier and Ken Cruikshank, "The War on the Squatters: Hamilton's Boathouse Community and the Re-creation of Recreation on Burlington Bay, 1920–1940," *Labour/Le Travail* 51 (Spring 2003): 9–46.

48. Presbyterian Church in Canada, Board of Social Service and Evangelism, the Department of Temperance and Moral Reform of the Methodist Church, and the Community Council of Hamilton, *Report of a Preliminary and General Social Survey of Hamilton* (April 1915); "Say Slum Conditions Exist in Hamilton," *Hamilton Spectator* (May 20, 1913); "Sounds Death-Knell of the Slum Districts" (July 23, 1915); "Social Survey of Hamilton in 1913," *Hamilton Herald* (January 6, 1914). See also the recounting of the city's early social planning by an activist who worked in the day, Ethel S. Ambrose, "Along the Road of Housing and Town Planning from the Earlier Days," (Unpublished manuscript, n.d., HPL. Special Collections); and Rosemary Gagan, "Mortality Patterns and Public Health in Hamilton, Canada, 1900–1914," *Urban History Review* 18 (February 1989): 161–175.

49. Noulan Cauchon Papers, National Archives of Canada, Ottawa, MG 30 v. 1, f. 38, *Reconnaissance Report on Development of Hamilton*, October 1917, 68; "How Hamilton Might Become Beautiful," *Hamilton Herald* (August 4, 1917); Brian Henley, "Cauchon Had Unique Vision for Hamilton," *Hamilton Spectator* (April 26, 1997).

50. Terpstra, "Local Politics and Local Planning," 121.

51. Cauchon Papers, vol. 2, 2–16, "The Ethical Basis of Town Planning," December 11, 1920; *Hamilton Spectator* (June 19, 1920).

52. On the creation of the Harbour Commission, see, Sproule-Jones, *Governments at Work*," 135–142.

53. F.I. Cousins, *Report on Harbour Front Development*, February 20, 1919; HCCM (1919), 549–556.

54. "A Bathing Beach?" *Hamilton Spectator* (May 3, 1926).

55. On parks board developments in this period, see John C. Best, *Thomas Baker McQuesten: Public Works, Politics, and Imagination* (Hamilton: Corinth Press, 1991), especially Chapter 5: "A Bachelor ... Whose Bride Is the City Parks System," 51–68; On McQuesten's relationship with Cauchon, see Mary J. Anderson, *The Life Writings of Mary Baker McQuesten, Victorian Matriarch* (Waterloo: Wilfrid Laurier University Press, 2004), 311, n. 573 and 312, n. 578.

56. For an overview of the land scheme, see John Weaver, "From Land Assembly to Social Mobility: The Suburban Life of Westdale (Hamilton) Ontario, 1911–1951," in *A History of Ontario: Selected Readings*, edited by Michael J. Piva (Toronto: Copp Clark Pitman, 1989), 214–241; Best, *Thomas Baker McQuesten*, 56–57. On earlier ideas about developing the area for industry, see, for example, "Coote's Paradise," *Hamilton Spectator* (September 13, 1877); "To Develop Marsh Lands on Big Scale," *Hamilton Spectator* (May 14, 1912); "Not Encouraging: Cootes Paradise Not Suitable for Factories," *Hamilton Times* (March 25, 1914); and Brian Henley, "Plan to Develop Cootes Raised a Ruckus," *Hamilton Spectator* (October 25, 1997).

57. R.C. Reade, "Hamilton Shows Toronto How," *Toronto Star Weekly* (November 16, 1929).

58. Leslie Laking, "Early Days at RBG," *PAPPUS* 11 (1992): 9–11; Best, *Thomas Baker McQuesten*, 59–60; John A. Scott, "A Short History of Cootes Paradise," *The Gardener's Bulletin* 14 (March 1970): 1–8; Thomas B. McQuesten High Level Bridge Scrapbook, vol. 1, HPL, Special Collections; Hamilton Naturalists Club, *Minute Book*, June 26, 1920, Archives of Ontario, MU 1285. For the larger context of nature conservation in Canada, see Janet Foster, *Working for Wildlife: The Beginnings of Preservation in Canada* (Toronto: University of Toronto Press, 1978), especially Chapter 6: "Protecting an International Resource," 120–154.

59. "Bird Sanctuary Law in Force," *Hamilton Spectator* (March 1, 1927); "Dundas Marsh Is Designated a Crown Game Reserve: Unlawful to Carry Arms on the Property," *Hamilton Herald* (February 12, 1927). Hunters apparently had to obtain these licences from provincial authorities in Toronto, rather than local authorities in Hamilton.

60. Some of them leased their land from area farmers, the city, and the Toronto Hamilton & Buffalo (TH&B) railway company for nominal rents of about $1.25 a month. Others were squatters, who held a tenuous legal claim to the land. See, HCCM, March 1, 1921; Bylaw no. 4188, "To Acquire Lands and Boat Houses Necessary for the Establishment and Laying Out of Longwood Road," Schedule A, "Parcels of Land Occupied by Certain Buildings and Boathouses Erected on City Property to the North of Desjardins Canal and West of York Street," March 31, 1931; Hamilton Board of Control Report 10, March 31, 1931, April 14, 28, 1931; "Would Remove Squatters on Marsh's Edge," *Hamilton Spectator* (February 14, 1939); Bouchier and Cruikshank, "The War on the Squatters."

61. Bouchier and Cruikshank, "Dirty Spaces"; On Stewart Park, see "Cecil Vanroy Langs," *Dictionary of Hamilton Biography*, vol. IV (Hamilton: W.I. Griffin, 1999), 149.

62. "Publicity Helps to Stop Bay Pollution," *Hamilton Spectator* (June 22, 1926); "Oil in Bay," *Hamilton Spectator* (April 15, 1926); "Federal Officer Coming to Probe," *Hamilton Spectator* (April 16, 1926); "Acids and Oil Find Way to Bay," *Hamilton Herald* (June 13, 1931); "Oil Floating Again on Bay," *Hamilton Spectator* (April 21, 1932).

63. "Dye Dumping in Bay Must Stop," *Hamilton Spectator* (November 29, 1932).

64. "Bay Pollution," *Hamilton Spectator* (February 4, 1927).

65. "Wants Bay Water Analyzed," *Hamilton Herald* (July 24, 1924); "Untreated Sewage Is Polluting Bay Water," *Hamilton Spectator* (April 15, 1926); "Bay Pollution," *Hamilton Spectator* (February 4, 1927); "Blames Private Sewers," *Hamilton Herald* (July 6, 1933); "Bay Pollution Remedy Sought," *Hamilton Herald* (July 31, 1934).

66. "Provincial Expert Presents Report on Pollution of Bay," *Hamilton Spectator* (December 23, 1943); on local reports to city council, see HCCM (1942) "Report Addressed to W.L. McFaul, City Engineer" (June 25, 1942), 299–300; HCCM, June 29, 1943, December 28, 1943.

67. As cited in *A Consolidated Report on Burlington Day* (Hamilton, 1958), n.p. See especially Chapter 4: "Sanitary Conditions," and Chapter 6: "Industrial Water."

68. "Bay Pollution Health Menace," *Hamilton Spectator* (August 10, 1946). Studying the dramatic rise in coliform bacteria between 1935 and 1960 along the beach strip, scientists determined that the major source of pollution in the western portion of Lake Ontario was from the Burlington Bay. See D.H. Matheson, *A Sanitary Survey Study of the Western End of Lake Ontario* (Hamilton: Department of Municipal Laboratories, 1964).

69. "Works Controller Alarmed over Pollution of the Bay," *Hamilton Spectator* (April 13, 1946).

70. "Our Deceptive Bay," *Hamilton Spectator* (August 17, 1946).

71. Ibid.

72. Lawrence Murphy and Philip Murphy, *Tales from the North End* (Hamilton, 1981).

73. Y.P., "City's Zoning and the Master Plan," *Hamilton Spectator* (April 1, 1946). On this phenomenon more generally, see John David Hulchanski, "The Origins of Land Use Planning in Ontario, 1900–46," Ph.D. dissertation, University of Toronto, 1981, 271–352.

74. Y.P., "City's Zoning."

75. Ibid.

76. Faludi, *A Master Plan*.

77. Hamilton, City Planning Committee, *Report on Existing Conditions Prepared as Base Material for Planning*, Town Planning Consultants, 1945.

78. Michael Quigley, "Flying High," *Hamilton Cue Magazine* (July 1983): 12. For other information on Brightside, see "Brightside," *Hamilton Spectator* (July 12, 1975).

79. HCCM (1930) Board of Control Report, November 11, 1930, 875–876.

80. Faludi, *A Master Plan*.

81. These acreages did not include areas like Brightside or the homes near Gage's Inlet, which, in Faludi's mind, already were industrial zones.

82. "Commission Outlines Plan to Fill in Extensive Tract," *Hamilton Spectator* (February 3, 1944).

83. Robert J. Hanley and Frank Oxley, "Port of Hamilton," *Hamilton Spectator* (May 27, 28, 29, 1957).

84. Yale Rabin, "Expulsive Zoning: The Inequitable Legacy of Euclid," in *Zoning and the American Dream*, edited by Charles M. Haar and Jerold S. Kaydon (Chicago: APA Planners Press, 1990), 100–103.

85. Weaver, *Hamilton*, 162–166.

86. "North-end Man Resents Slum Tag," *Hamilton Spectator* (March 7, 1978); see also "Demolition May Force Family Break-up," *Hamilton Spectator* (March 4, 1978).

Critical Thinking Questions

1. In what ways did the owners of the sawmills responsible for polluting the Ottawa River defend their actions and resist attempts to control their polluting practices?
2. Why were the mill owners able to resist legislative control over their activities for as long as they did? What finally tipped the scales in favour of a conservationist and environmentalist perspective?
3. Can you account for the changing government stance on pollution in the period covered by Gillis's article? Is it fair to say that, for a while at least, concerns over the health of the environment were marginalized by other, more definably political, concerns?
4. Why did formal urban planning fail to create environmental equity in Hamilton in the first half of the twentieth century? What forces combined to undermine the admittedly sensible idea that people should not live in polluted neighbourhoods nor too close to sources of pollution?
5. Cruikshank and Bouchier argue that "industrial and residential areas produced wastes, and containing those wastes to any one part of the city proved difficult." Why was this?
6. What efforts were undertaken to improve the environmental conditions in Hamilton? Were any successful and, if so, why in your view did they succeed?

FURTHER READING

Gilbert Allardyce, "'The Vexed Question of Sawdust': River Pollution in Nineteenth-Century New Brunswick," *Dalhousie Review* 52 (1972): 177–190. Reprinted in Chad Gaffield and Pam Gaffield, eds., *Consuming Canada: Readings in Environmental History* (Toronto: Copp Clark, Ltd., 1995), 119–130.

Although Allardyce's piece is much older than Gillis's reproduced here, the similarities in the narratives they construct are striking. In both cases the multiuse potential of waterways are limited as a result of the activities of one usage group. In both cases the environment is obviously damaged as a result of those same activities. Unlike the situation on the Ottawa

River, however, in southern New Brunswick the environmental damage was long term and led to a decline in the region's economy, a decline that was partially reversed by the creation, ironically, of Fundy National Park in the area.

Margaret Beattie Bogue, "To Save the Fish: Canada, the United States, the Great Lakes, and the Joint Commission of 1892," *Journal of American History* 79, no. 4 (March 1993): 1429–1454.

Bogue's article continues the historical theme found throughout this section, that of the development of responses to the increasingly clear environmental impact of industrial and resource-extraction activities in Canada in the late nineteenth century. In this case the degraded ecosystem is not riverine in nature but the Great Lakes, damaged by industrial pollution and by a grossly mismanaged fishing industry. But as in other cases, the story is one of stubborn resistance on the part of those whose activities damaged the environment. Bogue shows the ways in which the environmental degradation created tensions between the United States and Canada, but also agreement on the creation of what is one of the very first attempts to manage shared resources jointly.

Suzanne Zeller, "Darwin Meets the Engineers: Scientizing the Forest at McGill University, 1890–1910," *Environmental History* 6, no. 3 (July 2001): 428–449.

Zeller's piece shows the way in which the certainties of Victorian science confronted the challenges of defining and creating management strategies that could be applied to forestry in Canada. In her view those certainties were shaken to their core by the complexities evident in the Canadian forest ecosystems. The impact of their failure to understand and, therefore, to offer strategies to control the forests affected both the scientists and their discipline for decades. In many ways this article focuses on the limitations of reductionist science when it is applied to exceedingly complex ecosystems.

Jennifer Read, "'A Sort of Destiny': The Multi-jurisdictional Response to Sewage Pollution in the Great Lakes, 1900–1930," *Scientia Canadensis* 22–23, no. 51 (1998–1999): 103–129.

This article may be read in conjunction with Bogue's piece above. Read is primarily interested in the pollution of the waters of the Great Lakes as a result of urbanization and industrial activities along their shores. She, like Bogue, demonstrates the tensions that were created by the degradation of this shared resource, and also outlines the nature of the responses to the crisis. She points out that, of the various strategies suggested to eliminate the problem of water pollution and water-borne disease, only one was adopted, the Ontario Public Health Act amendments of 1912. In doing so she highlights, as Bogue did, the difficulties of creating an international response to even the most immediate and serious of environmental crises.

Richard Walker and Robert D. Lewis, "Beyond the Crabgrass Frontier: Industry and the Spread of North American Cities, 1850–1950," *Journal of Historical Geography* 27, no. 1 (2001): 3–19.

Walker and Lewis argue that conventional wisdom—that suburbanization and urban sprawl are post-war phenomena intimately linked with the increasingly widespread use of the automobile and the effects of urban decay—is in need of considerable revision.

They suggest instead that urban sprawl and suburbanization are in fact much older phenomena whose roots date back to the middle of the nineteenth century. They offer a view of a planning process and political partnerships that are more coherent than is the case suggested by Cruikshank and Bouchier in their study of Hamilton.

RELEVANT WEB SITES

Industrial Hamilton: A Trail to the Future
http://collections.ic.gc.ca/industrial/
   This historical site was created as part of the Canada's Digital Collections initiative. It narrates and analyzes in great detail the way in which Hamilton developed into a centre of Canadian industry in the 1800s. Although the environmental impact of this development is not *directly* addressed, it is possible to examine and understand the environmental effects of the industrialization of the region by examining the site.

Railways in Canada: A Brief History
http://imagescn.technomuses.ca/photoessays/railways/railways07.cfm
   This photo essay from the Canadian Science and Technology Museum is in reality a relatively comprehensive examination of the history, especially the early history, of railways in Canada. In addition, many contemporary photographs and descriptions of industrial development in Canada may be found in the "Industry" section of the museum's site.

When Coal Was King: Coal Mining in Western Canada
http://www.coalking.ca/index.html
   This site offers excellent coverage of the history of coal mining in western Canada. Sections focus on the history of coal mining, mining companies, technology, and a variety of subtopics. In addition, the social and economic impact of mining is discussed, as are its environmental effects and long-term legacies. The site offers a good range of pictures, videos, and audio tracks related to mining history and is published by the Heritage Community Foundation.

Sustainability and Conservation

The issues of sustainability and conservation reached the forefront of environmental debate in Canada in the twentieth century, and continue to be overwhelmingly important today. Each term is underpinned, in fact, by a very complex philosophy, a point that is developed in the last two readings in this section. The first, by Jennifer Read, illustrates the effect of popular environmental consciousness on corporations that manufactured environmentally unfriendly products in the 1960s. A response to alarms sounded by wildlife personnel and ecologists in the United States and Canada in the post-war period (the most famous of which was Rachel Carson's *Silent Spring*), environmentalism entered the consciousness of the baby boom generation as they reached adulthood. To a great extent this popular environmentalism was a part of a larger dissatisfaction exhibited by the 1960s generation, seen in youth support of (and often participation in) the civil rights movement and, later, in the more overt counterculture stances of the late 1960s.

Read shows how the baby boom generation affected pollution policy in Ontario, and contrasts it with the slightly earlier environmentalism exhibited by their parents' generation. The senior group enjoyed greater access to political power in the early 1960s, at local and regional government levels, but were unsure of how to wield it in such a way as to promote environmental action. Read reveals the halting attempts, well meaning but occasionally factually incorrect, of this generation to effect environmental change. She shows, too, that despite attacks from manufacturers and legislative delays, how these attempts were ultimately successful. A sharp contrast is then offered when Read examines the environmental activism of the younger generation. Although that activism flourished only five or six years later, differences in the political sophistication of the activists and their impact on public awareness could hardly have been more marked. Read leaves us wondering: are the enormous differences that mark environmental policy since the early 1970s in comparison to those from the early 1960s a consequence of vastly increased public awareness of environmental issues, or because of greater political skill on the part of environmental activists?

John Sandlos's "Where the Scientists Roam" and Alan MacEachern's "Changing Ecologies" both focus on the role of professional environmental management carried out within the Canadian national parks system from the 1920s to the 1960s. Each author highlights the dramatic shifts in policy, often in response to changing currents in scientific thought, which marked the history of the parks in the mid-twentieth century. Both Sandlos and MacEachern emphasize the environmental sincerity of parks personnel, but also the fact that they were hampered by tensions within their mandate: were parks designed to be sanctuaries for wildlife, as corners of "unspoiled" wilderness, or as locales for humans to experience nature's majesty? Furthermore, a question that raged throughout much of the period concerned the amount of resource exploitation (and this included wildlife as well as timber) that could be allowed within the parks system. As Sandlos points out, there were well-meaning but ultimately misguided attempts to cull the buffalo in Wood Buffalo National Park for commercial purposes, and Alan MacEachern outlines a similar situation in relation to logging in parks in Atlantic Canada.

But it is the tensions between the philosophies of management and exploitation, and between managers and scientists, that interest Sandlos and MacEachern the most. Each demonstrates clearly that Canadian parks policy was neither linear nor always entirely rational, and was shaped by scientists at some times, managers at others, and even on occasion by external political forces. The questions asked within that policy framework—for whom are parks created, people or the organisms that live there; what level of intervention is required in management activities; what level of commercial activity is acceptable, and so on—have not been settled satisfactorily, and remain central in the debate over the direction of Canada's national parks today.

Chapter Fifteen

"Let Us Heed the Voice of Youth":
Laundry Detergents, Phosphates, and the Emergence of the Environmental Movement in Ontario

Jennifer Read

You're glumping the pond
Where the humming-fish hummed!
No more can they hum,
for their gills are all gummed.
So I'm sending them off.
Oh, their future is dreary.

They'll walk on their fins
And get woefully weary
in search of some water
that isn't so smeary.
I hear things are just as bad
up in Lake Erie.[1]

Dr. Seuss's timely poem, *The Lorax*, highlighted a number of environmental issues prominent during the late 1960s and early 1970s. The hero of the poem, the Lorax, accuses the exploitative "Once-ler man" of habitat destruction, species extirpation, if not extinction, and air and water pollution. *The Lorax* is still in print today, 25 years later, and the issues retain ongoing significance. Despite the environmental movement's importance in shaping political debate since the 1960s, relatively little has been written about its origins in Ontario.[2] The controversy surrounding detergent pollution in the province, which extended through most of the 1960s, offers an excellent opportunity to examine the shift in attitudes that marked the emergence of environmentalism.

*****

Environmentalism emerged out of an intricate and evolving set of values reflecting an understanding of and concern about the human impact on nature as it relates to physical and spiritual human health. It was expressed initially in the efforts of the young, educated, and environmentally aware activists who demanded a decision-making role in areas that traditionally had been within the purview of scientifically trained, expert managers. In the United States environmentalists formed lobby groups that pressed governments at all levels to increase research and adopt the solutions presented by their own scientific experts.[3]

One of the earliest environmental issues to emerge in the Great Lakes basin was the detergent debate, which occurred in two stages during the 1960s. The first stage focused on the problem of excessive foaming, beginning early in 1963. In *Biodegradable: Detergents and the Environment* (1991), William McGucken examined the issue as it played out in the United States. The American industry solved the problem in 1965 by voluntarily changing to a biodegradable, non-foaming detergent.[4] While similar events occurred on this side of the border they have not yet been examined. Late 1968 marked the beginning of the second stage, when the phosphate content of detergents and its role in degrading water quality, especially in Lake Erie, captured significant media attention. McGucken's article, "The Canadian Federal Government, Cultural Eutrophication, and the Regulation of Detergent Phosphates, 1970" (1989), concentrated on the federal level, briefly touched on a number of environmental groups, but missed the debate's significance at the provincial level, especially the role played by Pollution Probe.[5]

Terence Kehoe used both stages of the detergent issue to highlight the shift in American business–government relations that occurred during this period. Traditional co-operation between the public and private sectors disintegrated as public input assumed increased importance in both the political process and business regulation. This was especially significant at the state and local levels. Kehoe stressed growing post-war affluence as a key factor in the change because growing concern about health and "the quality of life led to the creation of an extraordinary number of new laws and agencies charged with regulating the 'social conduct' of business firms."[6] Kehoe's concept of changing government–industry relations applies to the Ontario situation during the same period.

An examination of both phases of the detergent debate as they developed in Ontario will demonstrate the shift in thinking that marked the appearance of environmental values in this province. The first phase was distinguished by traditional business–government problem-solving strategies, which rejected non-expert input despite a significant outcry from municipal governments across the province. Phase two was markedly different. By 1969, public values had changed significantly, enabling non-governmental environmental groups, specifically Pollution Probe, to challenge closed-door decision making. A comparison of the effectiveness of the tactics used in both phases will demonstrate the changing milieu created by the emergence of environmental values in the province.

Called "the prosperous years," the 25 years after the Second World War marked a fundamental change in the social and economic composition of Ontario society. The provincial economy grew steadily until the first post-war recession, beginning in 1957; the economy recovered in 1963 to continue expanding into the 1970s. Throughout the period, the unemployment rate rarely rose above 4 percent. Between 1941 and 1961 Ontario's population increased from 3.7 million to over 6 million. The province attracted some 600,000 immigrants during the period, while the baby boom produced an unprecedentedly high annual birth rate of 25–26 per 1,000 with a high of 28.9 in 1947.[7]

The burgeoning population increasingly concentrated in cities, especially the Toronto-centred area extending along the Lake Ontario shoreline from Niagara in the west to Oshawa in the east, known as the "golden horseshoe." By the 1960s, this region was home to over 50 percent of the province's population. It supported a manufacturing sector that employed roughly 30 percent of the provincial workforce and created significant employment in related industries and services. Here, Ontario firms produced almost half of the nation's manufacturing output and were responsible for over 80 percent in areas as diverse as automobiles, soaps and washing compounds, leather tanning, agricultural implements, and prepared breakfast foods.[8]

The province's lakes and rivers were particularly vulnerable to this post-war industrial surge. Already degraded from Depression-era neglect and the considerable industrial measures undertaken during the Second World War, water bore the brunt of the post-war boom.[9] The heavy industrialization of this period ensured the introduction of a wide range of effluents into the Great Lakes; none were as visible as synthetic detergents. Wartime developments in the petroleum sector enabled manufacturers to produce more effective and cheaper synthetic cleaning agents than had previously been possible. This coincided with pent-up consumer demand unleashed by the return to peace. Thousands of new laundry machines and dishwashers readily accommodated the synthetic detergents.

A detergent is any agent that, when added to water, thoroughly saturates accumulated soiling particles such as dirt and oil (wetting), separates the particles from the item (dispersing), then links the particles to water molecules (emulsifying) and carries them away from the item. There are primarily two types of detergents, soaps and synthetics. Soaps are made from animal or vegetable fat and are most useful in soft water, which is low in dissolved minerals such as calcium and magnesium. Both soaps and synthetic detergents have components called surface-active agents, or surfactants, which interface directly with distinct surfaces. Detergents are able to clean because the surfactant inserts itself between dirt and the item being washed and holds the dirt suspended in water.[10] In addition to the surfactant, synthetic detergents also contain a builder, most often a phosphate, which softens hard water by drawing suspended minerals out of the solution. In the early 1960s, surfactants counted for 10–15 percent of synthetic detergents, while the builder made up another 60 percent by weight. The remaining volume was concerned with aesthetic aspects, such as smell.[11] Due to the versatility and new affordability of synthetic detergents, they were favoured for both domestic and industrial cleaning.

A molecule called alkyl benzene sulphonate (ABS) served as the surfactant of the synthetic laundry detergents introduced in the immediate post-war years under names such as Fab, Tide, and Surf. ABS was a long, asymmetrical molecule; one end attracted dirt, and the other attracted water. Between the two ends the molecule branched several times. This structure created two significant and related problems. Because of its design, ABS caused visible and long-lasting foam in concentrations as low as 1 mg per litre in sewage effluent. Other synthetic detergents, such as those used for industrial applications, did not branch and yielded much less foam, which readily broke down in sewage treatment. Detergent manufacturers used the foaming property of ABS as a sales strategy appealing to housewives who, for years, had been encouraged to equate plentiful suds with cleanliness. Ultimately, the excessive sudsing problem was an artificial creation of detergent marketers caught up in a campaign over who had the longest-lasting suds.[12]

The other related problem with ABS was its lack of biodegradability. Easily biodegradable substances can be broken down into harmless materials through the bacterial action of normal biological processes. This is the basis of many sewage treatment plants. Because of its branched construction, ABS was not biodegradable and the long-lasting suds, which so appealed to consumers, piled up along those of the province's fast-running rivers and lake shores receiving treatment plant effluent. University of Toronto professor P.H. Jones noted: "The major hazard created by these foam banks was the break down in public relations between the large soap companies and the customer."[13]

Although complaints about foam in sewage treatment plants and rivers began in the early 1950s, the detergent industry insisted that ABS compounds were only partly responsible for the increasing mess. Surely, they argued, the small concentration of synthetic detergents could not be the primary cause of foaming in plants and in the waters receiving plant effluents.[14] In June 1962, the provincial agency responsible for water quality, the Ontario Water Resources Commission (OWRC), hosted a conference on "Problems in the Use of Detergents," which included representatives of the major detergent manufacturers, Colgate-Palmolive, Lever Brothers, and Procter & Gamble. D.F.

Carrothers, representing the Canadian Manufacturers of Chemical Specialities Association, concluded: "While detergent materials can be a contributing factor to some of the problems met in sewage treatment and water pollution, they are by no means the only or necessarily the most important factor. Thus, we consider that much of the publicity blaming detergents specifically is unwarranted."[15] Industry representatives insisted that problems "aggravated" by ABS were restricted to "a few areas in the world and do not yet exist in Canada."[16] Nevertheless, they agreed to co-operate with the OWRC and to look into the Ontario problem.

One area that experienced visible pollution problems was Wentworth County at the head of Lake Ontario. This included the heavily industrialized area surrounding Hamilton Harbour, including the city of Hamilton and, beyond that, Wentworth County. "Sewage, detergents, sludges, chemicals, oil ... they all pour into the harbour," a *Hamilton Spectator* headline moaned in November 1962. Among the pollution problems explored in the accompanying article, detergents received a significant airing. Detergent residues persisted after sewage and water treatment and, in some American cities, the article noted, "the tap water already comes with a foamy crest." The American and German situations offered examples to be avoided in Canada. The author of the article pointed out that German manufacturers had to be ordered to change the formulation of their detergents in order to address the dilemma there. While the problems in Ontario were not as severe as in Germany, both the local medical officer of health and the director of Hamilton's municipal laboratories predicted that "the probable outcome will be much the same as Germany's, as syndets [synthetic detergents] build up in the water." To emphasize the point, the article featured photographs of an Ontario sewage treatment plant enveloped by foam.[17]

Disturbed by the problem of excessive form, the Wentworth County Council, consisting of municipal representatives from Stoney Creek, Dundas, Waterdown, and the surrounding Hamilton area townships, unanimously passed a resolution on December 18, 1962 urging the provincial government to ban the use and sale of synthetic detergents. Citing its alarm over the pollution of Ontario waters by laundry detergents, the council noted

> that the basic cause of detergent pollution arises from the fact that most detergents marketed in Ontario have a mineral base (i.e., phosphorus) which cannot be broken down and purified by natural or artificial purification methods;
>
> AND ... the pollution from mineral based detergents does not dissipate but rather has a cumulative effect causing such serious problems as the algae buildup in many lakes and other inland waters with its consequent ill effects;
>
> AND ... other jurisdictions (i.e., Germany) have solved the detergent pollution problem by prohibiting the use of mineral based detergents;
>
> AND ... detergents can be produced with equivalent cleansing properties by using an organic base instead of a mineral base (i.e., German and some parts of the U.S.)[18]

In addition to urging a ban on detergents, the council insisted that the province and the OWRC alert the public to the seriousness of detergent pollution. The council sent a copy of its resolution to every municipal council in Ontario, the chair of the OWRC, the premier, and the leaders of the provincial Liberals and New Democrats.

Thomas Beckett, a Hamilton lawyer who went on to chair the Hamilton Region Conservation Authority as well as to become a member of the Conservation Council of Ontario (CCO), proposed the resolution.[19] The impetus for Beckett's move is unclear. In all likelihood it was a combination of press coverage, similar to the *Spectator* article, and a meeting he had attended that fall at Hamilton's Royal Botanical Gardens with OWRC secretary, William McDonnell.[20] There, McDonnell had admitted that pollution from detergents was a serious problem of which the public remained unaware. McDonnell pointed out that, unless the public demanded change, the detergent industry would not undertake it voluntarily.[21]

The Wentworth resolution struck a chord with municipal governments across the province and soon letters supporting it flooded the OWRC offices. Some 277 municipalities from all regions of the province endorsed the resolution between January and March 1963. On March 11, the Ontario Association of Rural Municipalities approved the resolution at its annual meeting.[22]

The OWRC hastened to reassure the municipalities approving the Wentworth resolution that it had the situation in hand. Remedial activity included OWRC meetings with industry representatives and an ongoing commission investigation into the detergent problem. Although the OWRC admitted that detergent foaming somewhat interfered with the operation of treatment plants, and partially contributed to the growth of algae in rivers and lakes, it maintained that detergents were merely a nuisance.[23] Commission personnel had investigated the assertions in the Wentworth resolution and noted that "the authors of the resolution are mixed up in the causes and effects.[24] In fact the document did confuse phosphorus builders, which readily broke down in sewage treatment plants and provided nutrients to support algae growth, with the ABS surfactant, which did not break down and caused the persistent problems with foaming. To commission scientists, the Wentworth Council had clearly confused the scientific and technical aspects of the detergent debate and therefore should not be taken seriously. David Caverly, OWRC general manager, believed the commission's best response to the resolution would be to widely publicize the fact that it remained actively involved in the search for a solution. His attitude was clear: leave problems to the people best able to deliver answers, scientifically trained resource managers.[25]

For the detergent industry, the Wentworth resolution raised the unwelcome spectre of government intervention. The potential for provincial legislation was heightened by European precedents, notably in Germany, and legislation then under consideration in some American states. In the United States, the industry had been searching for a non-foaming substitute for ABS since the 1950s, as pressure from Congress and state legislatures threatened to push detergent manufacturers toward a solution more quickly than they wished. The industry decided to find and introduce alternatives to ABS on its own terms rather than have change dictated by politicians. Detergent manufacturers also recognized that negative publicity had an unfavourable impact on sales. In June 1963, therefore, the American detergent industry announced that it intended to produce more readily biodegradable detergents within two years.[26]

In the meantime, Canadian representatives of the detergent multinationals worked to avert government intervention in Ontario. In May, delegates of the Canadian

Manufacturers of Chemical Specialities Association met with the OWRC and reasserted the industry's position that detergent foaming was not yet a serious problem in Canada. They resisted the commission's demand for expensive research into the problem, arguing that they had access to the extensive work being done in Europe and the United States through their international affiliates. They insisted that coercive legislation would only interfere with their research program and would result in more expensive and inefficient alternatives by rushing the industry to a less than ideal solution. Legislation that emphasized only one aspect of a complex situation, they pointed out, might cloud the real cause of the problem.[27] A few months later, Canadian detergent producers refused outright to implement the detergent formulation changes just announced by the American industry.[28]

By the summer of 1963, the OWRC finally recognized that its co-operation with the industry was not as smooth as had been intimated to the public. The commission admitted internally that foaming was a problem and that it did not accept the industry's position. Behind closed doors, the OWRC applied pressure for change, but it continued to support detergent manufacturers in public.[29] The OWRC's strategy resembled government–business relations typical of the conservation era. Commission personnel believed that more could be accomplished through dynamic interaction between their experts and business representatives than by offering antagonistic public ultimata. As long as the technical people were at work on a solution the public did not need to know the details.

Press coverage of the situation generally favoured the industry's stance and ridiculed the Wentworth resolution.[30] Many writers found the industry explanation convincing and echoed the OWRC position that blamed the current confusion on the technical and scientific ignorance of the municipal officials who had drafted the resolution. These comments infuriated Thomas Beckett, the author of the Wentworth resolution, who responded to a particularly scathing article in the *Globe and Mail* by firing off a letter to William McDonnell, commission secretary. Beckett accused the OWRC of supporting the detergent industry against the municipalities. Referring to McDonnell's speech at the Royal Botanical Gardens, Beckett reminded Secretary McDonnell that he himself had acknowledged the level of detergent pollution in Ontario waters to be of "serious proportions." All the same, Beckett continued, "it is quite apparent that an attempt is being made to minimise the important [*sic*] of this problem and suggest that the municipalities were badly informed." Beckett was angered that an agency that had been created to meet the needs of Ontarians now acted like "an ally of the soap manufacturers," ignoring "the pleas of several hundred municipalities."[31]

Ultimately, the OWRC continued to ignore the objections of the local governments and allowed the detergent industry to implement its own solution. After a thorough investigation of the matter, the commission determined that the situation did not require the degree of intervention called for in the Wentworth resolution. In part, the OWRC based its decision on a 1963 American Water Works Association investigation. The study concluded that U.S. legislators should not address the problem of foam at that time. Various state hearings in the matter

had found that the industry was making significant headway toward alleviating the situation. The report also declared that none of the proposed legislation adequately addressed the situation and that forcing a change on the industry, before it was prepared to switch, would only drive up the cost of detergents. The U.S. industry's announcement of a voluntary change reinforced the report's conclusions. The British House of Commons' decision not to regulate the U.K. detergent industry also offered strength to the OWRC's decision.[32]

The OWRC believed that the American shift to biodegradable detergent, scheduled for 1965, would affect Ontario as well. Given the relatively small number of North American companies manufacturing detergent components, it was unlikely that the material required to produce non-degradable detergents would be available to Canadian manufacturers once U.S. manufacturers had completed the large task of converting to an ABS substitute. "It seems reasonable, therefore, not to insist upon regulation of an industry which is already heavily committed toward regulating itself," David Caverly concluded.[33] Indeed, late in 1963, Canadian manufacturers reversed their initial decision to maintain existing detergent formulations and declared that Canadian detergents would be changed to solve the problem of foam. The new detergents replaced the branched ABS surfactant with an unbranched derivative—linear alkylate sulphonate (LAS). Detergents formulated with LAS surfactants foamed much less and were readily degradable with existing sewage treatment technology.[34] Industry and government newsletters stressed the voluntary nature of the decision and the industry's responsibility in tackling the issue.[35]

By 1965, the issue of foaming appeared to be solved. OWRC attempts to influence the industry had been resisted successfully by the detergent manufacturers and they had been able to address the problem on their own terms. Despite the industry's concern about the Wentworth resolution, concentrated negative reaction to foaming materialized only at the municipal level. The provincial legislature appeared more concerned about declining Great Lakes levels than about pollution problems associated with detergent foaming. The issue was raised during the annual debate on the estimates for the Department of Energy and Resources Management in 1963, but it did not appear to concern Premier John Robarts. In the House of Commons, the possibility of diverting Great Lakes water into the United States dominated federal discussion. In pollution debates, detergent foaming generated little concern.[36]

Oddly, the OWRC did not take the opportunity presented by the Wentworth resolution to use the argument of overwhelming public concern to demand changes in detergent formulae. This stemmed from its reluctance to acknowledge the relevance of the scientifically inaccurate resolution. The press also appeared inclined to trust the scientific experts and to question the credibility of the Wentworth County Council. At this point, the general public demonstrated little interest in environmental issues. For a visible and messy problem, the issue of foaming raised relatively little public complaint in comparison with the later response to pollution problems.

Although the detergent industry had addressed the problem of foaming successfully, another issue associated with detergents soon took its place. The trouble was algae. Whether by accident or design, the Wentworth resolution had addressed the appearance of algae in the province's waters, but had tied it to detergent foaming. However, algal blooms had nothing to do with the ABS surfactant. Phosphate builders were at the root of the new problem.

Algae are rootless water plants that, like all living things, require energy and nutrients to grow. The plants get their energy from sunlight and the nutrient fuel they require from carbon, hydrogen, oxygen, nitrogen, and phosphorus. All these are available naturally to algae, with the exception of phosphorus, which therefore determines the extent of growth and is called the limiting nutrient.[37] When organic matter, including algae, dies in water, it is broken down by aerobic and anaerobic bacteria. The aerobic bacteria require oxygen to convert the material into simpler organic substances, some of which are then used for food. The more organic substance there is to decompose, the faster oxygen is consumed by the bacteria.[38] Thus, a large amount of dying and decaying algae threatens fish and other aquatic life, which require dissolved oxygen to live.

The phosphorus producing the algal blooms entered waterways from three sources: the spring runoff of manure applied to frozen fields during the winter; partially treated sewage effluent rich in the nutrient; and detergents. Sewage treatment plants could remove only the limited amount of phosphorus required in the bacterial treatment process; any amount in excess of that entered the receiving waters with the treated sewage effluent. Even before the introduction of synthetic detergents, sewage contained more of the nutrient than treatment plants could use. After the advent of the new cleaning agents, the phosphate content of sewage more than doubled, causing two to three times the amount to enter North American lakes and rivers than before the Second World War. Scientists called this rapid, human-generated enrichment of water cultural eutrophication to distinguish it from the natural, long-term process.[39]

All this phosphorus created the luxurious blooming of many algae, but particularly one called cladophora, which grew in large, filamentous green clumps. In September 1964, some 800 square miles of algal bloom coated the surface of Lake Erie. At the same time, close to 43 miles of shoreline between Toronto and Presqu'ile Point were covered by accumulated cladophora. The following summer, Lake Erie and the southeast section of Lake Ontario were again subject to extensive bloom. Algal growth interfered with recreational and commercial boating and fishing, affected water intake pipes and treatment plants, and created "obnoxious odours" when it washed up and decayed along the shoreline.[40] Many scientists attributed the growth of cladophora to cultural eutrophication.[41]

In 1964, the Canadian and American federal governments asked the International Joint Commission (IJC) to investigate the pollution of Lakes Ontario and Erie, and the international section of the St. Lawrence River. This was the fourth time the IJC had been asked to assess Great Lakes pollution since its creation under the 1909 Boundary Waters Treaty. Consisting of six commissioners, three each appointed by the federal governments of Canada and the United States, the IJC was assigned investigatory powers under the treaty. The commission's previous pollution investigations had been limited to the Great Lakes connecting channels—the St. Mary's, Rainy, St. Clair, Detroit, and Niagara rivers.[42]

With increasing instances of algal blooms, fish kills, and oil spills, the two governments asked the IJC to examine pollution in the lakes themselves. For research such as this, the commission supplemented its staff by seconding federal, provincial, and state civil servants, and occasionally private consultants, to serve on its technical advisory boards. All technical surveys required during the course of the investigation were carried out by government water agencies. The federal departments of Energy, Mines, and Resources and of Health, as well as the OWRC, contributed both personnel and facilities to the project. This served to tie personnel and research from both levels of government closely together. By December 1965 the IJC had completed its first interim report and sent it to the two governments. The report outlined the eutrophication problem and recommended that both American and Canadian federal authorities co-operate immediately with provincial and state governments to ensure maximum removal of phosphates from municipal and industrial waste being discharged into the lakes and their tributaries.[43]

The IJC report coincided with an outpouring of public concern about the pollution problem in the Great Lakes in general, and about algae in particular. Citing the commission's findings, in February 1966 the *Globe and Mail* called for swift federal action.[44] When the House of Commons resumed sitting later that month, opposition MPs took up the cry and urged more federal spending on pollution research and control.[45] Growing concern over environmental issues reflected a shift in societal attitudes toward the natural world. By the mid-1960s, Ontarians had come to expect available and abundant outdoor recreation space. Suddenly, Great

Lakes beaches, where they had raised their children or grown to adulthood themselves, were a mass of stinking algae and dying fish and no longer the beautiful recreation spots they had once been.

Letters began to flow into the OWRC from a variety of sources over the fall and winter of 1965–1966. Among those expressing concern about phosphate pollution were women's groups, the United Auto Workers, and private citizens. One young writer, alarmed by the widely reported death of Lake Erie, wrote: "Why cannot Ontario, which covers half the lake's shoreline, co-operate with other border states to get tough on the sources of this sewage. Our generation will look back either with appreciation to your generation's foresight in this matter, or with disappointment at your inability to deal decisively with this important problem."[46] Such letters reflected the emergence of environmental values in the province, although at this point the concern still lacked focus.

Not surprisingly, commission personnel reacted defensively to the criticism. OWRC general manager David Caverly scathingly attacked those he called publicity-seeking, scientifically ignorant "Johnny-come-latelies" both inside and outside the government:

> They pay no attention to actual figures or to verified statistics. With their pet theories, and their preconceived notions, they belong to that *"my mind is made up, don't confuse me with the facts"* group of people who are a part of any society. The result is that they have stirred up "John Q Citizen" to the point of almost hysteria. The old pollution fighters have been pushed into the background, and our task has been made more difficult.[47]

Caverly's comments echoed the disdain he and his colleagues had shown for the Wentworth County Council. His initial reaction to the new environmental values suggests a continued belief in the superiority of scientific training and expertise. Caverly and his colleagues were also reacting to the growing changes in Ontario society, which encouraged citizens to question government pronouncements rather than quietly accept them.

Detergent manufacturers remained complacent and oblivious to the societal changes occurring around them, relying on the "real spirit of co-operation," which they maintained had developed during the detergent foaming controversy.[48] At the Canadian Council of Resource Ministers conference, "Pollution and Our Environment," held in October 1966, they argued that society would be better served by more effective sewage treatment than by any alteration to their detergent formulae. This technological "fix" would provide the most efficient elimination of nutrients at the lowest cost. It would also ensure that Canadian sanitary standards would remain high, something the industry widely predicted to be in jeopardy if formulae were drastically changed.[49] Clearly the manufacturers expected this phase of the debate to play out much like the detergent foaming stage, with the domination of their agenda and public support from their government "partners." The manufacturers did not count on the changes to Ontario society, which would make the traditional business–government relationship suspect, then impossible, before the decade was over.

As the "baby boom" generation came of age in the late 1960s, the province, along with most of the Western democracies, entered a new moral, intellectual, and political era. […] The protest organizations emerging in Canada during this period originated, for the most part, on university campuses. Here, a minority of radical students encouraged their more moderate cohorts to address issues of social justice, racial equality, and, late in the decade, sexual equality. They reoriented political debate and presented the agendas of the new movements to both the public and the government, in the process shifting the focus of political discourse from the traditional political parties to pressure groups. These organizations were able to rally otherwise nebulous public concern and translate it into demand for government action.[50]

These issues triggered Pollution Probe, an environmental group that emerged at the University of Toronto in February 1969. Concerned with social justice, the students were motivated by a sense of outrage at their voicelessness and by the desire to force patronizing politicians to hear their opinion. Pollution Probe's core came from the university's Department of Zoology and initially organized in response to the controversy surrounding the CBC documentary "The Air of Death." Pollution Probe soon broadened its scope. Its mandate grew to include investigating all environmental pollution, determining its effects on human health and mobilizing public opinion on specific measures.[51] According to a Probe pamphlet, the group represented a "grassroots movement with professional expertise which gives form and strength to the public concern over environmental quality" through research, education, and action. "We are fighting not for an antiseptic world, but for a healthy environment. There is a difference."[52]

Several environmental activists, widely recognized as pre-eminent in their fields, began their work with Probe. For instance, University of Toronto Zoology professor Donald Chant chaired the advisory board, which also included professors Ralph Brinkhurst, Henry Regier, John Pales, Phil Jones, and Marshall McLuhan and broadcaster Stanley Burke. These people lent Probe legitimacy and their expertise when the organization tackled a problem falling within their purview. From the beginning, though, Probe's strength came from its student members, such as Monte Hummel, now head of the World Wildlife Fund (Canada). The students' youthful enthusiasm and idealism propelled the organization. They orchestrated Probe's publicity events, such as the mock funeral held for the "dead" Don River. They canvassed door to door and took every opportunity to present their message city-wide, even nationally, through the CBC and Toronto newspapers. It was to this group of energetic and dedicated young people that Chant referred when he urged "Let us heed the voice of youth."[53]

Pollution Probe had been in existence for eight months when the IJC released the *Report to the International Joint Commission on the Pollution of Lake Erie, Lake Ontario and the International Section of the St. Lawrence River* in October 1969.[54] The report recommended comprehensive phosphorus reduction for the lakes, to be achieved by an immediate lowering of the phosphorus content of detergents to the minimum practicable level. In addition, the IJC advocated cutting the nutrient content of municipal and industrial effluent discharged directly into Lake Erie and Lake Ontario by no less than 80 percent. The commission urged both federal governments to begin research to control agricultural runoff. To avoid further nutrient loading of the already taxed waters in the Great Lakes basin, the IJC also recommended the immediate regulation of all new uses of phosphorus. The report and the possibility of restricting phosphate-based detergents received wide discussion in the press.[55]

Just prior to the release of the IJC report, the federal minister of Energy, Mines, and Resources, J.J. Greene, appointed a departmental task force to investigate detergent pollution. When the report came out, the minister directed the task force to consider how to implement the IJC's recommendations on detergents. As part of its investigation, the members of the task force and OWRC representatives visited the Procter & Gamble research facilities in Cincinnati, Ohio, in December 1969. At the meeting, company spokespeople reiterated the position they had taken the month before with the minister. Although they acknowledged concern about the potential for negative publicity in connection with phosphates, they were unwilling to admit that their search for a phosphate substitute had been prompted by the eutrophication problem. Instead, Procter & Gamble spokespeople insisted that their researchers were looking for a substance to enhance product performance.[56] Naturally the government representatives were disappointed with the industry's stance. No doubt the OWRC people experienced a sense of *déjà vu*. As with the issue of foaming, the detergent producers refused to acknowledge a problem until they had developed their own solution.

On December 23, the Task Force on Phosphates and Pollution from Detergents submitted its findings to the minister. It advised a multiple-stage solution to curb accelerated eutrophication. The six main recommendations echoed those of the IJC's October report, and included improved sewage treatment to be financed through amendments to the Canadian Mortgage and Housing Corporation Act. The task force also advised the government to issue a directive ordering replacements for phosphate builders and urged federal research into possible phosphate substitutes. Finally, the report recommended the development of a water quality plan for the Great Lakes basin that would require federal co-operation with the provinces, the United States, and the chemical industry.[57]

Before the minister could announce federal policy, however, the IJC held public hearings on the October report between January 20 and February 6, 1970. In the Great Lakes states, the commission met at Toledo, Erie, and Rochester, while in Ontario meetings were convened in London, Hamilton, and Brockville. Many people representing industry, local citizens' groups, various agencies from local, state, provincial, and federal governments, as well as concerned individuals, presented briefs at the hearings.[58] Detergent manufacturers resisted the boards' recommendation to replace phosphate builders in synthetic detergents. They explained that housewives expected a certain level of cleaning performance and would only use more detergent to achieve the expected results, thereby counteracting the efficacy of the reduction. As no effective phosphate substitute then existed, the best solution was improved sewage treatment facilities.[59] In contrast, Pollution Probe argued that improvements to sewage treatment facilities would take much too long to implement. It urged instead an immediate reduction in the consumption of phosphate-based detergents by the introduction of an immediate ban on their manufacture, sale, and use. To achieve this goal, Probe envisaged a two-part approach—strong consumer demand to convince industry to replace phosphates, combined with public pressure to force governments to legislate a ban on phosphate detergents. "The state of our lakes demands immediate action," Probe asserted.[60]

Probe had already begun its campaign to see both steps carried out. In the early winter, Phil Jones, a University of Toronto civil engineering professor, and Pollution Probe volunteers tested samples of all the major detergents and soaps for phosphorus content by weight. When Probe appeared before the IJC, the results were already complete and on February 8 it broke the story on CBC's "Weekend." Over the next few days, the list and accompanying news release were carried in most Canadian daily newspapers. Probe urged concerned citizens to write to Prime Minister Trudeau and Premier Robarts, the federal and provincial Cabinet ministers responsible for pollution control, and their MPs and MLAs.[61] Brian Kelly, one of Probe's student leaders, appeared on CBC's "Take Thirty" on February 13 and the Larry Solway show on CHUM radio soon after. By March, Probe had received over 7,000 requests for the phosphate content list and it had been reprinted and distributed across the country. John Bassett, publisher of the Toronto *Telegram*, helped Probe's campaign by supplying space for free advertisements created by Vickers and Benson, a Toronto advertising agency.[62] Probe's campaign, coinciding with action on the part of both the provincial and federal governments, helped to educate the public and keep enthusiasm high.

On February 6, the federal minister of Energy, Mines and Resources, J.J. Greene, announced in the House of Commons that the Canada Water Bill, then being considered by a parliamentary committee, would be amended to allow the federal government to regulate the phosphate content of laundry detergents.[63] On March 24, the coordinators of both the Ottawa and Toronto Probe branches, Phil Reilly and Peter Middleton, along with ecologist Ralph Brinkhurst and limnologist Michael

Dickman, appeared before the Commons Committee on National Resources and Public Works. Probe's brief on the Canada Water Bill reflected the organization's belief that its demand for action must be supported by scientific evidence, and thus Reilly, Brinkhurst, and Dickman stressed their expertise as biologists. They emphasized the need for swift federal action on phosphates. Although admitting that more research needed to be done, Dickman and Brinkhurst insisted that the government already possessed enough information to act. They explained that advanced sewage treatment technology existed and should be installed in the Great Lakes basin as soon as possible. Brinkhurst countered industry claims that there was no viable alternative to phosphate builders by reminding the committee that similar objections had been raised over the issue of foam, and it had been resolved responsibly. "I think nothing will work faster than requiring somebody to use their ingenuity," he declared.[64]

On February 9, the Ontario Department of Energy and Resources Management had announced that the province would introduce legislation to restrict detergent formulations gradually over five years.[65] Concerned that phosphate builders would not be reduced quickly enough, Probe submitted a 10-point brief to Premier Robarts in April. The brief called for provincial legislation limiting the maximum level of phosphorus in detergents to less than 1 percent by January 1971, rather than the graduated plan announced by the province in February.[66] Probe's worry was addressed when, after intense negotiations with the federal government, the provinces agreed that the best approach to the problem would be national phosphate restrictions listed under the Canada Water Act.[67] Greene acknowledged the intense federal–provincial consultation that had taken place when he introduced the nutrient-loading amendment in the House of Commons. With the difficult aspect of the process over, the Canada Water Act quickly passed its third reading and received royal assent by the end of June. The minister announced the phosphate regulations under the Act a month later. As of August 1, 1970, the phosphate content of detergents was limited to 20 percent by weight and further reduced to only 5 percent by the end of 1972.[68]

Probe cannot be given sole credit for the new provisions included in the Canada Water Act and the first regulations listed under its auspices. As is clear from internal Water Resources Branch memoranda, the minister's advisers had already determined that the federal government should act on the IJC report, and the provincial government had also considered action. Nevertheless, Probe helped to concentrate public concern and kept the issue before the government while the parliamentary committee considered the legislation. Probe's effective use of the news media was perhaps its greatest strength. Certainly the detergent industry felt the impact of Probe's activity. During the first five months of 1970, national sales of synthetic detergent declined by 5 percent while soap flakes and chips rose by 50 percent over the same period of 1969.[69]

In contrast to the issue of foam, the detergent industry was unable to set its own agenda when phosphates became a concern in the late 1960s. This was due, in part, to public receptiveness to the issues, reflecting a dawning wariness of big business and its influence on government, as well as a growing concern about pollution.[70] Probe focused the public debate and suggested actions that the average citizens could undertake, from writing to their MP to

buying soap instead of synthetic detergent. Because of its scientific expertise, Probe's recommendations to government were reliable and allowed it to be more than just "another alarmist group." The phosphate issue captured attention in the House of Commons and the provincial legislature so that support, indeed pressure, for government initiatives emanated from that direction as well. Although they did not like criticism levelled by environmental activists, civil servants were less certain that the industry would police itself at this phase of the debate. They joined with citizen groups and urged their political bosses to take coercive, legislative action. In contrast to the detergent foaming phase, the phosphate debate was marked by a convergence of public interest, press, and government pressure, reflecting the new environmental values.

When the Wentworth County Council circulated its resolution urging the provincial government to ban foaming detergents, the issue did not engender substantial public or political support beyond the level of municipal governments. This failure can be attributed partly to the council's lack of scientific credibility, but also to the fact that societal attitudes had not yet shifted to favour environmental issues. The press accepted the assessment of OWRC personnel, which paid little regard to either the resolution or the council. Without significant media promotion, the Wentworth resolution did not garner the public support necessary to influence the provincial government.

In comparison, Pollution Probe's phosphate campaign proved to be much more effective. The group piqued public interest and support by challenging the problem-solving style of the traditional wise-use conservation experts. Pollution Probe's strength lay in its use of scientific expertise to educate the public and offer well-considered alternative solutions to those suggested by the government scientists and manufacturers. This enabled the group to mobilize the public, drawing on emerging environmental concern and focusing on specific issues. Clearly, Probe was more effective than the Wentworth County Council—so much so that the *Financial Post* concluded: "But for the most part, Probe's aims and achievements have become almost as respectable as motherhood, so drastically has public opinion changed [regarding] the need to curb pollution."[71] In fact, it was the OWRC that found itself disconnected from public opinion and unable to adjust to the emerging environmental attitudes. In response to these new values, the provincial government created a Ministry of the Environment, which, in turn, absorbed the commission in 1972.

NOTES

1. Adapted by Pollution Probe from Dr. Seuss, *The Lorax* (New York, 1971).
2. Ontario historians have examined the emergence and impact of the Algonquin Wildlands League in its efforts to preserve Ontario's remaining wilderness areas. See Gerald Killan and George Warecki, "The Algonquin Wildlands League and the Emergence of Environmental Politics in Ontario, 1965–1974," *Environmental History Review* 16 (Winter 1992): 1–27; also Gerald Killan, *Protected Places: A History of Ontario's Provincial Parks System* (Toronto, 1993). Stephen Bocking, "Fishing the Inland Seas: Great Lakes Research, Fisheries Management, and Environmental Policy in Ontario," *Environmental History* 2 (January 1997): 52–73, examines the impact of environmental attitudes on Great Lakes fisheries management policy.
3. Samuel P. Hays and Barbara Hays, *Beauty, Health and Permanence: Environmental Politics in the United States 1955–1985* (Cambridge, 1987), 531–534. See also Robert Paelke, "Environmentalism," in *Conservation and Environmentalism: An Encyclopedia*, edited by Robert Paelke (New York, 1995), 260–261.
4. William McGucken, *Biodegradable: Detergents and the Environment* (College Station, 1991), 10.
5. William McGucken, "The Canadian Federal Government, Cultural Eutrophication, and the Regulation of Detergent Phosphates, 1970," *Environmental Review* 13 (Fall/Winter 1989): 155–166.
6. Terence Kehoe, "Merchants of Pollution?: The Soap and Detergent Industry and the Fight to Restore Great Lakes Water Quality, 1965–1970," *Environmental History Review* 16 (Fall 1992): 21–46.
7. J.K. Rea, *The Prosperous Years: The Economic History of Ontario, 1930–1975* (Toronto, 1985), 14–15, 193–222. See also Doug Owram, *Born at the Right Time: A History of the Baby-Boom* (Toronto, 1996), 4.
8. Ontario Department of Economics, *Ontario: Economic and Social Aspects Survey* (Toronto, 1961), 143–154; also Rea, *The Prosperous Years*, 14–34; Regional Development Branch, Department of Treasury and Economics, *Design for Development: The Toronto-Centred Region* (Toronto, 1970), 2–4.
9. This account prefaced almost every Ontario Water Resources Commission speech given before groups such as the Ontario Municipal Association, the University Women's Club of Welland, various mining and industrial associations, the Petroleum Association, the Department of Agriculture, the Dairy Branch Field Men's Conference, the London Progress Club, the Consumers' Association, the International Labour Council, the Niagara Regional Development Council, the Engineering Institute of Canada, the Smith Falls Water Commission, the Long Point (Norfolk) Ratepayers Association, and various Rotary Clubs. Ontario Archives (AO), RG 84,

OWRC, Central Records, "Ontario Municipal Association," and "Public Relations: Speaking, General, 1966–67."

10. McGucken, *Biodegradable*, 12–13; also AO, RG 84, OWRC, Subject Files, "Detergents: Miscellaneous Information," David Caverly to OWRC Management Committee, March 29, 1963. See also the Water Management Committee of the Canadian Manufacturers of Chemical Specialities Association, "Detergents and the Aquatic Environment," *Pollution and Our Environment: Background Papers*, vol. 2 (Ottawa, 1966), 2–3.

11. McGucken, *Biodegradable*, 16–17. See also Tom Davey, "Eutrophication and Detergents: An Interview with P.H. Jones," *Water and Pollution Control* 106 (September 1968): 23.

12. P.H. Jones, "Does LAS Spell 'Pollution Free'?" *Water and Pollution Control* 105 (August 1967): 24; also William Ashworth, *The Late, Great Lakes: An Environmental History* (Toronto, 1986), 134–136; McGucken, *Biodegradable*, 21.

13. Jones, "Does LAS Spell 'Pollution Free'?" 24; also Ashworth, *The Late, Great Lakes*, 136.

14. Water Management Committee of the Canadian Manufacturers of Chemical Specialities Association, "Detergents and the Aquatic Environment," 4–5; McGucken, *Biodegradable*, 22–23.

15. D.F. Carrothers, "Household Detergents in Water and Sewage." AO, RG 84, OWRC, Subject Files, "Algae and Detergents."

16. Ibid., "Detergents: Miscellaneous Information," the Canadian Manufacturers of Chemical Specialities Association, "A Brief to the Ontario Water Resources Commission," June 12, 1963.

17. *Hamilton Spectator* (November 8, 1962).

18. *Wentworth County. Proceedings of the Municipal Council of the County of Wentworth for the Year 1962*, December Session, December 18, 1962 (Dundas, Ont., 1962).

19. AO, RG 84, OWRC, Central Records, "Public Relations: 1968, Jan.–June, General Information," OWRC memo, April 10, 1968. The Conservation Council of Ontario (CCO) was an umbrella organization founded by Francis (Frank) Kortright in 1952 for groups and individuals with an interest in conservation. It served a lobby/watch dog function. CCO Minutes, 1952–1953, vol. 1, Conservation Council of Ontario Library and Archives.

20. One editorial noted that the Germans had banned non-foaming detergents by 1965 as well as citing OWRC findings that algal blooms were caused by detergents. *Globe and Mail* (December 3, 1962).

21. AO, RG 84, OWRC, General Managers' Files, "Detergents: Miscellaneous Information," Thomas Beckett to William McDonnell, June 12, 1963.

22. Ibid., OWRC, Central Records, "Wentworth County Resolutions, 1963." An interesting exception was the Sarnia City Council, which chose not to endorse the resolution after Dr. Duncan Cameron, a researcher at Imperial Oil, warned them that "50 percent of all detergents is used by industries and if it was prohibited this would mean shutting down industry." Cameron suggested that endorsing the resolution would prove embarrassing to the Sarnia council. *London Free Press* (February 5, 1963).

23. One of the inquiries came from the minister of Transport, James Auld. AO, RG 84, OWRC, Subject Files, "Detergents: Miscellaneous Information," OWRC memo to the Hon. J.A.C. Auld from David Caverly, March 13, 1962. See also ibid., Central Records, "Wentworth County Resolutions, 1963," OWRC press release, December 14, 1962.

24. Ibid., Central Records, "Wentworth County Resolutions, 1963," OWRC memo to David Caverly from F.A. Voege, March 27, 1963; also ibid., Subject Files, "Detergents: Miscellaneous Information," OWRC memo to Management Committee from David Caverly, "Re: Technical Aspects of the Recent Detergent Problem," March 29, 1963. OWRC research into algae growth began during the summer of 1958. In 1963, the commission's Research Division confirmed

phosphorus as the limiting nutrient. OWRC, *Third Annual Report, 1958* (Toronto, 1958), 66–67, and *8th Annual Report, 1963* (Toronto, 1963), 99–100.

25. AO, RG 84, OWRC, Subject Files, "Detergents: Miscellaneous Information," OWRC memo to Management Committee from David Caverly, "Re: Recent Controversy on Detergents," March 29, 1963.

26. McGucken, *Biodegradable*, 66–91. See also Kehoe, "Merchants of Pollution?" 26–27.

27. AO, RG 84, OWRC, Subject Files, "Detergents: Miscellaneous Information," "A Brief to the Ontario Water Resources Commission, Presented by the Canadian Manufacturers of Chemical Specialities Association," May 1963.

28. Ibid., "Meeting with Detergent Industry, August 14th, 1963."

29. Next to the association's assertion, on the OWRC copy of an industry brief, that "no such situation [similar to Europe] has been created in Canada by the use of synthetic detergents," someone wrote: "We dispute this." See "A Brief to the Ontario Water Resources Commission, Presented by the Canadian Manufacturers of Chemical Specialities Association," May 1963. See also AO, RG 84, OWRC, Subject Files, "Detergents: Miscellaneous Information," "Meeting with Detergent Industry, August 14th, 1963."

30. For sample newspaper items, see *The Financial Post* (March 30, 1963); *The Globe Magazine* (May 18, 1963), reprinted in Conservation Council of Ontario, *Bulletin* 10 (May 1963): 3. Also *Globe and Mail* (June 21, 1963).

31. AO, RG 84, OWRC, Subject Files, "Detergents: Miscellaneous Information," Beckett to McDonnell, June 21, 1963.

32. Ibid., OWRC, Central Records, "Wentworth County Resolutions, 1963," David Caverly to Wentworth County Council, May 21, 1964.

33. Ibid.

34. Jones, "Does LAS Spell 'Pollution Free'?" 24.

35. Canadian Institute on Pollution Control, *Newsletter* (1964): 19; AO, RG 84, OWRC, Central Records, "Resource Ministers Council, 1962," Pollution and Our Environment newsletter, *Resources 2* (December 1965). See also "Detergents Made Biodegradable," *Water and Pollution Control* 104 (February 1966): 27; and Water Management Committee of the Canadian Manufacturers of Chemical Specialities Association, "Detergents and the Aquatic Environment," 4–5.

36. For example, see Ontario. Ontario Legislature. *Debates*, December 19, 1962, 473; March 11, 1963, 1576–1587; January 31, 1964, 347–358; February 24, 1964, 883; April 21, 1964, 2253–2326; March 18, 1965. A good gauge of the pollution debate at the federal level is the annual proposal for an amendment to the *Criminal Code* to make water pollution nuisance punishable under the *Criminal Code*. W.L. Herridge, NDP member for Kootenay West, first proposed the amendment in 1961. He made the proposal annually between 1961 and 1968 while he sat in the House. For sample debates, see Canada, House of Commons, *Debates*, June 2, 1961, 5793–5801; February 13, 1962, 822–827; February 1, 1963, 3366–3375; July 7, 1964, 1943–1950. Select newspaper stories and editorials on general pollution issues: *Toronto Telegram Magazine* (August 31, 1963); *Toronto Telegram* (May 6, 1964); *Simcoe Reformer* (May 1, 1964); *Globe and Mail* (May 5, 6, 12, and 20) and *Globe Magazine* (June 13, 1964); *The Montreal Star* (May 13 and 22, 1964); *Windsor Star* (May 5, 1964); and *Toronto Star Weekly* (June 13, 1964).

37. Davey, "Eutrophication and Detergents," 22–25; J.M. Appleton, "'Fertility Pollution': The Rapidly Increasing Problem," *Water and Pollution Control* 106 (June 1968): 26–27 and 44; and Ashworth, *The Late, Great Lakes*, 129–136.

38. Gilbert Masters, *Introduction to Environmental Engineering and Science* (Englewood Cliffs, 1991), 116–118; also Ashworth, *The Late, Great Lakes*, 126.

39. Davey, "Eutrophication and Detergents," 22–25; Appleton, "'Fertility Pollution,'" 26–27 and 44; also Jones, "Does LAS Spell 'Pollution Free'?" 24–25. See also Masters, *Introduction to Environmental Engineering and Science*, 134–146.

40. International Joint Commission (IJC), *Interim Report of the International Joint Commission United States and Canada on the Pollution of Lake Erie, Lake Ontario and the International Section of the St. Lawrence River* (Ottawa, 1965), 3–5. The OWRC had been investigating the appearance of cladophora since 1958, the year of the first significant bloom after the commission's creation. See OWRC, *Third Annual Report, 1958*, 66–67, and annual reports through to the 1970s; also OWRC, *A Report on Algae Cladophora* (Toronto, 1958); OWRC, *Cladophora Investigations — 1959 — A Report of Observation on the Nature and Control of Excessive Growth of Cladophora sp. in Lake Ontario* (Toronto, 1959); and Duncan McLarty, *Cladophora Investigations — 1960 — A Report of Observation on the Nature and Control of Excessive Growth of Cladophora sp. in Lake Ontario and Lake Erie* (Toronto, 1960). Until the late 1960s, commission investigations focused on controlling algae through the application of chemical algicides and through mechanical means of collecting inshore growth.

41. IJC, *Interim Report, 1965*, 6. Eutrophication is a gradual, natural process whereby organic wastes wash into a lake, decompose, and consume oxygen.

42. The IJC also had quasi-judicial powers under the treaty, being the arbiter of boundary water diversion. For its earlier pollution findings, see IJC, *Final Report on the Pollution of Boundary Waters* (Ottawa, 1918); *Final Report on the Pollution of Great Lakes Connecting Channels* (Ottawa, 1951); *Report of the International Joint Commission United States and Canada on the Pollution of Rainy River and the Lake of the Woods* (Ottawa, 1965).

43. IJC, *Interim Report, 1965*, 16. The report also recommended that the construction of combined sanitary and storm sewers be prohibited and that the process of separating combined sewers then in existence be started. During heavy rainfall or spring runoff, combined sewers often outstripped treatment facility capacity and spilled untreated effluent into the lakes and rivers, increasing the phosphorus load, and risking bacterial contamination of the receiving waters.

44. *Globe and Mail* (February 3, 1966). The *Toronto Star* had raised the issue several years previously (February 3, 1962).

45. Canada, House of Commons, *Debates*, February 8, 1966, 934–935.

46. AO, RG 84, OWRC, Central Records, "Great Lakes: Public Enquiries, 1964–76," Greg McConnell to Premier John Robarts, January 24, 1966. Other letters in the file include: Ora Patterson, Hamilton Local Council of Women to OWRC, December 30, 1965; Fred Palmer to OWRC, December 28, 1965; George Burt, Canadian director, Canadian Region, UAW to Prime Minister Lester B. Pearson and Premier John Robarts, February 23, 1966. See also ibid., "Public Relations Information, 1966."

47. D.S. Caverly, "What Are We Doing about Pollution?" *Water and Pollution Control* 104 (September 1966): 50. Emphasized text is Caverly's. For examples, see AO, RG 84, OWRC, Central Records, "Public Relations Information, 1966," OWRC to Pierre Berton and Charles Templeton, May 12, 1966; "Public Relations—Public Speaking, D.S. Caverly," Caverly address to CCO, May 16, 1966; "Public Relations, 1966," OWRC press release, May 26, 1966; "Public Relations, 1967," OWRC press release, March 20, 1967.

48. Water Management Committee of the Canadian Manufacturers of Chemical Specialities Association, "Detergents and the Aquatic Environment," 1–11.

49. Canadian Council of Resource Ministers, *Proceedings: Pollution and Our Environment* (Ottawa, 1966), 151–172.

50. Owram, *Born at the Right Time*, 216–247. See also A. Paul Pross, *Group Politics and Public Policy*, 2nd ed. (Toronto, 1992), 1–17.

51. Interview with Donald Chant, Toronto, February 5, 1997. Pollution Probe Foundation Library and Archives, "'Air of Death' Pollution Probe Brief to CRTC," March 5, 1969. Also see Donald Chant, "Pollution Probe: Fighting the Polluters with Their Own Weapons," *Science Forum* 14, no. 3 (April 1970): 19–22; and AO, F1O58, Pollution Probe Foundation Papers, MU 7328, "Pollution Probe History," n.d. In response to ongoing complaints about fluoride pollution in Port Maitland and Dunnville, and the CBC program, the provincial government appointed a three-person committee to study the problem. The report criticized the producers of the CBC program for exaggerating, even falsifying, some of the evidence of fluoride poisoning. George E. Hall, W.C. Winegard, and Alex McKinney, *Report of the Committee Appointed to Inquire into and Report upon the Pollution of Air, Soil, and Water in the Townships of Dunn, Moulton, and Sherbrooke, Haldimand County* (Toronto, 1968).

52. AO, F1O58, Pollution Probe Foundation Papers, MU 7328, "Aims, Objectives, Policies," n.d.

53. Interview with Donald Chant, Toronto, February 5, 1997. Donald Chant, *Pollution Probe* (Toronto, 1970), v; also AO, F1O58, Pollution Probe Foundation Papers, MU 7328, "Advisory Board," n.d. For the Don River funeral, see *Toronto Star* (November 17 and 18, 1969); *Toronto Telegram* (November 17, 1969); *Globe and Mail* (November 17, 1969); University of Toronto *Varsity* (November 19, 1969).

54. The International Lake Erie and Lake Ontario-St. Lawrence River Water Pollution Boards, *Report to the International Joint Commission on the Pollution of Lake Erie, Lake Ontario and the International Section of the St. Lawrence River* (Ottawa, 1969).

55. Ibid., 10–11. Also see Canada, National Archives (NA), RG 89, Water Resources Branch, vol. 509, file 7875-2, Pt. I, "Report of the Task Force on Phosphates and Pollution from Detergents," December 23, 1969.

56. Ibid., Memo to A.T. Davidson, ADM (Water) Department of Energy and Resources Management from A.T. Prince, director, Inland Waters Branch, December 10, 1969. See also McGucken, "The Canadian Federal Government," 160–161.

57. NA, RG 89, Water Resources Branch, vol. 509, file 7875-2, Pt. 1, "Report of the Task Force on Phosphates and Pollution from Detergents," December 23, 1969.

58. See UC, Library and Archives, Docket 83-2-4: 1–6. Also see the *Hamilton Spectator* (February 3, 1970).

59. IJC, Library and Archives, Docket 83-2-4: 2, "Briefs: Erie," Dr. Frank H. Healey, January 20, 1970, and W.R. Chase, January 20, 1970; also Docket 83-2-4: 5, "Briefs: Hamilton," Alan Rae, February 2, 1970, and John Dixon, February 2, 1970. The condescending tone of these briefs angered several housewives present, who indicated that they were more interested in the future of the environment than in how white they could get their family's laundry. See the *Hamilton Spectator* (February 3, 1970). CCO noted: "Any current emphasis upon 'whiteness' appears to originate from industry-sponsored advertising campaigns, and informed housewives have left no doubt that they would be willing to sacrifice both some cost savings and some 'whiteness' to stem the deterioration of our waters." IJC, Library and Archives, Docket 83-2-4: 5, "Briefs: Hamilton," the Conservation Council of Ontario, February 2, 1970.

60. Ibid., "Briefs: Hamilton," Pollution Probe, February 2, 1970. Other environmental groups that presented briefs at Hamilton included: Bryan Kingdon for CHOP — Clear Hamilton of Pollution; Stewart Hilts for Pollution Probe, London; Committee of a Thousand; and the CCO. These dealt with pollution more broadly.

61. *Toronto Daily Star*, the *Toronto Telegram*, and the *Hamilton Spectator* (February 9, 1970); *Globe and Mail* (February 10, 1970).

62. A0, F1058, Pollution Probe Foundation Papers, MU 7346, "Probe Newsletter 1969–1972," *Probe Newsletter* 2 (March 31, 1970). See Chant, "Pollution Probe," 20–21, and *Business Week* (August 8, 1970). For the press release, see Pollution Probe Foundation Papers, MU 7346, "Press Releases,

1970," Probe press release, February 9, 1970. It is probably no coincidence that Probe's tactics closely mirrored those of the Algonquin Wildlands League, whose leader, Douglas Pimlott, was also a biology professor in the Department of Zoology at the University of Toronto. See Killan, *Protected Places*, 155–204.

63. Canada, House of Commons, *Debates*, February 6, 1970, 3293–3295.

64. Canada, House of Commons, Committee on National Resources and Public Works, *Standing Committee on National Resources and Public Works, Minutes of Proceedings and Evidence* (Ottawa, 1970), 14: 1–14: 47.

65. The *Toronto Telegram* (February 10, 1970).

66. AO, F1O58, Pollution Probe Foundation Papers, MU 7346, "Probe Newsletter 1969–1972," *Probe Newsletter* 2 (March 31, 1970): n. 3, 3–5.

67. OWRC disquiet over the initial provisions of the Canada Water Bill stemmed from concern about duplication of programs, the fight for a limited number of trained water resource personnel, and the federal proposal to adopt river basin organization rather than the regional approach favoured by Ontario. These concerns were addressed during the amending process. See AO, RG 84, OWRC, Central Records, "Legal Acts, Canada Water Act, 1969," OWRC memo, August 20, 1969 and ibid., "Legal Acts, Canada Water Act, Jan.–June, 1970," Minutes of meeting held in Quebec City, January 27, 1970; OWRC memo, January 28, 1970; and OWRC memo, April 30, 1970.

68. NA, RG 89, Water Resources Branch, vol. 52, file 7709-1-2, "Canadian Initiatives Concerning the Eutrophication of the Lower Great Lakes," n.d., 1–2. See also ibid., ACC 88-89/059, box 24, file 7354-1, Pt. 5, "Phosphorus Concentration Control Regulations"; and AO, F1058, Pollution Probe Foundation Papers, MU 7346, "Probe Newsletter, 1969–1972," *Probe Newsletter* 2, no. 3. See also McGucken, "The Canadian Federal Government," 163.

69. NA, RG 89, Water Resources Branch, ACC 88-89/059, box 24, file 7354-1, vol. 6, Memo to J.P. Bruce, dir. Canadian Centre for Inland Waters, from T.R. Lee, September 3, 1970.

70. By March 1970, 91 percent of Ontarians polled had heard about pollution; 78 percent believed the situation was "very serious," while a further 19 percent believed it "fairly serious." Canadian Institute of Public Opinion, *The Gallup Report: Canada's Only National Opinion Poll with Publicly Recorded Accuracy* (Toronto, March 25, 1970). In December 1970, 65 percent of Canadians polled wished to see the government devote resources to reducing air and water pollution, while the second-place option, reducing unemployment, received the support of 59 percent. Canadian Institute of Public Opinion, *The Gallup Report* (December 2, 1970).

71. *Financial Post* (October 16, 1971).

Chapter Sixteen
Where the Scientists Roam
Ecology, Management, and Bison in Northern Canada

John Sandlos

On October 3, 1893, Royal Northwest Mounted Police Commissioner L.W. Herchmer wrote urgently to the comptroller in Ottawa that "I have the honour of drawing to the attention of the Department the imminent danger of the total extermination of the small band of wood bison, at present ranging in the country in the vicinity of Great Slave Lake."[1] Herchmer's immediate concern was the reports he had received of deep snow during the last winter, which had allowed Native hunters easy access to the wood bison and precipitated a decline in the populations to an estimated total of 150 animals, but his correspondence marked the beginning of a series of debates concerning the condition and fate of this last free-roaming herd of bison in Canada. The debate has continued to the present day. Herchmer's letter, combined with reports of a steep decline in the wood bison that had appeared in the popular writings of famous hunter-naturalists such as Caspar Whitney and Warburton Pike, prompted the Dominion government to enact a closed season on buffalo through the Unorganized Territories Game Preservation Act of 1894. In 1897, the federal government established a North West Mounted Police detachment in Fort Smith to enforce the restrictions on buffalo hunting, and the first two convictions against Native hunters for game offences were successfully prosecuted in 1898.[2]

To assess the effectiveness of the new game laws, the government made its first tentative steps to gather scientific information about the northern bison from the formal and popular reports of geologists, amateur naturalists, police officers, and zoologists. Most notably, the three journeys of Major A.M. Jarvis, RNWMP, the American naturalist E.A. Preble, and the famous author-naturalist Ernest Thompson Seton to the Salt River and Little Buffalo River regions in 1907 produced an estimate of 300 animals, a number that was widely circulated in Seton's popular travelogue *The Arctic Prairies* (1911). A ground survey of the buffalo range by the zoologist Francis Harper in 1914 produced an estimate of 500 bison, and the geologist Charles Camsell calculated a total herd of 300 bison in 1916 during expeditions conducted under the auspices of the Canadian Geological Survey. These early naturalists never acknowledged the absurdity of attempting to estimate buffalo numbers based on a few short excursions into a huge range that included thousands of square kilometres of forest cover and wetlands. Nonetheless, they used their suspect assumption that the northern bison herds were in decline—for the most part due to Native hunting, according to Jarvis and Seton—as a justification for establishing a more active federal wildlife administration in the Northwest Territories.[3] The Forestry Branch of the Department of the Interior, for example, took responsibility for the wildlife in dominion forests in 1911 and established a regular game warden service operating out of Fort Smith. One year later, Maxwell Graham, who was chief of the Parks Branch's Animal Division, began a bureaucratic campaign for a large bison preserve in the Slave-Athabasca region that resulted in the creation of Wood Buffalo National Park in 1922. The practice of wildlife "science" had for the first time identified a wildlife crisis in northern Canada, and the state had responded by creating a preserve for the threatened animals.[4]

But the question of what to do with the wood bison of northern Canada did not end with the creation of the national park; the park was merely the precursor to a program of intense bison management instituted by federal wildlife conservationists. From the mid-1920s to the early 1970s, bison have been at various times relocated, rounded up, herded, corralled, vaccinated, and slaughtered. From 1925–1928, 6,670 plains bison were transferred from the overcrowded range at Buffalo National Park near Wainwright, Alberta, to the supposedly understocked range in Wood Buffalo National Park, a process that resulted in hybridization

between the plains buffalo and wood buffalo and the infection of the northern herds with tuberculosis and brucellosis. An annual hunt of roughly 20 animals per year was inaugurated in 1929 to supply meat to local missions and hospitals. This program was greatly expanded beginning in 1951, as a large-scale test and slaughter program designed to eliminate disease and to produce meat for commercial markets resulted in the killing of between 120 and 900 bison annually over the next two decades. The outbreak of anthrax among bison herds outside the northeast boundary of the park in 1962 (and within the park in 1964) resulted in new management programs designed to herd bison away from infected areas using helicopters and to corral and inoculate as many animals as possible each year. Attempts were also made to depopulate the herds in the Grand Detour area to create a buffer zone between bison in the park and those in the anthrax-infected Hook Lake area farther to the northeast in November 1964 and March 1965. Clearly, the animals Maxwell Graham rhapsodically described as "the last of their species living to-day under absolutely free and wild conditions" have been managed as if they were nothing more than a herd of beef cattle.[5]

*****

How can we explain the variable position of science in the debates over the fate of the wood bison? Certainly, on the most basic level, the shifting relationship between scientists and wildlife managers within Wood Buffalo National Park can be understood in the context of broader intellectual and historical trends in North American wildlife management. Evolving ecological concepts such as the "balance of nature," carrying capacity, and optimal herd productivity were all used to defend both non-interventionist and intrusive approaches to bison management. Shifting institutional arrangements both inside and outside the federal government also influenced the values of the scientists working with the northern bison. By the 1920s, for example, several zoology programs had been established at Canadian universities. The programs provided a relatively independent institutional base for critics of federal bison policy during this period.[6] Conversely, the creation of a federal government agency solely responsible for wildlife science in 1947 tended to stifle criticism of bison policy as broad institutional alliances evolved among wildlife researchers and park managers during the post-war period. Other factors more specific to their particular time and place have also influenced scientists who worked with the northern bison — the impact of northern development programs on wildlife policy, the immense ontological significance of the last free-roaming herd of wood bison, and perhaps even the individual personalities involved in the controversies have all contributed to the shifting and variable political position of science during the debates over the fate of the wood bison. Certainly, all these social influences on the practice of wildlife ecology in the Canadian North indicate, in accordance with the critics of ecology, that science is not a politically neutral or "value-free" institution. Yet the widely divergent political relationship of scientists to the bureaucratic managers of the wood bison also suggests that wildlife ecology is not captive to one particular set of political interests, but instead possesses more of a capacity for internal debate, self-criticism, and independent thought than the critics of science have thus far allowed.

FINDING RANGE FOR BUFFALO

By the beginning of the 1920s, the Canadian government could justifiably proclaim that they were world leaders in the conservation of American bison. The government had purchased one of the last herds of plains bison from the Montana rancher Michael Pablo in 1907, and between 1909 and 1914, 748 plains bison were transferred to Buffalo National Park near Wainwright, Alberta. But by the early 1920s, park managers faced the dilemma common to all buffalo conservation programs on the Southern Plains: what to do with irrupting bison populations on fenced ranges that were surrounded by land designated for agricultural purposes. The government did attempt to control the bison population through the issue of hunting licences and the provision of breeding stock for ranchers, but when these programs failed to reduce the bison population, the Parks Branch began to slaughter surplus animals in 1921. There was immediate public opposition to the slaughters (particularly among humane societies), which reached a peak in 1923 due to several reports in major newspapers that a Hollywood film company intended to make a movie of the annual kill using "authentic" costumed Natives to hunt down the animals from horseback with spears. The negative publicity prompted wildlife officials within the federal government to consider a management scheme that would forever tarnish their image as "saviours" of the bison and earn them the enmity of zoologists and bison conservationists throughout North America.[7]

The impetus to find a "humane" alternative to the slaughter program at Buffalo National Park appears to have originated in the office of W.W. Cory, the deputy minister of the Interior. On May

26, 1923, he wrote Parks Commissioner James Harkin with the suggestion that "instead of slaughtering [the surplus bison at Wainwright] it would be a good idea to transfer some of the healthy young stock to the Wood Bison Reserve administered by the Northwest Territories Branch."[8] A meeting was held on May 30 to discuss the idea with Cory, Harkin, O.S. Finnie (director of the Northwest Territories and Yukon Branch), Dr. F. Torrance (veterinary director general with the Department of Agriculture), A.G. Smith (superintendent of Buffalo National Park), and Maxwell Graham (chief of wildlife, Northwest Territories and Yukon Branch) in attendance. Graham, the individual most responsible for the creating the buffalo preserve, had the most immediate professional and personal interest in the fate of the wood bison. In the days leading up to the meeting, Graham expressed concern to his superiors about the possible spread of tuberculosis from the infected Wainwright herd to the wood bison. He also hinted at possible objections from scientific circles: "[I]t is a question," he wrote, "whether the authorities, such as Doctor W.T. Hornaday, would approve of plains bison being introduced and interbred with the 'wood-buffalo.'"[9]

Whatever his concerns in the days leading up to the meeting, Graham emerged a convert to the transfer proposal. It is not clear what swayed his opinion, but Torrance's suggestion that separating yearling animals from the main herd and testing them for tuberculosis before shipment would reduce the risk of disease transmission appears to have provided the degree of "scientific" legitimacy to the proposal that Graham and the rest of the committee was seeking.[10] Graham became the key proponent of the transfer program both publicly and within the federal

government. His tenacious commitment to the program—some at that time might have termed it narrow-mindedness—left little room for compromise. To achieve his goals, Graham was willing to dismiss the opinions of leading zoologists, misrepresent the views of his colleagues, and ignore expert advice he had received from within the civil service. In one glaring example, Graham continually cited Torrance's view that shipping young animals would reduce the risk of tuberculosis transmission to negligible levels, but he neglected to mention that a key condition for Torrance's approval of the transfer plan was the separation *and* testing of the animals for tuberculosis and that a decision had been made to dispense with these tests due to potential cost overruns.[11] W.J. Rutherford, dean of Agriculture at the University of Saskatchewan, forwarded a critique of the transfer proposal authored by veterinarian Seymour Hadwen, suggesting Torrance's comments had been misinterpreted by departmental officials, and that the only way to eliminate the risk of TB transmission was to separate "transfer" calves from cows at birth. Graham replied that the danger of tuberculosis spreading among animals who were widely dispersed in the vast open range of Wood Buffalo Park was "extremely remote."[12] If Graham provided no evidence for his assertion, perhaps it was because he was fully aware of the risks associated with disease transmission. In an internal memo, he offered a very different interpretation of Torrance's recommendations than the one he trumpeted in public:

It would seem therefore in Doctor Torrance's opinion we must face a certain risk of infection from the introduction of even young, tested, buffalo coming from the infected herd at Wainwright.... Since Dr. Torrance has given his opinion it is hardly proper for me to say more on the matter of possible infection.[13]

For Graham, no level of risk was going to derail the transfer project. The letter concludes with a discussion of the "perfect feasibility" of the project from a practical point of view.

*****

There can be little doubt that the debate over the bison transfer in northern Canada proceeded on as many of the same terms as the larger, continent-wide debate between zoologists who advocated less intrusive forms of wildlife management and government bureaucrats devoted to maximizing the productive potential of wildlife herds. When Graham drafted an article for the December 1924 issue of the *Canadian Field-Naturalist*, where he claimed the transfer project was meant to "save" a Wainwright calf crop "which otherwise cannot, apparently, be saved," the response in subsequent issues of the journal was universally negative.[14] Respected naturalist W.E. Saunders wrote a letter calling for abandonment of the scheme, and Cornell zoologist Francis Harper publicly challenged the Canadian government to refer the transfer scheme to a vote at the annual meeting of the American Society of Mammalogists in 1925.[15] Invoking the familiar theme of an inviolable nature, Harper wrote that the transfer project "raises anew the old question of man's interference with nature, which, in too many cases, is alike unnecessary and unjustifiable." Harper chastised his opponents in government for ignoring the expert advice of zoologists, arguing that "too many serious mistakes have been made in the past through failure on the part of legislators and other

government officials to consult zoological authorities in conservation matters."[16]

At the same time American scientists were challenging the federal government, William Rowan, a zoologist at the University of Alberta, was organizing colleagues around Canada to oppose the bison transfer. Rowan was no stranger to controversy: since 1920 he had feuded with his university's president, Henry Tory, who pressured Rowan to conduct much of his path-breaking work on migratory birds at home and on his spare time because it did not accord with the program of applied and economically useful scientific research Tory hoped to build at the university.[17] Certainly, the defence of pure scientific research as a necessary prerequisite to intrusive management programs was at the forefront of the arguments used in the letters from prominent zoologists Rowan assembled and forwarded to Ottawa in April 1925. Scientists such as A.K. Cameron at the University of Saskatchewan, B.A. Bensley at the University of Toronto, and C. McLean Frazer at the University of British Columbia argued passionately that the wood bison subspecies (*Bison bison athabascae*) identified by Rhoads in 1898 would be lost to science forever if allowed to interbreed with the introduced Plains bison. Nevertheless, the preservationist argument did not rest solely on the erudite taxonomic value of the wood bison; the intrinsic public value of the species was also cited as an argument against the transfer. Cameron suggested that the wood bison were "a natural asset in which all Canadians must be interested," and Rowan argued that "the proposed introduction is to be greatly regretted as it means the permanent loss to Canada and to the world of the largest known living bison." A second major concern of the zoologists was the risk of the Wainwright animals infecting the northern bison with tuberculosis. They urged a cautionary approach to the problem of overcrowding at Wainwright and at times were critical of the very idea of an intrusive managerial approach to wildlife ecology. [...]

The do-nothing stance promoted by the zoologists was an almost completely alien idea to the federal wildlife bureaucrats, who supported the transfer project. Indeed, the pragmatic wildlife officials within the Department of the Interior had routinely promoted intensive management programs designed to increase the numbers of northern wildlife herds—not as a tourist or sport-hunting attraction, as in the United States, but as a potential food supply. In 1919, for example, the minister of the Interior appointed a Royal Commission (whose membership included Parks Commissioner James Harkin) to investigate the feasibility of establishing a meat industry in the Canadian North based on domesticating musk oxen and importing reindeer from Alaska. The final report described wild caribou as a "very valuable national asset, and one the value of which could be greatly enhanced under a definite policy of conservation and development."[18] Despite their smaller numbers, the northern bison were included in this push toward resource development and agricultural expansion. C. Gordon Hewitt, an entomologist in the Department of Agriculture and a founding father of wildlife conservation in the Canadian government, wrote in his seminal 1921 volume on wildlife conservation that "the greatest value of the buffalo ... lies in the possibility of its domestication." He went on to praise recent experiments to cross "surplus" bison with domestic cattle and yaks at the Dominion Experiment Station in southern Alberta, for their potential to add to the agricultural economy of the

Great Plains.[19] Even Maxwell Graham, the "saviour" of the last herd of free-roaming wood bison, readily applied the principles of agricultural production to wildlife management. One of the key arguments he used to promote the transfer project was that "the introduced plains bison, under the leadership and protection of the adult wild ones now in the southern range of the Wood Buffalo Park, should so multiply that a future source of food supply may be assured the natives in the surrounding district. While in the immediate future this project holds out the promise of re-stocking vast areas suitable for the propagation of bison at comparatively little cost."[20]

Graham also responded to the criticism of the transfer in a more direct way by insinuating that the zoologists who opposed the project were ivory-tower critics who were unaware of actual conditions on the ground in Wood Buffalo National Park. He continually cited Charles Camsell's 1916 geological survey, F.V. Siebert's boundary survey in 1922, and his own observations on the same journey with Siebert, all of which suggested that there were two distinct herds in the park—one in the north near the Salt River and one in the southern portion near the Peace River—and that these animals never intermingled because they were separated by a 20-mile band of muskeg. Since the Wainwright animals were only to be released in the southern range, Graham reasoned, the wood bison would remain inviolate in the northern part of the park. If this argument failed to convince the skeptics, Graham claimed that the extinction of a species was not at issue because the wood bison were distinct only due to geographic factors. The Plains bison at Elk Island National Park had, according to Graham, adopted some of the characteristics of the wood bison (darker pelage, larger size, and so on) after only 10 years of habitation in a wooded environment. The Plains bison introduced into the northern range would thus effectively become wood bison, according to Graham, and erudite taxonomic concerns of zoologists were hardly worth consideration. Graham wrote in his "Statement" on the transfer that "a subspecies is merely a geographic race, and the preservation of a subspecies, as such, is more an academic than a practical problem."[21] Graham also insisted that his critics had misunderstood the practical administrative reasons for the transfer. He argued that an expansion of Buffalo National Park was impossible due to the prohibitive expense of land in a settled area, and he rejected Rowan's suggestion that the surplus bison at Wainwright could be relocated farther south in the Birch Mountain area because of the additional costs associated with hiring a warden service when one already existed in Wood Buffalo National Park.[22]

Considering the disparate political, technical, and ethical positions between the rival factions in the debate over the transfer proposal, it is not surprising that the tensions between the two camps boiled over in the summer of 1925. In the days leading up to the first shipment of Wainwright bison, the project was denounced in a desperate and almost belligerent article that appeared in the July issue of the *Canadian Forum*. It is not clear who wrote the article, but in what is perhaps a reflection of the chilly climate of the period for pure researchers who criticized government authorities, it was authored anonymously by a "Canadian Zoologist." The article firmly contested many of Graham's key arguments about the low-risk nature of the transfer. It suggested that the risk of TB transmission was acute no matter what precautions were

taken, and that even if there were separate herds in the park, the introduction of large numbers of bison to the southern range would surely encourage a crossover to the north in a search for food. What clearly vexed the author of the article the most was Graham's assertion that "practical" bureaucrats should dismiss the arguments of trained zoologists merely because those arguments were speculative and theoretical. The author cited an article from the May 5 edition of the *Edmonton Journal* in which, he reported, the Department of the Interior acknowledged the protests of leading zoologists in North America but decided to disregard them since "their own experts are better qualified to judge the policy, because of experience and practice, than are zoologists at a distance." The Canadian Zoologist countered this argument with the suggestion that "it is greatly to be regretted [that] when a government department makes reference to its experts it refrains from divulging their names." The author goes on to question the zoological acumen of the government agents who have conducted biological investigations in the North:

[W]e know that numerous government geologists have visited the wood bison park; but, even if a thousand geologists had investigated conditions there, their collective opinions on the race *athabascae* would not be of equal value with the conclusions derived from a single skull by a single trained zoologist, even if he had made his investigation in Honolulu or at the South Pole.

The article concludes with a condemnation of the transfer project that was perhaps unnerving for a dedicated conservationist such as Graham: "[N]ever before in the annals of conservation, as far as we are aware, have the last survivors of a unique race of animals been knowingly obliterated by a department of conservation."[23]

In the end, the arguments of the zoologists were not enough to sway a bureaucracy that was determined to solve the problem of overcrowding they faced at Buffalo National Park by stocking a supposedly underutilized northern range. On June 21, 1925, the first shipment of 196 Plains bison arrived by barge and were released at Labutte along the Slave River. (Two died in transit.) By the beginning of August, 1,926 Plains bison had been relocated to Wood Buffalo National Park. Over the four summers from 1925–1928, a total of 6,673 bison were shipped by rail and barge and were discharged to their new northern range. In the short term, many hailed the transfer project as a practical success. The vast majority of the bison survived the arduous transportation process, and the park wardens reported that the animals were responding well to their new environment. Most important for supporters of the transfer, many of the Plains bison began to migrate immediately upon arrival toward the muskeg country south of the Peace River. This raised fears that the buffalo might be trying to return to Wainwright, but also confirmed, for supporters of the transfer, that the Wainwright animals were not mixing with the northern herd of wood bison.[24] Laudatory press reports on the transfer appeared in newspapers across North America, and even some former critics were persuaded to soften their stance toward the program. The deputy minister of the Interior, W.W. Cory, for example, wrote to Finnie from Washington in January 1926 to report that he had managed to persuade famous conservationist William T. Hornaday and the president of the American Bison

Society, Edmund Seymour, of the sound reasoning behind the transfer program. M.S. Garretson, secretary of the American Bison Society, also wrote to Finnie later that year to say he was pleased that the Wainwright bison were "doing nicely."[25]

If criticism was somewhat muted in the wake of the transfer, it did not dissipate entirely. An article by Dr. James Ritchie appeared in the February 1926 issue of the prestigious journal *Nature* denouncing the project and recommending that specimens of pure wood bison be secured by museums for posterity. A year later, a brief article in the same periodical reported that the wood bison and the Plains bison had begun to cross-breed, and that Rowan's recent examination of a buffalo skeleton had proven definitively that the wood bison were a distinct subspecies.[26] Rowan had travelled to Wood Buffalo Park in the summer of 1925, and he produced a report on his investigations that was sharply critical of the federal government's administration of the wood bison range. In addition to reiterating the reasons for his opposition to the introduction of the Wainwright bison, Rowan noted a lack of published information on the wood bison since 1908 and the lack of available data with which to compute accurate numbers of wood bison in the region. In a pointed critique of the scientific ineptitude of the federal administration, Rowan refuted the idea that the transfer had been conducted with expert advice. He recommended that a general zoologist be appointed to the Department of the Interior and that trained zoologists control all decisions pertaining to wildlife.[27]

There is some irony attached to Rowan's comments—it was, in the end, his ability to rally a scientific community independent from the management imperatives of the federal bureaucracy that provided the most trenchant critique of the transfer proposal. Faced with an overcrowding problem at Wainwright, adverse publicity due to large slaughters of bison, and huge financial costs associated with any effort to expand Buffalo National Park, federal bureaucrats were willing to construct any biological argument (for example, the two-herd theory) and ignore any internal warnings they did receive (for example, Torrance's reservations about tuberculosis) in order to promote a scheme that was, in essence, an administrative solution. Because the zoologists who opposed the transfer were detached from a bureaucratic structure that demanded manageable solutions to complex problems, they were able to freely criticize the transfer proposal by exposing the fallacies and inaccuracies embedded in the federal government's arguments. Of course, the intense opposition of the zoologists to the bureaucratic managers of Wood Buffalo Park also ensured that the zoologists were marginalized from the decision-making process.

*****

Although a non-interventionist approach to ecology and environmental management has come under intense criticism in recent years—particularly for its historical role in promoting fire suppression in pristine wilderness areas—hindsight has clearly vindicated the critics of the transfer proposal.[28] The original wood bison of Wood Buffalo National Park have become infected with tuberculosis and are today considered by some zoologists to be hybrids with the introduced Plains bison.[29] The problem of overcrowding at Wainwright was never solved—large annual slaughters continued until the park bison were liquidated in 1939, and Buffalo National Park was deleted from the system

in 1947. Harvard scientist Thomas Barbour effectively summed up the character of the entire transfer program when he wrote in 1932 that it was "one of the most tragic examples of bureaucratic stupidity in all history."[30] Perhaps stung by such strong and persistent criticism, the federal government moved, in the years following the transfer, to build its own capacity to conduct scientific research on northern wildlife. Never again would the boundary between science and management be so clear in Wood Buffalo National Park.

THE WILDLIFE INVESTIGATIONS OF J. DEWEY SOPER

The immediate priority for officials in the Department of the Interior in the years following the bison transfer was to secure some measure of the success of the program. To this end, the department devoted a great deal of effort to determining the number of bison in the park. Park warden diaries for the years following the transfer, for example, reported meticulously on the numbers of buffalo seen in each patrol district. By February 1931, the district agent at Fort Smith, J.A. McDougal, had gathered enough data to wire O.S. Finnie in Ottawa with an estimate of 12,000 bison in the park with a 20 percent annual increase in the herd.[31] Nonetheless, as early as 1929, Finnie thought it would be prudent to acquire the services of a trained biologist to conduct an extensive survey of the bison range and accurately assess the numbers of buffalo in the park. A report on the first aerial survey ever conducted in the park in February and March 1931 confirmed Finnie's suspicion that no accurate data yet existed on the bison population in Wood Buffalo National Park. A mere 1,467 buffalo were counted on the survey. The number produced such a huge discrepancy with the warden's estimates that the department was left with no precise population numbers with which to conduct a public-relations campaign in support of their approach to bison management.[32]

Finnie chose biologist J. Dewey Soper, who had been conducting ornithological investigations on Baffin Island for the Department of the Interior since 1923, to carry out the bison study. Despite his position within the federal bureaucracy, Soper's scientific interests were not restricted to the narrow concerns of his administrative superiors. In fact, from the very beginning of the bison study in 1931, there was a distinct tension between Soper's desire to produce a wide-ranging and significant biological study of Wood Buffalo Park and the desire of administrators (primarily Deputy Minister H.H. Rowatt and the chair of the Lands Board, H.E. Hume) for Soper to restrict his work to accurate reporting on the population of the wood bison. In his proposal for the research program, Soper insisted that the study include detailed work on topography and plant cover, as well as a scientific collection of all birds and mammals known to exist in the park. Hume's instructions to Soper did allow for judicious study of species other than bison, but they nevertheless clearly identified the buffalo census as the first priority of the research project. Soper was also instructed by Hume to use the census data to fulfill a management objective by reporting on the advisability of establishing an abattoir in the park to produce buffalo meat for relief purposes and for sale to White settlers and mining communities.[33]

The study began well enough from a practical perspective with a journey from Fort Smith toward Pine Lake on June 1, 1932, but Soper was soon mired in a series

of conflicts with his superiors that would persist throughout the investigation. He continually frustrated Hume and Rowatt by refusing to speculate on the total numbers of bison in the park and by forwarding detailed reports on non-economic species of wildlife. In just one example, Soper was clearly enthralled by the variety of migratory waterfowl he observed on a trip to the Peace-Athabasca Delta in the early summer of 1933. He sent a detailed report on his field notes to Hume. The report provided no recorded sightings of buffalo, but stated that "the trip represented one of the most interesting and fruitful from the standpoint of wildlife observations."[34] When Hume and Rowatt, near the beginning of autumn, pushed Soper to provide an estimate of the bison population, he pleaded for more time and noted the difficulties inherent to an accurate assessment of a large wildlife population spread unevenly over a huge territory. To assuage the deputy minister, Soper provided frequent and detailed reports outlining the vast areas of the park that he was able to cover during his investigations—an extraordinary 8,015 miles from June 1932 to May 1934. Above all, Soper asserted that his fidelity to the rigours of natural science took precedence over the administrative priorities of the department.

*****

Soper did finally provide an estimate of the bison population in May 1933. His figure of only 7,500 animals must have come as a shock to his superiors, as it was less than the total number of introduced Plains bison (6,673) combined with the estimated number of wood bison in the park at the time of the transfer (1,500 to 2,000). Soper did, however, reassure Hume that the estimate was conservative, and revised the figure upward to 7,700 animals with a maximum population of 8,000 to 9,000 (based on the possibility of 10 percent error) in his final report in May 1934.[35] By the time Soper had completed the manuscript for his monumental publication, "History, Range and Home Life of the Northern Bison," he concluded that the bison population had likely increased to 12,000 animals. On a subsequent investigation to Wood Buffalo Park in 1945, Soper was characteristically vague when he suggested that Warden Dempsey's estimate of 20,000 animals was within the range of possibility, but that no conclusions could be made from his own "superficial examination."[36]

Nevertheless, Soper's relationship as a scientist to the operative management paradigm for bison in the post-transfer era was, in the final analysis, complex and contradictory. Soper was clearly not a yes man. When asked by Hume to comment on Barbour's critique of the transfer proposal, Soper expressed complete agreement with the idea that the project was a case of bureaucratic stupidity. He criticized the federal administration for ignoring expert opinion, claimed that the band of muskeg separating the two main herds was a myth, and suggested there was little doubt that the Wainwright bison had infected the wood bison with tuberculosis. Soper declared that he was simply offering the "independent viewpoint of another 'mammalogist,'" and that he "would be doing less than my duty were I not above-board and perfectly honest in doing so."[37] Yet Soper's reluctance to extend this notion of duty to the public realm suggested that his position within the federal bureaucracy placed certain limitations on his ability to act as an independent scientist. In fact, when Soper presented the idea of publishing a scientific monograph on the northern bison to his superiors, he offered to confine all controversial material from his research to a confidential supplementary report. In just one example of the inconstancies between the two documents, Soper stated in the internal document that the northern bison were infected with tuberculosis; in the public monograph he declared that "the wild herds of Wood Buffalo Park are undoubtedly in a healthy condition for they apparently exhibit as little sign of disease as the great herds of the southern range."[38] Soper also encouraged expanded wolf-control operations—despite a scientific and philosophical awareness of the important role that predators played in a natural ecosystem. This suggests that he was not completely averse to management programs designed to increase the bison population.[39]

In spite of such inconsistencies, there is little doubt that Soper's fidelity to the principles of scientific accuracy and attentiveness to non-economic wildlife during his investigations represented at least a form of tentative resistance to the sustained-yield management model that federal bureaucrats had envisioned for the bison. Soper refused to feed administrators the large numerical projections for the bison population that were necessary to respond decisively to critics of the transfer program and justify commercial meat production. Indeed, Soper's continual assertion of scientific doubt and uncertainty about the status of the bison population constituted an implicit refusal to participate in the management process. For Soper, wildlife science was, in the most ideal terms, an independent check on the potential oversights of managers. Nowhere is this idea more clear than in Soper's comments on the bison transfer, where he argued that "it is always unsafe to make an important decision in matters pertaining to wildlife against the judgment of specialists in this line; their lives are exclusively devoted to such things."[40] This plea for science to play a more prominent role in wildlife management decisions went largely unheeded until after the Second World War, when Canadian wildlife scientists gained a measure of respect in management circles. Ironically enough, it was at the precise historical moment that Soper's devotion to natural science as an independent endeavour unfettered by administrative controls was rejected by a new generation of scientists working in Wood Buffalo National Park.

THE SCIENTIFIC MANAGEMENT ERA: 1950 TO 1968

The Canadian government took very little interest in the bison of Wood Buffalo National Park in the later years of the Great Depression and during the Second World War. Small annual slaughters continued for relief purposes, but other than a casual survey of the buffalo in 1944 by B.I. Love, superintendent of Elk Island National Park, there was no systemic scientific study of the bison and no management program for the animals until the years after the war. Soper's report of a serious decline in the park facilities—overgrown roads, downed telephone lines, and collapsing warden cabins—on his return visit in 1945 suggests that the park had been subject to neglect in the years since his first investigation.[41]

Yet the conclusion of the war brought significant changes to the administration of wildlife in Canada. These changes greatly increased scientific activity in Wood Buffalo National Park and furthered the integration of pure research and applied management programs. The most important development was the creation

of the Dominion Wildlife Service in 1947 (later the Canadian Wildlife Service), which centralized much of Canada's wildlife research infrastructure within the federal government. It was clear from the very beginning that the new service would not function as a pure research body, but would also serve to bolster the productive-management aims of the federal government. At the Dominion Provincial Wildlife Conference of 1945, the first director of the DWS, Dr. Harrison Lewis, called for increases to the scientific research capacity of governments, museums, and universities because "we must realize the importance of Canada's wildlife resources in the post-war development of the country. Ninety per cent of the area of Canada supports wildlife and on two-thirds of the area it is the most important permanent crop."[42]

The new emphasis on resource production among Canadian wildlife scientists in the post-war period was part of a larger international movement among ecologists away from the non-interventionism of an earlier generation. According to Donald Worster, ecologists in the 1930s were already rejecting the organic holism associated with the balance of nature school and moving toward a mechanistic model that was more amenable to production-oriented and managerial interests. [...] By the end of the war, the new emphasis on interventionist science had taken hold throughout the federal wildlife bureaucracy. In the national parks, for example, scientists openly advocated stocking species that had declined and culling others that had become overpopulated. Maintaining balance in wildlife populations, it seemed, could no longer be entrusted to the vagaries of nature but required the active participation of the natural scientist.[43] Moreover, in the eyes of federal officials, the by-products of these wildlife management programs had the potential to produce an economic return. In a paper presented to the 1945 wildlife conference delegates, Dr. Lewis argued that surplus wildlife that was cropped to prevent disease and overgrazing should be utilized as fully as possible to produce revenue for the federal government. In addition to a renewed interest in the economic value of wildlife, the federal government was also fully committed to developing the economic potential of the Canadian North. It was thus abundantly clear by the early 1950s that the protected bison of Wood Buffalo Park would not be excluded from the development rush that had gripped both wildlife scientists and the federal administration.[44]

The momentum for an expanded cropping program in Wood Buffalo Park began to build in government circles shortly after aerial surveys of the bison herds, conducted by the resident CWS mammalogist William Fuller in 1949 and 1951, revealed a large and potentially harvestable population of roughly 12,000 animals.[45] Dr. F.E. Graesser, the veterinary inspector of the Health of Animals Branch, participated in the annual hunt of 1950 and discovered an infection rate for tuberculosis of approximately 25 percent in the 75 animals he examined. Graesser was appalled at the conditions associated with the field slaughter, and recommended improvements to sanitation and haphazard butchering techniques.[46] The reports of increased bison numbers, higher infection rates for tuberculosis, and the apparent need to professionalize the slaughter led to proposals within the Department of Resources and Development to construct portable abattoir facilities at Hay Camp and kill 500 animals the following season. The needs of wildlife science and economic

efficiency could therefore both be served by the expanded slaughter; the infection rate for tuberculosis could be assessed more accurately and, as one memo suggested, "through sound management practices in connection with the care of this herd and the utilization of the meat therefrom, within a period from five to six years, sufficient revenue could be obtained annually from the sale of meat products, etc., to balance this cost of maintaining the Wood Buffalo Park."[47]

There is little evidence to indicate that the CWS scientists who worked in the park during this period were uncomfortable with the commercialization of the northern bison. Fuller openly supported the expanded slaughter and claimed that the thinning of the population would reduce the spread of tuberculosis.[48] Harrison Lewis also supported the proposal at the November 1951 meeting of the interdepartmental Advisory Board on Wildlife Protection in spite of some initial reservations.[49] It was in many ways the scientists working within the Canadian Wildlife Service who provided the momentum behind the slaughter program. In a report authored in 1954, CWS biologist W.E. Stevens articulated four principal aims for bison management: the preservation of the herds, keeping the population within the capacity of the range, reducing the incidence of disease, and the provision of "a cheap source of meat for the resident population of the Northwest Territories and Wood Buffalo Park."[50] These recommendations provided the conceptual basis for bison management in Wood Buffalo National Park throughout the 1950s and 1960s. In the program, the commercial sale of meat was justified by the testing and slaughtering of tubercular animals (and later those infected with bovine brucellosis) to reduce the incidence of disease. Between 1951 and

1967, more than 4,000 bison were killed for commercial sale in park abattoirs at Sweetgrass and Hay Camp. The meat was sold to local missions, the Indian Affairs Branch, the Hudson Bay Company, and to specialty restaurants in southern Canada. In the final year, more than 150 bison were slaughtered to provide meat for sale as quintessentially Canadian buffalo burgers at Expo '67 in Montreal.[51]

There is some indication that CWS scientists and Wood Buffalo Park field staff had begun to have doubts about the slaughter program by the early 1960s. A report produced by CWS biologist Nick Novakowski in February 1961 maintained that the combined effect of the slaughters and severe flooding in 1960 had "decimated" the bison population. He criticized the bison management program for a lack of focus, arguing that "at present no policy exists."[52] Novakowski's superiors quickly advised him that his comments were ill considered, and he was instructed to conduct another survey of the bison range in the Egg Lake area in April. The new census revealed similar results to the first, but a full test and slaughter program proceeded in September 1961, and approximately 400 bison were killed.[53] Widespread flooding the following year rendered the abattoir at Hay Camp unusable and caused a further reduction in the number of bison around the Sweetgrass site. When officials from the Department's Industrial Division requested 50,000 to 100,000 pounds of meat for the southern specialty market, park superintendent B.E. Olson responded by noting that a kill of that size would require the slaughter of prime breeding stock, and "to do so would not be management, or herd reduction, but sheer mass murder." Olson went on to caution that "we have here a fundamental conflict between a scientific and practical

management policy of the buffalo herds in the park and the production of first class meat for the sophisticated 'outside market.' In a normal year there need not be a dangerous conflict between these two but in a year such as this the conflict is irreconcilable."[54]

The emerging debate over the contradictory objectives of the buffalo management program was interrupted by the outbreak of anthrax in the bison herds just outside the northeastern park boundary in the Grand Detour and Hook Lake areas in 1962. More precisely, the introduction of a disease that was thought to represent an acute and deadly threat to the bison had the effect of diminishing further criticism of the intensive management paradigm. When anthrax spread to the park herds in 1964 and 1967, wildlife managers considered the outbreak as an extreme crisis. Immediate and drastic steps were taken to contain the disease. A multi-agency anthrax committee was formed, which subsequently recommended a program to depopulate the herds in the Grand Detour area to create a buffer zone between the infected area and the park bison. Attempts to herd buffalo by helicopter met with only limited success, as animals from outlying areas dispersed into the infected area. Slaughter operations, which resulted in the death of 554 bison, were carried out in November 1964 and March 1965. Buffalo herds within the park were further thinned and attempts were made to inoculate as many animals as possible in massive roundups at the Sweetgrass and Hay Camp abattoirs.[55]

Behind many of these management programs was the recently expanded Veterinary Pathology Section of the Canadian Wildlife Service. Indeed, the almost exclusive focus on disease control in the park afforded CWS veterinary scientists such as L.P.E. Choquette and Eric Broughton the opportunity to take a leading role in the bison management programs.[56] Many of the techniques they employed, such as depopulation, mass inoculation, and the fencing of infected areas, were borrowed directly from agricultural disease eradication programs and had limited or no success with wild animals. Moreover, the narrow focus on disease eradication as the *raison d'être* for bison conservation was firmly tied to the agricultural model of bison management that had been practised by the federal government at least since Maxwell Graham's day. Choquette and Novakowski asserted in a 1968 report, for example, that the bison range could hold 14,000 animals, but was at present "understocked" with only 10,000 animals. The two scientists argued that the preservation of the bison was not the key issue. They wrote that "the low productivity, caused chiefly by brucellosis and malnutrition, and the high mortality do not significantly endanger the population, but at the same time, there is a substantial waste of a resource." Choquette and Novakowski recommended a five-year disease-eradication program involving the fencing of bison herds at various points in the park, the testing and slaughtering of all reactors for tuberculosis and brucellosis within the confined areas, and the eradication of all the rest of the buffalo in the surrounding area.[57]

In many respects, Novakowski and Choquette's management plan represents the logical end-point of the scientific management era. The authors made no attempt to prove that disease threatened the bison with extinction at the population level—the report clearly asserted that saving the bison from the ravages of disease was less important than producing a population that was large enough to

provide meat for relief purposes and the southern specialty market. Furthermore, the report stressed the importance of managerial technique rather than scientific research, highlighting the fact that all the half-measures designed to improve the health of the bison had thus far failed. It was therefore imperative, according to Choquette and Novakowski, that the buffalo be herded into more manageable areas, or destroyed, so the complex ecology of the bison range could be reduced to the fundamental preoccupations of the agricultural scientist—the herd, the disease, the surplus. In the end, Choquette and Novakowski's management plan was never implemented due to apprehension about the costs and possible negative public reaction.[58] The free-roaming bison herds had survived an era of extreme manipulation with their wild status intact. What was lost during the era, however, was Soper's expectation that wildlife ecologists within government would act as a check on the worst excesses of administrators. Scientists and managers had become, for the time being, practically indistinguishable.

A RETURN TO CONTROVERSY: THE ARMAGEDDON OPTION

The disease-eradication program proposed by Choquette and Novakowski was in many ways one of the last echoes of a policy regime within Wood Buffalo National Park that emphasized intrusive management of the bison herds. The transfer of authority over the park from the Northern Administration Branch to the National and Historic Parks Branch in 1965 resulted in a return to management philosophies that, by the early 1980s, emphasized preservation and non-interference with wildlife populations.

The final commercial slaughter of bison in Wood Buffalo Park was carried out in 1967; the last slaughter for relief purposes took place in 1974. New wildlife policies emphasized monitoring and research and permitted management intervention only if the long-term survival of the population was threatened.[59]

But the question of what to do about the bison did not dissolve with the introduction of the new "hands off" approach to wildlife management in the park. By the late 1980s, the imminent eradication of tuberculosis and brucellosis from domestic livestock in Canada and a precipitous decline in the bison population in Wood Buffalo National Park provoked renewed interest in the disease issue. An intergovernmental steering committee that included federal, territorial, and provincial representatives was established in 1986 to discuss the national implications of disease in the park bison. The committee subsequently presented four possible courses of action: maintenance of the status quo, fencing of the park boundary, a combination of fences and buffer zones near the park boundary, and the complete eradication of the park bison with replacement by a disease-free herd of wood bison. In February 1988, these recommendations were forwarded to a Federal Environmental Assessment and Review Panel (EARP) for consideration. The panel invited submissions from interested government agencies, public groups, and individuals, held "issues scooping" hearings in the spring of 1989 and formal public hearings in January 1990. The panel concluded that there was a risk of disease transmission from bison to domestic cattle and recommended the Armageddon option—the complete eradication and replacement of the diseased park bison.[60]

The intense public debate following the panel recommendation stood in stark contrast to the one that emerged around the transfer proposal in 1925. First and foremost, it is clear that the opposition to the slaughter-and-replacement option was not restricted to university-based wildlife scientists. A more critical press, for example, was able to arouse broad public concern over the environmental implications of the depopulation scheme, and opposition from politically organized Aboriginal residents living near the park added a voice from a constituent group that was nearly absent during the debate over the transfer proposal.[61] In addition, there were more cracks within the federal bureaucracy in 1990 than during the transfer proposal debates. In fact, no single administrative perspective prevailed during the controversy. Agriculture Canada clearly supported the slaughter program and acted as a proponent during the environmental review process, but Environment Canada remained ambivalent and presented a brief that the report described as "a thinly veiled version of the status quo."[62] The Wood Buffalo National Park staff, however, unequivocally opposed the proposal. They sent a critique of the EAP report to all Environment Canada employees that suggested the risk of disease transmission to cattle was overstated, that it was impractical to attempt the slaughter of the entire park bison population and that the project was a precedent-setting justification for management interventions in other parks to satisfy external economic interests.[63]

The position of scientists in the debate over the EAP report was equally variable and contradictory. Even among scientists who supported the eradication proposal there were a variety of perspectives and motivations. Some, such as the veterinary pathologists Stacy Tessaro and Gary Wobeser, argued in journal articles and popular publications that the bovine disease reservoir in Wood Buffalo National Park represented a clear threat to domestic livestock in Canada. Tessaro reasoned that the slaughter proposal should go ahead because "the national and international effort to eradicate bovine tuberculosis and brucellosis has played a major role in reducing animal and human suffering and hardship."[64] Although there was not a rigid separation between the agricultural and conservation scientists, several wildlife ecologists argued that the hybrid and diseased bison of Wood Buffalo Park were a threat not just to livestock, but also to the disease-free herds of wood bison in the Mackenzie Bison Sanctuary. Several studies had postulated that the decline of the bison population in Wood Buffalo Park from 12,000 animals in the 1950s to approximately 3,000 in 1990 was attributable to the impact of disease on reproductive and mortality rates, a circumstance that could, it was argued, be replicated elsewhere if disease were allowed to spread.[65] Indeed, Cormack Gates, the chair of the Wood Bison Recovery Team and a scientist with the Department of Renewable Resources, NWT, claimed that the hybrid and diseased animals threatened the entire wood bison recovery effort by limiting the amount of secure range available for the reintroduction of disease-free and taxonomically pure animals. The image of worthless diseased and hybrid bison that emerged out of these scientific studies was incorporated into the EAP report, which concluded "the presence of infectious diseases [in the WBNP buffalo] reduces the heritage value of the free-roaming bison herd."[66]

Nevertheless, scientific support for the eradication proposal was not unanimous, and an almost immediate critique of the EAP report emerged from the scientific community in the wake of its release. University of Calgary professor Valerius Geist published a study in 1991 that claimed the extermination of the hybrid bison in Wood Buffalo Park could not be justified taxonomically because the wood bison was an ecotype rather than a subspecies of the wood bison.[67] Geist maintained that the effort to save wood bison would result in a genetic bottleneck: "the view that WBNP bison be killed off and replaced by genetically impoverished 'wood bison' from EINP [Elk Island National Park] would destroy the largest continuous, well-tested gene pool of the species *B. bison* and give priority to the impoverished gene pool of an inbred phantom subspecies."[68] A second major objection to the report was the implicit assumption of the panel that disease represented a limiting factor, or threat, to bison populations. University of Guelph ecologist Thomas Nudds released a study in 1993 that employed allometry—a technique used to estimate the population density of herbivores from their body mass—to suggest that the decline in the Wood Buffalo Park bison was not an overall conservation problem but merely an adjustment from the ecological anomaly of the transfer to a more optimum population of roughly 2,000 animals. Nudds concluded that management to eliminate disease was unjustified if this factor was not the ultimate cause of the population decline. Moreover, if it could not be clearly demonstrated that the spread of disease represented a general threat to other bison populations, "there appears to be little reason to kill Bison for the purposes of conserving Bison in the park."[69] In the same year, L. Carbyn, S. Oosenbrug, and D. Anions published an expansive investigation of the Wood Buffalo Park bison through the Canadian Circumpolar Institute at the University of

Alberta. The authors noted that disease was spread evenly throughout the park bison, but that the population decline was far greater in the Peace-Athabasca Delta. They examined a range of factors that might be contributing to the population decline, including predation, disease, changing snow depth, hydrological changes in the Peace-Athabasca Delta, and food availability. They concluded that the increased wolf predation since the cessation of predator control in the 1970s was the key factor leading to population declines, in concert with habitat losses that had forced the bison to congregate into fewer areas, thus making them more vulnerable to wolf attacks.[70]

In the end, the broad opposition to the EAP report had a very different outcome than the debate over the transfer project: the negative response from the public and the scientific community prompted a deferral of the slaughter proposal. Yet in many ways, the terms of the debate over disease management were not so different than those that governed the controversy over the Wainwright transfer. Once again, wildlife scientists tended to criticize a far-reaching and intrusive management project that primarily served an administrative and economic agenda. On the other side of the debate stood a coalition of bureaucrats and veterinarians within the departments of agriculture and environment, who were convinced that, despite a lack of scientific research on the precise impact of diseases on the bison herds, the animals represented enough of a dire threat to Canada's agricultural economy that they had to be eliminated.[71] The environmental assessment panel adopted much of this same rhetoric in their final report, arguing that scientific research on the prevalence and impact of the disease on the bison was not a necessary

prerequisite to the slaughter program because the animals were clearly a threat to domestic livestock no matter what the rate of infection. In response, wildlife scientists such as Geist, Nudds, and Carbyn were able to argue effectively that not enough was known about bison ecology to justify such an extreme intervention in the Wood Buffalo Park ecosystem.

In other respects, however, the controversy surrounding the slaughter was very different than the debate between supporters of pragmatic interventionism and the advocates of non-interference with the northern bison. Ironically, many of the wildlife ecologists who supported the depopulation scheme invoked the very images of a balanced nature that an earlier generation had used to oppose intensive management of the bison. Slaughter advocates such as Cormack Gates claimed that the presence of diseased and hybridized bison on the northern range inhibited efforts to re-establish the pure wood bison on much of its prehistoric range. According to Gates, the wood bison were a keystone species that occupied and maintained a unique habitat of sedge and grass meadows. One of the larger goals of the disease elimination program was thus "the restoration of wood bison to near their former status in a healthy, productive, and intact ecosystem within a large portion of the boreal forest region." If a forceful moment of human intervention was necessary to achieve this goal, the result would be the recovery of a self-sustaining population of wood bison that supposedly existed in a state of harmony and stable equilibrium with their environment prior to the northward expansion of the European fur trade in the late eighteenth century.[72]

Ironically, criticism of this equilibrium model from within the larger field of ecology had begun to seep into the scientific literature opposing the slaughter proposal. Since the mid-1970s, chaos theorists such as Robert May had carried out a frontal attack on the notion of ecological balance. The year the slaughter controversy erupted, ecologist Daniel Botkin published a popular text calling for a new ecology based on disturbance and disequilibrium as the operative metaphors for nature. On one level, Botkin's rejection of steady-state ecology as a naive fiction based on dogmatic assertions of a pristine nature left little room for a critique of managerial intervention. Indeed, Botkin celebrated human management of the natural world as both inevitable and desirable—he argued for a shift away from the preservationist rhetoric of the environmental movements toward a discussion of the rates and kinds of change that were desirable in the natural world. Nevertheless, his rejection of steady-state ecology did not provide a free rein for environmental managers—Botkin argued that, in the absence of any doctrinal master narrative of nature, the new ecology must conduct specific and meticulous research on the dynamics of diverse and ever-changing ecosystems as a precondition to any management intervention.[73] At least some scientists working with the northern bison incorporated these criticisms of the stability model into their arguments against the disease eradication program. They argued that the Wood Buffalo National Park ecosystem was dynamic and fluid—the process of isolating and removing disease as the single factor inhibiting the return of the bison to their stable historical status ignored the multitude of ecological changes that had taken place in the park over time. Ludwig Carbyn embodied much of this thinking when he revised his predation-disease-habitat hypotheses and asserted that there were too many dynamic elements in a complex and changing ecosystem such as Wood Buffalo National Park to isolate one single factor as the ultimate cause of the decline in bison numbers. According to Carbyn, factors other than disease (fire suppression, changing flood regimes due to dam construction on the Peace River, agricultural expansion outside the park, shifting predator control policies, and variable precipitation patterns) may have contributed to the decrease in the bison population. He called for a multidisciplinary approach to research that would consider the interactions between a broad range of ecological factors because "too many changes have taken place in the delta to permit facile explanations or to justify narrowly focused research."[74] In a similar manner, Markus J. Peterson argued for a research approach that incorporated uncertainty, complexity, and the concept of multiple causation when determining if diseases represented an overall threat to the bison populations. He also directly invoked Botkin's critique of the unspoken assumptions underlying equilibrium ecology to suggest that a cultural awareness of the values that frame scientific debate and careful research based on extensive hypothesis testing are necessary prerequisites to any sound management intervention.[75]

If the reaction to a recently released report of a five-year research project conducted by University of Saskatchewan biologists Damien Joly and Francois Messier on the effects of disease on the northern bison is any indication, Carbyn's and Peterson's arguments against an overly simplistic and intuitive approach to wildlife management may be taking hold at senior levels of the Parks Canada bureaucracy. The report once again advocated the eradication and replacement of the entire bison population in Wood Buffalo National Park, but Parks Canada concluded from the responses to an academic peer review process that the investigation failed to support the recommended management solution because it did not demonstrate conclusively that diseases represent a threat to bison at the population level.[76] If this decision holds, then the practice of wildlife management in Canada has clearly changed a great deal from the days of the bison transfer, when a small group of civil servants could push through a radical intervention in the ecology of a national park while ignoring criticism and contrary evidence from the scientific community. It appears as if J. Dewey Soper's ideal of the careful natural scientist who is willing to sound a cautionary note in response to ill-considered management interventions has survived among wildlife ecologists in Canada.

CONCLUSION

The past seven decades of bison management in Wood Buffalo National Park have provided fertile ground for debate among wildlife scientists and administrators in Canada. Certainly, the tremendous importance attached by conservationists to the largest free-roaming herd of wood bison in the world has ensured that the debates about what to do with these animals have been intense and unwavering. Science has also clearly played a variable role in the controversy, ranging from complete resistance to managerial objectives to an almost total integration of technical knowledge with an interventionist agenda. Clearly, as the history of meat production in Wood Buffalo Park demonstrates, a dominant paradigm in the history of wildlife management in

Canada—particularly in the North—has been an agricultural model based on the maximum sustained yield of ungulates to serve economic interests. Aldo Leopold's dictum that game management is the art of producing a crop of wild animals suggests that the roots of the productionist agricultural model run very deep in the history of North American wildlife science.[77]

But the history of bison management in Canada also suggests that scientists have not always aligned themselves with the dominant productionist model. A recurring theme in the debates has been the resistance of zoologists and ecologists to management interventions they considered detrimental to the bison. The historical controversy suggests that wildlife science is capable of both challenging state interests that are intent on controlling nature, and partaking in self-criticism and internal debate. As the history of northern bison management suggests, wildlife ecologists in Canada have at times been able to articulate a critical intellectual practice that rejects the utilitarian resource management models critics such as Merchant, Livingston, and Evernden see as inherently tied to the scientific endeavour. Indeed, the case of bison management in Wood Buffalo National Park strongly suggests that more study is needed of the extent to which administrators within government bureaucracies in Canada (and elsewhere) have contributed to the development of managerial approaches to nature, as opposed to more abstract explanatory models that point to the influence of a generalized scientific world view. As Peterson argues, science is not a monolithic intellectual activity, and much of the source of the debate within the field of wildlife science can be attributed to the conflicting values associated with disparate disciplinary and institutional affiliations—

that is, wildlife ecologists who have advocated a wider ecosystem approach to the bison versus agricultural scientists and veterinarians who have advocated intrusive disease management programs.[78] Indeed, the historical practice of wildlife science in Wood Buffalo National Park suggests that the disagreements between scientists and administrators, and among different scientific and managerial factions, have not progressed as mere technical disagreements from within a unitary philosophical framework, but instead proceeded as fundamental philosophical debates about the proper role and place of humans in the natural world. Programs to relocate surplus plains bison to the northern range, or to eradicate disease by slaughtering and replacing all the bison in Wood Buffalo National Park, prompted scientific and philosophical questions that lie at the root of so many environmental debates. To what extent should humans interfere with nature? To what extent is it desirable to leave well enough alone? If wildlife ecology provided no clear answers to these questions, it has at least provided a rich testament not only to what human beings might manage and control, but also to what we might cherish and defend.

NOTES

1. L.W. Herchmer to comptroller, Royal Northwest Mounted Police, October 3, 1893, RG 18, vol. 489, file 381-15, National Archives of Canada.

2. See RG 18, vol. 489, file 381-15, NAC. Warburton Pike, *The Barren Ground of Northern Canada* (New York: Macmillan, 1892), 145. Caspar Whitney, *On Snow-Shoes to the Barren Grounds: Twenty-eight Hundred Miles after Musk-Oxen and Wood-Bison* (New York: Harper and Brothers, 1896), 116–117.

3. A.M. Jarvis, "Report of Inspector A.M. Jarvis, C.M.G. on Wood Buffalo in the Mackenzie District," Appendix N, Report of the Royal Northwest Mounted Police, *Sessional Papers for 1907*, 122–129; Edward A. Preble, *A Biological Investigation of the Athabasca-Mackenzie Region. North American Fauna* no. 27 (Washington: G.P.O., 1908); Ernest Thompson Seton, *The Arctic Prairies* (New York: Harper and Row, 1911); Francis Harper, "Mammals of the Athabasca and Great Slave Lake," *Journal of Mammalogy* 13, no. 1 (1932): 19–36; Charles Camsell, "The Wood Buffalo Range of Northern Alberta." RG 85, vol. 1390, file 406-13, pt. 1, NAC.

4. For the events leading up to the creation of Wood Buffalo National Park, see RG 85, vol. 1390, file 406-13, pt. 1, NAC. For an assessment of the inaccuracies associated with Seton's and Jarvis's condemnation of Native hunters, see Theresa A. Ferguson, "The 'Jarvis Proof': Management of Bison, Management of Bison Hunters and the Development of a Literary Tradition," a paper presented at the Fort Chipewyan/Fort Vermilion Bicentennial Conference, September 23, 24, 1988.

5. For background material on the bison transfer, see RG 85, vol. 1391, file 406-13, NAC. For detailed records on the bison slaughters and disease control program, see RG 85, vol. 157, file 472-73, NAC, and RG 84, vol. 2237, file WB 232, NAC. For an overview and political analysis of bison management in the park, see Patricia McCormack, "The Political Economy of Bison Management in Wood Buffalo National Park," *Arctic* 45, no. 4 (December 1992): 367–380.

6. University scientists were not, however, completely free from political pressures applied by their superiors in senior administration during this period. See Marianne Gosztonyi Ainley, "Rowan vs. Tory: Conflicting Views of Scientific Research in Canada," *Scientia Canadensis* XII, no. 1 (Spring–Summer 1988): 3–21.

7. For an overview of the problems associated with the bison irruption at Buffalo National Park, see RG 84, vol. III, file BU232, NAC. For an analysis, see Barry Potyandi, "Wood Buffalo

National Park: An Historical Overview and Source Study," unpublished manuscript, Wood Buffalo National Park Library, Fort Smith, NWT, 1979, 86. See also "Kill Buffalo as Movie Men Make Pictures," *Ottawa Journal* (October 27, 1923), and "To Kill Buffalo in a Humane Way," *Ottawa Citizen* (October 17, 1923), where government officials are quoted as having assured the humane societies that the buffalo were not killed during the making of the film and that only blunt spears were used.

8. Cory to Harkin, May 26, 1923, RG 85, vol. 1390, file 406-13, NAC.

9. Graham to Finnie, May 28, 1923.

10. It was finally agreed that construction on corrals and testing facilities at Buffalo Park should be arranged by Harkin and Smith immediately, with shipments to begin by rail and barge the following spring. A summary of the meeting is contained in a document titled "Memorandum for Fyle," July 5, 1923, RG 85, vol. 1390, file 406-13, NAC. Perhaps conscious of the potential controversy, Finnie asked for Torrance to send his views in writing to the deputy minister. See Torrance to Cory, May 30, 1923, RG 85, vol. 1390, file 406-13, NAC. Harkin expressed reservations about the program in a letter to Cory dated November 23, 1923. He wrote, "there is one subject that has not been discussed in the department ... that is, the possibility of the transfer of Wainwright buffalo to the wood buffalo range resulting in the introduction of tuberculosis to the wild herd. You will remember that last Spring we had a conference with Dr. Torrance upon this subject and that the Doctor would not commit himself concerning this aspect to [*sic*] the situation." RG 85, vol. 1390, file 406-13, NAC.

11. Sec Graham's discussion of tuberculosis with Finnie in response to a letter from Dean Rutherford, Faculty of Agriculture, University of Saskatchewan, April 9, 1925, RG 85, vol. 1391, file 406-13, NAC. For the decision to dispense with tuberculosis testing, see A.G. Smith to Harkin, April 16, 1924, RG 85, vol. 1390, file 406-13, NAC.

12. M.J. Rutherford to Charles Stewart, minister of the Interior, May 23, 1925, RG 85, vol. 1391, file 406-13, NAC; Maxwell Graham, "Statement Concerning the Transfer of Buffalo from the Park at Wainwright, Alberta, to the Southern Range of the Wood-Buffalo in Northern Alberta, Prepared from Memoranda Supplied by the Parks and Northwest Territories Branches," June 25, 1925, RG 85, vol. 1391, file 406-13, NAC. See also the unpublished article drafted by Graham, titled "Canada's Repatriation of the Buffalo," RG 85, vol. 1391, file 406-13, NAC.

13. Graham to Finnie, December 20, 1923, RG 85, vol. 1390, file 406-13, NAC.

14. See Maxwell Graham, "Finding Range for Canada's Buffalo," *The Canadian Field-Naturalist* 38 (December 1924): 189.

15. Francis Harper, "Letter to the Editor," *The Canadian Field-Naturalist* 39 (February 1925): 45; W.E. Saunders, "Letter to the Editor," *The Canadian Field-Naturalist* 39 (May 1925). See A. Brazier Howell to Harkin, April 13, 1925, RG 85, vol. 1391, file 406-13, NAC. A resolution opposing the transfer was passed by the society at their April 1925 meeting, only one year after the predator control controversy erupted. The resolution was printed in its entirety in the *CFN* (next to Saunders's letter), and a copy of the resolution was forwarded to Harkin and to "organizations interested in wild-life conservation."

16. Harper, "Letter to the Editor," 45.

17. See Ainley, "Rowan vs. Tory," 3–21.

18. See John Rutherford, James McLean, and James Harkin, *Report of the Royal Commission to Investigate the Possibilities of the Reindeer and Musk-Ox Industries in the Arctic and Sub-Arctic Regions of Canada* (Ottawa: The King's Printer, 1922), 31. A herd of domestic reindeer was imported from Alaska to the Mackenzie Delta region in 1929. This experimental effort to transform Inuit hunters into Old World herders met with only limited success. One of the difficulties associated with such an intensive resource management project was keeping track of the wide-ranging reindeer

herds. See A.E. Porsild, "Reindeer and Caribou Grazing in Northern Canada," *Transactions of the North American Wildlife Conference* 7 (1942): 381–391, and C.H.D. Clarke, *Report on Development of the Reindeer Industry—Mackenzie District* (NWT Lands, Parks and Forests Branch, Department of Mines and Forests), mimeographed unpublished report.

19. C. Gordon Hewitt, *The Conservation of Wild Life in Canada* (New York: Charles Scribner's Sons, 1921), 136–142.

20. Graham, "Finding Range for Canada's Buffalo," 189. For a discussion of the utilitarian orientation of wildlife conservation in Canada, see Thomas Dunlap, *Nature and the English Diaspora: Environment and History in the United States, Canada, Australia, and New Zealand* (Cambridge: Cambridge University Press, 1999), 168–174. For a critique of the agricultural emphasis in bison conservation programs in Canada and the United States, see Andrew Isenberg, *The Destruction of the Bison: An Environmental History, 1750–1920* (Cambridge: Cambridge University Press, 2000), 164–192.

21. See Graham, "Statement Concerning the Transfer of Buffalo." See also Graham's memo to Finnie, April 9, 1925, RG 85, vol. 1391, file 406-13, NAC. The debate surrounding the taxonomic status of the wood bison surfaces again in the final section of the paper in the context of the proposal to eradicate disease in the park bison by slaughtering and replacing all the hybrid bison in the park.

22. Graham to R.A. Gibson, May 1, 1925, RG 85, vol. 1391, file 406-13, NAC. See also the "Statement" and the memo to Finnie cited in note 21.

23. A Canadian Zoologist, "The Passing of the Wood Bison," *The Canadian Forum* (July 1925): 301–305.

24. For reports on the Wainwright bison in their new habitat, see Warden Report Extracts, July 1926, KG 85, vol. 1391, file 406-13, NAC. For a report of the transplanted bison migrating to Wainwright, see "One Thousand Buffalo Roam at Will over Great North-West," *Montreal Standard* (August 22, 1925). For an account of how the buffalo were settling in to their new home and not migrating south to Wainwright, see "Alberta Buffalo Sent Far North, Like New Home," *Calgary Herald* (February 1, 1926).

25. For favourable press reports, see "Buffalo Now Secure," *Ottawa Citizen* (March 20, 1926); Earl Gaye, "Plains Buffalo Take Kindly to New Home around Fort Smith," *Edmonton Journal* (July 28, 1928); John F. Ariza, "A Vanishing American Comes Back," *Baltimore Sun* (August 17, 1930). For the positive reaction from former critics, see Cory to Finnie, January 28, 1926 and Garretson to Finnie, March 24, 1926, RG 85, vol. 1391, file 406-13, NAC.

26. James Ritchie, "The American Bison: a Questionable Experiment," Supplement to *Nature* (February 20, 1926); "*Bison bison athabascae*, Rhoads," *Nature* 119, no. 2985 (January 15, 1927): 95–96.

27. A summary of, and quotations from, Rowan's report appear in "Notes on Professor Rowan's Report, Re: Wood Buffalo National Park," RG 85, vol. 1391, file 406-13, NAC.

28. For a critique of the balance of nature idea, see Daniel Botkin, *Discordant Harmonies: A New Ecology for the Twenty-First Century* (New York: Oxford University Press, 1990).

29. See C.G. Van Zyll de Jong, *A Systematic Study of Recent Bison, with Particular Consideration of the Wood Bison (Bison bison athabascae Rhoads 1898)* (Ottawa: National Museums of Canada, Publication in Natural Sciences, no. 6, 1989). For a contrary view, see Valerius Geist, "Phantom Subspecies: The Wood Bison *Bison bison 'athabascae'* Rhoads 1897 Is Not a Valid Taxon, but an Ecotype," *Arctic* 44, no. 4 (December 1991).

30. For the buffalo slaughters at Wainwright and the liquidation of the park bison, see RG 84, vol. 2239, file WB 299, NAC. See also "History of the Buffalo of Buffalo National Park," unpublished manuscript, 1961. For Barbour's statement, see "Review of *Wild Beasts To-day*, by Harold Shepstone," *Science* 76, no. 1978 (November 25, 1932): 491.

31. For the warden patrol reports, see RG 85, vol. 152, file 420-2, NAC. Estimates of a 20 percent herd increase were likely derived from reports of herds with a 20 percent calf crop on patrols by wardens Taylor and Dempsey in July and August 1930. For the report of bison numbers to Ottawa based on the warden reports, see McDougal to Finnie, February 23, 1931, RG 85, vol. 1391, file 406-13, NAC.

32. Finnie's interest in hiring a biologist is expressed in a memo to deputy minister of the Interior H.H. Rowatt, October 22, 1931, RG 85, vol. 1200, file 400-15-1, NAC. The report on the aerial census is contained in a report sent by Warden Dempsey to McDougal, March 12, 1931, RG 85, vol. 1391, file 406-13, NAC.

33. Soper's proposal is contained in a memo to Finnie, November 3, 1931, RG 85, vol. 1200, file 400-15-1, NAC. Soper's instructions are located in the same file and are not dated. Rowatt's comments on the draft instructions to Soper state clearly that "the buffalo herd at Ft. Smith is being developed primarily as a food supply for the native population of the Northwest Territories. If it reaches proportions where there is a surplus beyond the needs of the native population then the sale of buffalo meat could be considered along the lines prevailing in the Buffalo Park, Wainwright, which is eminently satisfactory." Rowatt to Hume, April 12, 1932, RG 85, vol. 1200, file 400-15-1, NAC.

34. Soper to Hume, July 5, 1933.

35. Soper to Hume, November 30, 1933, and Soper to Hume, May 12, 1934.

36. J. Dewey Soper, "History, Range and Home Life of the Northern Bison," *Ecological Monographs* 2, no. 4 (October 1941): 394; J. Dewey Soper, "Report on Wildlife Investigations in Wood Buffalo Park and Vicinity, Alberta and Northwest Territories, Canada," unpublished report (Winnipeg: National Parks Bureau, 1945), 16–17.

37. Soper to Hume, February 22, 1933, RG 85, vol. 1391, file 406-13, NAC.

38. Soper wrote to Parks Commissioner James Harkin asking for permission to publish the monograph, but stated that "all reference of a controversial nature as to wood bison and the Wainwright bison should be rigorously excluded, so far as possible, and the work confined to a straight forward [sic] account of the history of the park, bison life habits therein, local migrations, trails, wallows, disposition, traits, breeding, young, etc., etc." Soper to Harkin, November 5, 1935, RG 85, vol. 1200, file 400-15-1, NAC. The published study of the bison appeared as J. Dewey Soper, "History, Range and Home Life of the Northern Bison," *Ecological Monographs* 2, no. 4 (October 1941): 396. The unpublished report exists as an appended summary to the manuscript for the published monograph. Sec RG 85, vol. 1200, file 400-1 S-1, NAC.

39. Soper's extensive report on wolves is contained in a report to Hume, March 4, 1933, RG 85, vol. 1200, file 400-15-1, NAC.

40. Soper to Hume, February 22, 1933, RG 85, vol. 1391, file 406-13, NAC.

41. See B.I. Love, "Survey of Animals and Surroundings, Wood Buffalo Park, NWT, August 4–27, 1944," RG 85, vol. 1392, file 406-13, NAC. For a report of Soper's shock at the condition of Wood Buffalo Park, see Harrison Lewis to A.L. Cumming, December 10, 1945, RG 85, vol. 1392, file 406-13, NAC. For an overview of the minimal exploitation era in Wood Buffalo National Park, see McCormack, "The Political Economy of Bison Management," 370–371.

42. For Lewis's comments, see the minutes for the Tenth Dominion Provincial Conference on Wildlife, February 22–24, 1945, RG 22, vol. 4, file 13, NAC.

43. For a discussion of wildlife management practices in the Canadian national parks during this period, see Alan MacEachern, *Natural Selections: National Parks in Atlantic Canada, 1935–1970* (Montreal: McGill-Queen's Press, 2001), 190–203.

44. For an account of the broader northern resource development program, see Morris Zaslow, *The Northward Expansion of Canada, 1914–1967* (Toronto: McClelland & Stewart, 1988), 234–270. The Northern Affairs Branch, which administered Wood Buffalo Park until 1964, opened the park

to logging and commercial fishing in Lake Claire in the early 1950s, and considered reducing the park boundary to allow mining and petroleum exploration in the region in the 1960s. For records of timber berths in Wood Buffalo National Park, see RG 84, vol. 2237, file 206-503m, NAC. For records of the Lake Claire fishery and proposals to reduce the park boundary to open areas for resource development, see RG 85, accession 1997–98/076, box 73, file 406-13, NAC.

45. For count numbers from Fuller's report, see "Buffalo Report—Wood Buffalo Park," n.d., RG 85, vol. 157, file 472-3, NAC.

46. See also F.E. Graesser, "Report—Investigations of Buffalo, Wood Buffalo Park," December 29, 1950, RG 85, vol. 157, file 472-3, NAC.

47. J.W. Burton, acting chief, Conservation and Management Services to J.G. Wright, acting chief, Northern Administrative Division, July 6, 1951, RG 85, vol. 157, file 472-3, NAC.

48. Fuller to Lewis, October 29, 1951, RG 85, vol. 157, file 472-3, NAC.

49. See the minutes for the Advisory Board on Wildlife Protection, November 6, 1951, RG 22, vol. 16, file 69, NAC.

50. W.E. Stevens, "Bison Report—September 1954," unpublished report available at the Wood Buffalo National Park Library, NWT.

51. See "Summary of Buffalo Reduction by Park, 1951–1966," RG 84, vol. 2239, WB 299, NAC. For the development of the meat production program, see RG 85, vol. 157, file 472-3, NAC and RG 85, vol. 158, file 472-3, NAC. For shipment of buffalo meat to Expo, see R.T. Flanagan, head, Parks Management Section, memo for file, January 13, 1967, RG 84, vol. 2239, WB299, NAC.

52. N.S. Novakowski, "Estimates of the Bison Population in Wood Buffalo Park and the Northwest Territories Based on Transect and Total Counts," February 1961, RG 84, vol. 2237, file WB232, NAC.

53. For a record of Novakowski's dressing down, see J.R.B. Coleman to B.G. Sivertz, April 4, 1961, RG 84, vol. 2237, file WB232, NAC. The resurvey is recorded in N.S. Novakowski, "Total Counts of Bison in the Higher Density Areas of Wood Buffalo Park," April 1961, RG 84, vol. 2237, file WB232, NAC. For a report on the slaughter, see B.F. Olson, "Wood Buffalo National Park: Buffalo Round-up, Testing and Slaughter Program, 1961," RG 84, vol. 2239, file WB299, NAC.

54. B.E. Olson to C.L. Merrill, October 28, 1962, RG 84, vol. 2239, file WB299, NAC.

55. The decision to depopulate the Grand Detour herds was taken at a meeting of the Anthrax Committee on July 17, 1964. See RG 84, vol. 487, file WB210-1, NAC. See also N.S. Novakowski, "Slaughter Report—Grand Detour 1964, 1965," RG 84, vol. 2239, file WB299, NAC. See also L.P.E. Choquette and Eric Broughton, "Anthrax in Bison, Wood Buffalo Park and the Northwest Territories—Report for the Year 1967," RG 84, vol. 2237, file WB210-1, NAC.

56. For the development of the Pathology Section of the CWS, see J. Alexander Burnett, "A Passion for Wildlife: A History of the Canadian Wildlife Service, 1947–1997," *The Canadian Field-Naturalist* 113, no. 1 (January–March 1999): 61–64.

57. N.S. Novakowski and L.P.E. Choquette, "Proposed Five-Year Management Plan for Bison in Wood Buffalo National Park," RG 84, vol. 2237, file 210-1, NAC.

58. See McCormack, "The Political Economy of Bison Management," 373.

59. See Ministry of the Environment, *Wood Buffalo National Park: Management Plan Summary* (Ottawa: Minister of Supply and Services, 1984). For an overview, see McCormack, "The Political Economy of Bison Management," 373. The movement toward non-interventionism in the national parks began in the late 1950s, but was slow to take hold in the development-minded Northern Affairs Branch. For a discussion of the wider movement toward non-interventionism in the Parks Branch, see MacEachern, *Natural Selections*, 220–228.

60. Environmental Assessment Panel, *Northern Diseased Bison* (Ottawa: Minister of Supply and Services, 1990).

61. For articles critical of the slaughter proposal, see Ed Struzik, "The Last Buffalo Slaughter," *Canadian Forum* 69, no. 794 (November 1990): 6–11; Andrew Nikiforuk, "If Armageddon Arrives, the Bison Are Dead Meat," *The Globe and Mail* (October 4, 1990): A21. For a survey of Aboriginal perspectives on the slaughter proposal, see Theresa A. Ferguson and Clayton Burke, "Aboriginal Communities and the Northern Buffalo Controversy," in *Buffalo*, edited by John Foster, Dick Harrison, and I.S. MacLaren (Edmonton: University of Alberta Press, 1992).

62. Environmental Assessment Panel, *Northern Diseased Bison*, 28.

63. This memo was leaked to the press. See Struzik, "The Last Buffalo Slaughter," 8–9.

64. Stacy Tessaro, "The Existing and Potential Importance of Brucellosis and Tuberculosis in Canadian Wildlife: A Review," *Canadian Veterinary Journal* 27 (1986): 119–124; Stacy Tessaro, "Bovine Tuberculosis and Brucellosis in Animals, Including Man," in Foster, Harrison, and MacLaren, eds., *Buffalo*, 207–224; Gary Wobeser, "Disease in Northern Bison: What to Do?" in Foster, Harrison, and MacLaren, *Buffalo*, 179–188. Wobeser was a member of the EAP.

65. Brucellosis causes spontaneous abortions in females, and tuberculosis weakens older animals. See Tessaro, "The Existing and Potential Importance"; S. Tessaro, L.B. Forbes, and C. Turcotte, "A Survey of Brucellosis and Tuberculosis in Bison in and around Wood Buffalo National Park, Canada," *Canadian Veterinary Journal* 31 (1990): 174–180; Eric Broughton, "Diseases Affecting Bison," *Bison Ecology in Relation to Agricultural Development in the Slave River Lowland*, edited by H.W. Reynolds and A.W.L. Hawley, Canadian Wildlife Service Occasional Paper Series Number 63 (Ottawa: Minister of Supply and Services, 1987), 34–38; Jack Van Camp and George W. Calef, "Population Dynamics of Bison," in Reynolds and Hawley, eds., *Bison Ecology*, 21–24.

66. See Environmental Assessment Panel, *Northern Diseased Bison*, 21; C. Gates, T. Chowns, and H. Reynolds, "Wood Buffalo at the Crossroads," in Foster, Harrison, and MacLaren, eds., *Buffalo*, 139–165; Stacy V. Tessaro, Cormack C. Gates, and Lorrey B. Forbes, "The Brucellosis and Tuberculosis Status of Wood Bison in the Mackenzie Bison Sanctuary, North West Territories, Canada," *Canadian Journal of Veterinary Research* 57 (1993): 231–35. The wood bison of the Mackenzie Bison Sanctuary were relocated to the Fort Providence area in 1959 when a small herd of relatively pure wood bison were discovered in the Nyarling River area of Wood Buffalo National Park. See A.W.F. Banfield and N. Novakowski, "The Survival of the Wood Bison *(Bison bison athabascae* Rhoads) in the Northwest Territories," National Museum of Canada Historical Paper 8 (1960): 1–6.

67. The irony is that Maxwell Graham used the same idea to dismiss the critics of intensive buffalo management. As mentioned earlier, Graham countered the criticism that the transfer project would result in the hybridization between distinct subspecies with the suggestion that the wood bison were merely a geographic race. See Graham, "Statement Concerning the Transfer of Buffalo."

68. Geist, "Phantom Subspecies," 297. The questions surrounding the taxonomic status of the wood bison have not yet been resolved within the scientific community. For a contrary view to Geist's, see Van Zyll de Jong, *A Systemic Study of Recent Bison.*

69. Thomas D. Nudds, "How Many Bison, *Bison bison,* Should Be in Wood Buffalo National Park?" *Canadian Field-Naturalist* 107, no. 1 (1993): 119.

70. L. Carbyn, S. Oosenbrug, and D. Anions, *Wolves, Bison and the Dynamics Related to the Peace-Athabasca Delta in Canada's Wood Buffalo National Park* (Edmonton: Circumpolar Institute, 1993).

71. See Environmental Assessment Panel, *Northern Diseased Bison*, 19.

72. Gates, Chowns, and Reynolds, "Wood Buffalo at the Crossroads," 157.

73. Robert M. May, "Biological Populations with Nonoverlapping Generations: Stable Points, Stable Cycles, and Chaos," *Science* 186 (November 15, 1974): 645–647; Botkin, *Discordant Harmonies,*

4–6. For an analysis, see Donald Worster, *Nature's Economy: A History of Ecological Ideas*, 2nd ed. (Cambridge: Cambridge University Press, 1994), 406–420.

74. Ludwig N. Carbyn, Nicholas J. Lunn, and Kevin Timoney, "Trends in the Distribution and Abundance of Bison in Wood Buffalo National Park," *Wildlife Society Bulletin* 26, no. 3 (Fall 1998): 463–470.

75. M.J. Peterson, "Wildlife Parasitism, Science, and Management Policy," *Journal of Wildlife Management* 55, no. 4 (1991): 782–789. Peterson argues that various value-laden symbolic filters lay at the root of the conflict between scientists conducting research on diseased bison. Veterinary scientists, for example, tended to focus their questions on the health of individual bison, while wildlife scientists framed their research in terms of the larger ecological system—with variable results, as we have seen. Peterson writes that "recognizing that culturally-validated assumptions underlie all human enterprise (including science) may encourage wildlife managers to explore the assumptions upon which research is based, thus determining whether those assumptions are ecologically sound, before accepting the recommendations of any study or proposal as appropriate justification for future policy" (see 788).

76. Damien Joly and Francois Messier, *Limiting Effects of Bovine Brucellosis and Tuberculosis on Wood Bison within Wood Buffalo National Park*, unpublished report, Wood Buffalo National Park, Heritage Canada (2001). Information on the peer review process comes from a personal communication with Professor Thomas Nudds, University of Guelph.

77. For a discussion of the agricultural roots of wildlife management in North America, see Thomas D. Nudds and Robert G. Clark, "Landscape Ecology, Adaptive Resource Management and the North American Waterfowl Management Plan," *Proceedings of the Third Prairie Conservation and Endangered Species Workshop*, edited by G.L Holroyd, H.L. Dickson, M. Regnier, and H.G. Smith, Provincial Museum of Alberta, Natural History Occasional Paper no. 19 (1993): 180–190. For a discussion of the agricultural model as it was applied to bison conservation, see Isenberg, *The Destruction of the Bison*, 164–192.

78. Peterson, "Wildlife Parasitism," 787.

Chapter Seventeen

Changing Ecologies
Preservation in Four National Parks, 1935–1965

Alan MacEachern

The period 1935 to 1965 is an interesting one for studying Canadian national parks, if only because so little is supposed to have happened then. Most park histories offer only a pause between the 1930 Parks Act and the rise of environmental interest in parks in the 1960s. Kevin McNamee's 25-page "From Wild Places to Endangered Spaces: A History of Canada's National Parks," for example, gives these years two small paragraphs.[1] A central reason for this neglect is that the system did not grow much in these decades: the only new parks were the four Atlantic Canada ones. There was a sense within the Parks Branch itself that the system was now complete, demanding only eternal vigilance to maintain.[2] Much the same attitude was present in the United States national park system, which saw the addition of only five new natural areas between 1940 and 1959. An American park historian has aptly titled this period the "we think we've done it all" era—and, also aptly, she then spends little time discussing it.[3]

The general neglect of this period by historians is surprising when one considers that in both Canada and the United States the 1960s is seen as one of the most important decades in park history. In that decade in both countries, the size of the park system mushroomed, there came to be greater public respect for the parks' preservationist philosophy, and skyrocketing park attendance fostered debate on how to curb development. The continent-wide environmental movement is seen as so central to all of this that the park systems' own histories are presumed to be irrelevant. I would argue that in fact the decades leading up to the 1960s were instrumental to the shaping of park preservationist policies. In its treatment of fish, wildlife, and vegetation, the Canadian National Parks Branch took an increasingly hands-on approach during the 1940s and 1950s, until—recognizing the ecological, philosophical, and political damage of its actions—it sought to become less interventionist in the 1960s. At both stages, it justified its approach by invoking the name of science.

*****

WILDLIFE

The Parks Branch [exhibited a] general lack of interest in the wildlife that the new Atlantic parks contained, a lack evident in the reports on the planned parks in the 1930s. [...] Why was there so little interest in the wildlife of Maritime parks? The branch believed that no place in the region had wildlife that satisfied the national park ideal. To an agency used to relying on bear, moose, elk, deer, buffalo, and even large predators such as wolves and mountain lions to attract tourists and to prove that it hosted real nature, the animals in the Atlantic provinces hardly compared. Small animals such as foxes, skunks, and porcupines (not to mention insects, birds, and amphibians) might be abundant in these parks, but they would hardly draw tourists. There were just not enough different kinds of big mammals, and too few of the kinds that there were.[4]

There were other reasons wildlife played a minor role in the creation of the first Maritime parks. For one thing, wildlife management held a weak administrative position for much of the branch's history. The Dominion Wildlife Division had been established in 1918 as a small agency within the Parks Branch, but its staff was very small and preoccupied with enforcement of the Migratory Bird Regulations Act; indeed, staff had been hired for their expertise in ornithology.[5] Although the Wildlife Division could and did communicate the latest scientific information to the Parks Branch, it was administrators within the branch itself who then formulated policy. An example of this on the national level was the Parks Branch's management of predators such as wolves and mountain lions in the 1930s. Reading animal population studies from the United States, the Wildlife Division's supervisor

of wildlife protection, Hoyes Lloyd, grew convinced that predators offered a natural and necessary check on prey populations.[6] He sought to convince Commissioner Harkin that the practice of killing predators in Canadian parks should be discontinued, and wrote lengthy reports to that effect in the early 1930s. Harkin used Lloyd's reports to fight proponents of increased kills, though he did not take the advice to stop killing predators altogether.[7] The Wildlife Division had little direct power, but because it existed there was no need to hire biologists or other natural scientists as senior administrators. As a result the Parks Branch was headed by foresters, engineers, and career bureaucrats who lacked training in wildlife matters. The men who initially inspected the Maritime parks and assessed the variety and range of wildlife there were the engineers Cautley and Williamson, the architect Cromarty, and the forester Smart.

The Wildlife Division itself leaned toward non-intervention in the 1930s, as a result of simultaneously following the most established and the most innovative strands of biology. Following the tradition of natural history, the division saw its primary goals to be classifying all park species, and obtaining information about them. Following the latest tenets of ecology—the twentieth-century science that studies relationships between organisms—the wildlife officers believed that all species act together in concert, and that actions affecting one population might seriously affect others.[8] This was why, for example, Hoyes Lloyd opposed predator extermination policies: the results of intervention were unforeseen and so by their nature unwanted.[9] Lloyd won an ally on this issue with the promotion of F.H.H. Williamson as James Harkin's replacement in 1936. More so than Harkin, Williamson was aware of and responsive to the latest findings in ecological research, and used cases of predator eradication leading to prey irruptions as cautionary tales against tinkering with nature. This is not to say that the park system under Williamson practised complete non-intervention: some predators were still killed, animals were collected for museums and zoos, and parks were stocked with "useful" wildlife. But it may be said that when the Cape Breton Highlands and Prince Edward Island national parks were established, North American ecologists were wary of human attempts to regulate wildlife numbers, and had convinced the National Parks Branch to think likewise.[10]

Administrators were at the same time changing their idea of wildlife's role in the parks. In the first decades of the century, Canadian national parks had often been justified as sanctuaries, where game species would have the freedom to grow, prosper, and then wander outside to be killed by hunters.[11] Fish and game organizations came to accept the argument that sanctuaries of all sorts ultimately helped to maintain healthy wildlife breeding stock.[12] In the 1920s and 1930s, fish and game groups in New Brunswick, Nova Scotia, and Prince Edward Island were important lobbyists for parks in their provinces. But by this time parks were promoted more in terms of tourism, and the Parks Branch was less interested in selling them as sanctuaries. If sportsmen wanted a reserve, they should ask their province to set one up. The presumed absence of wildlife in the proposed Maritime parks, while unfortunate, was thus not catastrophic. A park had to have remarkable scenery; it could, if necessary, import wildlife. Of course, the branch was not likely to tell fish and game organizations doing much of the local legwork in promoting parks that

the park-as-sanctuary idea did not hold the importance it once had. Staff instead complained to Ottawa that the locals did not understand what national parks were all about. [...] As each of the first three Maritime parks was established, local fish and game clubs reacted identically: first with delight that their work had borne fruit, then with surprise that the branch was acting so slowly to increase game populations, and finally with anger when it was clear that the park was uninterested in improving hunting outside its boundary.

In sum, scientific, administrative, political, and aesthetic positions might reasonably have been expected to keep the Parks Branch from an interventionist wildlife policy in Cape Breton Highlands and PEI national parks in the years following their creation. The point is moot, however. The arrival of the Second World War in 1939 allowed little thought or opportunity for wildlife management during most of the next decade. Since wildlife in the first two Maritime parks was deemed relatively insignificant and not even threatened by large predators, for the time being the act of setting up a park boundary and hiring wardens was considered management enough.

*****

The only interventionist policy practised in these parks during the war was the attempted reintroduction of moose and beaver to Cape Breton Highlands. At the Park Branch's request, two colonies of beaver were captured, crated, moved, and released by the Nova Scotia Department of Lands and Forests in 1938. One of the colonies survived the move, though the other one, like a group of moose transferred the same year, was not so fortunate.[13] The justification for these reintroductions was that these animals had been native to the park area, but had been extirpated by the humans there in the past century. They should be returned, then, to restore the park to its pristine condition. This might seem to be an endorsement by the Parks Branch of what was still a radical idea in park wildlife management. In the United States, biologist Joseph Grinnell in the 1920s and the authors of the 1933 *Fauna of the National Parks of the United States* had met resistance for proposing that parks be returned to their original state, with native species reintroduced and exotic ones removed.[14] But the Cape Breton Highlands case was not revolutionary. Though it was true that beaver and moose were native but absent, they were chosen for reintroduction because they were attractive to tourists and to the branch's idea of what wildlife a park should have.[15] Bringing them back certainly intruded on existing biological conditions, but this was more of a blip in the wildlife policy of the time than a sign of change. These were one-time-only interventions, and once released the new park residents were left on their own. It was taken for granted that nature would help its own, after this initial push from park staff.

Biologist C.H.D. Clarke's arrival at Cape Breton Highlands in 1942 to report on its wildlife was the first real sign of branch interest in the animals of the Maritime parks. Yet his report also shows the conservative nature of wildlife research at the time. Clarke, hired as the Wildlife Division's mammalogist in 1938 (and its first staffer not trained in birds), offered an essentially hands-off plan in keeping with the time. He wrote that the population fluctuations resulting from the creation of the new park were of great scientific interest, but they should only be monitored and not directed. "They are absolutely

natural phenomena and we have no concern in trying to interfere with them."[16] Instead, the branch's management of wildlife should be limited to protecting all species (predators included), obtaining all possible information on their populations, and restoring vanished species—though only moose and caribou were mentioned. Clarke was careful to note that only the native variety of woodland caribou still present in Gaspé, not the Newfoundland caribou, should be considered for introduction: "For one thing, survival would be doubtful; for another, it is not desirable to introduce exotic species into the parks."[17]

*****

The National Parks Branch began in the late 1940s to take a much more activist role in preservation issues regarding wildlife. Post-war prosperity allowed for an exponential increase in funds for wildlife management, as it did for federal projects generally. The budget for wildlife matters jumped from $200,000 in 1948 to $400,000 in 1954, to $700,000 by 1960, and to $3.9 million by 1969.[18] To handle this increased responsibility, the Canadian Wildlife Division was replaced in 1947 by a new Dominion Wildlife Service, renamed the Canadian Wildlife Service in 1950. The idea was that as a separate agency the Wildlife Service would have greater autonomy and a more objective, credible role in managing the nation's wildlife. Thanks to enlarged responsibilities and loosened purse-strings, the Wildlife Service grew from a professional staff of seven in 1947 to ninety in 1969. In turn, the availability of scientists and funds encouraged more proactive projects in wildlife management.[19]

The Canadian Wildlife Service did not take over the management of park wildlife.

Parks Branch controller James Smart wrote all the superintendents early in 1948: "You are aware that during the recent reorganization of the Department and this Branch a Wildlife Service was set up which will act as our technical division to advise us on wildlife management. This does not mean that we give up any responsibility for looking after wildlife in parks."[20] But it did mean that the Parks Branch would be drawing on more and better funded scientists for future policy advice.

The creation of the Canadian Wildlife Service signalled not only a new scale for wildlife science in Canada, but a new direction as well. Hoyes Lloyd, Harrison Lewis, and others who were the backbone of the old division retired around the time that the service was created. A 1971 official account, *Scientific Activities in Fisheries and Wildlife Resources*, hints at the perceived difference between the old breed of wildlife scientists and their successors. The former "were a group of keen, hard-working individuals who contributed a great deal to our knowledge of the occurrence and distribution of the wildlife of Canada. After their retirement, the Wildlife Service replaced them with men who were trained along more formal lines."[21] The difference was as clear as that between natural history and science. The Wildlife Service was now staffed with young men schooled in the United States in the latest precepts of biological science. Ecology in particular was drawing attention in this era because it theoretically studied a whole system at once rather than a single species, and so was seen as a science of unlimited promise. Hugh Keenleyside, deputy minister of the Department of Mines and Resources, proudly announced plans in 1949 for a system-wide inventory of national park wildlife: "It will be no mere cataloguing of plants and animals, a great deal of which has already been done, but will be concerned with the community of living things and with the manner in which the various forms of life affect one another. This will be what scientists term an ecological survey."[22]

Ecology was not a value-free science — or, more accurately, like any science it was not practiced in a value-free way. It accommodated different interpretations. This may be seen in the application of its best-known discovery of the 1920s: that predators were essential to habitat health. Knowing that deer and wolf populations were dependent on each other for healthy populations could have taught ecologists that both species had to be left alone — and, indeed, some ecologists did take this lesson. But more interpreted it to mean that to manage deer effectively, both wolf and deer populations had to be actively managed. Likewise, ecology of the 1940s and 1950s tended to be taken to justify interventionism. Historian Gail Lotenburg writes of the period: "most federal wildlife administrators in Canada interpreted ecological concepts as a means towards securing traditional management goals."[23] It is worth stressing that this did not demand a contrived misapplication of ecological theory: it was (in most cases) the product of honest interpretation. Ecology had since the 1930s been moving away from an organic model to a mechanistic one, from a study of individuals and their places in a community to movements of physico-chemical properties within a system; it was natural that this change in metaphor distanced ecologists further from the subject of their inquiry.[24]

The direction taken by ecology would greatly affect preservationist policies of the Canadian national park system. Parks Branch staff relied on the Canadian Wildlife Service to explain ecology and to recommend how it should be applied within the parks (though they made the final decisions—unless upper levels of the department overruled them, of course).[25] As a result, wildlife policies in the national parks would become considerably more interventionist in the 1950s. Rather than merely fencing off an area and letting wildlife thrive, staff now considered it their responsibility to manage wildlife numbers. As early as 1949, the new philosophy—outlined, notably, by the head of the Canadian Wildlife Service, Harrison Lewis—was as follows: "It should, perhaps, be emphasized that it is not the established policy to administer the wildlife of the National Parks by simply letting it alone, letting nature take its course, and trusting in the idea that is commonly referred to as the 'balance of nature'. That would not be practicable, because the park areas are already more or less disturbed by human activities and they are surrounded by areas that are even more altered by man."[26] By the mid-1950s, this idea went still further: animal populations were bound to erupt in the unnaturally natural conditions of a national park.[27] According to a 1957 policy statement, "This brings in its wake such evils as starvation, disease, range destruction, damage to forest regeneration, displacement of desirable plants and soil erosion. Unless these surpluses migrate naturally out of the Parks, they must be removed without hesitation either by careful killing or live-trapping.... All of the National Parks of Canada are potential danger areas for the development of excessive populations of game, predators and fur-bearers."[28] The Parks Branch had created the environment in which animals could overpopulate; to leave them alone to do so would therefore in itself be a form of management, and an immoral one. There was a real arrogance in taking this stand.

Animals were living and dying because of past human choices, so humans must take responsibility by continuing to make choices to ensure their general survival. This reflected the Parks Branch's opinion not only of animals, but of human society as well; just as in past decades the agency had proven it was more progressive than the general public by not interfering in animals' lives, it would now demonstrate its relative intelligence by interfering efficiently.

The swing toward interventionism is evident in the wildlife policies chosen for the new Fundy National Park in the early 1950s. Unlike that at Cape Breton Highlands and Prince Edward Island, Fundy's wildlife received attention immediately upon establishment. By 1951, there were already four reports discussing Fundy's animal population. The early consensus was that the area's potential for wildlife was excellent. Wildlife Service staffer John Kelsall reported that in places where lumbering had been practised, there was unlimited young growth for moose and deer to feed on. He did warn, however, that moose were dying in winter, probably from a combination of moose ticks and difficult travelling in deep snow.[29] Three years later, the chief mammalogist of the Wildlife Service, A.W.F. Banfield, reiterated the hopes that moose, which currently numbered about 120, would increase naturally to the area's "carrying capacity." (This wildlife management term refers to the maximum population of a species that a given area can sustain. Perhaps inevitably, the idea of carrying capacity came to suggest that ecology could determine the right number of *all* species in *all* places, so that managers could maintain populations at just those numbers.)[30] Banfield noted that "[w]ithout a moose reduction programme, we can expect continuing heavy moose tick infestations and winter mortality."[31] In other words, Banfield believed the moose population was below its natural maximum, and yet still needed pruning. This is not really surprising, because winter deaths always bothered park staff. Although such mortality in the animal world is as natural as death by predation or old age, it always seemed needless and avoidable.[32] It was not difficult for wildlife officers at Fundy to convince themselves that the ungulate population needed their help.

By 1953, there were reports of "serious overbrowsing" at Fundy.[33] Apparently, the park's character as a sanctuary had permitted moose numbers to balloon just in the park's first four years. At wildlife officer J.S. Tener's recommendation, the Wildlife Service approved the killing of 20 moose "for the purpose of game management."[34] It is not clear from the files whether this reduction actually took place; before it could, Superintendent Saunders warned his superiors that the province felt it owned the moose by virtue of the New Brunswick Game Act. On the advice of the attorney-general's office, the Parks Branch explained that the federal Parks Act overrode the provincial act, thanks to the clause that gave Canada the right to "all profits, commodities, hereditaments and appurtenances whatsoever thereunto belonging or in anywise appertaining" to the park—including moose.[35] In 1955, with the park moose population around 150 and the carrying capacity now estimated to be only 80, another culling program was set in motion. This moose kill demonstrates just how dubious the branch's claims to scientific objectivity really were. Though the reduction was supposedly intended to lower an overextended and unhealthy population, it was done in such a way as to make such results impossible. Old bull

moose—the least likely animals in the herd to affect long-term population—were targeted. The nine that were killed were found to be "in good condition" and almost all had fat on their quarters, though they had been killed because they were supposedly running out of food.[36]

*****

The reductions implemented in Cape Breton Highlands, [at] Fundy, and elsewhere were both too much and too little: purposeful turns away from hands-off preservation as a result of questionable scientific reasoning, yet involving such small numbers that they were unlikely to have any impact on the stated desire to regulate populations. In fact, Fundy would experience a moose overpopulation and crash within five years of its 1955 reduction program. Other choices could have been made. "Surplus" animals could have been removed to other places in the province that would be glad to have them (although, it is true, they would then be targets for hunters rather than wardens). Strips of forest could have been opened up for browse (which, admittedly, would have meant destroying vegetation instead of wildlife). The other option was to do nothing, and see if the anticipated overpopulations actually occurred.[37]

Though the Parks Branch tried to maintain a fixed philosophy on park wildlife, its policy decisions were often shaped by the political and pragmatic needs of the moment, and by the perceived value of the wildlife in question. Such was the case with the hunting of wildfowl in Prince Edward Island National Park. Hunting has traditionally been the most forbidden activity in parks. Whereas parks throughout their history accepted the removal of trees, minerals, and fish from their borders, the removal of animals was always taboo.

When the PEI National Park was established, local politicians complained on behalf of area duck hunters over the loss of a favourite shooting spot.[38] But by the mid-1940s, hunters were resigned to the park's existence and lobbied instead that it do everything to build up bird populations so that the birds would spill outside the park to be shot. However, the discovery of a mistake in R.W. Cautley's original survey of the park forced the Parks Branch to re-evaluate how it dealt with local hunters. Cautley had mistakenly set a park boundary at Stanhope to "the line of mean high tide" on swampland not influenced by tide; as a result, the park did not have title to two pieces of land that hunters coveted. The park chose for a time to condone hunting in the contested areas, and even an acting superintendent hunted there.[39] The Parks Branch sought to rectify this situation permanently in 1957 by arranging a land swap with Gordon Shaw, owner of the local Shaw's Hotel. The park received a parcel it had long thought it owned, and Shaw received a piece of land that he could offer his visitors as a private shooting ground. Provincial hunters, however, felt they had lost out on the deal and again had their politicians complain. In researching their claim, locals discovered that Cautley had defined part of the park's shore boundary as the "high water mark of the Gulf of St Lawrence": hunters could presumably shoot on the beach at low tide. Trying to make the best of the situation, Superintendent Browning suggested that though it would seem that the PEI National Park did not own the beaches, "Perhaps we could claim squatters rights."[40] The Parks Branch believed that its position was stronger than this, but to avoid disagreements with

locals, staff were told by Ottawa to let duck hunting continue in the disputed areas.[41] In this case, the prime directive of national park wildlife policy was as flexible as the park's boundary.

The Parks Branch was not above advocating the killing of wildlife thought to be pests. The most prolonged example of this was the branch's battle against insects, which will be discussed later. But larger pests were also targeted. Porcupines were causing damage at Fundy headquarters, so staff requisitioned "Good-Rite zip" a chemical deer repellent, and were later given a more direct deterrent, a shotgun.[42] Five hundred muskrat were authorized to be killed in PEI National Park since they had reached "nuisance-value proportions"; nonetheless, only 64 could be found to be killed.[43] Reductions were not always even aimed at any perceived need, but simply so the Parks Branch could demonstrate it was being a good neighbour. In 1950, Wildlife Service biologist John Kelsall discussed killing foxes at Prince Edward Island Park: "The shooting would not be done with a view to eliminating or controlling the fox population but rather to assure any possible sources of complaint that the Park authorities are cognizant of the high fox population and are doing something about it."[44] In this instance the Parks Branch chose not to initiate a reduction program, not because it violated park philosophy, but because it was decided there were not enough complaints to take this step. The most perverse manifestation of the zeal for control arose when the Prince Edward Island National Park staff dealt in 1959 with a very large deceased pest: a beached whale. They blew it up. This may have been the best decision under the circumstances, but Superintendent Kipping's description of the event suggested a certain perverse enjoyment of the spectacle. He corresponded with CBC entertainers "Gentleman and Olga," who had reported on the incident, telling them that staff had to blow the whale "to heaven with 600 pounds of dynamite."[45] And when Ottawa asked Kipping for details, he sent pictures of what he called "the explosion which transported the whale out of the realm of our jurisdiction."[46]

That was exactly the sort of image the head office wanted to avoid. When the Parks Branch felt obliged to kill and remove some park wildlife, it sought to do so with discretion and decorum. On the subject of controlling porcupines at Fundy, for example, Ottawa ordered that "[a]ny control activity should, of course, be carried out as inconspicuously as possible in the interest of good public relations" and "[t]hese operations must be carried on with discretion, in order to avoid offense to the public."[47] Keeping the reduction programs quiet became such a standard policy that staff even convinced themselves that such discretion had a managerial purpose. The minister in charge of national parks in 1962, Walter Dinsdale, told the president of the New Brunswick Fish and Game Association that park hunting was not permitted because "the game species involved would become much more retiring, thus reducing the opportunity of worthwhile recreation for a large segment of the public." For the same reason, he said, when staff had to kill certain animals they were careful not to harass or frighten others.[48]

In reality, it was the human animals that the Parks Branch did not want to disturb. The branch knew that the public was unlikely to understand why different rules applied for park staff than for park users. Why was hunting forbidden in the park, yet proper game management might demand the shooting of some animals? If management was needed, why were locals not allowed a limited hunt? Why, for that matter, were farmers' cattle impounded if they strayed into Cape Breton Highlands, while Highland cattle were kept by Keltic Lodge into the 1950s?[49] Such awkward questions, typical of those prompted by park development throughout the world, were raised repeatedly about these four parks in this period. There were some relatively valid answers to some of these contradictions. National parks are intended to be places free of human effect; however, the parks need humans to help enforce this ideal (as well as to make the visitor's communion with the park as pleasurable, yet as passive, as possible). Moreover, the Parks Branch argued that nature had been affected by humans before the park's existence. Humans were therefore needed to re-establish the natural state of the park, which demanded management. All this is sensible if we accept the original idea that parks have an inherent logic, and if we accept that staff's decisions for the park are free of human interest.

FISH

Writing of the American case, Richard West Sellars states, "In its management of fish, more than of any other natural resource, the Park Service violated known ecological principles."[50] The same could be said of the Canadian National Parks Branch. As with wildlife, fish in Canadian national parks were deemed deserving of preservation, and as with wildlife, some fish deserved more preservation than others. But it was accepted that tinkering with the population and distribution of fish species could be tolerated to a degree that would not be tolerated for air breathers. Of course, the ultimate tinkering was that fishing was not merely tolerated but encouraged. National parks were promoted as having some of the best sport fishing lakes and rivers in Canada. The Parks Branch never tried to justify or sugar-coat fishing, which in itself suggests how ingrained the logic of allowing it in parks was. As with wildlife, the amount of intervention by park staff on fishing matters increased during the 1950s, before abating somewhat in the 1960s.

Fishing was supposed to have an especially important role in the first Maritimes national parks for a number of reasons. First, the Parks Branch believed the very name "Maritimes" promised tourist opportunities for fishing (that it suggested salt-water fishing rather than parks' traditional fresh-water fishing was not insurmountable). Second, it was hoped that an abundance of fish in Eastern parks would compensate for their insufficient wildlife. Finally, inland fisheries were still under federal control in the Maritimes, having been delegated to the provinces elsewhere. The Parks Branch therefore would benefit from particularly close affiliation with the federal Department of Fisheries in managing park fish populations.[51]

Nova Scotia was renowned for its fishing, and the Cape Breton site's potential for fishing was an important factor at park establishment. Cape Breton's Margaree River was already famous among North American anglers, and the Parks Branch hoped that it could likewise put the Cheticamp River on the map. Even before Cautley's inspection in 1934, Commissioner James Harkin stated that one of his conditions for a northern Cape Breton park would be "to co-operate with the Dominion Government in doing away with all net fishing of salmon off the mouth of Cheticamp River. I am informed that at the present time the salmon fishing in both Margaree and

Cheticamp rivers has been practically ruined by the mouths of both rivers being netted by shore fishermen."[52] Harkin was not concerned about depletion of the fishing stock per se, but felt that net fishing was fundamentally unsportsmanlike and kept anglers from catching the same fish. Cautley's 1936 report came to the same conclusion, and suggested that if the park was to go through there, fish ladders could be installed to make access up the Cheticamp easier for salmon. He even hoped that, if legally practicable, the Parks Branch could restrict net fishing outside the park boundary.[53]

Fish were also important in the planning of Prince Edward Island National Park. Williamson and Cromarty recommended that the water off the north shore of PEI be made a fish preserve, "in the same way as the Western National Parks act as reservoirs for game."[54] However attractive this idea, and however seemingly responsive to contemporary concerns for the viability of lobster stocks, it had more to do with creating a credible park than with preserving an endangered species. The North Shore waters were of interest as a reserve because of their location off a possible park site; Williamson and Cromarty had no idea whether or not this was an especially good place to find and protect fish. And it would be a fish sanctuary "except for taking of fish by hook and line as under Park regulations." When the Department of Fisheries concluded that the fish preserve would serve no purpose, and Island politicians complained about how it would affect North Shore fishing, the idea was scrapped.[55]

Once the two parks were established, the Parks Branch investigated the fish stock they had and tried to determine how they could generate more. The findings were not encouraging. Prince Edward Island

National Park lakes did not contain trout, as was hoped, but "a serious enemy," white perch.[56] As well, other lakes outside the park supplied much better angling. Cape Breton Highlands was inconvenienced by both white perch in its lakes and netters outside its main river, the Cheticamp. The natural solution was to stock the park rivers and lakes with fish. The Cheticamp had been stocked with salmon from the Department of Fisheries' Margaree River hatchery since 1916; the Parks Branch simply dumped in more fish. Cape Breton Highlands was stocked with 170,000 fingerlings in the first year of its existence, 180,000 in 1938, up to 250,000 by 1941. After a late wartime lull, by the late 1940s the park was again being "planted" at rates up to 100,000 per year.[57] Around the same time, Prince Edward Island National Park also began to be stocked with trout, even though its lakes were not considered likely to provide good fishing.[58]

*****

The Parks Branch relied on stocking because it was straightforward and seemed sensible. Staff needed to be doing something, since the Maritime parks' appeal to tourists depended so much on fishing.[59] Though R.W. Cautley gave the eventual New Brunswick park low marks for fishing potential in 1930—a 50 out of 100, with the comment "There are practically no lakes"[60]—by 1950 the opening ceremonies brochure for Fundy National Park showed two photos of fishermen hard at play. James Smart directed limnologist V.E.F. Solman to sell American fishermen on Prince Edward Island's perch by using the more impressive name "silver bass."[61] The Parks Branch then paid C.H.P. Rodman of the American magazine *Hunting and Fishing* to vacation in the Prince Edward Island

park; he did not catch a single silver bass, but he wrote a nice article anyway.[62] Cape Breton Highlands aggressively advertised its offshore swordfishing, though in 1948 Solman reported, "In spite of all this publicity not one tourist has arrived and demanded to be shown such fishing."[63]

It seemed that locals enjoyed the fishing in the new Eastern parks more than anyone. This hardly consoled the Parks Branch. In 1939, Commissioner F.H.H. Williamson said bluntly that fishing regulations did not "necessarily correspond with the Provincial dates as the parks policy is to cater more to the tourist trade than for the benefit of the local residents."[64] Tourists meant money, meant a park was succeeding; locals spent far less and then had the gall to catch the fish meant for tourists. The Parks Branch therefore adjusted the parks' fishing seasons to approximate the tourist season more closely. For example, the opening of the salmon season at Cape Breton Highlands was moved from the province-wide June 1 to June 15, since "practically all of the salmon have been caught by local residents before many of the tourists start arriving in the Park, which is usually from early June on."[65] This more than anything else demonstrates that the Parks Branch saw tourists and locals as fundamentally different users of park resources. Though ostensibly for the benefit, advantage, and enjoyment of all Canadians, parks were meant to be especially for tourists, Canadian or otherwise.

As with wildlife, fish management in the national parks became decidedly more interventionist in the early 1950s. This was due to a number of factors: the creation of the Canadian Wildlife Service with a resultant increase in scientific personnel, a move within ecology toward more extensive management projects, increased funding for such projects, and the Parks Branch's own dissatisfaction with fish stocking.[66] The most fundamental change was technical: the discovery and application of rotenone. When Prince Edward Island National Park had been established, foreman A.L. MacKay had noted that, instead of stocking the park's lakes, they could contaminate it, kill all its fish, and restock. He supposed, however, that such a task would likely leave the lakes "tainted for an indefinite period."[67] He was right for his time, but he did not know that the Fisheries Research Board of Canada was currently testing the purification of lakes using rotenone, a natural poison derived from the powdered root of a number of plants (most commonly, derris). In 1936 the fisheries staff succeeded in killing all fish along 25 miles of stream and in a 6-acre lake near the Cobequid Hatchery in Nova Scotia. Rotenone was found to wipe out all insect and fish life within half an hour, allowing for the introduction of preferred species, and there was no long-term effect to the water.[68] Rotenone poisoning became relatively commonplace in North American fish management in the 1940s, but it was not until the 1950s that the Canadian national parks began to use it.

★★★★★

Rotenone was a great success in park lakes. The branch's chief limnologist in 1960, Jean-Paul Cuerrier, noted that at Cape Breton Highlands, "The absence of competition has favoured the survival of hatchery trout which have provided satisfactory angling returns to Park visitors."[69] Nonetheless, there began to be rumblings of discontent about the drastic measure of poisoning lakes. After an inspection of Fundy in 1956, F.A.G. Carter, the secretary to the minister, questioned why the Wildlife

Service planned to drop rotenone in the unfortunately named Lake View Lake. "Apparently by the locals here," he wrote, "there is reasonable fishing (speckled trout) and no one understands why poisoning is necessary. It may be necessary; I do not know. I do know, however, that it would be most helpful if the Wildlife Service could explain the 'whys' to the wardens, etc., and to all concerned within the park—to try to win them over. This may have been tried and may be impossible. There is a considerable feeling throughout the Park that the Wildlife people are long haired experimenters. I have no doubt this is grossly unfair."[70] This nicely shows that to "locals"—a designation that apparently united ground-level park staff with people of the community—the methods and plans of biologists were a mystery. Carter, a federal bureaucrat, clearly felt the same way. However, it would be a mistake to infer from this that the Canadian Wildlife Service pushed the Parks Branch in managerial directions it did not wish to go. Two years later, it was the superintendent of Fundy who asked his superiors whether Wolfe Lake could be poisoned; his superiors considered the request, but opted for stocking it with fingerlings instead.[71]

The use of rotenone peaked and then disappeared in the mid-1950s; it would appear from available archival records that no lake or river in an Atlantic Canadian national park was again poisoned pure.[72] Perhaps this was because of complaints from tourists and locals or from staff within the Parks Branch over this ultimate form of intervention. Perhaps it was part of a general move away from active management of park resources around 1960. Or perhaps, though less likely, the lakes continued to be poisoned for some time, only without the fanfare and press releases. But the Parks Branch did continue to manage the fish populations under its domain, to a degree that would have been considered unacceptable for any other living things in the parks. Populations were introduced or killed off, with little attention to whether they were native or exotic. Staff knew and admitted this. As parks chief B.I.M. Strong explained to a superintendent in 1959, "As you know, fishing seasons in the National Parks are set not too much according to biological consideration of the game fish involved but rather according to the aims and objectives of the National Parks."[73]

VEGETATION

Neither fish nor wildlife preservation was ever completely practised within the national parks; it was always, which fish? how much wildlife? But the underlying philosophy, that national parks were places where living creatures should exist unbothered by humans, continued to hold even when the resulting policies changed. The theory of preservation was more directly challenged by vegetation. It was obvious that untouched vegetation was not necessarily "better" for itself or for the park; if left alone, it could choke out desired species, increase fire hazard, block views, attract disease, and be aesthetically unattractive. As well, thick forests, if cut, were financially lucrative, and not cutting them was wasteful. Finally, humans did not have the sort of relationship with vegetation that they had with animals. There was not, to the same extent, a feeling that plants had a natural right to existence.[74] For all these reasons, the National Parks Branch forever questioned the validity of a hands-off policy with regard to vegetation, particularly forests. In

the four parks studied here, the same trend is visible in forest management as in fish and wildlife—increased interventionism in the 1940s and 1950s, followed by a reversal in the 1960s—but more interesting is the volume of discussion about the underlying meaning of preservation.

The Parks Branch depended on the Canadian Forestry Service for expertise on forestry matters, just as it depended on the Canadian Wildlife Service for fish and wildlife matters, but its own knowledge of forestry was greater. Foresters were trusted with positions of general responsibility in the Parks Branch to a degree that wildlife biologists would never have been. James Smart, Parks Branch controller from 1941 to 1950 and director from 1950 to 1953, began his career with the Forestry Service after earning his forestry degree at the University of New Brunswick.[75] Prince Edward Island National Park's first superintendent, Ernest Smith, was chosen expressly because of his forestry experience. Fundy National Park's first superintendent, Ernest Saunders, had previously been in charge of the Acadia Forest Experimental Station, and was paid extra by the Parks Branch because it was understood he would have extensive silviculture work to perform.[76]

Forestry preservation was an issue from the very establishment of Cape Breton Highlands National Park. The Parks Branch had to decide whether people living near a new park should be allowed to continue their traditional use of its forests. James Smart, respecting foresters and their needs, pushed for small woodlots to be set up at out-of-the-way places throughout the park. Controller F.H.H. Williamson was less enthusiastic: "Settlers adjacent to a Park boundary, who have been dependent on timber in the Park area, form a problem, since satisfying their demands is usually opposed to Park regulations. The man who draws out such a plan as Mr. Smart proposes should be sufficiently experienced in National Parks' operations as to know what cuttings would constitute improvement from a Park standpoint, the only legitimate excuse we have for allowing timber to be taken out for settlers' needs."[77] In other words, the Parks Branch needed an excuse to show that helping fulfill local wood needs would be for the good of the park forests. Smart found one by having the proposed woodlots made "demonstration plots" to be cut on a sustained yield basis, with park staff choosing the trees to be targeted.[78] The Forestry Branch approved this idea, so the project went forward.

Allowing ex-residents to cut wood in the park was a noble idea, and a far cry from the usual treatment of their traditional rights once the parks were created. But it caused difficulties when put in practice, while reinforcing the Parks Branch's (and particularly James Smart's) tendency to see park resources as objects to be used. Staff found sustained yield difficult to manage, so they turned to exhausting one stand of trees and moving on—hardly an innovative forestry practice.[79] The Parks Branch and locals also disagreed about the types of trees to be cut. For building small boats, fishermen needed pine, which the Parks Branch specifically wished to preserve, since, in Smart's words, "this is a desirable species from a scenic point of view and there is very little of this species in the Park." Locals wanted hardwood for fuel, because it burned longer in woodstoves, but the Parks Branch was trying to preserve its hardwoods. (Staff even specifically chose softwoods when they needed to build guard rails, cabins, and such.) As Smart told the Cape Breton Highlands superintendent, "it is the opinion of the

Bureau and also of our foresters that the main species of wood to encourage on these areas are the hardwoods and any softwoods should be cut if they come within the size specified—in fact, it might be necessary to treat such as balsam as one which should be eliminated entirely from an area, as it is a very poor type of wood for any purpose and, being a very prolific seeder, it is inclined to take over the area at the expense of the hardwood species."[80] This is so interesting because it is so meaningless. The small sections of forest that the woodlots demanded would have had no appreciable effect on species composition for the park as a whole. More important, there should have been no reason to discuss a tree's "purpose," because it did not need one in the park. And if by purpose Smart was referring to softwoods' scenic value, he was admitting to a regional prejudice against the scrubby, coniferous look of Maritime forests; in any case, he was responding to a blatantly unnatural aesthetic the park could not possibly hope to satisfy. His description of succession—that softwoods tend to block out hardwoods—was just plain wrong. The opposite is in fact the case, and for a forester to say otherwise shows an irrational aversion to softwoods.[81] Smart's quotation demonstrates his uncertainty about how to manage the contradictory task of safeguarding and stabilizing a vulnerable and ever-changing resource. In trying to find a consistent position, the Parks Branch would fall back on questionable science, vague aesthetics, and the familiar values of capitalism.

*****

The 1950s was a decade in which the Parks Branch would continually find need for active forest management in the parks, and each time find justification. At PEI National Park, staff took to thinning and cropping trees that were obstructing tourist lodges' views of the North Shore. This was done to pacify park neighbours, but was not believed to contradict park policy, since, in the words of deputy minister R.G. Robertson, "trees are developing in a very overcrowded and unpleasant fashion. Keeping hands off these areas do *not* preserve them in their natural state because they are all second growth areas in any event."[82] At Fundy National Park, a sawmill (out of tourists' eyesight and knowledge) was maintained to meet the lumber needs of the Maritime parks. The mill on the Upper Salmon River owned by Judson Cleveland before expropriation was used by the park until it burned down in 1952. The Parks Branch considered implementing an intensive cutting plan for the entire park,[83] but in the end decided just to set up another small mill at Bennett Brook. Throughout the mid-1950s, staff yearly logged between 60 and a 100,000 board feet (about one one-hundredth the amount taken before the park's establishment), and in the 1960s supplied the new historic park at Louisbourg with spruce logs.[84] When asked by a New Brunswick government member about the park system's timber policy, Assistant Deputy Minister E.A. Cote noted, "I indicated to him that I thought our policy was in the state of evolution and that I could not give him an immediate answer."[85]

Spruce budworm infestation in New Brunswick, Nova Scotia, and, to a lesser extent, Prince Edward Island in the early 1950s cemented the belief that hands-off preservation was dangerous to the forests being preserved. This infestation was quite natural, as budworm populations cycle every 25 to 70 years.[86] Cape Breton Highlands reported the existence of budworm in 1951, and by 1955 they had spread widely, most noticeably around the park headquarters.[87] The insects feasted on balsam fir and spruce, and, with the aid of the balsam woolly aphid, left most of the park softwood forests, especially in the interior, dead or dying. The result was a park which, in the words of forest engineer D.J. Learmouth, had "a very ragged appearance.... Fortunately, much of the Park stands are mixed-wood stands and ... serve to camouflage much of the dead and deformed softwoods and maintain the aesthetic appearance of many of the fine views along the Cabot Trail."[88] But in the interior, "its present monetary, aesthetic or recreational value is practically nil."[89] The infestation affected so many trees that any sort of cleanup was out of the question. Fundy was similarly threatened, though the budworm that moved throughout New Brunswick in this period had not yet made it to the park. Nothing in the Parks Branch's theory and practice of park management prepared it for the spruce budworm outbreak. [...]

*****

The obvious solution was insecticide. Spraying insect pests was an accepted part of Canadian national park management. There was absolutely no consideration that insects deserved any of the protection provided to other living things in the parks; they were simply killed. For example, from the very first years of Prince Edward Island National Park, its stagnant ponds were annually sprayed with hundreds of gallons of furnace oil—the recipe for killing mosquito larvae.[90] And with the availability of war-surplus planes and the massive expansion of the pesticide industry in post-war North America, aerial spraying grew much more accepted and much more common.[91] It was in this climate that New Brunswick began a full-scale attack from the air on spruce budworm in the 1950s, in a program that would continue for decades.

In a rare act of restraint, though, the Parks Branch opted not to take on the Maritime parks' spruce budworm with aerial sprayers. The branch was likely wary about what such a massive interventionist move would mean to its public preservationist image. And it could hardly spray Cape Breton Highlands when the Nova Scotia government itself had chosen not to treat infested parts of the province, despite considerable pressure to do so. By the time the Parks Branch even considered spraying, it was advised by federal foresters that the infestation was too far developed to bother.[92]

There was another reason to avoid aerial spraying: in the words of Chief J.R.B. Coleman, "In a National Park, the effect of wide-spread spraying with DDT on the fish and wildlife populations must also be considered and may be reason enough to avoid such procedures on Park lands until more information on this aspect of insect control is available."[93] The chemical DDT—dichloro-diphenyl-trichloroethane—had been found by the American military during the Second World War to be a wonderful agent for insect control: it was inexpensive, very toxic, persistent, useful as either a contact or stomach poison, and yet of low acute toxicity to mammals, including humans. When it was made available for civilian use in 1945, it was heavily marketed as a sort of super-chemical that could cleanse the planet of pests.[94] Biologists were more cautious, and the Canadian Parks Branch took note of their concerns. In 1946, James Smart distributed to superintendents in all

the Canadian national parks a U.S. Fish and Wildlife circular, "DDT: Its effects on Fish and Wildlife," which discussed 12 studies showing the dangers of the chemical.[95] It was already proven that DDT tended to kill non-targeted populations such as birds and fish, though some of this seemed to be due to overspraying. Smart demonstrated both the national park experience with using DDT and awareness of its dangers when he wrote in 1948, "In using any chemicals we have, of course, to take into consideration the effect on our wildlife population and in particular on the fish as we have found the latter are very susceptible to DDT even in comparatively small quantities."[96] Insect control was nonetheless periodically carried out in Maritime national parks in the late 1940s and early 1950s—including in the first year of Fundy's existence—but it is not clear whether DDT was the pesticide of choice.[97]

When spruce budworm erupted in the 1950s, the park system knew enough about DDT to be distrustful of spraying it by airplane. And yet, action seemed necessary. Cape Breton Highlands superintendent Doak, who happened to be a great advocate of DDT, informed Ottawa that the public was constantly criticizing the park about its trees, and he pushed for a spray program.[98] More important, senior politicians became involved in the matter. The assistant deputy minister of northern affairs and national resources, C.W. Jackson, told the director of the Parks Branch, James Hutchison, that the ministry was under pressure—presumably from the provincial government or federal Nova Scotia politicians—to act. Though the Forest Branch had suggested that the infestation would have to run its course, this was now deemed unacceptable. As Hutchison relayed to Chief Coleman, "I can tell you now the Minister is not prepared to accept such a stand unless it is a matter of last resort." A more activist response was needed.[99]

The Parks Branch decided that a localized DDT spraying program would take place around public parts of the Cape Breton Highlands Park. Vegetation along roadsides, campground and picnic areas, and park headquarters would be targeted. Of course, by spraying only public places, the Parks Branch was trying to keep its budworm problem out of sight. Moreover, it was giving the impression that trees close to tourists were of greater value to the park than distant ones, and that their appearance was worth an increased risk to local fish, wildlife, and humans.[100] Learmouth, a forester, assured the Parks Branch that he had discussed this "informally with mammalogists, ornithologists, and limnologists of the Canadian Wildlife Service" who did not think it would be a hazard, since it would only be a light spray on a small area.[101]

The spraying of spruce budworm at Cape Breton Highlands took place in the summer of 1957. It was a largely pointless exercise, since the bulk of infestation had already swept through the area in past years and was gone. Three thousand gallons of 25 percent DDT emulsion was mixed on site with 12,000 gallons of water to form 15,000 gallons of 5 percent DDT, which was then distributed through a chemical sprayer. Superintendent Doak, though warned that people tended to overdose DDT, complained to Ottawa that the small spray coming out of the machine could not possibly be doing the job.[102] The following spring, it was reported that the control program had preserved the appearance and limited the mortality of the sprayed trees, but not to the degree expected. This was not considered a failure, though, since the exercise was constantly referred

to as having "limited control objectives" and as being "somewhat experimental in nature."[103]

*****

Insect control soon became normalized. What had originally been used for epidemic infestation became the weapon against everyday black fly and mosquito populations. Ex-staff told me of walking around the campgrounds a half-hour before spraying, warning campers to cover their food. They then hooked up the fogger to a 45-gallon drum filled with a diesel oil and DDT mix and tied it in the back of a truck. Staff worked without masks, and when they drove around, the fogging operation drew crowds of children who would run along behind. In the words of one staff member, the spray would "kill all the birds and squirrels, there'd be nothing left."[104] The Parks Branch's proactive insect control program became a point of pride. In a St. John's *Evening Telegram* story on the new Terra Nova National Park, the community of tourist cabins was portrayed as offering Newfoundlanders unprecedented freedom from insects: "Among the appealling [*sic*] features of this little snug town nestled in deep woods is that there are no flies to bother the inhabitants. They'd be there in hordes but for the 'treatment' given the area three times weekly by Ben Roper. Warden Roper sprays the region with DDT. Dense clouds of it bellow from a compressed air gun and floats over the cabin area eliminating any flies which might be found."[105] The park was sufficiently proud of its new machine that Warden Roper posed for a picture of the fogger in action.

In the same period, however, pesticide use in general and DDT use in particular were beginning to draw widespread opposition. Wildlife biologists were finding evidence that chemicals of low toxicity may become concentrated and cause indirect poisoning further up the food chain; how often this reached humans was unclear. It became a matter of public debate in 1962 with the publication of a series of *New Yorker* articles by Rachel Carson, and the subsequent release of her book *Silent Spring*. Carson's message was that today's pesticidal contamination of the environment might be having permanent, irreparable effects on the health of the planet and its residents. Carson even specifically discussed how the aerial spraying of DDT to combat spruce budworm had killed salmon stock on New Brunswick's Miramichi River.[106] Just as Carson's work was becoming a public issue, the Parks Branch received its first written complaint about its spraying program. An American chemist who had visited Cape Breton Highlands wrote,

> There was the indiscriminate fogging of the areas around the camp ground with a DDT-oil fog. We were informed by the park warden doing the fogging that a stronger fog was used last year and resulted in the death of many of the birds in the area. Even though he was using a fog reduced in strength by some 37.5% this year I feel the practice is both foolish and dangerous. It is foolish because the fog had only momentary effect on the black flies which were causing all of the campers much discomfort. The inexpensive insect repellents now on the market proved to be more effective, and without any cost to the National Park. It is dangerous for reasons which I am sure you are familiar.... As a chemist and as a citizen I condemn the widespread and indiscriminate use of any insecticide and urge that your department discontinue the practice we witnessed on Cape Breton.[107]

The superintendent disagreed with this assessment, saying that staff did not feel there was significant bird mortality—especially now that they had reduced the DDT formula to the recommended 5 percent level.[108] But there were more complaints, and the Parks Branch began to experiment with different types of chemicals. As a narrator of a 1965 CBC television show on the PEI park explained, the program to spray mosquitoes was ultimately discontinued when people noticed that "the song birds disappeared with the mosquitoes.... Now mosquitoes feed on people, and the songbirds feed on mosquitoes: Nature's perfect balance restored."[109] A happy ending.

*****

Just as the Parks Branch had, for a variety of reasons, originally chosen certain locations for its national parks over others that would probably have been equally suitable, it chose some parts of the parks' constituent living things over others in managing them. Effect followed effect, often beyond the branch's understanding. The preference for attractive forests filled with healthy trees led park staff to cut the forests in a way reminiscent of traditional use. It also led to the killing of insects, and secondarily to the killing of birds, small mammals, and fish. Even the most innocent gestures produced unforeseen results: When staff selectively cut down and removed trees killed by spruce budworm, they felt they were doing the forest a favour; instead, they were removing the homes of insects that were budworm predators, and so helping the depredation continue. [...] [But] park staff slowly grew aware of the changes they were bringing to the parks, and grew more assiduous in at least considering the potential consequences of their actions.

But the Canadian National Parks Branch as an agency has never really acknowledged the extent to which it was responsible for decades of rampant intervention, and how it served as an example to the Canadian public. Today, a sign at Fundy National Park concerning its peregrine falcon population reads, "Toward the 1950s peregrine numbers declined drastically. The widespread use of DDT, an insecticide which accumulated in the food chain, was responsible for the decrease." Also, "A pair of peregrines nested on a seaside cliff a few kilometres east of Point Wolfe in 1948, the year that Fundy National Park was established. They were not seen again and decades would pass before others of their species would replace them."[110] The sign never quite mentions that at one time pesticides were used so widely in North America, and thought so safe, that they were dispersed abundantly in this very national park.

*****

The culminating moment for the Canadian park system in the 1960s was Arthur Laing's address on park policy to the House of Commons in September 1964. Park historians have called it "a very significant milestone" and one that "established the preservation of significant natural features in national parks as its 'most fundamental and important obligation.'"[111] The ingredients for drama were certainly there. A quiet revolution was already under way within the park system, as both staff and the public sought nature-friendly policies. The Parks Branch had been labouring for six years on a declaration explaining how it was attempting to reconcile preservation and use. Now, the agency felt pressure to make this statement

public: Businessmen at Banff and Jasper were pressing their right to develop wherever they wanted in the townsites; a National and Provincial Parks Association of Canada had recently formed to protect the parks from development; and in 1963 *Maclean's* magazine published a scathing article, "Beauty and the Buck," that painted the park system as being on the verge of collapse.[112] Author Fred Bodsworth had pulled no punches: "We are losing them because a lax and indecisive parks policy, particularly toward business and political pressures, has allowed many national parks to deteriorate into commercialized, honky-tonk resorts where the major aim is no longer park preservation but rather separating tourists from their money."[113] It was hoped that Laing's speech would answer such criticisms by showing that the Parks Branch was already correcting past mistakes; it might also discourage developers from planning further incursions in the parks. Most of all, it would do away with the system of ad hoc, unwritten policy directions which had taken shape since 1930, replacing this with a single, decisive policy document.

It is surprising then that to today's reader, Laing's speech is not very decisive at all. The closest he comes to a pointed declaration of park priorities is to say:

National park policy cannot contribute to a solution of the crisis if it is based on one of the two extremes, maximum preservation on one hand, or maximum public use and development on the other. One would deprive the public of the benefits they receive from national parks; the other would destroy the special enjoyment and pleasure the public receives from lands kept in a near natural state. The objective of national park policy must be to help Canadians gain the greatest long term recreational benefits from their national parks and at the same time provide safeguards against excessive or unsuitable types of development and use.[114]

It helped, of course, for this position to be clarified for the Canadian public, and for Parks Branch staff to have such a ministerial statement from which to refer.[115] But rather than providing a solution, Laing's address simply restated the problem: that parks were always under a dual mandate of preservation and use. This was entirely appropriate. Though this was a period when park staff and the public leaned toward less interventionist policies—and, indeed, Laing's speech was interpreted as a victory for these beliefs—the Parks Branch could not and should not codify non-interventionism. A perfect balance between preservation and use must always be the park system's goal, even though it is a goal which will never be perfectly achieved.

NOTES

1. Kevin McNamee, "From Wild Places to Endangered Spaces: A History of Canada's National Parks," in Philip Deardon and Rick Rollins, eds., *Parks and Protected Areas in Canada: Planning and Management* (Toronto: Oxford University Press, 1993), 28.
2. As early as 1938, Commissioner F.H.H. Williamson noted, "In Canadian Parks possibly the development has reached a point where we should call a halt on new work and confine ourselves simply to the improvement, completion, and maintenance of existing works." Williamson to Gibson, January 4, 1938, RG 84 vol. 2101, file U172 (6), NAC.
3. Susan Power Bratton, "National Park Management and Values," *Environmental Ethics* 7 (Summer 1985), 119 and 126–127.
4. There is a large literature on human perceptions of animals. A standard text, though defined more by its subtitle than its title, is Keith Thomas, *Man and the Natural World: Changing Attitudes in England 1500–1800* (London: Penguin, 1983). On animal conservation, see John A. Livingston, *The Fallacy of Wildlife Conservation* (Toronto: McClelland and Stewart, 1981); on animal rights, Peter Singer, *Animal Liberation: A New Ethic for Our Treatment of Animals* (New York: Avon, 1977); and on the history of environmental ethics, Lisa Mighetto, *Wild Animals and American Environmental Ethics* (Tuscon: University of Arizona Press, 1991).
5. For the Wildlife Division and the origin of wildlife research in Canada, see W.F. Lothian, *A History of Canada's National Parks*, 4 vols. (Ottawa, Parks Canada, 1977-1981), 4: 55–59; D.H. Pimlott, C.J. Kerswill, and J.R. Bider, *Scientific Activities in Fisheries and Wildlife Resources*, Special Study #15 (Ottawa: Science Council of Canada, 1971), 112; Ian McTaggert Cowan, "A Naturalist-Scientist's Attitudes towards National Parks," *Canadian Audubon* 26 (May 1964), 93–96; and Gail Lotenburg, "Wildlife Management Trends in the Canadian and US Federal Governments, 1870–1995," unpublished report for Parks Canada, 1995.
6. Lloyd's formal training was as a chemist, but he came to the Wildlife Division because of his interest in ornithology. On Lloyd, see Janet Foster, *Working for Wildlife: The Beginning of Preservation in Canada*, 2nd ed. (Toronto: University of Toronto Press, 1998 [1978]), 159–161.
7. Alan MacEachern, "Rationality and Rationalization in Canadian National Parks Predator Policy," Chad Gaffield and Pam Gaffield, eds., *Consuming Canada: Readings in Environmental History* (Toronto: Copp Clark, 1995), 204–205.
8. For ecology in this period, see Donald Worster, *Nature's Economy: A History of Ecological Ideas*, 2nd ed. (Cambridge: Cambridge University Press, 1985 [1977]). On ecology in the national parks, see Richard West Sellars, *Preserving Nature in the National Parks: A History*. (New Haven: Yale University Press, 1997); and Thomas R. Dunlap, "Wildlife, Science, and the National Parks, 1920–1940," *Pacific Historical Review* (May 1990): 187–202.

9. North American biologists were greatly affected by the lessons learned in the Kaibab National Forest in Arizona. U.S. Biological Survey staff had cleansed this game preserve of all predators by 1920, and in subsequent years the deer population exploded. All foliage was soon picked clean, and by 1925 starvation wiped out much of the herd. In *Discordant Harmonies*, Botkin offers a valuable critique of the simplistic lessons learned from the Kaibab. Essentially, he points out that deer and predator populations do not simply move up and down in relation to one another. This ignores all other variables in their environment, and factors their "value" strictly by their absence or presence in the system. See also Thomas R. Dunlap, "That Kaibab Myth," *Journal of Forest History* 32 (1988), 60–68.

10. More research is needed on the effect that the Depression had on Canadian wildlife preservation policies. In the United States, it has been said of Franklin Roosevelt's 1930s that "No other decade or administration did so much to save wildlife." Donald Worster, *An Unsettled Country: Changing Landscapes of the American West* (Albuquerque: University of New Mexico Press, 1994), 77. See also Theodore W. Cart, "A 'New Deal' for Wildlife: A Perspective on Federal Conservation Policy, 1933–1940," *Pacific Northwest Quarterly* 62 (July 1972), 113–120.

11. Parks commissioner James Harkin was seen as Canada's greatest advocate of sanctuaries, in national parks as well as in provincial and public reserves. For Harkin's interest in sanctuaries, see Foster, *Working for Wildlife*, especially 88, 198, and 206, as well as his own "Wildlife Sanctuaries." On sanctuaries in general, see Ira N. Gabrielson, *Wildlife Refuges* (New York: Macmillan, 1943); and Worster, *Nature's Economy*, 259–260; and *An Unsettled Country*, 76–77.

12. Some hunters mistakenly believed that parks were sanctuaries *for* hunting rather than sanctuaries *from* them. In 1942, Ernest Smith, superintendent of Prince Edward Island National Park, thought this belief so prevalent that he asked his superiors if he could advertise in local papers that hunting was forbidden in the park. See RG 84, vol. 23, file PE1300, NAC. When Fundy was established, Egbert Elliott of Alma wrote the National Parks Branch asking if he could set up a tourist business that would cater to hunters. See RG 84, vol. 1024, file F16, NAC.

13. See RG 84 ,vol. 140, file CBH272, NAC, for the beaver reintroduction. The moose transfer is discussed in Williamson to Sarty, RG 84, vol. 1002, file CBH300, vol. 1, pt. 3, NAC.

14. Sellars, *Preserving Nature*, 96–98.

15. Reintroduction of species killed off by locals in the past was also a way of asserting the legitimacy of the park's takeover of the land and its resources.

16. C.H.D. Clarke report, March 23, 1942, 5, RG 84, vol. 1002, file CBH300, pt. 3, NAC.

17. Ibid., 7. In another case of rejecting exotic but otherwise desirable species during this period, the Parks Branch declined the offers of Pheasants Unlimited on PEI and the provincial fish and game association of Nova Scotia to let pheasants loose in the respective parks because they were not native birds. Harrison Lewis of the Wildlife Division even suggested that if pheasants began to arrive from outside the parks and adversely affected native wildlife, they should be controlled. Lewis to Spero, May 3, 1946, RG 84, vol. 182, file PE1301, NAC. Smart to Frank Nolan, president, Fish and Game Protective Association, December 3, 1948, RG 84, vol. 139, file CBH301, NAC.

18. Canada, Departments of Resources and Development, Northern Affairs and National Resources, and Indian Affairs and Northern Development, *Annual Reports* (1948–1969). Since there was a departmental reorganization in 1947, and more items were targeted by this expenditure from then on, it is difficult to know exactly how much of an increase this was.

19. Frank B. Golley's *A History of the Ecosystem Concept in Ecology: More Than a Sum of the Parts* (New Haven: Yale University Press, 1993) is helpful in demonstrating how the size and budgets of scientific studies can shape the direction that their work takes.

20. Smart to superintendents, January 6, 1948, RG 84, vol. 2102, file U172, vol. 7, NAC.

21. Pimlott, Kerswill, and Bider, *Scientific Activities*, 112.

22. Press release, Deputy Minister Hugh Keenleyside, June 11, 1949, RG 22, vol. 153, file 5.0.1.35, vol. 6, NAC.

23. Lotenburg, "Wildlife Management Trends." Discussing the Canadian situation generally, Stephen Bocking writes in "A Vision of Nature and Society: A history of the Ecosystem Concept," *Alternatives* 20, no. 3 (1993), 12–18, that "most Canadian ecological research has been tied more or less closely to immediate resource management concerns" (16).

24. See chapter 14 of Worster, *Nature's Economy*, and Bocking, "A Vision of Nature and Society."

25. Lothian, *A History*, 4: 59. Because the Parks Branch was ultimately responsible for management decisions, it is credited with the policies discussed in the remainder of this chapter. Of course, it should be noted that the Canadian Wildlife Service often played a critical role in decision making on park wildlife issues, just as the Forestry Service did in park forestry matters — and politicians did to any matter in which they had an interest.

26. H.F. Lewis to George J. Keltie, president, Western Canada-Yukon Fish and Game Council, March 31, 1949, RG 84, vol. 39, file U300, vol. 16, NAC. See also Lewis to Smart, March 4, 1949, ibid.

27. Slaughters of elk, moose, and buffalo had been taking place in Western parks since the 1930s, and became annual affairs during the Second World War. These were justified (time and again) in local, practical terms; not until the 1950s did wildlife managers speak of overpopulation as the inevitable result of a park's existence.

28. "A Policy Statement Respecting Wildlife in the National Parks of Canada," Coleman to Hutchison, January 21, 1957, RG 84, vol. 2140, file U300, pt. 18, NAC.

29. John P. Kelsall report, "Mammal and Bird Survey, New Brunswick National Park, June 4 to July 4, 1948," to Smart, October 29, 1948, RG 84, vol. 141, file F300, NAC.

30. R.Y. Edwards and C. David Fowle's "The Concept of Carrying Capacity," *Transactions of the Twentieth North American Wildlife Conference* (Washington, DC: Wildlife Management Institute, 1955), 589–602, discusses how a supposedly scientific term as this can lose precise meaning owing to its popularity and apparent universal applicability. In 1955 they were already writing that "most definitions of carrying capacity are vague and that some are almost meaningless" (589).

31. A.W.F. Banfield, "Fundy National Park Wildlife Investigations, March 12–17,1951," RG 84, vol. 1002, file CBH300, vol. 1, pt. 2, NAC. Banfield did not yet recommend moose reduction.

32. R. Gerald Wright, *Wildlife Research and Management in the National Parks* (Urbana: University of Illinois Press, 1992), 69.

33. J.S. Tener report, passed on by W.D. Taylor to Coleman, August 20, 1953, RG 84, vol. 1039, file F300, vol. 1, NAC.

34. J.A. Hutchison to Nason, legal adviser, October 22, 1953, RG 84, vol. 1024, file F2, vol. 5, NAC.

35. See RG 84, vol. 486, file F216, pt. l, NAC.

36. "Moose reduction program, Fundy National Park, January 1955," ibid.

37. Research in the U.S. has questioned wildlife managers' ability to judge browse quantity in the first place; what might seem to be overbrowsing may be quite natural for an area. Wright, *Wildlife Research*, 80.

38. There was discussion of this in the 1939 *Transactions of the Provincial-Dominion Wildlife Conference.*

39. F.A.G. Carter to Robertson, September 23, 1957, RG 22, vol. 476, file 33.21.1, pt. 4, NAC.

40. Superintendent E.C. Browning to Strong, May 7, 1958, RG 84, vol. 1778, file PE12, vol. 8, NAC.

41. Many locals may have disliked this concession. In 1960, Superintendent Kipping wrote that "indeed there was local derision directed at the Park staff because of their inability to cope with the hunting." Cited by Coleman to Côté, August 4, 1964, RG 84, vol. 1778, file PEI2, vol. 10, NAC.

42. Smart to superintendent, May 9, 1951, RG 84, vol. 141, file F300, NAC; and Strong to superintendent, December 15, 1959, RG 84, vol. 487, file F281, pt. 2, NAC.

43. C.O. Bartlett, wildlife biologist, Canadian Wildlife Service, cited by Superintendent Kipping to Strong, January 16, 1963, RG 84, vol. 1802, file PE1279, NAC.

44. John P. Kelsall to Harrison Lewis, October 9, 1950, RG 84, vol. 1802, file PEI275, NAC.

45. Superintendent Kipping to "Gentleman and Olga" of CBC "Long Shot," September 2, 1959, box 17, whales and seals file, PEINP files.

46. Kipping to Strong, November 13, 1962, ibid.

47. Strong to superintendent, December 15, 1959, RG 84, vol. 487, file F281, pt. 2, NAC; and Smart to superintendent, May 9, 1951, RG 84, vol. 141, file F300, NAC.

48. Dinsdale to Ralph H. Olive, June 8, 1962, RG 22, vol. 1083, file 300.52, NAC.

49. See RG 84, vol. 986, file CBHI6.I vol. 2, pt. 1, NAC.

50. Sellars, *Preserving Nature*, 123. See also Alfred Runte, *Yosemite: The Embattled Wilderness* (Lincoln: University of Nebraska Press, 1990), 65–66.

51. See Kenneth Johnstone, *The Aquatic Explorers: A History of the Fisheries Research Board of Canada* (Toronto: University of Toronto Press, 1977).

52. Harkin to Gibson, December 14, 1934, RG 84, vol. 983, file CBH2, vol. 1, no. 1, NAC.

53. Cautley report on Nova Scotia sites, 58, in R.H. MacDonald, *Transportation in Northern Cape Breton*, Appendix A (Ottawa: Parks Canada, 1979); and Cautley to Harkin, June 16, 1936, RG 84, vol. 520, file CBH296, pt. 1, NAC.

54. Cromarty and Williamson report on PEI sites, 11–12, RG 84, vol. 1777, file PEI2, vol. 1, pt. 1, NAC. They noted that this reserve should not include Rustico fishing grounds.

55. Wardle to William A. Found, deputy minister of fisheries, November 6, 1936, RG 84, vol. 1802, file PE1296, vol. 1, NAC.

56. Dr. A.H. Leim, director, Atlantic Biological Station, August 30, 1939, RG 84 ,vol. 1802, file PE1296, vol. 1, NAC.

57. Canada, Departments of Mines and Resources, and Resources and Development, *Annual Reports* (1939–1954). Salmon was the main fish stocked in early years, with Eastern brook trout increasingly stocked by the late 1940s. None of the Eastern parks had their own hatcheries, but the branch's philosophy on stocking can be seen in V.E.F. Solman, J.P. Cuerrier, and W.C. Cable's "Why Have Fish Hatcheries in Canada's National Parks," *Transactions of the Seventeenth North American Wildlife Conference* (Washington, DC: Wildlife Management Institute, 1952): 226–233.

58. Canada, Departments of Resources and Development, and Northern Affairs and National Resources, *Annual Reports* (1953–1954).

59. In the 1950 film *Father of the Bride*, Elizabeth Taylor's character breaks off her engagement because her fiancé wants a Nova Scotia honeymoon, complete with fishing. This nicely alludes to Nova Scotia's popularity with sport fishermen, while at the same time suggesting that the province might need to remarket itself if it hoped to attract post-war American tourism.

60. Cautley report on New Brunswick sites (1930), RG 84, vol. 1964, file U2.13, vol. 1, NAC.

61. Solman to Lewis, February 4, 1948, RG 84, vol. 1802, file PEI296, vol. 1, NAC.

62. Ibid. See also *Hunting and Fishing* (November 1952).

63. Solman to Lewis, February 11, 1948, RG 84, vol. 140, file CBH296.12, NAC.

64. Williamson to MacKay, March 28, 1938, RG 84, vol. 1802, file PE1296, vol. 1, NAC. See also MacKay to Williamson, June 10, 1938, ibid.

65. Superintendent MacFarlane to Smart, June 12, 1952, RG 84, vol. 200, file CBH3.1.1, NAC. In this case, the decision backfired. A group of Americans arrived early the following summer, and were very upset that the fishing season was two weeks away. In the parks studied here, the fishing seasons constantly changed as the Parks Branch tried to find a period which would attract the most tourists and yet correspond closely to the provincial season.

66. See Lothian, *A History,* 4: 60, 66–67; and Pimlott, Kerswill, and Bider, *Scientific Activities,* 62, for administrative changes in this period. See Bocking, "A Vision of Nature and Society," for changes in ecology.

67. MacKay to Williamson, March 15, 1938, RG 84, vol. 1802, file PE1296, vol. 1, NAC.

68. See Louis A. Krumholz, "The Use of Rotenone in Fisheries Research," *Journal of Wildlife Management* 12, no. 3 (July 1948): 305-317.

69. Cuerrier to Strong, August 4, 1960, RG 84, vol. 520, file CBH296, pt. 2, NAC.

70. Carter report, August 1, 1956, RG 22, vol. 472, file 33.6.1, pt. 4, NAC. According to the Wildlife Service, poisoning had not been considered. Coleman, August 23, 1956, ibid.

71. MacFarlane to Strong, June 5, 1958; and Strong to MacFarlane, June 17, 1958, RG 84, vol. 487, file F296, pt. 1, NAC.

72. In *A Walk in the Woods* (Toronto: Doubleday, 1997), Bill Bryson tells the story of the U.S. National Park Service's use of rotenone at Great Smokey Mountains National Park in 1957. Staff killed everything in Abram's Creek, including a fish thought extinct—which now was (97).

73. Strong to superintendent, June 5, 1959, RG 84, vol. 1802, file PE1296, vol. 2, NA.

74. Roderick Nash's *The Rights of Nature: A History of Environmental Ethics* (Madison: University of Wisconsin Press, 1989) discusses the evolution of rights over time. Though animal rights are quite well established, the idea of rights for other living things is in its infancy. See Christopher D. Stone's influential "Should Trees Have Standing? Toward Legal Rights for Natural Objects," *Southern California Law Review* 45 (Spring 1972): 450–501.

75. In bringing Smart to the Parks Branch in 1930, deputy minister of mines and resources W.W. Cory wrote, "[W]ith the growing needs and responsibilities for the protection of the Parks' forests and development of silviculture methods involved in the growth of the Parks system, it becomes a necessity to have a fully qualified Forester appointed to this Branch." Cory to William Foran, secretary, Civil Service Commission, April 10, 1930, RG 32, file 237, file 1888.02.29, NAC.

76. Smart to Gibson, January 14, 1948, RG 84, vol. 1039, file F200, vol. 1, NAC.

77. Williamson also warned that "there shall be no repetition in Cape Breton Highlands National Park of the timber disposal complications experienced at Riding Mountain Park" in Manitoba. Williamson to Gibson, December 4, 1939, RG 84, vol. 520, file CBH200, pt. 1, NAC. When Riding Mountain was created in 1930, it was decided that traditional use of park forests would be honoured, and sawmills were allowed to continue operation in the park. James Smart was the superintendent there in this period. Lothian, *A Brief History,* 78.

78. Smart to Williamson, January 30, 1940, RG 84, vol. 520, file CBH200, pt. 1, NAC.

79. W. Robinson to Smart, October 15, 1941, ibid.

80. Smart to superintendent, February 19, 1941, ibid. Also, D. Roy Cameron, Dominion forester, to J.C. Venness, Dominion Forestry Service, February 4, 1941, ibid.

81. Balsam have been known to take over areas wiped out by spruce budworm, and this may have been what Smart was referring to. He had to know, though, that this succession process was unusual. On foresters' poor understanding of succession in the early twentieth century, see Nancy Langston, *Forest Dreams, Forest Nightmares: The Paradox of Old Growth in the Inland West* (Seattle: University of Washington Press, 1995), 122–134.

82. Robertson to Coleman, August 20, 1957, RG 84, vol. 1786, file PE128, vol. 1, NAC. Complaints about park trees may be seen in North Shore Tourist Resort Operators to Department of Northern Affairs and Natural Resources, October 8, 1955, RG 22, vol. 317, file 33.21.I, pt. 3, NAC; and acting superintendent H.A. Veber to Coleman, June 30, 1954, RG 84, vol. 1786, file PE128, vol. 1, NAC.

83. Gibson to Timm, Dominion Forest Service, May 17, 1948, RG84 vol. 1039, file F200, vol. 1, NAC; and Smart to Superintendent Saunders, June 27, 1952, ibid.

84. See Superintendent MacFarlane to Coleman, May 25, 1956, ibid.; C.E. Mullen, *Literature Review: Fundy National Park* (Halifax: Parks Canada, 1974), 71; and Michael Burzynski, "Man and Fundy:

Story Component Plan," (Parks Canada, 1987), 43, internal document, Fundy National Park files. Those in charge of implementing the cut squeezed scientific benefit from even the most prosaic of forestry tasks. For example, at Bennett Brook, "It is proposed to lop and scatter the brush on part of the area and pile and burn on the remainder to provide cost data and esthetic comparisons between the two brush disposal methods." D.J. Learmouth, forestry engineer, to Coleman, October 20, 1955, RG 84, vol. 1039, file F200, vol. 1, NAC.

85. Côté to Jackson, September 11, 1956, RG 84, vol. 472, file 33.6.1, pt. 4, NAC.

86. On spruce budworm in the Maritimes, see Kari Lie, "The Spruce Budworm Controversy in New Brunswick and Nova Scotia," *Alternatives* 9, no. 2 (Spring 1980): 5–13; Charles Restino, "The Cape Breton Island Spruce Budworm Infestation: A Retrospective Analysis," *Alternatives* 19, no. 4 (1993): 29–36; and Elizabeth May, *Budworm Battles* (Halifax: Four East, 1982).

87. C.C. Smith report, December 1957, RG 84, vol. 520, file CBH181.1, pt. 3, NAC.

88. Learmouth report to Coleman, October 24, 1955, RG 84, vol. 520, file CBH180.3, pt. 1, NAC.

89. Learmouth report "A Fire Access Trail Program for Cape Breton Highlands National Park," 1956, RG 84, vol. 8, file CBH62, NAC.

90. Scribbler re Dalvay House information, PEINP files. For insect abatement in U.S. national parks, see Runte, *Yosemite*, 176–178.

91. For pesticide use in post-war America, see Thomas R. Dunlap, *DDT: Scientists, Citizens, and Public Policy* (Princeton: Princeton University Press, 1981). On the relationship between war and pesticide, see Edmund P. Russell, III, "'Speaking of Annihilation': Mobilizing for War Against Human and Insect Enemies, 1914–1945," *Journal of American History* (March 1996): 1505–29.

92. Learmouth to Coleman, October 17, 1956, RG 22, vol. 474, file 33.9.1, pt. 5, NAC.

93. Coleman to Hutchison, December 8, 1955, RG 22, vol. 473, file 33.9.1, pt. 4, NAC.

94. Dunlap, *DDT*, 37; and Russell, "Speaking of Annihilation," 1525. In a 1946 issue of the *Journal of Wildlife Management* devoted to DDT, a leading biologist of the time made the pronouncement, "No chemical substance has excited greater general interest than DDT." Tracy I. Storer, "DDT and Wildlife," *Journal of Wildlife Management* 10, no. 3 (July 1946), 181.

95. Clarence Cottam and Elmer Higgins, "DDT: Its Effects on Fish and Wildlife" (Washington: US Fish and Wildlife, Department of Interior Document, 1946). See Smart to superintendent, October 30, 1946, RG 84, vol. 1002, file CBH300, pt. 2 (1942–1952), NAC; and RG 84, vol. 23, file PE1300, NAC. Even prior to the civilian release of the chemical, the Canadian Provincial-Dominion Wildlife Conference passed a resolution stating, "Therefore, be it resolved, that this conference recommends a programme of scientific investigation of the complicated problems relating to DDT and wildlife before any large quantities of DDT are used for any large-scale commercial purpose, and also recommends that the use of DDT be kept under strict governmental control." February 1945, RG 22, vol. 4, file 13, NAC.

96. Smart to Harold Furst, manager, Seignory Club, Montebello, Quebec, November 8, 1948, RG 84, vol. 39, file U300, vol. 16, NAC. Smart wrote to find out how the Seignory Club kept insects away.

97. L.W. Ford to Coleman, February 23, 1955, RG 84, vol. 1786, file PE128, vol. 1, NAC; and Canada, Department of Mines and Resources, *Annual Report* (1949), 22.

98. Doak told Coleman, "If you should decide to carry out an aerial spray probably the Province of Nova Scotia would co-operate and have the entire Island sprayed." July 30, 1956, RG 84, vol. 520, file CBH181.1, pt. 3, NAC. In this letter, he advocated the use of DDT, even though he had just received a letter from a New Brunswick forestry engineer expressly recommending the use of another spray, if any. C.C. Smith, Forest Biology Lab, Fredericton, to Doak, July 20, 1956, ibid.

99. Jackson to Hutchison, August 1, 1956, ibid.; and Hutchison to Coleman, August 8, 1956, ibid.

100. There was as yet no firm evidence that DDT posed a threat to humans. However, it was certainly suspected. Dunlap, *DDT*, 88–89. There was constant discussion of the risks and benefits of DDT in the North American Wildlife Conferences of the 1950s. It is telling that in a 1952 defence of the chemical, David G. Hall of the U.S. Bureau of Entomology felt compelled to assure his readers, "DDT is not the human killer many people think it is." Hall, "Our Food Supply, Wildlife Conservation, and Agricultural Chemicals," *Transactions of the Seventeenth North American Wildlife Conference* (Washington, DC: Wildlife Management Institute, 1952), 30.

101. Lothian, assistant chief, to Hutchison, December 3, 1956, RG 22, vol. 474, file 33.9.1, pt. 5, NAC.

102. Doak to Strong, August 1, 1957, RG 84, vol. 520, file CBHI8I.I, pt. 3, NAC.

103. See, for example, ibid.

104. Interview with Neil MacKinnon, Pleasant Bay, August 4, 1994. Three ex-staffers from Maritime parks described DDT use to me, but asked not to be quoted.

105. Don Morris, "Park Paradise Beckons," St. John's *Evening Telegram*, August 6, 1960.

106. Rachel Carson, *Silent Spring* (Boston: Houghton Mifflin, 1962), 134–138.

107. Joseph G. Strauch Jr., Corvallis, Oregon, to David Munro, Canadian Wildlife Service, September 12, 1962, RG 84, vol. 1002, file CBH301, pt. 2, NAC.

108. Superintendent McCarron to Strong, November 22, 1962, ibid.

109. Narrator on "Diary of a Warden," CBC program "20/20," 1965. Thanks to Barb MacDonald of Prince Edward Island National Park for making this available to me.

110. Sign at Point Wolfe Beach, Fundy National Park. The collapse of the North American peregrine falcon population was critical in convincing the U.S. to ban DDT in 1972. See Dunlap, *DDT*, 130–132.

111. J. I. Nicol, "The National Parks Movement in Canada," J.G. Nelson and R.C. Scace, eds., *The Canadian National Parks: Today and Tomorrow*. Proceedings of a conference organized by the National and Provincial Parks Association of Canada and the University of Calgary, 9-15 October 1968, Vol. 1 (Calgary: University of Calgary Press, 1969), 47; and Kevin McNamee, "From Wild Places to Endangered Spaces: A History of Canada's National Parks," Philip Deardon and Rick Rollins, eds., *Parks and Protected Areas in Canada: Planning and Management* (Toronto: Oxford University Press, 1993), 30. It is not clear whom McNamee is quoting, since this phrase does not appear in Laing's policy statement.

112. On the park system in the early 1960s, see Leslie Bella, *Parks far Profit* (Montreal: Harvest House, 1986), 113–116; and McNamee, "From Wild Places to Endangered Spaces," 29–30. On the formation of the parks association (now the Canadian Parks and Wilderness Society), see Marilyn Dubasek, *Wilderness Preservation: A Cross-Cultural Comparison of Canada and the United States* (New York: Garland, 1990), 72–78. Fred Bodsworth, "Beauty and the Buck: A Holiday Through our Magnificent National Parks," *Maclean's* (March 23, 1963), 25 and 41–66. The Bodsworth article received a great deal of discussion within the publicly minded Parks Branch. RG 84, vol. 2103, file U172, vol. 12, NAC contains considerable correspondence on it.

113. Bodsworth, "Beauty and the Buck," 25.

114. Canada, House of Commons, *Debates*, September 18, 1964, 8192.

115. Staff often quoted or referred to Laing's speech in correspondence. For example, Coleman to Superintendent McAuley, May 21, 1965, RG 84, vol. 1802, file PEI181, vol. 2, NAC; and letter for Laing to Louis Robichaud, premier, New Brunswick, October 8, 1964, RG 84, vol. 1039, file F200, vol. 4, NAC.

Critical Thinking Questions

1. What sparked the public outcry over the environmental impact of detergents in Lake Ontario in the late 1950s and early 1960s? In your view, is it likely to have been the result of internal factors (that is, the increasingly obvious nature of the problem) or external factors (caused by the growing environmental consciousness of the time)?

2. Read makes the case that the younger generation of the latter half of the 1960s were able to affect environmental policy significantly because they seized control of the political agenda. Would such a thing be possible today, in your view?

3. To what extent was the opposition to the detergent industry generational, in your opinion? Are there similar instances of generational divisions today?

4. In what ways were scientific management practices in Wood Buffalo National Park compromised by their relationship with economic and political actors, according to Sandlos?

5. Are there ever legitimate reasons for controlled culls of wildlife, even of endangered species, in your view? Are any of those reasons not ecologically based?

6. What should be the foundations of sound wildlife management today, in your opinion?

7. MacEachern demonstrates the tension between conservation and access in Canada's national parks system. To what extent should National Parks policy take into account the needs and desires of the Canadian taxpayers, who pay for the parks after all? What restrictions should there be on parks access, and who should establish those restrictions, politicians or scientists?

8. How should we ensure that conservation policies are driven by sound science instead of political expediency?

FURTHER READING

Tina Loo, "Making a Modern Wilderness: Conserving Wildlife in Twentieth-Century Canada," *Canadian Historical Review* 82, no. 1 (2001): 92–121.

Loo traces the conservationist impulse in Canada through the twentieth century, and locates it firmly as forming because of the desires and needs of urbanites rather than as a result of concerns about environmental degradation *per se*. She shows, for example, that it was the desire to "get back to nature," created by the increasing pace and demands of urban life, that generated the impulse in a large part of the Canadian population to experience not just nature but *wilderness*. She argues that conservationism as it develops among certain segments of the population not only was but remains incompatible with scientifically based ecological conservationism, and that this has produced a deep tension over the questions of wilderness access, appropriate usage, and long-term sustainability.

D.H. Steele, R. Andersen, and J.M. Greene, "The Managed Commercial Annihilation of Northern Cod," *Newfoundland Studies* 8, no. 1 (1992): 34–68.

In the same way that Sandlos questions the coherence of scientifically based management processes, Steele, Anderson, and Greene argue that the supposedly scientific management of the northern cod stocks after 1949 were, in fact, bedevilled by social and especially political factors that had little to do with objective scientific reality. In the face of declining stocks, managers refused to accept that overfishing might be the cause, since this had potentially unpleasant social and political ramifications. Instead the bureaucrats and many of the scientists charged with managing the stocks failed to impose moratoria or meaningful limits on fishing activities, with the result that the stock crashed catastrophically in the early 1990s, producing precisely the social and political effects that managers and politicians had desperately tried to ignore. This piece is a harsh condemnation of management practices, but it also shows the limits of management, even supposedly scientific management, in the face of economic (and ecological) collapse.

I.D. Thompson, "The Myth of Integrated Wildlife/Forestry Management," *Queen's Quarterly* 94, no. 3 (Autumn 1987): 609–621. Reprinted in Chad Gaffield and Pam Gaffield, eds., *Consuming Canada: Readings in Environmental History* (Toronto: Copp Clark, Ltd., 1995), 213–224.

As his title suggests, Thompson rejects the policy of integrated wildlife and forestry management as it has been practised in Canada. He argues that our understanding of complex ecosystems is still too rudimentary and linear to form the basis of any coherent management framework that will last for the long term. He points out that the practice of setting aside protected areas is often self-defeating because those areas are simply too small to remain ecologically intact. He, like other authors in this section, is critical of the conservationist impulse that drives them, arguing that conservationism and rational-use policies as they have been practised in the twentieth century are all too often entirely incompatible with one another.

Alan MacEachern, "Rationality and Rationalization in Canadian National Parks' Predator Policy," *Canadian Papers in Rural History* 10 (1996): 149–164.

MacEachern shows the changing nature of predator management policy in Canada's national parks. He points out that predator management depended largely on the personal perceptions of those responsible for managing the parks system as a whole, and it oscillated between aggressive kill strategies to near-absolute protection strategies. This oscillation led to considerable internal contradictions in predator management policy and probably weakened the overall conservationist mandate of the parks system.

RELEVANT WEB SITES

Parks Canada Agency Sustainable Development Strategy
http://www.pc.gc.ca/docs/pc/strat/sdd-sds/index_e.asp

This Web site gives a good overview of Parks Canada's activities. It includes the Parks Canada Charter, the Canadian government's role in sustainability, and Parks Canada's commitments for sustainable development, including its Sustainable Development Strategy for 2004–2007. Interestingly, the title of that document is "Sustaining Ecological and Commemorative Integrity," suggesting at least the combination of two mandates that may not be entirely compatible with one another. The site includes information on all national parks, national historic sites, and national marine conservation areas of Canada. It is the best overview of Canadian sustainability and development on a national level.

Newfoundland Salt Fisheries: A Digital Exhibit
http://collections.ic.gc.ca/fisheries/

This comprehensive site presents a history of the Newfoundland salt cod industry from its beginnings in the 1500s to the middle of the twentieth century. It contains a very wide range of audiovisual items including film clips from the period after 1900, an impressive collection of photographs, and maps and manuscripts dating back to the earliest times of the fishery. Taken as a whole, the site represents a very well executed narrative of a way of life erased by industrialization and economies of scale.

Prince Edward Island: The Fisheries, Then and Now
http://collections.ic.gc.ca/peifisheries/index.asp

Another excellent site from the Canada's Digital Collections initiatives. This one covers the social, cultural, economic, and environmental history of the PEI fisheries from pre-contact times to the present. It emphasizes the role of management in the fisheries as well as the stresses created by the ever-changing impact of technological and market forces. It is a comprehensive and well-researched site with plenty of resources available.

Copyright Acknowledgements